Power Electronics and Variable Frequency Drives

Also of Interest from IEEE Press . . .

SENSORLESS CONTROL OF AC MOTOR DRIVES
Speed and Position Sensorless Operation
edited by Kaushik Rajashekara et al., *Delphi Energy and Engine Management Systems*
1996   Hardcover   528 pp   IEEE Order No. PC3996   ISBN 0-7803-1046-2

MODERN POWER ELECTRONICS
Evolution, Technology, and Applications
edited by Bimal K. Bose, *University of Tennessee, Knoxville*
1992   Hardcover   608 pp   IEEE Order No. PC2766   ISBN 087942-282-3

POWER ELECTRONICS
Converters, Applications, and Design, Second Edition
Ned Mohan, *University of Minnesota;* Tore M. Undeland, *Norwegian Institute of Technology;*
and William P. Robbins, *University of Minnesota*
Copublished with John Wiley & Sons, Inc.
1996   Hardcover   822 pp   IEEE Order No. PC5640   ISBN 0-471-58408-8

Power Electronics and Variable Frequency Drives

Technology and Applications

Edited by

Bimal K. Bose
Condra Chair of Excellence in Power Electronics
University of Tennessee, Knoxville

IEEE PRESS

IEEE Industrial Electronics Sociey, *Sponsor*
IEEE Industry Applications Society, *Sponsor*
IEEE Power Electronics Society, *Sponsor*

The Institute of Electrical
and Electronics Engineers, Inc., New York

This book and other books may be purchased at a discount
from the publisher when ordered in bulk quantities. Contact:

IEEE Press Marketing
Attn: Special Sales
445 Hoes Lane, P.O. Box 1331
Piscataway, NJ 08855-1331
Fax: (908) 981-9334

For more information about IEEE PRESS products,
visit the IEEE Home Page: http://www.ieee.org/

ISBN 0-7803-1061-6
IEEE Order Number: PC4150

Printed in the United States of America

10   9   8   7   6   5   4   3   2   1

Library of Congress Cataloging-in-Publication Data
Power electronics and variable frequency drives: technology and
  applications / edited by Bimal K. Bose.
     p. cm.
  Includes bibliographic references and index.
  ISBN 0-7803-1084-5
  1. Power electronics. 2. AC drive technology.
I. Bose, Bimal K.
TK7881.P69 1996
621.46–dc20   96-6092
                                                       CIP

Contents

Chapter 2   **Electrical Machines for Drives 36**
             G.R. Slemon, University of Toronto, Canada

Contents

FOC

Chapter 7   **High Power Industrial Drives   332**

H. Stemmler, Swiss Federal Institute of Technology, Switzerland

Preface

It is with pride and a sense of accomplishment that I am presenting this book to the power electronics community of the world. Unlike my previous books, this one can be considered unique. As you know, power electronics and variable frequency drives are fast-growing multidisciplinary fields in electrical engineering, and it is practically impossible for one individual specialist to write a book covering the entire area. Therefore, I have identified the key specialists in the field and invited them to contribute a chapter in their respective areas of expertise. In fact, the chapters in this book are modified and expanded versions of papers contributed to the special issue of the *Proceedings of the IEEE* published in August 1994.

We are indeed proud to be members of the power electronics profession. I myself have seen this field grow inch by inch like a gigantic tree, with the evolution from thyristors to transistors to DC drives to microprocessors to advanced converters to AC drives to sensorless control and estimation to AI-based control and more. In my own lifetime, I am proud to have seen modern power electronics technology take root, grow into a blossomed tree, and receive wide acceptance in the marketplace. In the twenty-first century, we will watch with greater pride the large impact of power electronics on industrial automation, transportation, utility systems, energy conservation, and environmental pollution control in a broader perspective. We are on the verge of completing 100 years in the evolution of power electronics, which originally started as industrial electronics. We wish it had been born another hundred years ago so that we could experience its full maturity and applications.

Let me now give you a broad historical perspective. As you know, the history of industrial civilization in the world started around 200 years ago with the invention of the steam engine by James Watt of Scotland in 1785. It was the beginning of an industrial renaissance, when muscle power was gradually being replaced by mechanical power. The advent of the internal combustion engine in the late nineteenth century gave further momentum to this trend. Around the same time (1888), Nickola Tesla invented the commercial induction motor. The DC and synchronous machines were introduced with a few years time lead and lag, respectively. The introduction of electrical machines along with commercial availability of electrical power started the new electrical age. Then, in 1948, the invention of the transistor by Bardeen, Brattain, and Shockley marked the start of the modern electronics age. Gradually, the technologies of solid-state power electronics, computers, robotics, communication, and the like were born. Solid-state power electronics is often considered to have brought in the second electronic revolution. Interestingly, power electronics blends the technologies brought in by the mechanical age, electrical age, and electronics age. It is truly an interdisciplinary technology.

While we are on the verge of the twenty-first century industrial revolution and human society is at its peak of glory to reap all the fruits of modern science and technology, we need to realize that behind this glamour and affluence, there is the dark side of suffering humanity. It is ironic that out of the 5.5 billion people on earth, more than 85% suffer immensely from poverty, starvation, and disease. How can we say that the world is a happy and healthy place when a handful of people live in affluence yet the majority suffer so much? The resources of mother earth appear to be abundant, particularly when using modern science and technology, and can comfortably support all the people for a healthy living. I appeal to the whole power electronics community to come forward with courage to pay back selflessly, and with kindness, sympathy, compassion, and love, the society to which we owe so much for the success of our ambitious careers. Let us unite under the banner of Power Electronics for Universal Brotherhood to fulfill this sacred mission.

Let me now give you a broad perspective of the book. The book can be considered as a state-of-the-art review of the interdisciplinary field of power electronics and variable frequency drives that encompasses power semiconductor devices, electrical machines, converter circuits, pulse width modulation techniques, AC machine drives, simulation techniques, estimation and identification, microprocessors, and artificial intelligence techniques. Our goal is to make the book a guiding star of technology that should be on the shelf of every engineer in this field. It has been assumed that the readers have basic knowledge in power electronics and variable frequency drives.

The "Introduction to Power Electronics and Variable Frequency Drives," contributed by me, gives a fast-running portrait of the whole technology and, in a sense, acts as the cementing bond for all the chapters. I have made no attempt to discuss any fundamentals. In fact, considering the scope of this book, fundamentals have hardly been dealt with in the chapters.

Chapter 1, "Power Semiconductor Devices for Adjustable Speed Drives," by Baliga, gives a review of power MOSFETs, IGBTs, MOS-controlled thyristors, power rectifiers, and smart power control chips particularly related to drives application. The MOS-gated devices are emphasized because of their importance in the recent drive applications. The losses are analyzed and compared for different devices in an inverter leg that use the conventional hard-switching technique. With soft switching, the losses will decrease, particularly for MCT-like devices. Although the present devices are based solely on silicon material, the next-generation material silicon carbide shows the promise for considerable reduction of losses.

Chapter 2, "Electrical Machines for Drives," by Slemon, gives a review for different machines, such as induction, permanent magnet synchronous, permanent magnet trapezoidal, synchronous reluctance, permanent magnet reluctance, switched reluctance, and wound field synchronous motors used for variable frequency drives. The features of DC machines that are traditionally popular for variable speed drive are also discussed. The application aspect of each type of machine has been amply reviewed. The chapter concludes with the comment "the optimum choice for a particular drive depends on a detailed assessment of the system design criteria..., such as initial cost, life-cycle cost, dynamic performance, ease of maintenance, robustness, weight, and environmental acceptability."

Chapter 3, "Power Electronic Converters for Drives," by van Wyk, emphasizes the historical perspective of converter evolution related to drives rather than summarizing possible converter topologies and describing their functional characteristics. It starts with the generic issues of switching function and covers the broad topological aspects, including the modern resonant link converters. It also covers snubbers, electromagnetics, cooling, protection, and converter integration aspects. The comprehensive bibliography at the end is useful.

Chapter 4, "Pulse Width Modulation for Electronic Power Conversion," by Holtz, describes modern PWM techniques. It reviews space vector principles and then discusses PWM performance criteria, such as harmonic currents and their spectra, harmonic torques, switching losses, and dynamic performance. The PWM techniques have been broadly classified as open loop, or feedforward, method and closed loop, or feedback, method. In open loop methods, suboscillation, modified suboscillation, space vector modulation, synchronized carrier modulation, selected harmonic elimination, and optimal subcycle methods are discussed. In closed loop methods, hysteresis current control, suboscillation current control, space vector current control, predictive current control, trajectory tracking control, and the like are discussed.

Chapter 5, "Motion Control with Induction Motors," by Lorenz, Lipo, and Novotny, starts with the introduction of voltage-fed and current-fed inverters used for induction motor drives. It heavily emphasizes the modern vector or field-oriented control methods in the servo drives. The basic current control PWM methods commonly used in vector control are reviewed. The flux and torque estimation and their control are discussed, and the modern self-commissioning of drives and continuous tuning techniques are presented.

Chapter 6, "Variable Frequency Permanent Magnet AC Machine Drives," by Jahns, is a comprehensive treatment on the control of permanent magnet trapezoidal and sinusoidal machines. The principles of the machines, converters, control, and sensor elimination are reviewed extensively with some application examples. Finally, the future trend of the technology is discussed.

Chapter 7, "High Power Industrial Drives" by Stemmler, reviews the essential types of highpower drives, such as cycloconverter-fed synchronous motor drives, slip power recovery wound rotor induction motor drives, voltage source, and current source inverter drives with induction and synchronous machines. It considers the selection of machine and converter types and discusses the design, performance, and application aspects of the drives. The future trend of each class is also reviewed.

Chapter 8, "Simulation of Power Electronic and Motion Control Systems," by Mohan, Robbins, Aga, Rastogi, and Naik, discusses first the need for simulation in the power electronics environment and then reviews the classical simulation tools. The widely used digital simulation programs, such as SPICE, EMTP, and MATLAB/SIMULINK, are reviewed. Particularly interesting are the simulation programs of the vector-controlled induction motor drive using SIMULINK, PSPICE, and ATP (a version of EMTP). Finally, the modeling aspects of power semiconductor devices are discussed.

Chapter 9, "Estimation, Identification, and Sensorless Control in AC Drives," by Ohnishi, Matsui, and Hori, is a vital topic for R&D today. It deals with estimation of feedback control signals (and the corresponding sensorless control) and identification of machine parameters in variable frequency AC drives using both induction and PM synchronous machines. The estimation of mechanical disturbance torque and control are also covered.

Chapter 10, "Microprocessors and Digital ICs for Control in Power Electronics and Drives," by Le-Huy, is a comprehensive treatment of digital control principles, microcontrollers, advanced microprocessors, and ASIC technology. It reviews the hardware design, software design, and system evaluation of practical systems. Application examples on PM synchronous motor drive for EV propulsion and induction motor drive with transputers are also given. It ends with a discussion on future trends.

The last chapter, Chapter 11, "Expert System, Fuzzy Logic, and Neural Network in Power Electronics and Drives," has been contributed by me. The "intelligent control" using artificial intelligence techniques is possibly the most dynamic field in power electronics. The principles of expert system, fuzzy logic, and neural network are discussed in simple language and then are followed by several application examples in each. The design methodology of each technique is discussed. Since the technology is somewhat new to the power electronics community, a comprehensive glossary has been added at the end. It is my opinion that the neurofuzzy control techniques will penetrate fast and deeply in the power electronics area in the coming years.

Bimal K. Bose
University of Tennessee

Acknowledgments

I must express my sincere thanks to the chapter contributors for their enthusiastic response, earnest cooperation, and timeliness, without which this book could not have been possible. If the book is well received, it will be updated periodically, with a cooperative effort from the contributors to furnish the latest technology to the readers. I would like to thank my graduate students, particularly Marcelo G. Simoes (now at the University of São Paulo, Brazil) and Sunil Chhaya (now at General Motors), for their help. Savoula Amanatidis at IEEE Press has been particularly helpful in bringing out this book. Thanks are also due to the reviewers for their painstaking efforts in assessing the entire manuscript on behalf of the IEEE. Finally, I am thankful to my family for their cooperation.

Bimal K. Bose
University of Tennessee

Bimal K. Bose

Introduction to Power Electronics and Drives

The area of power electronics and variable frequency drives has recently grown as a major and extremely important discipline in electrical engineering. Power electronics, basically, deals with conversion and control of electrical power for various applications, such as DC- and AC-regulated power supplies, heating and lighting control, electrical welding, electrochemical processes, induction heating, active harmonic filtering and static reactive power generation, control of DC and AC machines, and so on.

Electrical machine drives (often defined as *motion control*) are particularly a very fascinating and challenging area in power electronics because of their spectrum of applications, such as computer peripheral drives, machine tool and robotic drives, pump and blower drives, textile and paper mill drives, electric vehicle and locomotive propulsion, ship propulsion, cement mill and rolling mill drives, and so on. With the present trend of global industrial automation, the application of power electronics and variable frequency drives is expected to grow enormously in the future. Very recently, another important role of power electronics is becoming visible, that is, energy saving that can control the environmental pollution. It has been estimated that roughly 10–15% of generated electrical energy can be saved by widespread use of power electronics in applications, such as variable speed drives, high-frequency electronic ballasts for fluorescent lamps, and so on. In addition, power electronics–intensive environmentally clean photovoltaic and wind energy resources also show a bright future. The world has a vast potential for wind energy. A rough estimate indicates that all the electricity needs of the world can be

1

met even if 10% of the available wind energy is tapped. Photovoltaic energy is yet very expensive.

Power semiconductor device is the heart and soul of a modern power electronics apparatus. The present age of power electronics began by the introduction of thyristor (or silicon controlled recitifer) in the late 1950s. Gradually, other types of devices were introduced, such as triacs, gate turn-off thyristors (GTOs), bipolar power transistors (BPTs or BJTs), power MOS field effect transistors (MOSFETs), insulated gate bipolar transistors (IGBTs), static induction transistors (SITs), static induction thyristors (SITH), and MOS-controlled thyristors (MCTs). Along with this evolution, the voltage and current ratings and other electrical characteristics of these devices began improving significantly. Historically, the evolution of power electronics has generally followed the power semiconductor device evolution. Again, power semiconductor devices have followed the evolution of solid-state electronics. The power electronics community is extremely grateful to solid-state researchers who have worked relentlessly to improve semiconductor processing, device fabrication, and packaging to produce today's high-density, high-performance, high-reliability, and high-yielding microelectronic chips at such an economical price. These efforts have also contributed to the successful evolution of today's power semiconductor devices. The age of solid-state power electronics is often called the *second electronics revolution*. The first electronics revolution ushered due to the advent of transistors and integrated circuits.

Traditionally, diodes and phase-controlled thyristors have been the workhorses in power electronics from the beginning. The enormous growth of diode and phase-controlled converters on utility systems is now causing serious power quality problems mainly due to distortion of line current waves. The recent IEEE and IEC harmonic standards tend to severely limit such harmonic distortion. Of course, active type power line conditioners, however expensive, can filter the line harmonics as well as compensate the power factor condition. The industry favors MOS-gated, self-controlled, high-frequency power switches, such as power MOSFET, IGBT, and MCT. MOS gating not only permits high-power gain of the switch, but also has the advantages of control and protection integration on the same chip (often called *smart power*). Power MOSFET is a universally popular device for low-voltage, low-power, high-frequency applications, such as switching mode power supply, portable brush, and brushless DC drives, and has no fear of competition by other devices in the future. The static induction transistor (SIT) is basically a solid-state vacuum triode tube and has the characteristics of high-voltage, high-current junction FET. It is used in high-frequency, high-power induction heating, AM/FM transmitters, and so on. For motor drive and other applications in the low- to medium-power range (up to several hundred kilowatts), IGBT has found wide acceptance and may eventually extend up to a one megawatt power range. The MCT has the advantage of thyristor-like low-conduction drop, but compared to IGBT, its safe operating area (SOA) is somewhat lower, indicating its superiority for high-efficiency soft-switched applications. Power device research is extremely expensive and time consuming. Historically, a device introduced in the beginning always has a limited capability, but it is refined continuously by evolutionary process as more knowledge is gained. The power electronics community eagerly awaits a device like the MOS-GTO to replace the conventional GTO. Whether an MCT-like device can fulfill this goal in the long run is yet to be seen. At present, momentum is gathering for research in silicon carbide (and diamond) power semiconductors which have the potential for high-power, high-temperature, high-frequency, low-conduction-drop, and high-radiation hardness characteristics. Eventually, they will challenge the conventional silicon-based devices, but we may have to wait a number of years for their commercial availability. A silicon carbide power MOSFET will possibly be the first device to replace a GTO. Of course, modern high-power IGBTs are already challenging the lower end of GTOs. The history of power converter evolution has also been very fascinating. The introduction of a new device or the quantum improvement of its capability has always spurred topology development to optimize use of the device capability, although the primitive Graetz bridge circuits are still so common today. As mentioned before, the traditional diode and phase-controlled converters will be gradually deemphasized because of their power quality problems. Except for limited retrofit applications, active power line conditioners for harmonics and power factor correction are too expensive. Economical hybrid power line conditioners that combine the active and passive filters have been proposed particularly for large converter loads. The active line current wave shaping and power factor correction with self-controlled MOS-gated devices built in within the line-side converter are very convenient. For nonregenerative applications, the front-end diode rectifier can be followed by a PWM boost (or buck-boost) chopper for shaping sinusoidal line current at unity power factor. On the other hand, for regenerative applications, a PWM voltage-fed or current-fed line converter can be used. A variable frequency drive, of course, needs an additional PWM inverter on the machine side. With the present trend of decreasing converter cost, double-sided PWM voltage-fed and current-fed converters will find increasing applications for four-quadrant variable frequency drives. Of course, the voltage-fed topology is superior to current-fed topology in the overall figure-of-merit consideration. Among the different PWM techniques proposed for voltage-fed converters, the sinusoidal PWM (SPWM) technique has been popularly accepted. The space vector PWM technique is somewhat complex but gives superior performance, and it will be increasingly accepted in the future. The double-sided, two-level, voltage-fed converters can be extended to multilevel (three levels or higher) topology (defined as neutral-point-clamped topology), permitting higher power at higher voltage and improved PWM quality within the constraint of switching frequency. GTO-based three-level PWM converters are already used in multi-megawatt capacity. As knowledge for high-power PWM converters improves and high-power, self-controlled, high-frequency devices become available, the phase-controlled cycloconverters that are so popular for large power drives will gradually become obsolete. For very-high-power drives, wound field synchronous motor with thryistor load-commutated inverter and phase-controlled rectifier is difficult to beat. Multiphasing of the rectifier can minimize the line harmonic problem, but external VAR correction is needed to restore the power factor to unity. Although considerable effort has been spent in the development of matrix converters, hardly any attention has been paid by the industry so far because of unavailability of high-frequency, self-controlled AC switches and the need of line-side AC capacitors. Fabrication of AC power switch using the conventional unipolar devices is very uneconomical. The same reasoning is valid for high-frequency AC resonant link converters, which will be discussed shortly.

In recent years, soft- (or relaxed-) switching converters have been proposed in place of the conventional hard- (or stressed-) switched converters, and the published literature in this area has become large. The soft switching at zero voltage or zero current provides the following advantages: (1) minimization of switching loss and, therefore, improvement of converter efficiency; (2) relieving stress on the device, that is, improving its reliability; (3) elimination of snubbers; and (4) reduced EMI. Soft switching is possible by resonant link DC (also AC), resonant pole, or quasi-resonance technique. Resonant-link DC converters use zero voltage switching (ZVS) where the constant DC link voltage is converted to discrete voltage pulses through a resonant circuit. AC resonant link converters use high-frequency AC in the link and soft switching occurs at natural zero voltage of the link. Resonant-pole converter uses zero current switching (ZCS) with the help of auxiliary circuit connected to each pole or phase leg of the converter. Quasi-resonant converters permit resonant voltage notch at selected instants to improve PWM quality instead of free resonance in resonant link converter. Double-sided PWM converters can easily be constructed using these principles. In spite of the advantages of soft switching, penalties are often paid in the form of voltage or current derating, additional device count, or extra losses in the resonant circuit and load. The regenerative snubber is definitely an alternate attractive method of improving converter efficiency, particularly for higher power range. Converter technology has practically reached the stage of saturation region of S-curve, and only small and incremental activities are expected to follow in the future. Intelligent converter modules, where the power, control, and protection circuits are embedded in the same package and interface directly with the PWM logic signals, will be increasingly popular in the future. Custom-designed converters with automated design tools similar to VLSI design techniques will save expensive man-hours. This area, along with integrated packaging aspects, will get increasing attention in the future. For variable frequency drives, the converter module and the control and protection module, should be mounted directly on the machine frame ("smart" machine) in the low- to medium-power range. Electrical machines have been used in the industry for more than a century, but their evolution has been very slow, unlike that of power semiconductor devices, converters, and control of drives. Yet today, an overflowing number of papers are published in national and international conferences in this area. The machine is a very complex device electrically, mechanically, and thermally. The primitive 60 Hz (or 50 Hz) AC machine was very bulky and had poor performance. The advent of modern digital computers, improved machine modeling, and CAD programs, and the introduction of new materials have significantly contributed to high-power density, efficiency, reliability, and improved mechanical and thermal design of machines in the recent years. The studies related to machine modeling, state estimation, parameter identification, measurement, simulation, and interaction with the converters are still exciting R&D topics today. Although, traditionally, DC machines have been used for variable speed drives and today still count as majority, in the next century they will gradually fall into oblivion. Cage-type induction machines are the workhorse for the majority of today's variable frequency drive applications, and possibly will remain so in the future. The doubly-fed (or wound rotor) machines with slip power recovery control principle have been used in large power pump- and fan-type drives within limited speed range, but the undesirable features of this drive (that is, high cost of machine, poor line-power factor, and harmonics) coupled with the decreasing cost of self-controlled converters have already made this class of converters of diminishing importance. It is interesting to note here that once Thomas Alva Edison, the inventing wizard of the nineteenth century, was a strong proponent of DC machines, and believed that the induction machine had no future. The drives with synchronous machines are often in close competition with those of induction machines. Very large multi-megawatt-power-level wound field synchronous machines are used in pumps and compressors, cement and rolling mill drives, ship propulsion, and so on. The machine has better efficiency and can operate at unity power factor (thus demanding less power rating of the converter), although it is more expensive. An additional advantage is that the simple load-commutated thyristor converter (with slightly leading power factor) can be used in the drive if a very fast response is not demanded. Recently, permanent magnet synchronous machine drives are being accepted in the lower range of medium power (up to 100 kW). The drive is more expensive, needs an absolute shaft position sensor, and is difficult to operate at extended speed field-weakening region. However, higher efficiency (no rotor copper loss) makes its life-cycle cost somewhat favorable. Brushless DC drives (that use PM trapezoidal synchronous machines with position sensor controlled inverter) with ferrite magnet are already very popular up to a few kilowatt power range. The high-energy neodymium-iron-boron magnet is gradually replacing ferrite magnet, in spite of higher cost particularly in applications where low rotor inertia and small size have more advantage. Synchronous reluctance motors have been disfavored in preference to PM machines because of poor power factor and bulky size, but are now showing some limited preference because of improved design. In cost and performance, they are now often comparable to induction motors. The step motors (variable reluctance or hybrid) have been traditionally popular for open loop digital drives but have poor performance for speed control applications. Another type of variable reluctance machine, the switched reluctance motor (SRM) is receiving wide attention in the literature, and great effort has been made to commercialize it in competition with the induction motor. The machine is simple and economical, but an absolute position sensor is mandatory. The pulsating torque and acoustic noise are problems in this drive. Of course, various position estimation and pulsating torque elimination methods have been attempted recently. A converter is indispensable in SRM drive (no bypass operation is possible), multiple machine parallel operation is not possible and feedback signal estimation from machine voltage and currents is extremely difficult. It is doubtful whether it will ever be competitive in general-purpose industrial applications.

It is expected that in future, every electrical machine will have a variable frequency converter module in the front-end generating sinusoidal line current, irrespective of whether the drive is variable speed or fixed speed. For a fixed speed 60 Hz drive, the variable frequency starter will find increasing acceptance as the converter cost decreases and harmonic standards are strictly enforced.

The control, estimation, and identification related to variable frequency drives are a fascinating area of research and are at present receiving wide attention in the literature. Although many scalar and vector* (or field-oriented) control techniques have been proposed in the literature, it appears that industry has accepted the simple and economical open loop volts per hertz control for low-performance applications and the complex vector control for high-performance drives. With vector control, AC machines are controlled like a separately excited DC machine and, therefore, have all the virtues (often in excess) of DC machine control performance. Fortunately, the advent of modern microcomputers has made the vector control viable for industrial applications. Many problems related to direct and indirect vector-controlled drives, such as parameter estimation and slip gain tuning, speed and flux estimation, and complete autocommissioning of drive based on on-line parameter extraction are receiving wide attention in the literature. Although rotor flux-oriented vector control gives decoupling and has been widely accepted, the stator flux-oriented direct vector control has the advantages of a small parameter variation problem and ease of speed sensorless control from zero speed. In fact, with the modern DSP-based control, the complexity of speed and flux sensorless vector control is transparent to the user and makes it marginally different from the open loop volts per hertz control architecture. It can be safely projected that vector control will appear as universal controller for the AC drive.

A machine operating at the rated flux gives optimum transient response, but at light-load operation, the efficiency is nonoptimum because of excessive core loss. The flux can be weakened at light load using a flux-torque function generator or on the basis of real time search so that the input power becomes minimum. At any transient condition, however, the rated flux can be restored, thus combining both the efficiency optimization and transient response optimization features in the drive.

The robust performance against variation of machine parameters and load torque disturbance is very important in servo type drives, and it has been a subject for research for a long time. In an ideal drive, it is desirable that the load torque (transient or steady) and plant parameters (such as moment of inertia) should not affect the control response. Various adaptive control methods, such as self-tuning regulation (STR) based on parameters and disturbance torque estimation,

---

*The author strongly recommends standardizing the term "vector" instead of "field-oriented" because the former term is truly general and is applicable to machine as well as other control applications.

sliding mode or variable structure system (VSS) control, model referencing adaptive control (MRAC), and H-infinity control have been proposed in the literature. With microprocessor-based inner loop vector control, the implementation of such robust control in the outer loop is somewhat easy because of DC machinelike plant model.

Recently, AI (artificial intelligent) techniques—such as expert system, fuzzy logic, and artificial neural networks (ANN)—are showing much promise for intelligent adaptive control (often called *self-organizing* or *learning control*) and estimation of drives. Sophisticated on-line diagnostics and fault-tolerant control are other areas where adequate attention has not yet been paid to improve the drive reliability. For example, three-phase induction motor can be operated in single-phase mode (if the load condition permits) in case of device failure. Although there is a loss of torque and there is other performance degradation, the continuity of drive operation can be maintained for critical process control and transportation drives. Since sensors add cost and unreliability, there is a tremendous urge to develop sensorless high-performance drives, as mentioned before. The machine terminal voltages and currents can be sensed and processed to calculate speed, position, flux, torque, power and other feedback signals with the help of a microprocessor. The inaccuracy of estimated signals due to machine parameter variation always remains a problem. Accurate identification of machine parameters to compensate the estimation error is a challenging task. In addition, direct integration of machine voltages near zero speed (i.e., near zero frequency) to calculate the flux is another problem because of offset problem of the integrator. The flux of a machine can be conveniently estimated from the stator voltage model in higher speed range and rotor current model (Blaschke equation) in low-speed range. The vector flux signal gives the magnitude for feedback flux control and unit vectors for vector transformation of forward and feedback signals. The speed can be estimated by MRAC method by matching the flux estimation from stator voltage model and speed-dependent rotor current model. Instead of using direct integration for flux estimation, the closed loop, observer-based computation can be used for better estimation accuracy particularly at low speed. The speed adaptive flux observer has been proposed for improved estimation of speed. The stator flux-oriented direct vector control with cascaded low-pass filter-based flux synthesis has been proposed, which operates from zero speed with insensitivity to machine parameters.

The rapid advancement of microprocessor and application specific IC (ASIC) technologies has been amazing in recent years. Indeed, today's high-performance drive control and estimation techniques could remain a dream without advancement of microprocessor technology. The first generation of microprocessors began with the introduction of Intel 8080 family. As the demand for real-time control grew, the microcontrollers, such as the 8051 and 8096 families, were introduced. The modern digital signal processors (DSPs) with Harvard pipelined architecture (such as TMS320C14, TMS320C25, TMS320C30, and so on) have very high speed and functionality. The reduced instruction set computer (RISC)-type microprocessors (such as Intel 860, 960, etc.) are essentially optimized for data processing applications. The transputer-type microprocessors (such as T414, T800 etc., by

INMOS) are adapted for concurrent processing of modular control functions. Various ASIC chips, such as PWM modulator, vector rotator, or full-function controllers, are widely used in power electronics systems. Hybrid (analog and digital) ASIC chips with microprocessor core are expected to find wide applications in power electronics. In such a chip, the peripheral control hardware along with the frozen control functions will be implemented in ASIC, whereas the programmable control and estimation software particularly influenced by plant parameter variation will be implemented in the processor. Many functional elements in a power electronic control system will be implemented by neural network and embedded in the ASIC chip. Eventually, the whole power electronic control system will be based on a single ASIC.

Chapter 1

B. Jayant Baliga

Power Semiconductor Devices for Variable Frequency Drives

1.1. INTRODUCTION

Improvements in the performance of variable frequency drives have been directly related to the availability of power semiconductor devices with better electrical characteristics [1, 2]. It has been found that the device performance determines the size, weight, and cost of the entire power electronic system. For power-switching applications, an ideal power device must be able to support a very large voltage in the off-state with negligible leakage current, carry high current in a small area with a low on-state voltage drop, and be able to switch rapidly between the on- and off-states. In addition, it is preferable for the device to be able to regulate the rate of rise of current when it is turned on and to limit the current in the circuit under faulty operating conditions without the aid of external circuit elements. Although much progress has been made in achieving these goals, the ideal device continues to elude the power semiconductor designer, thus providing strong motivation for further research and development in this area.

The introduction of the power thyristor in the marketplace in the 1950s marked the beginning of the revolution in solid-state-device–based power electronics. Since then, a steady growth in the ratings of the thyristors and their operating frequency has enabled extension of their application to motor control. The growth in thyristor ratings can be traced with the aid of Figure 1-1. Starting with the introduction of 500 volt devices, the blocking voltage capability has been scaled to exceed 6500 volts. This very high voltage blocking capability has been achieved by the ability to produce very uniformly doped high-resistivity N-type silicon by neutron transmutation doping (NTD). The NTD process allows the conversion of a silicon isotope into phosphorus by the absorption of a thermal neutron. A very uniform doping concentration can be produced throughout the wafer if a uniform neutron flux distribution can be achieved. The sudden increase in blocking voltage capability for thyristors in the 1980s indicated in Figure 1-1a can be traced to the availability of NTD silicon wafers. There has been a concomitant increase in the current-handling capability for the power thyristors as indicated in Figure 1-1b. Since the on-state current density for the device has not changed over the years due to the relatively unchanged thermal impedance of the device package, this increase in current ratings has been possible by the availability of larger-diameter silicon wafers. Single devices are now manufactured with wafers of 4 inches in diameter.

Following the introduction of power thyristors in the 1950s, many bipolar power devices were introduced with improved electrical characteristics during the next 20 years. Among these, the power bipolar transistor was the key to extending the operating frequency of power systems above 1 kHz. Bipolar power transistor ratings were continuously increased over the years until devices with 600 volt block-

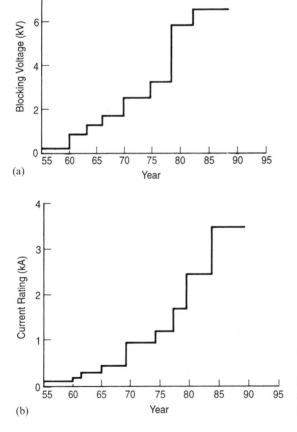

(a)

(b)

Figure 1-1. Growth in power thyristor ratings (a) blocking voltage, (b) on-state current.

ing capability became available with current-handling capability of several hundred amperes. These devices have been extensively used for motor control applications. In addition, bipolar power transistors with 1500 volt blocking capability have been developed for TV deflection circuits. A basic limitation of the power bipolar transistor is that it is a current-controlled device. Although much effort has been performed on device optimization to maximize the current gain, the current gain at typical on-state current densities is only about 10. This creates a significant problem in driving the power bipolar transistor. The control circuits must be assembled from discrete components leading to a bulky and expensive system that is often difficult to manufacture. For this reason, the power bipolar transistor has been replaced by the insulated gate bipolar transistor (IGBT) in the 1990s.

The development of the gate turn-off (GTO) thyristor was the key to extending the power rating of many systems to the megawatt range. It has found widespread application to traction drives (electric streetcars and locomotives). Devices have been developed with the capability for blocking 8000 volts and switching 1000 amperes. As in the case of the power bipolar transistor, the GTO is a current-controlled device. A very large gate current is required to enable turn-off of the anode current. When optimized for blocking high voltages, the current gain of the GTO is found to be less than 5. In addition, large snubbers are required to ensure turn-off without destructive failure. Consequently, the cost of the system is increased by the need for large snubbers and control circuits.

In the 1970s, devices called the static induction transistor (SIT) and the static induction thyristor (SITH) were introduced in Japan. Research on these devices was also performed in the United States on these devices under the name of vertical junction field effect transistors (VFET) and the field-controlled diode or thyristor (FCD, FCT) [5]. Although these devices were found to have some attractive characteristics, such as superior radiation tolerance and a very high cutoff frequency, they have not found much commercial acceptance for variable frequency drives. The two principal reasons for this are, first, these devices have normally on characteristics with poor high-temperature blocking, and second, a large gate drive current (nearly equal to the anode current) is need to achieve rapid turn-off.

Modern power semiconductor technology evolved by the assimilation of the metal-oxide-semiconductor (MOS) technology originally developed for CMOS integrated circuits. This first led to the advent of the power MOS-gated field effect transistors (MOSFETs) in the 1970s. Although these devices were initially touted by the industry to exhibit the characteristics of the ideal device, their performance was found to be unsatisfactory for medium-power applications where the operating voltages exceed 300 volts, thus relegating them to the low-voltage, high-frequency applications arena.

The introduction of the insulated gate bipolar transistor in the 1980s was aimed at providing a superior device for the medium-power applications by attempting to combine the best features of the bipolar power transistor and the power MOSFET. Although the IGBT has been widely accepted for motor control applications, its characteristics do not approach those of the ideal device. This has motivated research on MOS-gated thyristor structures, which have recently been commercially introduced for high-power applications. Further, theoretical analysis has shown that power devices made from silicon carbide could have electrical characteristics approaching those of the ideal device. Over the long term, these devices could completely displace silicon devices used for power electronics.

The objective of this chapter is to provide an overview of power semiconductor devices from the point of view of their application to adjustable speed motor drives. The intention is to concentrate on the impact of the modern MOS-gated power devices upon system efficiency. The commercially available switches that have been considered in this analysis are the power MOSFET, the insulated gate bipolar transistor, and the MOS-controlled thyristor (MCT). To assess the impact of ongoing research on silicon device technology, the analysis includes the base resistance controlled thyristor (BRT) and the emitter switched thyristor (EST). It is also illustrated in this chapter that the development of power rectifiers with improved high-frequency switching characteristics is critical to the development of systems operating above the acoustic frequency limit, which is important for reduction of ambient noise, particularly in commercial applications. In the future, it is anticipated that silicon carbide-based switches and rectifiers will significantly enhance the performance of variable speed drives. For this reason, the power loss reduction that can be achieved by replacement of silicon devices with silicon carbide-based devices is also analyzed.

1.2. BASIC VARIABLE SPEED DRIVE

The most commonly used adjustable speed drive technology is based upon the pulse width modulated (PWM) inverter [1, 2] in which the input AC line power is first rectified to form a DC bus. The variable frequency AC power to the motor is then provided by using six switches and flyback rectifiers. The switches are connected in a totem pole configuration as illustrated in Figure 1-2. The power delivered to the motor is regulated by adjusting the time duration for the on- and off-states for the power switch. By using a sinusoidal reference waveform, a variable frequency output current can be synthesized by using a switching frequency well above the motor operating frequency. To reduce the acoustic noise from the motor, it is desirable to increase the switching frequency for the transistors to above the acoustic range (preferably above 15 kHz).

A typical set of current and voltage waveforms in the transistor and rectifier during one switching cycle are shown in Figure 1-3. These waveforms have been linearized to simplify the power loss analysis. It is assumed that the motor current is initially flowing through the flyback rectifier at the bottom of the totem pole circuit. At time t_1, the upper transistor is switched on with a controlled rate of rise of current. During the time interval from t_1 to t_2, the motor current transfers from the rectifier to the transistor. Unfortunately, the PiN rectifiers that are used in these circuits are unable to recover instantaneously from their forward conduction state to their reverse blocking state. Instead, a large reverse current flow occurs with a peak value I_{PR} prior to the rectifier becoming capable of supporting reverse voltage. This reverse recovery current flows through the transistor. It is worth pointing out that

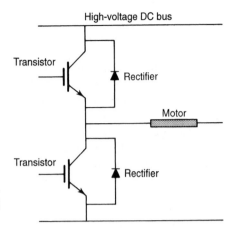

Figure 1-2. Typical totem pole configuration used in a pulse width modulated motor drive.

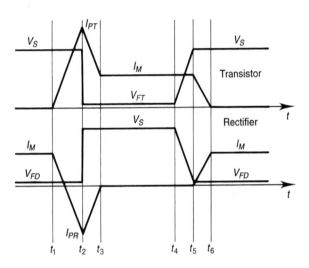

Figure 1-3. Linearized current and voltage waveforms for the switch and rectifier.
V_s-DC supply voltage
V_{FT}-on-state voltage drop for the transistor
V_{FD}-on-state voltage drop for the rectifier
I_M-motor current (on-state current in the transistor and rectifier)
I_{PR}-peak reverse recovery current for the rectifier
I_{PT}-peak transistor current during its turn-on.

the net transistor current during this time interval is the sum of the motor current I_M and the diode reverse recovery current. Further, during this time, the transistor must support the entire DC bus voltage because the rectifier is not yet able to support voltage. This produces a significant power dissipation in the power switch when it is being turned on. This will be referred to as the turn-on power loss for the transistor. Further, the transistor is subjected to a high stress due to the presence of a high current (I_{PT}) and a high voltage (V_S) simultaneously. This can place the transistor in a destructive failure mode if the stress exceeds its safe operating area limit. At time t_2, the rectifier begins to support reverse voltage, and its reverse current decreases to zero during the time interval from t_3 to t_2. During this time, the transistor current falls to the motor current, and a large power dissipation occurs in the diode because it is supporting a high voltage while conducting a large reverse current.

The other important switching interval is from t_4 to t_6 during which the transistor is turned off and the motor current is transferred to the rectifier. During the first part of this time interval from t_4 to t_5, the voltage across the transistor rises to the bus voltage while its current remains essentially equal to the motor current because of the large inductance in the motor winding. In the second portion of this time interval from t_5 to t_6, the current in the transistor decreases to zero while it is supporting the bus voltage. Since there is a large current and voltage impressed on the transistor during both these time intervals, there is significant power dissipation in the transistor during its turn-off.

Due to relatively long turn-off time for bipolar power transistors and the first IGBTs introduced into the marketplace, the emphasis has been on reducing the power loss in the switches during their turn-off. It has been found that methods employed to reduce the turn-off time for the switches is usually accompanied by an increase in their on-state voltage drop, which increases their on-state power dissipation during the time interval t_2 to t_4. It has been customary to compare power switches by calculating the sum of the on-state and turn-off power losses as a function of frequency [3] as given by

$$P_L = \delta\, I_F V_F + t_F I_F V_S f \tag{1.1}$$

where δ is the duty cycle, I_F is the on-state current, V_S is the DC bus voltage, t_F is the turn-off time for the switch, and f is the switching frequency. The results of such calculations for the devices discussed in this chapter are given in Figure 1-4. The calculations were performed for the case of a motor current of 15 amperes, a DC bus voltage of 400 volts, and a duty cycle of 50%. The device characteristics used for the calculations are provided in Table 1-1. Unfortunately, this method for comparison grossly underestimates the power losses in the motor drive circuit because it does not take into account the turn-on losses and the power dissipation incurred in the

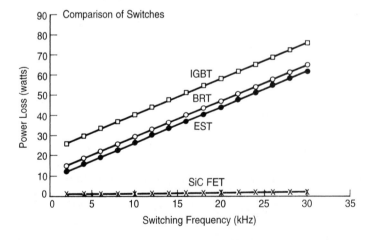

Figure 1-4. Comparison of power switches based upon a simple analysis, including on-state and turn-off losses only. The DC bus voltage is assumed to be 400 volts, and the motor current is 15 amperes.

TABLE 1-1   CHARACTERISTICS OF POWER SWITCHES USED
FOR POWER LOSS CALCULATIONS*

Device	Forward drop	Current density	Turn-off time
IGBT	3.0 V	100 A/cm^2	0.3 μsec
BRT/MCT	1.1 V	100 A/cm^2	0.3 μsec
EST	1.5 V	100 A/cm^2	0.3 μsec
SiC FET			
Case 1	0.1 V	100 A/cm^2	0.01 μsec
Case 2	0.1 V	100 A/cm^2	0.05 μsec

*All the devices are assumed to have a breakdown voltage of 600 volts.

rectifier. It is, therefore, important to perform the power loss analysis by considering the entire waveform for the transistor and rectifier as shown in Figure 1-3 for the complete switching period. The power losses obtained with this method will be provided in this chapter to illustrate the significance of the power rectifier reverse recovery behavior upon the power dissipation. Before such calculations are presented in this chapter, the basic structures and electrical characteristics of various switches and rectifiers will be discussed. Since the most commercially successful power devices with a high input impedance have been based upon MOS-gated structures, this review will not include the bipolar power devices but will focus upon the characteristics of power semiconductor devices with MOS-gated structures.

1.3. POWER MOSFET

The development of the power MOSFET was based upon the technology created for the fabrication of CMOS logic circuits [4]. The metal-oxide-semiconductor structure has an inherent insulator between the gate electrode and the semiconductor resulting in a high input impedance under steady state conditions. The structure of the power MOSFET is shown in Figure 1-5a,b. As in the case of bipolar discrete power devices, a vertical device architecture was adapted to obtain a high current-handling capability by placing the high-current electrodes (source and drain) on opposite surfaces of the wafer. This avoids the problems associated with interdigitation of the source and drain metal. In addition, a refractory gate (usually polysilicon) is used to enable fabrication of the device by using the double-diffused process to form the DMOS structure shown in the figure. This process allows fabrication of devices with submicron channel length by adjustment of the relative diffusion depth of the P-base and N+ source regions without the need for expensive high-resolution photolithographic tools.

The vertical DMOS device structure can support a large voltage across the P-base/drift region junction. The maximum blocking voltage is determined by the onset of avalanche breakdown of this junction or at the edge termination. A larger blocking voltage can be obtained by using a thicker, higher resistivity drift region.

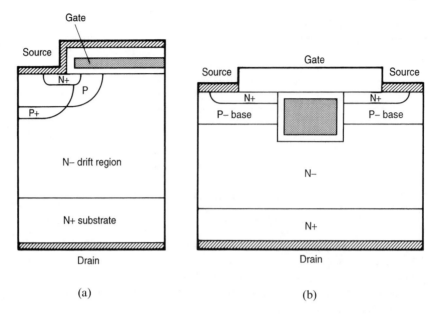

Figure 1-5. (a) Power MOSFET structure fabricated using the DMOS process; (b) Power MOSFET structure fabricated using the UMOS process.

When a positive gate bias is applied, an inversion layer is formed at the surface of the P-base region. This forms a channel between the source region and the drift region through which current can flow. The on-state current is conducted using only majority carriers (electrons for the n-channel structure). This feature of the power MOSFET provides very fast switching performance as compared with previous bipolar transistors.

The current rating of the power MOSFET is determined by the joule heating produced in the internal resistances within the device structure. There are many resistance components in the power MOSFET structure [5]. Among these, the most significant are the channel resistance (R_{ch}); the resistance between the P-base diffusions, which is called the JFET resistance (R_J); and the resistance of the drift region (R_D). During the last five years, power MOSFETs with trench-gate structures (see Figure 1-5b) have been investigated with the goal of reducing the on-state resistance. In this structure, the JFET resistance is eliminated. Further, the trench-gate process allows increasing the MOS-channel density by a factor of 5 times leading to reduction in the on-resistance. This is particularly significant for power MOSFETs with low (< 100 volt) blocking voltage capability. In the case of high-voltage-power MOSFETs, the resistance of the drift region becomes dominant for both the DMOS and the UMOS structures, and its value increases very rapidly with increasing breakdown voltage. This results in the on-state voltage drop of silicon power MOSFETs, as determined by the product of the on-state current and the on-resistance, becoming unacceptable for the types of variable speed motor drives being discussed in this chapter. Thus, in spite of its many other attractive features, the power MOSFET is not considered a viable device for high-voltage variable speed drive applications. This will be evident from the power loss calculations presented later in the chapter.

1.4. INSULATED GATE BIPOLAR TRANSISTOR

To provide a high-input impedance device for high-voltage applications, the insulated gate bipolar transistor was proposed in the 1980s [6]. In the IGBT, an MOS-gated region is used to control current transport in a wide-base high-voltage bipolar transistor. This results in a device with the attractive characteristics of the high-input impedance of a power MOSFET combined with the superior on-state characteristics of bipolar devices. In fact, it has been shown that the on-state characteristics of the IGBT are superior to those for a high voltage transistor and approach those of a thyristor [5].

A cross-sectional view of the IGBT structure is provided in Figure 1-6 together with its equivalent circuit. Its structure is similar to the power MOSFET with the exception that a P+ substrate is used instead of the N+ substrate. This structural similarity has allowed rapid commercialization of the device because the process for the fabrication of the IGBT is nearly identical to that for the power MOSFET. However, its operating physics is quite different as indicated by the equivalent circuit. This circuit indicates the presence of a parasitic thyristor between the collector and emitter terminals. The latch-up of this thyristor results in loss of gate control and destructive failure of the IGBT. In commercial devices, the latch-up has been suppressed by using a variety of techniques such as the P+ region under the N+ emitter to reduce the resistance R_P. These methods are aimed at preventing the

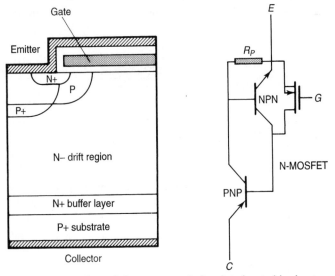

Figure 1-6. Cross section of the asymmetric insulated gate bipolar transistor structure and its equivalent circuit.

activation of the NPN transistor so that the IGBT can operate as a wide-base PNP transistor driven by an integrated MOSFET.

Both forward and reverse blocking capability are inherent in the IGBT structure because the voltage can be supported by the reverse biasing of the P-base/N-drift region junction in the forward quadrant and the P+ substrate/N-drift region junction in the reverse quadrant [5]. However, commercial devices are available with only forward blocking capability because they are made with a N+ buffer layer interposed between the N-drift region and the P+ substrate, as shown in Figure 1-6. This structure is referred to as the punch-through (PT) structure because the depletion layer extends throughout the N-drift region during forward blocking and punches-through to the N+ buffer layer. The buffer layer structure provides superior on-state and safe operating area characteristics for DC circuit applications. Some manufacturers have been developing devices using the nonpunch-through (NPT) design shown in Figure 1-7 but have optimized the device only for DC applications. This structure can support a large reverse voltage if the lower junction is properly terminated [6]. In the future, the symmetrical blocking structure may become available due to its demand for AC circuit applications.

The IGBT can be turned on by the application of a positive gate bias to form an n-channel at the surface of the P-base region. With a positive collector potential, current flow can now occur across the forward-biased P+ substrate/N-drift region junction. As shown in Figure 1-6, the equivalent circuit for the IGBT consists of a wide-base PNP transistor being driven by a MOSFET in the Darlington configuration. This produces a low on-state voltage drop in the IGBT during current conduction, while retaining the gate-controlled current saturation properties of the power MOSFET. Since the current transport in the IGBT occurs by the injection of a high concentration of minority carriers into the N-drift region, its switching behavior is determined by the rate at which these carriers are removed during turn-off. For applications at low frequencies, such as off-line appliance controls, as-fabricated

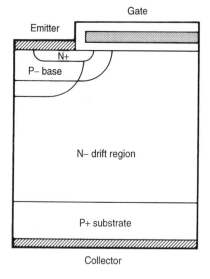

Figure 1-7. Cross section of the symmetric insulated gate bipolar transistor structure.

devices with relatively long minority carrier lifetime (10 microseconds) are acceptable. However, for adjustable speed motor control, where turn-off times of less than 0.5 microseconds are desirable, it is necessary to reduce the minority carrier lifetime by using electron irradiation [7]. Unfortunately, in all bipolar power devices, devices with shorter turn-off times have a larger on-state voltage drop. A trade-off curve between the on-state voltage drop and the turn-off time can be generated for any device structure by using lifetime control to alter the switching speed. A comparison of the trade-off curve for the asymmetric and symmetric IGBT structures can be performed with the aid of Figure 1-8. It can be seen that the trade-off curve for the asymmetric structure is superior for the blocking voltage of 600 volts. Careful optimization of the IGBT structure by adjusting the lifetime in the drift region and the buffer layer doping for the asymmetric structure has led to commercial devices with on-state voltage drops of 3 volts and turn-off times of 0.3 microseconds. Devices are now commercially available with ratings of several hundred amperes with blocking voltages as high as 1500 volts, and it can be anticipated that the ratings of IGBTs will continue to grow by an increase in both the die size and the blocking voltage rating.

The availability of high-performance IGBTs has made them the device of choice for variable speed drives. A calculation of the power losses incurred during device operation with a 50% duty cycle is shown in Figure 1-9 when the IGBT is used with a PiN rectifier. The electrical characteristics of the devices are given in Tables 1-1 and 1-2. It can be seen that the power loss increases rapidly with increase in operating frequency, indicating that the switching losses are more important than the conduction losses. It is interesting to analyze the power losses in the IGBT during various phases of operation. The power losses in the IGBT during the on-state, turn-off, and turn-on are shown in Figure 1-10 together with the total power loss in the IGBT. Note that although the on-state power loss in the IGBT is dominant at switching frequencies below 5 kHz, the switching power losses become dominant at higher frequencies. More important, the turn-on power loss is seen to be larger than the turn-off power loss. This occurs because of the large reverse recovery current for the PiN rectifier which the IGBT switch must conduct during its turn-on (see Figure

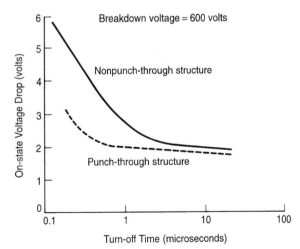

Figure 1-8. Comparison of the trade-off curve between on-state voltage drop and turn-off time for the symmetric and asymmetric IGBT structures.

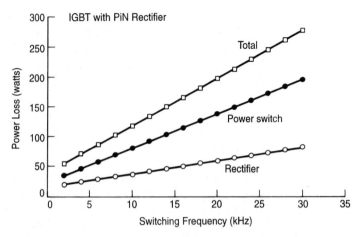

Figure 1-9. Power loss for drive with IGBT and PiN rectifier. The DC bus voltage is assumed to be 400 volts, and the motor current is 15 amperes.

1-3). From these charts, it can be concluded that a reduction in the power losses can be achieved by improving the reverse recovery performance of the power rectifier. Thus, progress in power rectifier technology has been essential to obtaining high performance variable frequency drives.

1.5. POWER RECTIFIERS

With the advent of high performance power switches, such as IGBTs, it has become very important to develop power rectifiers with good high-frequency switching characteristics. In addition to high reverse blocking capability, the important characteristics for power rectifiers are (1) low on-state voltage drop, (2) short reverse recovery switching time, (3) small peak reverse recovery current, and (4) soft reverse recovery behavior. The last requirement stems from the high-voltage spikes that can occur if the reverse recovery di/dt is large.

TABLE 1-2   CHARACTERISTICS OF POWER RECTIFIERS USED
FOR POWER LOSS CALCULATIONS*

Device	Forward drop	Current density	Reverse recovery time	Peak recovery current
PiN	2.0 V	100 A/cm^2	0.5 μsec	45 amps
MPS/SSD	1.0 V	100 A/cm^2	0.25 μsec	10 amps
SiC				
Case 1	1.0 V	100 A/cm^2	0.01 μsec	0 amps
Case 2	1.0 V	100 A/cm^2	0.05 μsec	0 amps

*All the devices are assumed to have a breakdown voltage of 600 volts.

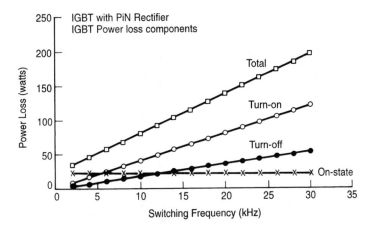

Figure 1-10. IGBT power loss components for drive with IGBT and PiN rectifier. The DC bus voltage is assumed to be 400 volts, and the motor current is 15 amperes.

Many high-voltage power rectifier structures have been proposed and investigated to achieve the above goals [8–13]. Three of these structures, called the merged pin/Schottky (MPS) rectifier, self-adjusting P emitter efficiency diode (SPEED), and static shielded diode (SSD), are illustrated in Figure 1-11 for comparison with the PiN rectifier. In all these structures, the goal is to reduce the amount of stored charge within the N-drift region during the on-state by reducing the injection efficiency of the upper $P+/N$-junction in the rectifier. A reduction in the stored charge is desirable because this results in a shorter reverse recovery time with a smaller peak reverse recovery current. It has been found that although this can be accomplished by simply reducing the doping concentration of the $P+$ region, it results in a very high on-state voltage drop. This does not occur in the MPS and SSD structures because of the highly doped $P+$ region within the structures in spite of a reduction in the stored charge by a factor of up to eight times. A careful comparison between the electrical characteristics of all the proposed rectifier structures has been performed with the same parameters for the drift region [14]. The trade-off curve between on-state voltage drop and the stored charge (which is a measure of the reverse recovery behavior) is shown in Figure 1-12. It can be seen that the MPS and SSD structures provide the best method for reducing the stored charge without a severe increase in the on-state voltage drop. With these structures, it is possible to obtain rectifiers with on-state voltage drop of about 1 volt with one-half the reverse recovery time and one-third the peak reverse recovery current of the PiN rectifier.

The impact of replacing the PiN rectifier with the MPS rectifier is illustrated in Figure 1-13, which provides the total power losses in the drive and the power losses in the IGBT and power rectifier. By comparison with Figure 1-9, it can be seen that the total power loss has been reduced by nearly a factor of 2 by replacement of the PiN rectifier with the MPS rectifier. This is due to not only a smaller power loss in the rectifier but also in the IGBT. This is evident from Figure 1-14 where the power loss

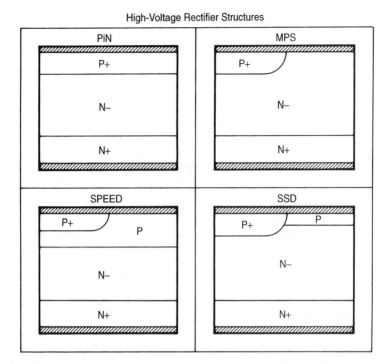

Figure 1-11. Power rectifier structures.

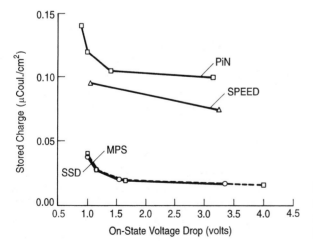

Figure 1-12. Comparison of the trade-off curve between on-state voltage drop and stored charge for various power rectifier structures.

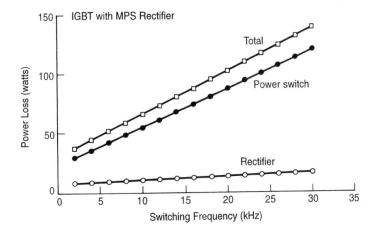

Figure 1-13. Power loss for drive with IGBT and MPS rectifier. The DC bus voltage is assumed to be 400 volts, and the motor current is 15 amperes.

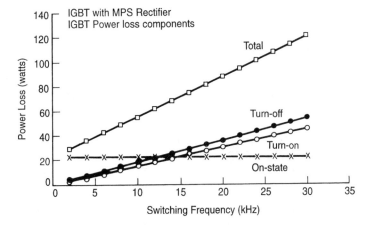

Figure 1-14. IGBT power loss components for drive with IGBT and MPS rectifier. The DC bus voltage is assumed to be 400 volts, and the motor current is 15 amperes.

components in the IGBT are provided. By comparison of the power loss components in this case to those obtained for operation of the IGBT with the PiN rectifier (see Figure 1-10), it is evident that a significant reduction in the turn-on losses for the IGBT is responsible for the improved performance. These charts demonstrate the importance of the rectifier characteristics on the system performance.

1.6. MOS-GATED THYRISTORS

Although the on-state voltage drop of the IGBT is much superior to that for the power MOSFET at high blocking voltages, its on-state voltage drop increases with increasing switching speed [7]. It is well known that power thyristors have superior on-state characteristics when compared with bipolar transistors. Consequently, there has been considerable interest in the development of thyristor structures that can be switched on and off under MOS-gate control. The MOS-gated turn-on of a vertical thyristor structure was first demonstrated in 1979 [15], and this method is now incorporated in all MOS-gated thyristor structures. The ability to turn off the thyristor current using an MOS-gated structure is much more difficult due to the regenerative action occurring within the thyristor. Several promising methods to achieve this capability have been proposed and demonstrated experimentally. Among these approaches, the most promising are the MCT, the BRT, and the EST structures.

The MOS-controlled thyristor (MCT) structure [16–18] is shown in Figure 1-15 with its equivalent circuit. The thyristor that carries the on-state current is formed by the coupled N+PN- and PN-N+P+ transistors. The MCT can be turned on by using the n-channel MOSFET formed across the NPN transistor [15] and has excellent on-state voltage drop (approximately 1.1 volt). No current saturation is observed in the on-state; that is, this device does not exhibit any forward-biased safe operating area (FBSOA). This can be a problem for drives that rely upon the device for providing short-circuit protection (as done quite commonly by using IGBTs). However, the regenerative action of the thyristor can be broken by short circuiting the N+ emitter/P-base junction by gating on the p-channel MOSFET

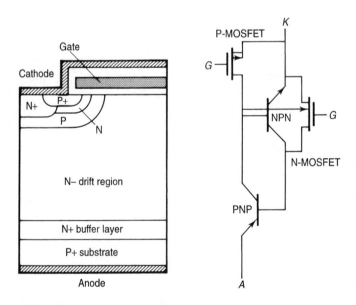

Figure 1-15. Cross section of the MOS-controlled thyristor (MCT) and its equivalent circuit.

integrated into the P-base region. It has been shown that devices with blocking voltages of 3000 volts can be made with good on-state voltage drop. Although the maximum controllable current density for small test devices has been shown to be extremely large, for large multicellular power devices it is in the range of 200 amperes per square centimeter [19]. It has also been found that the device must be operated with snubbers because of a limited reverse biased safe operating area [17]. One of the drawbacks of this MCT structure is that its triple-diffused junction structure is more complex than the double-diffused junction structure for the IGBT making its manufacturing more difficult. In spite of these issues, an MCT with current rating of 75 amperes and 600 volt forward blocking capability has recently become commercially available.

To address this fabrication problem, two MOS-gated thyristor structures were proposed which rely upon the DMOS process for their fabrication [20–21]. In the base resistance controlled thyristor (BRT) structure shown in Figure 1-16, the MOSFET used to turn off the device is formed adjacent to the P-base region using the same process steps as used for IGBTs. In the on-state, the current flows via the vertical thyristor formed between the N+PN- and PN-N+P+ transistors. The device has been shown to exhibit excellent on-state characteristics with a forward voltage drop of 1.1 volt. The thyristor regenerative action can be stopped by diverting the holes flowing into the P-base region, via the lateral p-channel MOSFET, into the cathode electrode. Although it has been shown that a high current density can be turned off in small devices, this device suffers from the same drawbacks as the MCT in terms of lacking any significant FBSOA.

In comparison with the IGBT, the MCT and BRT have a lower on-state voltage drop for the same turn-off switching time (see Table 1-1). The impact of this upon

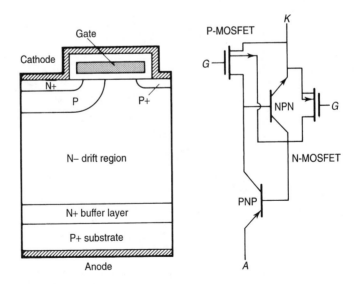

Figure 1-16. Cross section of the base resistance–controlled thyristor (BRT) structure and its equivalent circuit.

the power losses for the motor drive circuit is shown in Figures 1-17 and 1-18 when the device is used with either a PiN rectifier or an MPS rectifier. In comparison with the IGBT, although a reduction in the total power loss is observed at low operating frequencies due to the reduced on-state voltage drop, the impact is small at the higher operating frequencies. This indicates that the choice of the power switch is less critical than the choice of the power rectifier from the point of view of reducing the power losses in the drive.

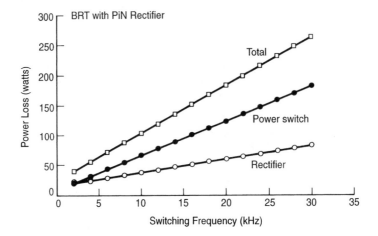

Figure 1-17. Power loss for drive with BRT (or MCT) and PiN rectifier. The DC bus voltage is assumed to be 400 volts, and the motor current is 15 amperes.

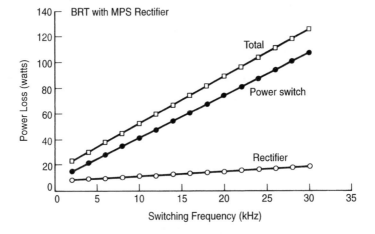

Figure 1-18. Power loss for drive with BRT (or MCT) and MPS rectifier. The DC bus voltage is assumed to be 400 volts, and the motor current is 15 amperes.

The emitter switched thyristor structure is shown in Figure 1-19 with its equivalent circuit. As implied by its name, in the EST, MOS-gated control is achieved by forcing the thyristor current to flow through a MOSFET channel. In the EST structure, this MOSFET is integrated into the P-base region of the thyristor. This not only provides MOS-gated–controlled turn-off capability but it also allows current saturation within a thyristor-based structure for the first time [22]. The FBSOA of the EST has been shown to be comparable to that for the IGBT [23]. However, this device has a slightly higher forward voltage drop than the MCT and BRT because the voltage drop across the MOSFET adds to the voltage drop across the thyristor because they are in series as indicated in the equivalent circuit. The typical on-state voltage drop of the EST is about 1.5 volts for a turn-off switching time of 0.3 microseconds as given in Table 1-1.

The power losses in the motor drive circuit for the case of an EST were calculated for the case of operation with a PiN rectifier and an MPS rectifier for the same load and bus voltage as the IGBT. These results are plotted in Figures 1-20 and 1-21 for comparison with the IGBT and BRT (MCT) cases. In each case, although the power loss is slightly greater than that for the BRT (MCT) case, the difference becomes small at higher operating frequencies. It can therefore be concluded that the EST may be a better choice than the BRT or MCT because of its excellent FBSOA.

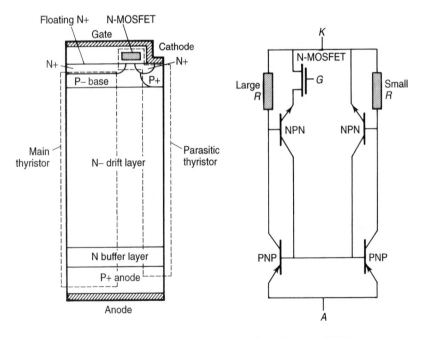

Figure 1-19. Cross section of the emitter switched thyristor (EST) structure and its equivalent circuit.

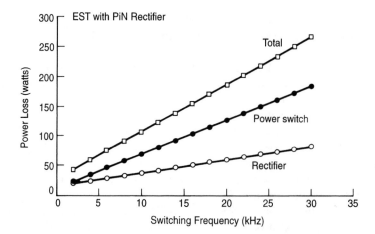

Figure 1-20. Power loss for drive with EST and PiN rectifier. The DC bus voltage is assumed to be 400 volts, and the motor current is 15 amperes.

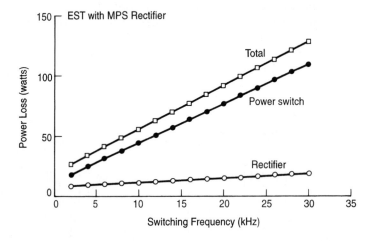

Figure 1-21. Power loss for drive with EST and MPS rectifier. The DC bus voltage is assumed to be 400 volts, and the motor current is 15 amperes.

1.7. NEW SEMICONDUCTOR MATERIALS

From the foregoing discussion, it can be surmised that further improvement in the power losses in the drive can be accomplished only by a reduction in the switching time for the power switch and the rectifier, as well as a reduction in the peak reverse recovery current for the rectifier. Recent theoretical analyses [24–25] have shown that very-high-performance FETs and Schottky rectifiers can be obtained by replacing silicon with gallium arsenide, silicon carbide, or semiconducting diamond. Among these, silicon carbide is the most promising because its technology is more mature than for diamond and the performance of silicon carbide devices is expected to be an order of magnitude better than the gallium arsenide devices. In these devices, the high breakdown electric field strength of silicon carbide leads to a 200-fold reduction in the resistance of the drift region. As a consequence of this low-drift region resistance, the on-state voltage drop for even the high-voltage FET is much smaller than for any unipolar or bipolar silicon device as shown in Table 1-1. These switches can be expected to switch off in less than 10 nanoseconds and have superb FBSOA. The analysis also indicates that high-voltage Schottky barrier rectifiers with on-state voltage drops close to 1 volt may be feasible with no reverse recovery transient. A plot of the on-state characteristics of SiC Schottky barrier rectifiers with different reverse blocking capability is shown in Figure 1-22. From these plots, it can be concluded that SiC Schottky barrier rectifiers with on-state voltage drop close to 1 volt will be possible with reverse blocking capability of up to 2000 volts. Recently, these theoretical predictions have been experimentally confirmed by the fabrication of Schottky barrier rectifier with breakdown voltages of 400 volts [26].

The low-drift region specific on-resistance for silicon carbide is also expected to allow the fabrication of power FETs with very low on-state voltage drop. A comparison of the specific on-resistance (on-resistance for 1 cm^2 device area) of silicon carbide drift regions for the three principle polytypes (3-C, 6-H, and 4-H) with silicon is provided in Figure 1-23. As pointed out earlier, an improvement in specific on-resistance by about two orders of magnitude is projected for all three polytypes. It is important to point out that in spite of the differences between the critical breakdown field strengths for the three polytypes, the specific on-resistance for the three cases is nearly identical because the difference in electron mobility provides a compensating factor. Using the projected drift region specific on-resistance, the on-state voltage drop of a SiC FET operating at an on-state current density of 100 amperes per square centimeter is calculated to be only 10 millivolts. If the additional resistances in the vertical FET structure are taken into account, the on-state voltage

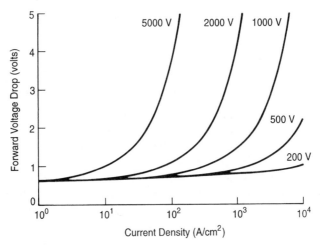

Figure 1-22. Calculated on-state characteristics of silicon carbide Schottky barrier rectifiers.

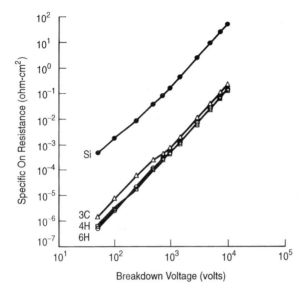

Figure 1-23. Comparison of the calculated specific on-resistance for three polytypes of silicon carbide drift regions with silicon.

drop can be anticipated to be less than 0.1 volts. For this reason, the power loss calculations for the SiC FET has been performed using an on-state voltage drop of 0.1 volts in this analysis. The most attractive SiC FET structure is the trench-gate device illustrated in Figure 1-5b. However, high-quality interfaces have not been obtained on thermally grown oxides on P-type SiC. Considerable research is now underway to explore methods for forming a gate insulator with good interface properties for fabrication of power MOSFETs.

If the silicon power switch and rectifier are replaced with the silicon carbide devices, it becomes possible to reduce the turn-off time to 10 nanoseconds because both the switch and the rectifier behave as nearly ideal devices. The power losses calculated for this case are shown in Figure 1-24 for comparison with the silicon devices. It is obvious that the power losses have been drastically reduced at all switching frequencies. Note that the power loss in the SiC rectifier is higher than that in the SiC FET because of its larger on-state voltage drop. This calculation assumes that the short switching time of the SiC devices can be utilized without encountering severe di/dt and dv/dt problems in the system. If this becomes a problem, it may be necessary to increase the switching time by adjusting the input gate waveform driving the SiC FETs. Calculations of the power losses in the drive when the turn-off time is increased from 10 nanoseconds to 50 nanoseconds have been performed and are given in Figure 1-25. In comparison with Figure 1-24, it can be seen that this results in an increase in the power loss in the switch, which doubles the total power loss. However, this power loss is still much smaller than that for the silicon devices.

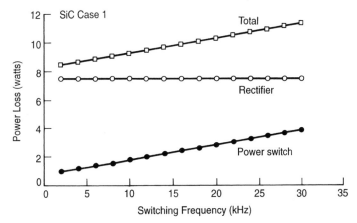

Figure 1.24. Power loss for drive with silicon carbide devices (Case 1 with 10 nanosecond turn-off time for switch). The DC bus voltage is assumed to be 400 volts, and the motor current is 15 amperes.

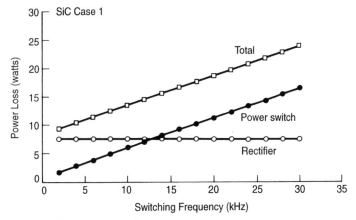

Figure 1.25. Power loss for drive with silicon carbide devices (Case 2 with 50 nanosecond turn-off time for switch). The DC bus voltage is assumed to be 400 volts, and the motor current is 15 amperes.

1.8. DEVICE COMPARISON

A comparison of the power losses for all the devices considered in the previous sections of this chapter can be performed with the aid of Figure 1-26 where the total loss for each case has been plotted. From this figure, it is clear that the impact of replacing the PiN rectifier with an MPS rectifier (or any other rectifier with faster reverse recovery time, reduced reverse recovery current, and comparable on-state

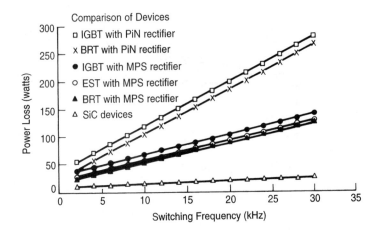

Figure 1-26. Comparison of total power losses for all the devices. The DC
         bus voltage is assumed to be 400 volts, and the motor current
         is 15 amperes. The results for the BRT are also applicable to
         the MCT.

voltage drop) is much greater than replacement of the IGBT with the BRT (or any
other MOS-gated thyristor-based device with on-state voltage drop of about 1 volt)
unless the operating frequency is relatively low. It is also obvious that the silicon
carbide devices are an extremely attractive choice for reducing power losses in variable speed motor drives.

1.9. SMART POWER CONTROL CHIPS

The development of the IGBT resulted in greatly simplifying the gate drive circuits
for motor drive applications. Since the IGBT gate drive circuit has much fewer
components than the base drive circuit for bipolar transistors and only relatively
small currents are needed to control the IGBT, it became possible to integrate the
gate drive circuit into a monolithic chip for the first time. This, in turn, created the
opportunity to add other functions (such as protection against adverse operating
conditions) and logic circuits to interface with microprocessors.

In a three-phase leg motor drive circuit, it is possible to partition the drive in
two basic ways. In one case, all the drive circuits for the lower switches in the totem
pole configuration are integrated together on one chip, while all the drive circuits for
the upper switches in the totem pole configuration are integrated on a second chip.
This avoids the need to integrate level shifting capability within the chip but requires
a separate high-voltage chip for providing this feature. Alternately, the drive can be
partitioned with the drive circuits for both the upper and lower switch in each phase
leg integrated on a monolithic chip. To achieve this capability, technology has been
developed to integrate the level shift circuits for the upper switch. This also requires
the ability for the drive circuit to rise in potential to above the DC bus voltage in

order to turn on the upper switch [27–28]. Three such control chips would be required in a three phase system.

In the smart power control chip, the sensing and protection circuits are usually implemented using analog circuits with high-speed bipolar transistors. These circuits must sense the following adverse operating conditions: overtemperature, overcurrent, overvoltage, and undervoltage. It is obvious that an overtemperature and overcurrent condition can cause thermal runaway leading to destructive failure, while an overvoltage condition can lead to avalanche injection-induced failure. The undervoltage lockout feature is also necessary because sufficient gate drive voltages are not generated at low bus voltages leading to very-high-power dissipation in the output transistors. An example of this condition is during system start-up. The bipolar transistors used in the analog portion must have a high-frequency response because of the high di/dt during short-circuit conditions. When the current exceeds a threshold value, the feedback loop must react in a short duration to prevent the current from rising to destructive levels.

The smart power chips used today are manufactured using a junction isolation technology. In these chips, the high-voltage-level shift transistors are usually lateral structures made using the RESURF principle to obtain a high breakdown voltage with thin epitaxial layers. It is anticipated that dielectric isolation (DI) technology will replace the junction isolation (JI) technology that is now being used for most smart power integrated circuits (ICs). Dielectric isolation offers reduced parasitics, a more compact isolation area, and the prospects for integrating MOS-gated bipolar devices that occupy less space that the lateral MOSFETs.

2.0. CONCLUSION

This chapter has been written with the point of view of reviewing power switch and rectifier technology of relevance to variable speed motor drive applications. By performing power loss calculations during a typical switching cycle, it has been shown that improvement in the reverse recovery behavior of power rectifiers is critical for reducing the total power loss. Although MOS-gated power switches with lower on-state voltage drop are useful for reducing the power loss in systems operating at low switching frequencies, they become of less importance for systems operating at higher frequencies of interest for motor control operation above the acoustic range. It has been shown that, in the future, power switches and rectifiers fabricated from silicon carbide offer tremendous promise for reduction of power losses in the drive. Thus, it can be concluded that advances in power semiconductor technology continue to look very promising for improving the performance of motor drives.

References

[1] Bose, B. K., "Power electronics and motion control—technology status and recent trends," *IEEE PESC Conf. Record*, pp. 3–10, 1992.

[2] Mokrytzki, B., "Survey of adjustable frequency drive technology—1991," *IEEE IAS Conf. Record*, pp. 1118–1126, 1991.

[3] Adler, M. S., and S. R. Westbrook, "Power semiconductor switching devices—a comparison based on inductive switching," *IEEE Trans. Electron Devices*, Vol. ED-29, pp. 947–952, 1982.

[4] Baliga, B. J., "Evolution of MOS-bipolar power semiconductor technology," *Proc. IEEE*, pp. 408–418, 1988.

[5] Baliga, B. J., *Modern Power Devices*, John Wiley and Sons Inc., New York, 1987.

[6] Baliga, B. J., M. S. Adler, P. V. Gray, R. P. Love, and N. Zommer, "The insulated gate transistor," *IEEE Trans. Electron Devices*, Vol. ED-31, pp. 821–828, 1984.

[7] Baliga, B. J., "Switching speed enhancement in insulated gate transistors by electron irradiation," *IEEE Trans. Electron Devices*, Vol. ED-31, pp. 1790–1795, 1984.

[8] Naito, M., H. Matsuzaki, and T. Ogawa, "High current characteristics of asymmetrical P-i-N diodes having low forward voltage drops," *IEEE Trans. Electron Devices*, Vol. ED-23, pp. 945–949, 1976.

[9] Shimizu, Y., M. Naito, S. Murakami, and Y. Terasawa, "High speed low-loss P-N diode having a channel structure," *IEEE Trans. Electron Devices*, Vol. ED-31, pp. 1314–1319, 1984.

[10] Tu, S. H. L., and B. J. Baliga, "Controlling the characteristics of the MPS rectifier by variation of area of Schottky region," *IEEE Trans. Electron Devices*, Vol. ED-40, pp. 1307–1315, 1993.

[11] Schlangenotto, H., J. Serafin, and F. Kaussen, "Improved reverse recovery of self-adapting P-emitter efficiency diodes (SPEED)," *Archiv fur Electrotechnik*, 74, pp. 15–23, 1990.

[12] Schlangenotto, H., J. Serafin, F. Sawitski, and H. Maeder, "Improved recovery of fast power diodes with self-adjusting P-emitter efficiency," *IEEE Electron Device Letters*, Vol. EDL-10, pp. 322–324, 1989.

[13] Nori, M., Y. Yasuda, N. Sakurai, and Y. Sugawara, "A novel soft and fast recovery diode (SFD) with thin P-layer formed by Al-Si electrode," *IEEE Int. Symp. Power Semiconductor Devices and ICs*, pp. 113–117, 1992.

[14] Mehrotra M., and B. J. Baliga, "Comparison of high voltage power rectifier structures," *IEEE Int. Symp. Power Semiconductor Devices and ICs*, pp. 199–204, 1993.

[15] Baliga, B. J., "Enhancement and depletion mode vertical channel MOS gated thyristors," *Electronics Letters*, Vol. 15, pp. 645–647, 1979.

[16] Temple, V. A. K., "MOS controlled thyristors," *IEEE Int. Electron Devices Meeting Digest*, Abstr. 10.7, pp. 282–285, 1984.

[17] Stoisiek, M., and H. Strack, "The MOS-GTO—a turn-off thyristor with MOS-controlled shorts," *IEEE Int. Electron Devices Meeting Digest*, Abstr. 6.5, pp. 158–161, 1985.

[18] Bauer, F., P. Roggwiler, A. Aemmer, W. Fichtner, R. Vuilleumier, and J. M. Moret, "Design aspects of MOS controlled thyristor elements," *IEEE Int. Electron Devices Meeting Digest*, pp. 297–300, 1989.

[19] Lendenmann, H., H. Dettmer, W. Fichtner, B. J. Baliga, F. Bauer, and T. Stockmeier, "Switching behavior and current handling performance of MCT-IGBT cell ensembles," *IEEE Int. Electron Devices Meeting Digest*, Abstr. 6.3.1, pp. 149–152, 1991.

[20] Nandakumar, M., B. J. Baliga, M. S. Shekar, S. Tandon, and A. Reisman, "A new MOS-gated power thyristor structure with turn-off achieved by controlling the base resistance," *IEEE Electron Device Letters*, Vol. EDL-12, pp. 227–229, 1991.

[21] Shekar, M. S., B. J. Baliga, M. Nandakumar, S. Tandon, and A. Reisman, "Characteristics of the emitter switched thyristor," *IEEE Trans. Electron Devices*, Vol. ED-38, pp. 1619–1623, 1991.

[22] Shekar, M. S., B. J. Baliga, M. Nandakumar, S. Tandon, and A. Reisman, "High voltage current saturation in emitter switched thyristors," *IEEE Electron Device Letters*, Vol. EDL-12, pp. 387–389, 1991.

[23] Iwamuro, N., M. S. Shekar, and B. J. Baliga, "A study of EST's short-circuit SOA," *IEEE Int. Symp. Power Semiconductor Devices and ICs*, pp. 71–76, 1993.

[24] Baliga, B. J., "Power semiconductor device figure of merit for high frequency applications," *IEEE Electron Device Letters*, Vol. EDL-10, pp. 455–457, 1989.

[25] Bhatnagar, M., and B. J. Baliga, "Comparison of 6H-SiC, 3C-SiC, and Si for power devices," *IEEE Trans. Electron Device*, Vol. Ed-40, pp. 645–655, 1993.

[26] Bhatnagar, M., P. K. McLarty, and B. J. Baliga, "Silicon carbide high voltage (400 V) Schottky barrier diodes," *IEEE Electron Device Letters*, Vol. EDL-13, pp. 501–503, 1992.

[27] Wildi, E., T. P. Chow, M. S. Adler, M. E. Cornell, and G. C. Pifer, "New high voltage IC technology," *IEEE Int. Electron Devices Meeting Digest*, Abstr. 10.2, pp. 262–265, 1984.

[28] Baliga, B. J., "An overview of smart power technology," *IEEE Trans. Electron Devices*, Vol. ED-38, pp. 1568–1575, 1991.

Chapter 2

Gordon R. Slemon

Electrical Machines for Drives

2.1. INTRODUCTION

The purpose of this chapter is to describe and discuss the various types of electric motors that are most usefully applicable in variable frequency, variable speed drives. Particular emphasis is placed on understanding the features and limitations of each motor type so that the designer of the drive system can make a rational choice of machine for each application.

Throughout the twentieth century, most of the drives for industrial processes, commercial equipment, and domestic appliances have been designed to operate at essentially constant speed, mainly because of the ready availability of economical induction motors operating on the available constant frequency AC power supply. For many mechanical loads, it has been recognized that a variable speed drive would provide improved performance, productivity, and energy efficiency. However, until recently, the provision of continuously variable speed has been considered too expensive for all but special applications for which constant speed was not acceptable. Examples of such drives are elevators, paper and steel mill drives, machine tools, and robots.

In recent decades, electric drives have been undergoing a major evolution as a result of two factors: (1) the availability and continuously decreasing cost of variable frequency electric supplies resulting from advances in power electronic switching devices and in microprocessor-based controls and (2) the increased concern about the present and future cost and availability of electric energy. Increasingly, variable speed drives are being designed into new systems and are being fitted to existing systems, permitting improved optimization of both the quality of the system's output and its energy efficiency.

In this chapter, attention is focused first on commutator or direct current motors — the traditional variable speed drive. These continue to be used but are gradually being superseded by other motor types. Next, consideration is given to induction motors — the current favorite for larger-power, variable speed applications. This is followed by a discussion of permanent magnet (PM) motors of both switched and synchronous types. These PM motors are increasingly popular and conceivably may become the dominant approach of the future. Other types currently being used, and with a promising future, are reluctance motors either of the switched or synchronous variety and hybrid motors combining PM and reluctance features.

2.2. MOTOR REQUIREMENTS FOR DRIVES

For each drive application, the mechanical system to be driven will have a specific set of desired criteria to be met. Those relating to torque and speed requirements can be appreciated with reference to Figure 2-1 [1, 2]. The base torque T_b (N) is normally the maximum continuous torque that can be provided at speeds up to the base speed ω_b (rad/s). For many drives, operation may be limited to the first or the forward drive quadrant of Figure 2-1. However, reversed speed and regeneration capability are frequently specified requiring operation in up to all four quadrants. Some applications call for continuous loading at base torque T_b. Others will specify a limiting duty cycle of loading. Values of torque in excess of T_b can be produced over limited

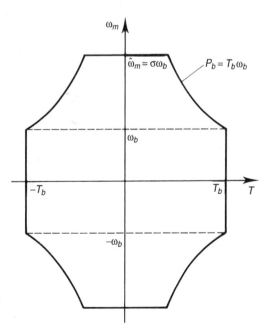

Figure 2-1. Operating region for drive with constant power range.

time periods, provided the motor temperature limit is not exceeded and certain limitations characteristic of the motor type are observed.

In some drives such as those used in traction, the power available from the electrical supply is limited. Rapid acceleration at low speed is desired, and in addition, operation is desired over a wide speed range. Assuming ideal efficiency, this supply power limit could be set at the product $P_b = T_b \omega_b$ (W). For speeds higher than the base speed, the drive can be operated at constant power up to a maximum value of $\hat{\omega} = \sigma \omega_b$ The required constant power speed-range ratio σ is frequently in the range 2 to 3 but may be as high as 6 in some locomotive drives.

For a high-performance drive such as in a robot, a maximum short-term torque $\hat{T}$ is typically to be maintained up to a base speed ω_b to provide rapid acceleration and deceleration in either direction of rotation. Depending on the duty cycle of the drive, this maximum torque may be several times the allowable continuous torque. A constant power speed range above ω_b may also be specified.

Other desired or limiting criteria that may be encountered in specifications are the maximum ratio of torque to rotor inertia (e.g., to provide maximum acceleration in high-performance drives), energy efficiency (e.g., for electric road vehicles), power-to-mass ratio (e.g., for airborne systems), torque ripple (e.g., for position control systems), acoustic noise, shape, volume, acceptability in hazardous environments, reliability, manufacturability, fail-safe features, initial cost, and present value of total lifetime cost, including the cost of energy.

2.3. COMMUTATOR MOTORS

Commutator motors, also known as direct current (DC) motors, have been widely used for variable speed drives for many years. Because of their long history and wide acceptability, it is convenient to begin with an examination of some key properties of these motors to establish a basis of comparison for the other variable frequency machine types to be discussed later.

Figure 2-2 shows a cross section of a commutator machine. A number of iron poles, typically four or six, project inward from a cylindrical iron yoke. A field coil encircles each pole, and these coils are normally connected in series with sequentially opposite polarity. Current in a field coil produces a magnetic flux in the air gap between the pole and the central rotating armature, the flux returning through adjacent oppositely directed poles. The armature is made of iron laminations and has axially directed slots in its outer surface to accommodate current carrying conductors. The coils of the armature winding are connected to a commutator or segmented mechanical switch mounted on the shaft. This commutator continuously switches the armature conductors so that those under each pole carry similarly directed currents.

Large commutator motors have additional narrow commutating poles located between the main poles and carrying armature circuit current. These aid in the switching of armature coils without sparking. Also, large motors may have a compensating winding consisting of conductors in slots in the pole faces. This winding

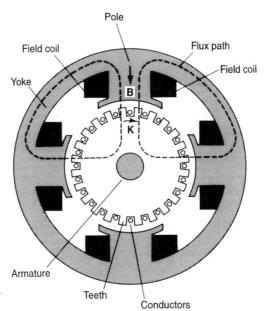

Figure 2-2. Cross section of a commutator motor.

ensures that the flux distribution across the pole face remains reasonably constant under load. These additional windings have little effect on the terminal properties of the motor. Typically, the main pole faces cover about 70% of the armature surface.

2.3.1. Torque Production

The interaction of the axially directed armature currents and the radially directed magnetic flux produces thrust and therefore torque on the shaft. The thrust force per unit of armature surface area is BK (N/m^2) where B (T) is the radial flux density in the air gap and K (A/m) is the linear current density around the armature periphery [1]. Thus, the torque exerted on the rotor may be expressed as

$$T = 2\pi r^2 \ell \gamma BK \qquad \text{N.m} \qquad (2.1)$$

where r is the rotor radius, ℓ is the stacked rotor length, and γ is the fraction of the rotor surface covered by the poles.

Typically, the armature slots occupy about half the circumference of the armature. The flux density in the armature teeth is limited to about 1.4–1.8 T by magnetic saturation. The gap flux density B under the pole is therefore limited to about 0.7–0.9 T.

2.3.2. Losses and Cooling

The continuous value of linear current density K is limited mainly by the maximum acceptable temperature of the armature winding insulation. An estimate of the allowable linear current density may be obtained by evaluating the winding power loss P_{SW} per unit of armature surface A,

$$\frac{P_{SW}}{A} = \frac{\rho K^2}{d_e} \qquad \text{W/m}^2 \tag{2.2}$$

where ρ is the resistivity of the armature copper at the operating temperature. The quantity d_e is the equivalent depth of armature conductor material if it was uniformly distributed on the armature surface and is typically about 0.2–0.3 of the armature slot depth.

The loss power that can be conducted away from the surface by the cooling air depends mainly on the air velocity. This cooling coefficient may be in the range 20–100 W/m^2 of surface area for each °C of temperature rise or difference between the surface and the cooling air [3].

Using these concepts in equation 2.1 and equation 2.2, a first approximation of the rotor dimensions can be obtained. For example, for a motor with a continuous rating of 50 kW at 2000 r/min (209 rad/s) with a cooling coefficient of 50 W/m^2.°C, a temperature rise of 60°C, and an equivalent depth of armature conductor of 10 mm, the linear current density could be about 40 A/mm. For an arbitrarily chosen ℓ/r ratio of 2, the rotor dimensions might be chosen at about $r = 95$ mm and $\ell = 190$ mm.

In addition to the loss in the armature winding, the commutator motor has losses in the iron laminations of the rotor core, conductor losses in the field coils and in the compensating and commutating pole windings, losses in the voltage drop of approximately 2 volts in the commutator, loss due to friction at the bearings and commutator, and loss due to windage. Typical values for the efficiency of fully loaded 5–150 kW motors might be 83–93% [2, 4]. Small motors may have an efficiency of less than 50%. Armature winding loss varies from 60% of the total loss for small motors to about 35% for large ones.

2.3.3. Equivalent Circuit

For a particular drive system, there will be a maximum or base voltage V_b available from the electrical supply which is usually a controlled rectifier or chopper (2). The number of turns in the armature winding and their series-parallel connections through the commutator to the brushes are chosen so that, at the base speed ω_b, and with maximum field flux, the voltage generated by the armature is approximately equal to the maximum or base supply voltage V_b. Then, ignoring losses, the rated or base supply current will be

$$I_b = \frac{T_b \omega_b}{V_b} \qquad \text{A} \tag{2.3}$$

The dynamic performance of the commutator motor may be predicted using the equivalent circuit of Figure 2-3, which includes the resistances and inductances of the armature and field circuits. The ratio of the torque T to the current i is a factor k_Φ, which is proportional to the flux and the number of armature turns. The same

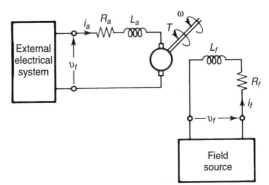

Figure 2-3. Equivalent circuit of a commutator motor.

constant relates the generated voltage to the speed. The field time constant $\tau_f = L_f/R_f$ (s) is much larger than the armature time constant $\tau_a = L_a/R_a$. With the field flux held constant at its maximum value, the torque response of the motor is limited only by the ability of its electrical supply to inject the required armature current i_a. The rotor inertia can be considered as part of the mechanical system.

Many drives require operation in all four quadrants of Figure 2-1. By appropriate design of the supply system, regeneration is achieved by reversal of the armature current direction and speed reversal by reversal of the supply voltage.

2.3.4. Constant Power Operation

The flux density in the air gap can be adjusted by controlling the current in the field coils. Normally, for all speeds below ω_b, this flux is kept constant at near its maximum value so that the maximum torque per ampere of armature current, and thus lowest loss in the winding resistance, can be achieved.

To operate the motor at the constant base power imposed by the supply power limitation at speeds above ω_b as shown in Figure 2-1, the field current is controlled so that the flux density is reduced in inverse proportion to the speed, thus keeping the generated armature voltage constant at about V_b while the supply current through the commutator to the armature winding remains at I_b. The torque at current I_b is proportional to the flux and thus reduces as the speed is increased.

2.3.5. Operational Limitations

Commutator motors have very desirable control characteristics, but their use is limited by a number of factors:

- A need for regular maintenance of the commutator
- A relatively heavy rotor with a high inertia
- Difficulty in producing a totally enclosed motor as required for some hazardous applications

- A relatively low maximum rotor speed limited by mechanical stress on the commutator
- Relatively high cost

For high-performance applications, the maximum torque that a commutator motor can produce is limited by the current that the commutator can switch without excessive sparking. Typically, this is in the range of two to five times the continuous rated current [4].

2.4. INDUCTION MOTORS

Most induction motors are designed to operate from a three-phase source of alternating voltage [1, 5, 6]. For variable speed drives, the source is normally an inverter that uses solid-state switches to produce approximately sinusoidal voltages and currents of controllable magnitude and frequency.

A cross section of a two-pole induction motor is shown in Figure 2-4. Slots in the inner periphery of the stator accommodate three phase windings a, b, and c. The turns in each winding are distributed so that a current in a stator winding produces an approximately sinusoidally distributed flux density around the periphery of the air gap. When three currents that are sinusoidally varying in time but displaced in phase by 120° from each other flow through the three symmetrically placed windings, a radially directed air gap flux density is produced that is also sinusoidally distributed around the gap and is rotating at an angular velocity equal to the angular frequency ω_s of the stator currents.

The most common type of induction motor has a squirrel cage rotor in which aluminum conductors or bars are cast into slots in the outer periphery of the rotor. These conductors are shorted together at both ends of the rotor by cast aluminum

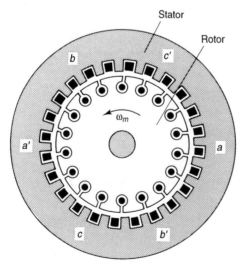

Figure 2-4. Cross section of an induction motor.

aluminum conductors or bars are cast into slots in the outer periphery of the rotor. These conductors are shorted together at both ends of the rotor by cast aluminum end rings, which also can be shaped to act as fans. In larger induction motors, copper or copper-alloy bars are used to fabricate the rotor cage winding.

2.4.1. Production of Torque

As the sinusoidally distributed flux density wave produced by the stator magnetizing currents sweeps past the rotor conductors, it generates a voltage in them. The result is a sinusoidally distributed set of currents in the short-circuited rotor bars. Because of the low resistance of these shorted bars, only a small relative angular velocity ω_R between the angular velocity ω_s of the flux wave and the mechanical angular velocity ω_m of the two-pole rotor is required to produce the necessary rotor current.

For low values of the relative angular velocity ω_R (normally called the slip velocity), the sinusoidally distributed linear current density of rms value K_r in the rotor conductors is effectively proportional to ω_R and is in space phase with the sinusoidally distributed air gap flux density of rms value B_g. The interaction produces a torque on the rotor of [1]

$$T = 2\pi r^2 \ell B_g K_r \qquad \text{N.m} \tag{2.4}$$

A magnetizing current is required in the stator windings to establish the flux in the machine. In addition, when the relative velocity of the rotor with respect to this flux wave causes a rotor current of linear density K_r to be established, the stator must set up an equal and opposite linear current density component so that the net effect of the magnetization will be retained. Thus, the stator can be considered to have two sinusoidally distributed space-wave components that are approximately in space quadrature with each other. The result is a linear current density K_s in the stator that is somewhat larger than that of the rotor.

The rms flux density B_g that can be maintained is limited by magnetic saturation in the teeth of both the stator and the rotor. Similar to the commutator motor armature, the stator tooth and slot widths are roughly equal, and the maximum flux density in the stator teeth is limited to the range 1.4–1.8 T. Thus, the maximum gap flux density is about 0.7–0.9 T, and the rms value of B_g is therefore usually in the range 0.5–0.65 T rms.

The continuous values of the rms linear current density K_r in the rotor and K_s in the stator are limited by many of the same heat dissipation considerations as discussed for the commutator motor. These limitations on flux density and current density are the main factors in establishing the base continuous torque T_b of the motor. The force per unit of rotor surface area and thus the rated torque are roughly comparable for induction and commutator motors of the same rotor radius r and effective length ℓ.

The base angular velocity ω_b of the motor will be fixed by the needs of the mechanical load to be driven. For the two-pole motor shown in Figure 2-4, the required base value of angular frequency of stator currents ω_s will differ from ω_b only by the angular frequency ω_R of the rotor currents.

Figure 2-5 shows the relation between the angular velocity and the torque for a number of values of stator frequency. For small values of the slip frequency ω_R, the torque is seen to be proportional to ω_R. For motoring action, ω_R has a positive value. To produce regeneration with reversed torque, the load will drive the mechanical angular velocity to a value greater than that of the stator angular frequency making the value of ω_R negative; that is, the sequence of the rotor currents is reversed. To reverse the direction of rotation, the sequence of the stator currents is reversed, that is, from *abc* to *acb*.

The flux density wave rotating at angular velocity ω_s induces sinusoidally time-varying voltages in the stator windings. The number of turns in each phase winding is therefore chosen so that, to a first approximation, the induced voltage E_s at the base value of frequency $\omega_s = \omega_b$ will be approximately equal to the base or maximum rms value V_b of the available stator phase voltage V_s. In most induction motor drives, the flux is maintained at or near its maximum value (limited by saturation) so as to obtain the maximum torque per ampere of stator current. To achieve this, the stator voltage is controlled to be approximately proportional to the stator frequency. More precisely, the induced voltage E_s, which differs from V_s by only the drop across the stator winding resistance R_s, is made proportional to ω_s, thus keeping the stator flux linkage phasor $\Lambda_s = E_s/\omega_s$ (Wb) constant in magnitude.

2.4.2. Equivalent Circuit Model

Just as the stator linear current density contains two components, similarly the stator current phasor I_s will have two components, a magnetizing component I_m lagging the stator voltage V_s by approximately 90° and a power component that is approximately in phase with the stator voltage for relatively small values of torque.

The rotor current and the stator current on the other side of the air gap set up a

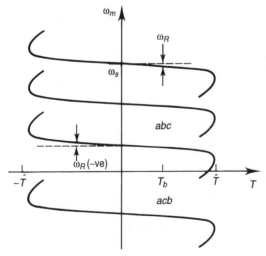

Figure 2-5. Speed-torque relations for induction motor.

leakage flux that follows a more or less circumferential path between these windings, partly in the air gap, partly across the tips of the rotor and stator teeth, and partly around the end windings [1]. This leakage reduces the flux entering the rotor, thereby reducing the induced rotor current and also shifting its space wave relative to the flux wave. Instead of the torque continuing to increase in direct proportion to the slip frequency ω_R between the angular velocities of the supply and the rotor, the torque reaches a maximum value $\hat{T}$ and then decreases for larger values of slip as shown in Figure 2-5.

The steady-state behavior of an induction motor as used in variable speed drives can be predicted with good accuracy by use of the equivalent circuit of Figure 2-6 [7] in which R_s is the resistance of a stator phase winding, L_m is the magnetizing inductance carrying the magnetizing component I_m of the stator phase current I_s, L_L is the leakage inductance, R_R is the rotor winding resistance as seen by the stator, and the load is represented by the resistance $R_R \omega_o/\omega_R$.

The magnetizing inductance L_m is essentially constant until magnetic saturation becomes significant in the stator and rotor teeth and possibly in the stator yoke. Normally, to make efficient use of the stator and rotor iron, the motor is designed for rated operation at a flux linkage level Λ_s, where saturation reduces the magnetizing inductance by up to about 30%. Typically, the magnetizing current is in the range of 20–50% of the rated stator current.

2.4.3. Number of Poles

While two-pole induction motors are used in some variable speed drives, an arrangement of four or more poles is common in conventional motors. For a p-pole motor, the sinusoidal distribution of each phase winding is repeated for $p/2$ complete waves around the stator. The mechanical angular velocity ω_m of the rotor at zero torque is then given by

$$\omega_m = \frac{2}{p}\omega_s \qquad \text{rad/s} \tag{2.5}$$

The term ω_o is used in the equivalent circuit of Figure 2-6 to denote the speed in electrical radians per second, that is, $\omega_o = (p/2)\omega_m$.

The number of poles does not affect directly the limiting values of either the air gap flux density B_g or the continuous rotor linear current density K_r. Thus, the torque capability of the motor as given in equation 2.4 is not changed with a change of pole number, and at the same speed, the rated mechanical power is relatively

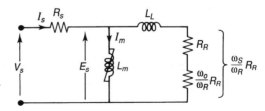

Figure 2-6. Steady-state equivalent circuit of induction motor.

unchanged. However, for the same speed, the required frequency of the supply will be increased in proportion to the number of poles.

An increase in the number of poles above two reduces the yoke thickness that is required outside the stator slots to accommodate the return paths of the radial tooth flux. This, in turn, allows for a larger rotor radius for a given overall frame size. From equation 2.4, it may be noted that, for given values of flux density, linear current density, and rotor shape (or ratio ℓ/r), the rated torque is proportional to the cube of the rotor radius. Thus, the power rating obtainable with a given frame size may be increased by increasing the pole number. Also, the required thickness of the rotor core inside the rotor slots is decreased. With a large pole number, a semi-hollow spoked rotor may be used reducing the rotor mass and inertia.

A disadvantage of increasing the number of poles is that, for a given shape ℓ/r, the magnetizing component of the stator current increases in proportion to the square of the number of poles. The power factor of the motor is therefore reduced, the loss in the stator windings is increased, and the required rating of the supply system is increased [8, 9]. Also, with increased numbers of poles, the coupling between the rotor and stator windings is somewhat decreased thus increasing the leakage inductance. Conversely, two-pole motors tend to have mechanical asymmetries that interact with the electromagnetic forces to produce rotor unbalance, shaft and bearing fluxes, and other parasitic effects. To avoid these effects, higher precision is required in machine fabrication than for motors with four or more poles. The optimum induction motor for a variable speed drive in the low- and medium-power range usually has either two or four poles. For higher-power, low-speed applications, higher pole numbers may be used together with shape ratios ℓ/r considerably less than 1.

2.4.4. Torque Expressions

Suppose the motor voltage V_s supplied by the inverter is controlled so that the voltage E_s induced in the stator winding is kept proportional to the supply frequency ω_s; that is, the stator flux linkage Λ_s is maintained constant at all values of speed. Using the equivalent circuit of Figure 2-6, the torque is equal to the power into the effective load resistance $R_R \omega_o/\omega_R$ for all three phases divided by the actual mechanical angular velocity ω_m. Alternatively, the torque can be evaluated as the total power entering the rotor circuit resistance $R_R \omega_s/\omega_R$ for all three phases divided by the synchronous mechanical velocity $(\frac{2}{p})\omega_s$. The value of the leakage inductance is typically in the range 0.15–0.25 per unit. With small values of the rotor frequency ω_R, the rotor circuit resistance will be in excess of 1.0 per unit. Then, the effect of the leakage inductance L_L can be ignored and the torque can be approximated by

$$T = 3 E_S^2 \frac{\omega_R}{\omega_s R_R} \frac{p}{2\omega_s} = \frac{3p}{2} \Lambda_s^2 \frac{\omega_R}{R_R} \qquad \text{N.m} \qquad (2.6)$$

Note the linear relation between torque and slip frequency. As the motor is loaded more heavily, the rotor frequency ω_R increases, decreasing the effective total resistance of the rotor circuit as seen by the stator, increasing the effect of the leakage inductance and shifting the rotor current distribution wave away from the flux wave. For constant flux linkage Λ_s, maximum power is transferred to the rotor when $\omega_R = R_R/L_L$ and the maximum torque is given by

$$\hat{T} = \frac{3p}{4} \frac{\Lambda_s^2}{L_L} \qquad \text{N.m} \tag{2.7}$$

This is just half of what it would have been at that rotor frequency if there were no leakage.

For regeneration, the slip frequency is negative. The same maximum reversed torque as in equation 2.7 is produced with a value of slip frequency $\omega_R = R_R/L_L$. In most motor designs, this maximum torque T will be in the range two to three times the continuous base torque T_b.

Most standard induction motors are made for operation on the standard utility constant voltage and constant frequency supply. To provide them with adequate starting torque, their rotors are frequently designed with deep bars or double cages of bars, making their effective resistance increase as the rotor frequency increases. This feature is not required when these standard motors are used in variable frequency drives since the rotor frequency is always small. When induction motors are designed specially for variable frequency operation, the rotor bars are made as large as can be accommodated to minimize the rotor resistance.

2.4.5. Losses and Efficiency

In the past, many motors, especially those in appliances, were designed to achieve minimal initial cost. Now, energy conservation has become a major concern. Because induction motors are so widely used, their efficiency has become an important public consideration.

The power losses in the stator and rotor windings can be determined using the resistances and currents in the equivalent circuit of Figure 2-6. Alternatively, the loss expression of equation 2.2 can be applied to both windings, including the end connections. Typically, at rated load, these winding losses each account for about 25–30% of the total loss [5]. In addition, there is the core loss in the laminated iron teeth and yoke of the stator that usually accounts for about 15–20% of the total. Further losses, usually called the stray losses, occur mainly because of the high-frequency local oscillations in fluxes in the stator and rotor teeth near the air gap. In larger motors, significant eddy current stray losses may occur in metal parts close to the stator end windings and the yoke. Most of the stray loss is proportional to the square of the stator current. Stray loss usually accounts for 10–15% of the total loss. There is also friction and windage loss, typically 5–10% of the total [10].

With the increased concern about the cost and eventual availability of electric energy, emphasis has been placed on the design and increased use of high-efficiency induction motors in contrast to the standard designs [11, 12]. These motors use more conductor material in both stator and rotor to achieve lower winding resistances and losses. They use a higher-quality iron at lower flux density in the stator to reduce core losses [13]. The air gap is increased to reduce the stray loss. For the same rating, these motors maintain the same frame size as standard motors but are made axially longer. The initial cost is increased by up to 25%. Typical values of efficiency for standard and high-efficiency, four-pole, 60 Hz induction motors are shown in Figure 2-7. It is seen that the losses are reduced by about one-third in the high-efficiency motors.

2.4.6. Dependence of Parameters on Size

The characteristics of the motor and the relative values of its parameters are dependent on the size of the machine. Suppose that all dimensions of a motor are multiplied by a factor k and that the shape, the speed, the flux density, and the cooling coefficient are unchanged. From equation 2.2, if only the dominant winding losses are considered, the power loss per unit of cooling surface area can be maintained constant if the linear current density K increases in proportion to $\sqrt{k}$. From equation 2.4, the torque and the power will then increase as $k^{3.5}$. The dominant winding losses will have increased by only k^2 and the iron losses by k^3. This, in part, explains the efficiency trend with rating of Figure 2-7.

Inductance parameters, being proportional to area over length, will increase by the factor k, while the winding resistances being proportional to length over area will be divided by k [14]. Thus, in larger motors, the leakage inductance will be increased. From equation 2.7, this increase in inductance tends to decrease the maximum available torque and also combines with the decrease in resistance to decrease the slip frequency at which this maximum torque occurs.

The magnetizing inductance will be increased in larger motors, and the magnetizing current will therefore be a smaller component of the supply current. The power factor of the motor increases with rating as shown in Figure 2-8. Note also the reduction in power factor as pole number is increased.

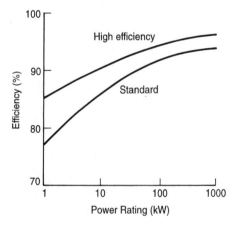

Figure 2-7. Efficiencies of standard and high-efficiency induction mo- tors over a range of power ratings.

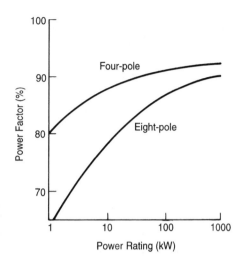

Figure 2-8. Typical power factor for four-pole and eight-pole induction motors over a range of power ratings.

2.4.7. Use in Industrial Drives

Induction motors are simpler in structure than commutator motors. They are more robust and more reliable. They require little maintenance. They can be designed with totally enclosed rotors to operate in dirty and explosive environments. Their initial cost is substantially less than for commutator motors and their efficiency is comparable. All these features make them attractive for use in industrial drives.

The majority of induction motors currently in use are in essentially constant speed applications. However, two factors have led to a reexamination of many of these applications: concern about process quality and productivity in manufacturing and concern about the cost of electric energy. An ability to adjust the speed of fans, pumps, and process drives addresses both these concerns. In most industrial applications, a capability for relatively slow adjustment of speed is adequate. Typically, a speed-sensing device provides a signal proportional to the angular velocity ω_o (in electrical rad/s) to which is added a signal representing the rotor angular frequency ω_R to give the desired torque with rated flux linkage. An inverter then supplies a frequency $\omega_s = \omega_o + \omega_R$ at a proportional value of voltage [15, 16]. The range of the rotor frequency $(\omega_R/2\pi)$ at rated load varies from about 5 Hz for small motors to less than 0.5 Hz for very large motors.

The use of high-efficiency rather than standard induction motors adds somewhat to the initial cost of the drive. This can, however, be more than compensated by the decrease in losses and the reduction in operating energy costs. For example, a typical increase in efficiency of 6% is obtained in a 10 kW, high-efficiency motor rather than one of standard design. If this motor is operated near rated load for 80% of the time, the annual saving will be 4200 kW.h per year having a value of $250 with electric energy at 6c/kW.h. This may be compared with a typical addition to the initial motor cost of only about $150. Thus, the added initial investment can be recovered in about eight months of operation. Alternatively, the present value of the energy saved over a 15-year lifetime with interest at 6% is about $2450, that is, about 16 times the added initial cost.

2.4.8. Constant Power Operation

Induction motors can be used to a limited extent for drives that require a constant power range such as that shown in Figure 2-1. Suppose the inverter voltage can be controlled up to some maximum value V_b. As the speed and the applied voltage are increased in proportion, rated flux linkage can be maintained up to the base supply frequency ω_b, or up to synchronous speed $(2/p)\omega_b$. As shown in Figure 2-9, values of torque up to the maximum value of equation 2.7 can be produced at speeds up to about this base value. For higher speed, the frequency of the inverter can be increased, but the supply voltage has to remain constant at the maximum value available in the supply. This causes the stator flux linkage Λ_s to decrease in inverse proportion to the frequency.

The maximum available torque is proportional to the square of the flux linkage as shown in equation 2.7. Figure 2-9 shows some speed-torque curves in the extended speed range. Typically, the induction motor is designed to provide a continuous torque rating T_b of about 40–50% of its maximum torque $\hat{T}$. Constant power can be achieved up to the speed at which the peak torque available from the motor is just sufficient to reach the constant power curve. A constant power speed range factor in the range of $\gamma = 2$–2.5 can usually be achieved. Within this range, the rotor frequency is increased until, at maximum speed, it is $\omega_R = R_R/L_L$. It may not be feasible to operate continuously over all this speed range because, at the limiting speed, the power factor will decrease to less than 0.7, increasing the stator

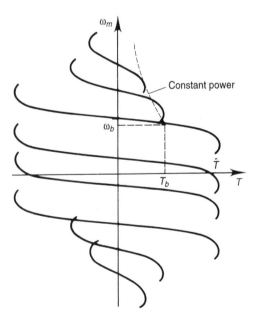

Figure 2-9. Torque-speed relations for constant power operation of induction motor.

current beyond its rated value. The motor heating may be excessive unless the duty factor is low.

In recent years, the drive choice for locomotives has frequently been the induction motor [17]. A similar choice has been made for some electric vehicle drives.

2.4.9. Use in High-Performance Applications

Induction motors are also being used extensively in applications requiring rapid and precise control of speed and position. Using an approach known as vector control [18], a transient response at least equivalent to that of a commutator motor can be achieved.

The capability of the induction motor in this high-performance role can be appreciated by use of the alternate, inverse-Γ form of equivalent circuit shown in Figure 2-10 [7]. The voltage, current, and flux linkage variables in this circuit are space vectors from which the instantaneous values of the phase quantities can be obtained by projecting the space vector on three radial axes displaced 120° from each other. (In much of the literature on induction motor dynamics, the real and imaginary components of the space vectors are separated, resulting in separate direct and quadrature axis equivalent circuits but with equal parameters in the two axes.)

The parameters of the circuit of Figure 2-10 are similar to and can readily be derived from those of Figure 2-6. The mechanical load is represented by a source voltage e'_o, which is proportional to both the speed and the rotor flux linkage. The instantaneous torque of the motor is given by

$$T = \frac{3p}{4} \, \lambda'_R \, i'_R \qquad \text{N.m} \tag{2.8}$$

Changes in the rotor flux linkage λ'_R can be made to occur only relatively slowly because of the large value of the magnetizing inductance L'_m of the induction motor. Vector control is based on keeping the magnitude of the instantaneous magnetizing current space vector i'_m constant so that the rotor flux linkage remains constant. The motor is supplied from an inverter that provides an instantaneously controlled set of phase currents that combine to form the space vector i_s. This consists of two components. The first is the magnetizing space vector i'_m, which is controlled to have constant magnitude to maintain constant rotor flux linkage. The second component is a space vector i'_R, which is in space quadrature with i'_m. This component is instantaneously controlled to be proportional to the demand torque.

To the extent that the inverter can supply instantaneous stator currents meeting these two requirements, the motor is capable of responding without time delay to a

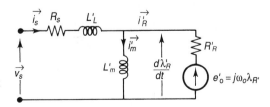

Figure 2-10. Transient equivalent circuit of an induction motor.

demand for torque. This feature, combined with the relatively low inertia of the induction motor rotor, makes this drive attractive for high-performance control systems.

Vector control requires a means of measuring or estimating the instantaneous magnitude and angle of the rotor flux linkage space vector λ'_R. Direct measurement is generally not feasible. Rapid advances are being made in devising control configurations that use measured electrical terminal values for estimation [19]. A number of these are discussed in Chapters 5 and 9.

2.4.10. Some Drive Design Considerations

For many drive applications, particularly those requiring relatively low power, inverters with high switching speed can produce variable voltage and variable frequency with little significant harmonic content. With these, either standard or high-efficiency induction motors can be used with little or no motor derating. However, the inverters used in larger drives have limits on switching rate that cause their output voltages to contain substantial harmonics of orders 5, 7, 11, 13, and so on. These, in turn, cause harmonic currents and additional heating in the stator and rotor windings. These harmonic currents are limited mainly by the leakage inductance. For simple six-step inverters, the additional power losses, particularly those in the rotor, may require derating of the motor by 10–15% [20].

If the motor is cooled by its rotor fans, it may also have to be derated for continuous low-speed operation. Typically, for a constant torque load, 80% rating might be acceptable for 20% speed and as low as 50% at very low speed. This derating may be avoided by use of forced ventilation from an auxiliary fan motor.

In large mill motors in the megawatt range, sets of phase-shifted stator windings and 6- or 12-step inverters may be used to reduce harmonic losses. Also, motors with relatively high leakage inductance may be chosen to reduce the harmonic currents in the rotor [21].

Existing constant speed drives frequently have an induction motor that is somewhat oversized. These can usually be converted to variable speed operation using the original induction motor. Most of the subsequent operation will be at lower load and lower loss than that for which the motor was designed.

The rugged construction of the induction motor allows for high values of maximum speed, limited by bearing life, windage losses, and the natural mechanical frequency of the rotor. A number of motors have been designed for the speed range 5000–50,000 r/min.

Many of the inverters supplying induction motors use pulse width modulation to produce a voltage wave with negligible lower-order harmonics. The wave consists of pulses formed by switching at relatively high frequency between the positive and negative sides of the DC link voltage supply. With larger motors that operate from AC supplies of 2300 V and above, the rapid rate of change of the voltage applied to the winding may cause deterioration and failure in the insulation on the entry turns of standard motors. The severity of these pulses can be enhanced by the cable between inverter and motor. For special variable speed designs, failure can be avoided by appropriate application of additional insulation.

2.4.11. Wound Rotor Motors

Wound rotor induction motors with three rotor slip rings have been used in adjustable speed drives for many years. In an induction motor, torque is equal to the power crossing the air gap divided by the synchronous mechanical speed. In early wound rotor induction motor drives, power was transferred through the motor to be dissipated in external resistances connected to the slip ring terminals of the rotor. This resulted in an inefficient drive over most of the speed range. More modern slip ring drives use an inverter to recover the power from the rotor circuit, feeding it back to the supply system [1, 2].

A speed range limited to about 70–100% of synchronous speed is adequate for many fan and pump drives. As the power of these drives is proportional to the cube of the speed, 70% speed produces 34% power. For these, the wound rotor system with rotor power recovery is frequently economic because of the reduced rating and cost of the inverter relative to that of a stator-fed induction motor drive. Disadvantages are the increased cost of the motor relative to that of a squirrel cage induction motor, the need for slip ring maintenance, difficulty in operating in hazardous environments, and the continued need for switchable resistors for starting.

2.5. SYNCHRONOUS PERMANENT MAGNET MOTORS

While induction and wound field commutator motors still dominate much of the drive market, much attention is currently being given to permanent magnet motors. These motors use magnets to produce the air gap magnetic field rather than using field coils as in the commutator motor or requiring a magnetizing component of stator current as in the induction motor. Significant advantages arise from the simplification in construction, the reduction in losses, and the improvement in efficiency.

The two most common types of PM motors are classified as (1) synchronous (i.e., with a uniformly rotating stator field as in an induction motor) and (2) switched or trapezoidal (i.e., with a stator field that is switched in discrete steps). The latter type is discussed in Section 2.6.

PM motors are made in a number of configurations. One of the simplest is shown in cross section in Figure 2-11. The rotor has an iron core that may be solid or may be made of punched laminations for simplicity in manufacture. Thin permanent magnets are mounted on the surface of this core using adhesives. Alternating magnets of opposite magnetization direction produce radially directed flux density across the air gap. This flux density then reacts with currents in windings placed in slots on the inner surface of the stator to produce torque.

PM motors are normally constructed with totally enclosed rotors. This protects the magnets from collecting iron dust from the cooling air and also excludes other

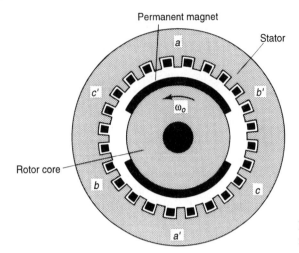

Figure 2-11. Cross section of permanent magnet motor.

In Section 2.3, it was noted that the maximum air gap flux density in induction motors is usually in the range 0.7–0.9 T because magnetic saturation limits the flux density in the stator teeth to the range 1.4–1.8 T and because optimum design usually results in nearly equal slot and tooth widths. Thus, a technically ideal magnet material for use in the synchronous PM motor of Figure 2-11 would be capable of producing an air gap flux density with peak value similar to that in an induction motor.

2.5.1. Permanent Magnet Materials

A review of the characteristics of the major permanent magnet materials provides a basis for appreciating the potential and limitations of PM motors. Magnetization characteristics of the most common magnetic materials currently used in motors are shown in Figure 2-12. All these PM materials have their magnetic domains well aligned, resulting in an essentially straight line relationship in much if not all of the second quadrant of the B-H loop. The slope of this characteristic is such that the magnet looks like a constant source current encircling a block of material with a relative permeability only slightly greater than that of air, that is, $\mu_r \approx 1.05$–1.07.

The magnet can be operated at any point on the linear part of the B-H characteristic and remain permanent. However, if the flux density is reduced beyond the knee of the characteristic (denoted as flux density B_D), some magnetism will be lost permanently. On removal of a demagnetizing field greater than this limit, the new characteristic will be a straight line parallel to but lower than the original.

Ferrite PM materials have been available for decades. Their cost is attractively low. However, their residual flux density B_r at only $0.3 - 0.4$ T is much lower than the desired range of gap flux density. Also, the types of ferrite that are made with a higher value of residual density B_r also have a higher value of minimum density B_D and thus are more easily demagnetized. Ferrites have high resistivity and thus have low core losses in situations with rapidly varying flux components. Ferrites can be

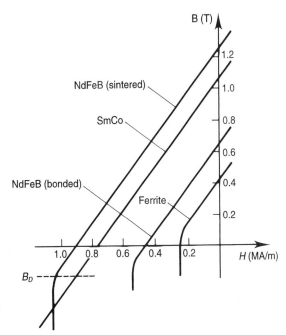

Figure 2-12. B-H characteristics for several permanent magnet materials.

higher value of residual density B_r also have a higher value of minimum density B_D and thus are more easily demagnetized. Ferrites have high resistivity and thus have low core losses in situations with rapidly varying flux components. Ferrites can be operated at temperatures up to about 100°C. An increase in temperature increases the residual flux density and decreases the knee point density [22]. Special care is needed in design to ensure that demagnetization does not occur when operating with high currents at very low operating temperature. Ferrite magnets are extensively used in low-cost PM motors. For these, the stator tooth width is made much smaller than the slot width.

Samarium cobalt magnets have a higher value of residual flux density (0.8–1.1 T). They are highly resistant to demagnetization having values of B_D well into the third quadrant of the B-H loop. The residual flux density B_r decreases somewhat with increase in temperature while the knee point density B_D increases, signifying an increased sensitivity to demagnetization as temperature is increased. The resistivity is about 50 times that of copper. The cost of these materials is relatively high, reflecting the cost of a rare earth element, samarium, and an expensive metal, cobalt. Until recently, samarium cobalt was the preferred magnet material for most high-performance PM motors.

The introduction of sintered neodymium-iron-boron (NdFeB) materials in 1983 provided a major impetus to the use of PM magnets in motors [23]. At room temperature, the residual flux density is in the range 1.1–1.25 T. This is adequate to produce a flux density of 0.8–0.9 T across a relatively large air gap, for example, a gap of 1 mm, with a magnet thickness of only about 3–4 mm. The residual flux density reduces by about 0.1% for each degree rise in temperature. The knee point flux density increases rapidly with temperature imposing a limit on the maximum operating temperature of NdFeB in the range 100–140°C, depending on the detailed composition of the material. The resistivity is about 85 times that of copper.

The cost of these sintered NdFeB materials is still relatively high, typically $150–175 per kilogram, largely because of manufacturing complexity of the sintering process. Experience with other materials would, however, lead to an expectation of lower cost in the future as volume of use is increased since two of the basic materials iron (77%) and boron (8%) cost relatively little and neodymium is one of the more prevalent of the rare earth elements.

Bonded NdFeB magnets have a residual flux density of about 0.6–0.7 T. They can be produced at about half the cost per kilogram of sintered NdFeB. Their density is about 75% of that of sintered magnet material. A significant advantage of this form of NdFeB over the sintered variety is that magnet sections can be made in a wider variety of shapes and in larger dimensions.

2.5.2. Equivalent Circuit

For the synchronous type of PM motor with surface magnets as shown in Figure 2-11, the stator is essentially identical to that of an induction motor. The three phase windings are each distributed to provide a near-sinusoidal distribution of linear current density around the air gap periphery so that a set of phase currents, sinusoidally varying in time, will set up a field that is sinusoidally distributed peripherally and is rotating at an electrical angular velocity ω_s, that is, a mechanical angular velocity of $(2/p)\omega_s$.

Each magnet may be viewed as equivalent to a resistanceless conducting loop with a current of about 800 B_r amperes for each millimeter of magnet thickness. This loop can be considered to encircle a space having a relative permeability μ_r only slightly greater than that of air. The air gap flux density B_g can therefore be approximated by [24]

$$B_g \approx B_r \frac{\ell_m}{\ell_g} \qquad \text{T} \tag{2.9}$$

where ℓ_m is the magnet thickness and ℓ_g is the gap between the iron faces of the rotor core and the stator with appropriate provision for stator slotting.

The flux density produced by the magnets is a rectangularly distributed wave of relatively constant magnitude over the magnet surface, alternating in polarity with adjacent magnet poles. To the extent that the stator windings are approximately sinusoidally distributed, only the fundamental space component of the magnet flux density wave can link with the stator windings. The magnitude of this component is proportional to $\sin\alpha$, where α is the half angular width of the magnet measured in electrical radians. A typical value for α for a synchronous PM motor is about $\pi/3$ rad, that is, a magnet coverage of two-thirds of the rotor surface. At this value, the rms gap flux density is 0.78 of the peak gap density.

When the rotor rotates at angular velocity ω_m, the effect of the rotor as seen by the stator can be represented as a sinusoidal current source with an angular frequency $\omega_s = (p/2)\omega_m$ for a p-pole motor. In the steady-state equivalent circuit of Figure 2-13, this current is represented by the phasor $I_F \angle 0$. The stator current phasor is $I_s \angle \beta$, where the phase angle β is the electrical angle by which the peak field of the stator leads the rotor magnet axis, which in turn is related directly to the angular position of the rotor [25].

The magnetizing inductance L_m for the PM motor is much smaller than that of an induction motor with the same stator because the effective magnetic gap ℓ_g between the stator iron and the rotor iron is much larger, the magnet material having a permeability approximately equal to that of air. Thus, the magnetizing current I_m is usually in the range 2–5 times rated stator current in contrast with the range 0.2–0.5 for induction motors. Typically, the current I_F has a magnitude of 2–5 times rated current. The product of I_F and $\omega_s L_m$ is such that the open circuit voltage is then equal to the rated value at rated speed.

2.5.3. Operating Characteristics

A PM motor provides continuous torque only when the speed is directly related to the supply frequency. The supply may be either a three-phase controlled voltage or controlled current. Many synchronous PM drives are fitted with a rotor position sensor from which the rotor angle can be obtained. The stator voltages or currents are then constructed by switching action of the inverter to have the desired waveshape and the desired relative instantaneous phase angle. For the current source drive, using the equivalent circuit of Figure 2-13, the torque can be derived as the total air gap power divided by the mechanical angular velocity and expressed as

$$T = \frac{3p}{2} L_m I_F I_s \sin \beta \qquad \text{N.m} \qquad (2.10)$$

Figure 2-14 shows the torque as a function of the angle β for a surface magnet motor with constant magnitude of stator current. For a given value of stator current, maximum driving torque can be obtained with a rotor angle of $\beta = 90°$. For this condition, the electrical variables are displayed in the phasor diagram of Figure 2-15a. This mode of operation gives the maximum torque per ampere of stator current and therefore a high efficiency. However, the supply power factor is seen to be less than unity. Operation in this condition is practical and preferred up to the speed at which the voltage limit of the inverter is reached. Above this speed, a more appropriate operating condition is one which provides for unity power factor thus maximizing the utilization of the inverter voltampere rating. This is achieved by

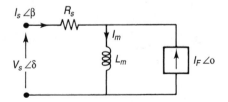

Figure 2-13. Equivalent circuit for PM motor.

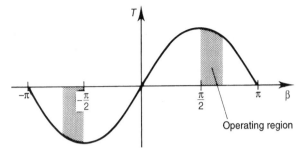

Operating region **Figure 2-14.** Torque-angle relation for PM motor.

increasing the current lead angle β as shown in the phasor diagram of Figure 2-15b. As shown in the operating region of Figure 2-14, this causes only a small reduction in torque. For regeneration, the rotor angle β is made similarly negative.

For a voltage-driven PM drive, the magnitude of the supply voltage and its frequency are made proportional to the motor speed. Operating conditions equivalent to those of Figure 2-15a and b can be employed by appropriate adjustment of the angle δ by which the stator voltage phasor leads the field axis. However, the motor can also be operated with a leading or capacitive stator current by increase of the angle δ as shown in Figure 2-15c. This condition is particularly desirable for some large drives since it allows use of a load-commutated inverter, that is, one where the switches are turned off by their currents going to zero rather than requiring auxiliary turn off means [2].

2.5.4. Magnet Protection

Demagnetization of a part of the magnet can occur if the flux density in that part is reduced to less than the knee point flux density B_D shown for various magnet

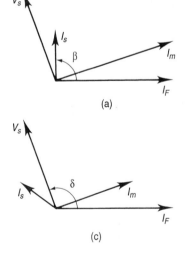

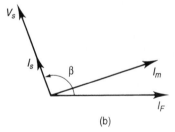

Figure 2-15. Phasor diagrams for PM motor (a) for maximum torque per ampere, (b) for unity power factor, (c) for leading power factor.

materials in Figure 2-12. This can result in permanent reduction in torque capability since it is usually not feasible to remagnetize the magnets without disassembly of the motor.

As seen in the phasor diagrams of Figure 2-15, the stator current in the driving mode is normally controlled to lead the effective magnet current by 90° or more. The effect of this stator magnetic field is to increase the air gap flux density near the leading edge of each magnet and decrease it near the lagging edge. With normal values of stator current, the effect of the air gap flux in producing torque is relatively unchanged. The limiting value of permissible stator current is that which brings the flux density down to the value B_D at the lagging edge. Most PM motors are designed to withstand considerable overload currents without danger to the magnets. The safe value of the stator current I_s usually can exceed the magnetizing current I_m without causing demagnetization [25]; that is, it can be typically 2–5 times the rated stator current.

The major danger to the magnet may arise from a short circuit on the stator terminals as a result of a failure in the inverter. Except for the effect of the stator resistance, which will be relatively small at high speed and frequency, particularly in larger motors, the steady-state short-circuit current will be equal to the equivalent magnet current I_F. The stator and magnet fields will oppose each other directly, causing the greatest reduction in flux density at the magnet center. If the limiting magnet flux density B_D is zero or negative, this steady-state current can usually be tolerated without damage [26]. However, the steady-state component of the short-circuit current space vector is accompanied by an initial transient space vector component of equal instantaneous magnitude. This transient decays at a time constant equal to the inductance L_m divided by the stator resistance R_s. In small machines, this time constant is so small in relation to a half period of the maximum fundamental frequency that it has little effect. However, in large motors, the time constant may be several periods in length and the demagnetizing effect of the short-circuit current may be nearly doubled. For these motors, one means of providing the required protection is to design for additional stator leakage inductance or to add external series inductance.

2.5.5. Losses and Efficiency

Synchronous PM motors are potentially much more efficient than induction motors. There is no equivalent of the loss in the rotor bars of the induction motor. Also, the stray losses that result mainly from magnetic interaction of the closely adjacent stator and rotor teeth in the induction motor are effectively eliminated by the large iron-to-iron gap of the PM motor. With modern magnet materials, the resistivity is such that losses due to tooth ripple variations in the air gap flux density produce little loss in the magnet material for most designs. This loss may, however, become significant for motors operated at very high frequencies and speeds. This leaves the stator core loss, the stator winding loss, and the windage and friction loss.

Let us compare PM and induction motors of the same frame size and shape. At the same speed, the friction loss in the bearings will be similar for the two motors.

The windage of the PM motor may be somewhat reduced due to the reduced need for an internal fan in a totally enclosed motor that has negligible rotor heating. For the same maximum flux density, the hysteresis component of the core loss in a surface magnet motor will be the same as that of an induction motor, but the eddy current component will be higher. This is mainly due to the rapid rise and fall in the flux density of the stator teeth as each magnet edge passes into or out of the space under a tooth [27]. The eddy current iron loss can be minimized by appropriate choice of the iron lamination material and also by designing with a smaller number of wider teeth per pole. In addition, the rate of change of tooth flux density can be reduced by beveling the magnet edges. Typically, the PM motor core loss may be in the range 1.25–2.0 times that of a machine with near-sinusoidal flux distribution.

The PM motor can be operated at unity power factor, while the induction motor will always have a lagging power factor, typically in the range 0.8–0.9 for four-pole motors and lower for larger numbers of poles (Figure 2-8). For the same input power rating, the ratio of the stator currents of the PM and induction motors will be this power factor. Thus, the ratio of the stator winding losses for the PM and induction motors will be the induction motor power factor squared.

For the same power ratings, the total losses in the PM motor will typically be about 50–60% of those of the induction motor [28]. Predicted values of PM motor efficiency are about 95–97% for ratings in the range 10–100 kW as compared with 90–94% for induction motors. Also, the reduction in PM motor efficiency with reduction in speed will be less than experienced with induction motors. The overall drive efficiency is further improved because the rating, and therefore the loss, in the inverter will be reduced roughly in proportion to the power factor.

With PM motors, the number of poles can be chosen to optimize the efficiency. Increasing poles increases the frequency and thus the core loss per kilogram of stator iron. However, the thickness and mass of the stator yoke is reduced in approximately inverse proportion to the pole number, partially compensating for the effect of increased frequency. Also, the decrease in yoke thickness allows an increase in the rotor radius for a given overall frame size. This allows the required torque to be produced by a shorter motor. An additional advantage of increased pole numbers is the shortening of the end turns of the stator winding reducing the stator resistance and therefore the loss. From an efficiency standpoint, the optimum number of poles is frequently in the range 8–12.

The efficiency of PM motors can be made sufficiently high that power loss per unit of surface area will be relatively low. The internal temperature of the motor can be maintained at a level where the temperature limit of the magnet material is not a limitation in many designs. Exceptions are special situations such as those with a very high ambient temperature.

2.5.6. Use in Industrial Drives

The surface PM motor has the potential to replace the induction motor in a number of industrial, commercial, and domestic variable speed drive areas based largely on its energy-saving capability. Its original cost will be higher than that of an induction motor mainly because of the cost of the magnet material. However, the present value of the energy saved by a 1% increase in efficiency is about $50 per kilowatt (kW) of motor rating, assuming electric energy at $0.06 per kW.h, continuous rated loading, interest at 6%, and a 15-year lifetime. Thus, the increase in efficiency of about 7–3%, which can potentially be achieved by PM motors with a range of ratings from 1 to 100 kW, may be valued at $350–$150 per kW of rating. This is many times the original trade cost of induction motors and well above the current estimated production cost of PM motors.

2.5.7. Constant Power Applications

In both commutator and induction motor drives, constant power operation is achieved by flux reduction. The surface PM motor considered in this section is essentially a constant flux motor. A limited amount of flux weakening can be achieved by increasing the lead angle β of the stator current. While the component of stator current proportional to $\sin \beta$ produces torque, the component proportional to $\cos \beta$ acts to reduce the flux. However, achieving a useful constant power range is not usually practical with surface magnet motors. A large demagnetizing component of stator current would be required to produce a significant reduction in magnet flux, and this would increase the stator loss substantially. Other PM motor structures with flux reduction capability are considered in Section 2.8 on PM reluctance motor drives.

In some traction applications, the surface PM motor may be a good choice even without flux weakening [29]. The motor size is determined by the required base torque and the losses are essentially independent of the number of stator turns. For this constant power mode of operation, the stator windings are designed so that the generated voltage is equal to the supply voltage at maximum speed rather than at base speed. At speeds up to the base speed ω_b of the constant power range, the efficiency of the motor is essentially the same as for one designed for rated voltage at speed ω_b. For operation above base speed, the stator current from the inverter is reduced in inverse proportion to the speed. This mode of operation in the high-speed range reduces the dominant stator winding losses relative to a machine in which the flux is reduced and the current kept constant. The losses in the inverter are, however, increased due to its higher current rating. For an electric road vehicle that must carry its energy store, the net energy saving may be sufficiently valuable to overcome the additional cost of the larger inverter. A further advantage of this approach is that, if the DC supply to the inverter is lost, the open circuit voltage applied to the inverter switches will be within their normal ratings, that is, V_b.

2.5.8. Use in High Performance Drives

Surface PM motors using NdFeB magnets are particularly suited for the high values of acceleration required in drives such as for robots and machine tools. These motors are often operated with high acceleration for a short time followed by a longer period of low torque. At such low values of load factor, the cooling capability is frequently not a limitation. The major interest is in obtaining the maximum acceleration from the motor. The short-term stator current of a surface PM motor is limited to the value required for magnet protection. Sample designs have shown that, with a duty factor of 5%, acceleration values greater than 100,000 rad/s^2 can be produced in a motor with a maximum torque capability of 40 N.m and an acceleration over 20,000 rad/s^2 is achievable for a 400 N.m peak torque motor [30]. These values of acceleration are significantly higher than can be achieved with either induction or commutator motors of similar maximum torque rating. PM motors have the additional advantage that their overall mass and volume can be made considerably less than that of other motor types.

2.6. SWITCHED OR TRAPEZOIDAL PM MOTORS

The other major class of PM motor drives is alternatively known as *trapezoidally excited PM motors*, or *brushless DC motors*, or simply as *switched PM motors* [31]. Normally, these have stator windings that are supplied in sequence with near-rectangular pulses of current.

2.6.1. Star-Connected Motor

A cross section of one type of motor is shown in Figure 2-16. In most respects, this switched motor is identical in form with the synchronous PM motor of Figure 2-11. For this two-pole motor, the rotor magnets extend around approximately 180° peripherally. The stator windings of this motor are connected in a star as shown in Figure 2-17a. These windings are generally similar to those of an induction or synchronous motor except that the conductors of each phase winding are full pitched; that is, they are distributed uniformly in slots over two stator arcs each of 60°.

The electrical supply system is designed to provide a current that can be switched sequentially to pairs of the three stator terminals. In the condition shown in Figure 2-16 and Figure 2-17a, current i has just been switched to enter into phase a and exit out of phase c. For the next 60° of counterclockwise rotation, two 120° stator arcs of relatively uniform current distribution are so placed with respect to the magnets as to produce counterclockwise or driving torque of relatively constant magnitude. As the leading edge of the upper magnet crosses the line between sectors b and a', the current i continues through phase c but is switched to enter through phase b rather than a.

The sequence of switching actions can be simply triggered by use of signals from position sensors (e.g., Hall generators) mounted at appropriate positions around the stator. Six steps of this switching cycle are required per revolution for a two-pole motor.

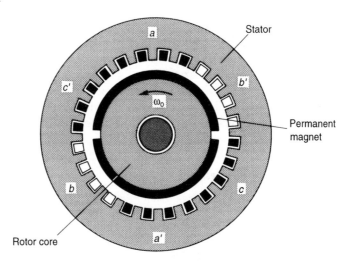

Figure 2-16. Cross section of a switched PM motor.

density K_{rms} is $\sqrt{(2/3)}$ of the maximum current density $\hat{K}$. Thus, the torque produced by this motor can be expressed as

$$T = \sqrt{\frac{2}{3}}\, 2\pi r^2 \ell B_g K_{rms} = k_\varphi\, i \qquad \text{N.m} \qquad (2.11)$$

where B_g is the air gap flux density.

The constant k_φ relating torque T to supply current i is proportional to the number of turns in each winding. From conservation of energy, it can be shown that the voltage induced in the two windings in series at any instant is equal to $k_\varphi \omega_o$, where ω_o is the angular velocity of the two-pole rotor. Thus, this motor behaves very much like a commutator motor. Its no-load speed is approximately

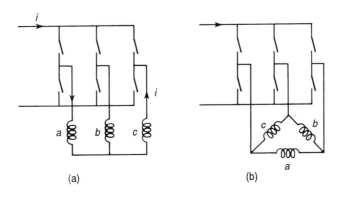

Figure 2-17. Supply and winding connections for switched PM motors: (a) star connected, (b) delta connected.

The constant k_φ relating torque T to supply current i is proportional to the number of turns in each winding. From conservation of energy, it can be shown that the voltage induced in the two windings in series at any instant is equal to $k_\varphi \omega_o$, where ω_o is the angular velocity of the two-pole rotor. Thus, this motor behaves very much like a commutator motor. Its no-load speed is approximately proportional to the applied direct voltage, and the speed reduces somewhat with torque due to the voltage drop across the winding resistances.

2.6.3. Losses and Efficiency

The power loss in the stator windings of a switched PM motor will be the same as for a synchronous PM motor if the rms values of linear current density are made equal. For the same rotor radius, the torque capability of the switched PM motor is about the same as that of a synchronous PM machine with 120° magnets. However, the total flux per pole is greater due to the wider magnet of the switched motor, and a thicker yoke is therefore required to accommodate this flux. The iron losses are somewhat higher because of the increased iron volume. There is, however, little significant difference in the efficiencies of the two motors.

2.6.4. Delta-Connected Motor

An alternate form of switched PM motor has magnets that are typically about 120° in angular width rather than 180° [31]. Its stator windings are connected in delta as shown in Figure 2-17b. Again, current is supplied in a switched sequence into one stator terminal and out another. At any instant, approximately two-thirds of this current flows through one phase winding while one-third flows through the other two phases in series.

For the same gap flux density, rms linear current density, and rotor dimensions, this motor produces about 13% less torque than the star-connected type. However, in the delta-connected motor, the flux is only about two-thirds of the flux of the star-connected motor, and the required thickness of its yoke is reduced proportionately. In this respect, it is similar to the synchronous PM motor that typically has a magnet width of about two-thirds or 120°. When the star and delta motors are compared using the same overall frame size, the air gap radius of the delta-connected motor is typically about 10% larger increasing the available torque as can be seen from equation 2.11. The net effect is that there is very little difference in the torque capabilities of the two motor types.

2.6.5. Design Features

For both synchronous and switched PM motors, the magnets are bonded to the rotor core using epoxy adhesives. The adhesive must have good aging properties and a sufficiently wide temperature range. The maximum speed of the rotor is limited by the action of the centrifugal force on the adhesive, a force proportional to $r\omega_m^2$, or alternatively to u^2/r, where u is the peripheral velocity of the rotor surface [9].

For larger high-speed motors, it may be desirable to band the rotor with glass or carbon fiber tape under tension. The gap between the stator iron and the magnet surface can be made quite large to accommodate this banding without significant reduction of air gap flux density, provided the magnet thickness is proportionately increased. For very high speed, canned rotors may be considered.

Theoretically, there appears to be no inherent limit on the maximum rating of torque and power for which PM machines can be built. Protection of the magnets from demagnetization in larger motors requires use of thicker magnets, an increased air gap, and increased numbers of poles. An increase in the magnet thickness requires a proportional decrease in the square of the limiting peripheral rotor velocity. If the ratio of rotor radius to magnet thickness is kept constant, the magnet volume and cost per unit of motor torque rating will either remain constant with constant values of B_g and K_{rms} or will decrease as the linear current density K_{rms} is increased. However, some practical matters may limit the motor size. For example, with increased size, the fixtures to contain the magnetic forces during motor assembly become more elaborate, and their cost may not be justified for small numbers of production units.

PM motors can produce a ripple torque arising from the interaction of the magnet edges and the individual stator teeth. In some designs this is eliminated by skewing the stator slots or the rotor magnets by one slot pitch. However, this increases the manufacturing complexity. Alternatively, the slot–tooth width ratio at the stator surface and the magnet width can be adjusted to minimize this source of ripple torque [32]. A further source of ripple torque occurs in switched PM motors arising from the finite switching time of the transition from one connection to the next. While this ripple can be made small, the overall ripple torque of switched PM motors tends to be larger than that of well-designed synchronous PM motors.

2.6.6. Operating Characteristics

Both switched and synchronous PM motors are capable of very fast dynamic response in high-performance drive applications. Because of the high resistivity of the magnet material, induced currents in the rotor are negligible except at very high frequencies. Thus, the motor does not have any significant internal time delay in its torque response to stator current. The response can be made essentially as rapid as the supply inverter can apply the appropriate phase currents.

The control system for switched PM motors is inherently simpler than that of the synchronous type. For all the PM motors, control is considerably less complex than the vector control discussed for induction motors.

set of sinusoidal stator currents flowing in the approximately sinusoidally distributed windings. The rotor is shaped with a small air gap in the direct (d) axis and a large gap in the quadrature (q) axis. In addition, the rotor of this motor is made of iron laminations separated by nonmagnetic material to increase still further the reluctance to flux crossing the rotor in the quadrature axis.

2.7.1. Equivalent Circuit

When the rotor of the reluctance motor rotates synchronously with the stator field so that the peak of the flux wave is aligned with the direct axis of the rotor, the magnetizing inductance as seen by the stator winding is L_d. When the peak of the flux wave is aligned with the quadrature axis, the magnetizing inductance is L_q. The ratio of these inductances is denoted as the saliency ratio, $\Delta = L_d/L_q$.

The steady-state operating characteristics of a reluctance motor can be adequately predicted by use of the equivalent circuit of Figure 2-19, where R_s is the stator resistance per phase and the inductance L_r is given by

$$L_r = \frac{L_q}{(1 - 1/\Delta)} \tag{2.12}$$

and where δ is the angle by which the rotating stator wave leads the rotor direct axis [1]. This circuit is seen to be similar in form to that of the induction motor shown in Figure 2-6. Therefore, similar operating characteristics are to be expected.

2.7.2. Torque Capability

Suppose the voltage applied to the stator is controlled so that the rms stator flux linkage Λ_s is maintained constant in magnitude at all values of speed. The torque of this motor can be derived from the power crossing the air gap divided by the synchronous mechanical speed and can be expressed as

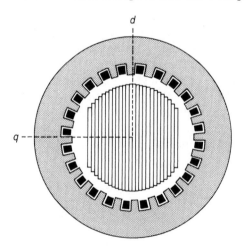

Figure 2-18. Cross section of reluctance motor.

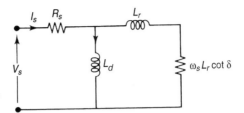

Figure 2-19. Equivalent circuit for reluctance motor.

and where δ is the angle by which the rotating stator wave leads the rotor direct axis [1]. This circuit is seen to be similar in form to that of the induction motor shown in Figure 2-6. Therefore, similar operating characteristics are to be expected.

2.7.2. Torque Capability

Suppose the voltage applied to the stator is controlled so that the rms stator flux linkage Λ_s is maintained constant in magnitude at all values of speed. The torque of this motor can be derived from the power crossing the air gap divided by the synchronous mechanical speed and can be expressed as

$$T = \frac{3p}{4} \frac{\Lambda_s^2}{L_r} \sin(2\delta) \qquad \text{N.m} \tag{2.13}$$

Comparison of this expression with that of equation 2.7 suggests that a reluctance motor can have a maximum torque similar to that of an induction motor if the saliency ratio Δ is made relatively large and if the quadrature axis inductance L_q is comparable with the leakage inductance L_L of the induction machine.

2.7.3. Operating Condition and Power Factor

The choice of the normal operating angle δ for a reluctance motor is conditioned by the effect of the power factor of the motor load on the inverter. The importance of the saliency ratio Δ in a reluctance motor is emphasized by noting that the maximum power factor obtainable from the reluctance motor is equal to $(\Delta - 1)/(\Delta + 1)$. To achieve a power factor of 0.85, a saliency ratio of $\Delta = 11$ is required.

Reluctance motors are normally designed for continuous operation at or near the condition of maximum power factor to minimize the rating or the required inverter. At maximum power factor, the torque T is related to the maximum available torque $\hat{T}$ by

$$\frac{T}{\hat{T}} = \frac{2\sqrt{\Delta}}{\Delta + 1} \tag{2.14}$$

As the saliency ratio increases, this torque ratio is reduced. At $\Delta = 11$, the torque ratio is 0.55. At $\Delta = 5$, the torque ratio is 0.75, but the power factor is only 0.67. These values may be compared with those of induction motors where the usual torque ratio is in the range 0.4–0.5.

2.7.4. Configurations

The type of rotor shown in Figure 2-18 can usually achieve an unsaturated saliency ratio in the range 6–12 [34]. The effective saliency ratio will, however, be decreased as the iron is magnetically saturated. This type of rotor can be made in a four- or six- pole configuration using appropriately bent laminations with nonmagnetic spacer material between laminations. A variety of alternate structures have been proposed.

Reluctance motors can be operated in a switched mode similar to that discussed for PM motors. Uniformly distributed stator windings are used, and pulses of current are supplied to the windings in sequence.

2.7.5. Losses and Efficiency

Synchronous reluctance motors can have an advantage in efficiency over induction motors of similar rating [35]. The rotor loss that accounts for about 20–30% of the induction motor losses is small or negligible in synchronous reluctance machines. If the saliency ratio is sufficient to produce a power factor equal to that of the induction motor, the stator winding loss will be the same. Also, the stator iron losses will be similar for the two motors.

2.7.6. Constant Power Operation

The reluctance motor is capable of operation in the constant power regime in much the same manner as described for the induction motor in Section 2.3.8. This would be expected in view of the similarity of their equivalent circuits as shown in Figures 2-19 and 2-6. Above the base speed where the supply voltage limit is reached, the flux linkage is reduced in inverse proportion to the speed, and, from equation 2.13, the torque is inversely proportional to speed squared.

2.8. PM RELUCTANCE MOTORS

The surface PM motor considered in Section 2.4 is not amenable to flux reduction as is desired for operation in the constant power range. However, if some of the properties of the reluctance motor are incorporated into the PM machine, effective flux reduction can be achieved. For example, a cross section of an inset-PM motor is shown in Figure 2-20. The magnets are inset into the rotor iron [36]. As a PM motor, the direct axis is aligned with the center of the magnet. Considered as a reluctance machine, its direct axis inductance L_d will be small because of the near unity relative permeability of the magnet material. The quadrature inductance L_q will, however, be relatively large because of its small iron-to-iron gap throughout the quadrature sector.

A typical relation between torque and rotor angle for a current-driven inset-PM motor is shown in Figure 2-21. It consists of a term proportional to sin β due to the

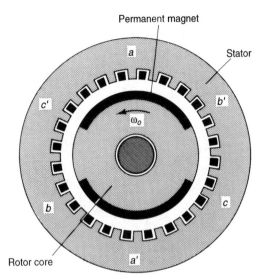

Figure 2-20. Cross section of inset PM motor.

direction to that produced by the magnets. The net flux is reduced to the point where the induced voltage is equal to the limiting value of supply voltage. The torque and the flux linkage are therefore controlled by simultaneous adjustment of the magnitude and angle of the stator current. Using the motor configuration of Figure 2-20, a constant power speed range of 2–5 can be achieved.

Another rotor configuration for a PM reluctance motor is shown in Figure 2-22. In this buried-PM motor, the magnets are directed circumferentially to supply flux to the iron poles, which then produce radially directed air gap flux. The central shaft is nonmagnetic. This type of structure has been widely used with ferrite magnets to achieve a satisfactory value of air gap flux density. The direct axis inductance of this motor is low because its flux path is largely through the low permeability mag-

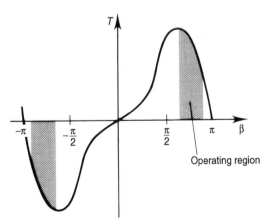

Figure 2-21. Torque-angle relation for inset PM motor.

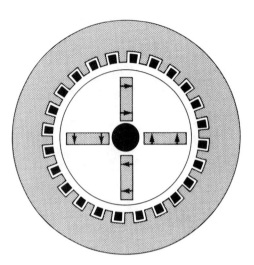

Figure 2-22. Cross section of circumferential PM motor.

tude and angle of the stator current. Using the motor configuration of Figure 2-20, a constant power speed range of 2–5 can be achieved.

Another rotor configuration for a PM reluctance motor is shown in Figure 2-22. In this buried-PM motor, the magnets are directed circumferentially to supply flux to the iron poles, which then produce radially directed air gap flux. The central shaft is nonmagnetic. This type of structure has been widely used with ferrite magnets to achieve a satisfactory value of air gap flux density. The direct axis inductance of this motor is low because its flux path is largely through the low permeability magnets, while the quadrature axis inductance is large because its flux can circulate in and out of each pole face. Many structural variants of buried-magnet machines have been proposed [37, 38, 39]. Some of these are well suited for high-speed applications. Both the structure and the control of PM reluctance motors are discussed further in Chapter 6.

2.9. SWITCHED RELUCTANCE MOTORS

The switched reluctance (SR) motor has a doubly salient structure [40]. The machine shown in cross section in Figure 2-23 has six stator poles and four rotor poles and is denoted as a 6/4 motor. Each stator pole is surrounded by a coil, and the coils of two opposite stator poles are connected in series to form each of the three phase windings. The three phase windings are connected in sequence to a power converter that is normally triggered by a set of shaft position sensors.

Consider the rotor position shown in Figure 2-23. If supply current is switched through the coils of phase *a*, there will be a counterclockwise torque acting to align the adjacent pair of rotor poles with the phase *a* stator poles. When aligned, if the

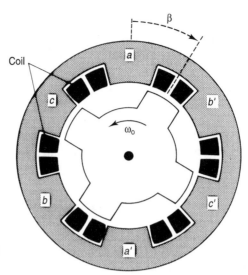

Figure 2-23. Cross section of switched reluctance motor.

current is now switched to phase b, the counterclockwise torque will continue. For the machine shown in Figure 2-23, the rotor moves $\pi/6$ radians for each step, and thus 12 switching operations are required per revolution. The direction of the coil current is not significant. The direction of rotation can be changed to clockwise by reversing the sequence of the currents to a, c, b.

Many other pole choices are useful, particularly 8/6 (four phase) and 10/8 (five phase) [41]. Increase in the number of poles increases the switching frequency and the losses in the iron. It decreases the ratio of aligned to unaligned inductance somewhat.

2.9.1. Torque Relations

If the motor is operated in the linear mode with no significant saturation in the iron, the torque can be expressed as

$$T = \frac{i^2}{2} \frac{dL}{d\beta} \qquad \text{N.m} \tag{2.15}$$

where i is the phase current, L is the phase inductance (where L is the inductance of the energized phase), and β is the rotor angle. To the extent that the inductance is proportional to the overlap angle, the torque can be essentially constant. The expression of Figure 2-15 is useful only for values of the current i for which the inductance is independent of current.

To achieve high torque and power from a given frame size, most switched reluctance motors are operated so that the poles are significantly saturated when aligned and energized. The average torque that these motors can produce may be assessed from the magnetizing characteristics relating the phase flux linkage λ to the coil current i when in the aligned and unaligned positions [1]. A typical set of relations for the 6/4 motor is shown in Figure 2-24.

It can be shown that the average torque over one step is equal to the heavily shaded area between the unaligned and fully aligned curves measured in joules (or Wb.A), divided by the rotational angle between steps (i.e., $\pi/6$ rad in Figure 2-23). To optimize this average torque, the slope of the linear part of the aligned characteristic ($\beta = 0$) can be increased by reduction of the air gap between the aligned poles while the slope in the unaligned position can be adjusted by changing the relative widths of the stator and rotor poles [42].

An approximate expression for the torque can be obtained by noting that the work W done in rotating from unaligned to aligned positions can be expressed as

$$W = k_r \hat{\lambda} \hat{\imath} \quad \text{J} \tag{2.16}$$

where the factor k_r is in the range from 0.45 for an unsaturated motor to about a maximum of about 0.75 for a highly saturated one. Assuming six stator poles, equal angular widths of stator and rotor poles of $\pi/6$ rad, N turns per coil, and a maximum flux density of B in the poles when aligned, the maximum flux linkage of a pair of coils in series can be expressed as

$$\hat{\lambda} = \frac{\pi}{3} r \, \ell N \hat{B} \quad \text{Wb} \tag{2.17}$$

The coil current $\hat{\imath}$ can be linked to the rms value of the stator linear current density K_{rms} around the motor by noting that each coil is on for one-third of the time.

$$K_{rms} = \frac{2\sqrt{3}N\hat{\imath}}{\pi r} \quad \text{A/m} \tag{2.18}$$

The torque can then be expressed as

$$T = \frac{k_r}{\sqrt{3}} \, \pi r^2 \ell \hat{B} K_{rms} \quad \text{N.m} \tag{2.19}$$

This may be compared with equation 2.11 for a switched PM motor recognizing that the maximum air gap flux density B for the SR motor can be made about $2\sqrt{2}$ times the rms flux density of a slotted stator motor. For this six-pole SR motor, the torque for the same rms linear current density will be somewhat less than the fraction k of that of a switched PM machine. Optimization of the pole widths can improve this ratio of SR to PM torque. To a first approximation, the continuous torque rating is independent of the number of poles [41].

2.9.2. Losses and Efficiency

When the poles are aligned, the stored energy in the magnetic field is represented by the shaded area between the aligned curve ($\beta = 0$) and the vertical axis of Figure 2-24. The power converter is usually designed to return this energy to the supply as efficiently as possible.

The main power loss for the switched reluctance motor is that in the stator coils. These coils have a relatively large cross section and the rms linear current density can be made relatively high.

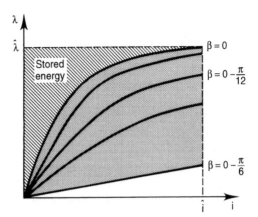

Figure 2-24. Flux linkage vs current characteristics for reluctance motor.

There is iron loss in both the stator and the rotor. These iron losses may be relatively low for low-speed motors but may contribute substantially to the total loss at high speeds because of the high ratio of switching frequency to speed. However, only a relatively small amount of iron is used in the SR motor.

Windage and friction losses are similar to or less than those of induction motors, but windage losses may be high at high speeds. An internal fan is generally not required because the rotor heating is small. Typical values of efficiency are comparable with those of high-efficiency induction motors of similar rating [43].

2.9.3. Design and Application Considerations

Switched reluctance motors are economical to manufacture. Precision may however be needed to achieve a suitably small air gap, particularly for small machines. A ratio of pole arc to air gap of about 25–30 is desirable. Maintenance is negligible. They are suitable for high-speed operation because of ruggedness of the rotor. The rotor inertia is low because of the spaces between the rotor poles. The maximum short-term torque is typically twice the continuous value. The dynamic response can be rapid.

The doubly salient structure of the switched reluctance motor produces relatively high-torque ripple. A constant current in the phase coils will not produce a constant instantaneous torque in a saturated motor as can be judged by comparing the four unequal shaded areas in the λ/i characteristics of Figure 2-24 between the five equally spaced rotor angular positions. When it is necessary to reduce this torque ripple, the coil current can be appropriately controlled during each step of angular motion [43].

Switched reluctance motors also tend to produce a high level of acoustic noise mainly due to the variations in the radial forces with rotor position [44]. These forces have high-frequency components that at certain speeds resonate with the several resonant frequencies of the stator core. Appropriate timing of the current pulses from the power converter can reduce the acoustic noise level associated with the mechanically resonant speeds.

For good performance, the design of a switched reluctance motor must be closely coordinated with the design of its power converter and control system [41]. This coordination is much more significant than for induction and PM motor drives. Switched reluctance machines can be built for a wide range of power ratings.

2.10. WOUND FIELD SYNCHRONOUS MOTORS

For variable speed mill drives in the multimegawatt power range, the machine of choice is frequently a synchronous motor with its magnetic flux provided by a field winding [45, 46]. For the medium-power range, the field current can be supplied through slip rings from a controllable direct current source. For larger machines, excitation can be provided through a rotating transformer and a rotor-mounted diode rectifier.

By adjusting the field current to produce a leading power factor, the supply inverter can be line commutated removing the need for forced commutation of the switching devices except at low values of speed.

Wound field synchronous motors are useful in traction applications where a wide speed range at constant power is required. The adjustment of the field current in this range is similar to that of a commutator motor.

2.11. LINEAR MOTORS

In transit applications and in many industrial situations, linear motion is required rather than rotational motion. This need has given rise to a number of linear machines [47].

The major properties of a linear induction motor can be appreciated by considering the evolution of the induction motor of Figure 2-25a. In this motor, the rotor consists of an iron core surrounded by a conducting ring of aluminum or copper. Suppose a cut is made along the dotted line and the machine is unrolled. The result is shown in Figure 2-25b, which has a two-pole stator or primary and the rotor or secondary has been extended indefinitely.

The synchronous speed of this linear induction motor is given by

$$\nu_s = \frac{z \omega_s}{2\pi} \qquad \text{m/s} \qquad (2.20)$$

where z is the wavelength of a two-pole section of the primary and ω_s is the angular frequency of the supply.

A linear induction motor typically has four to eight poles, but the number need not be an even number or even an integer. The velocity-torque relation for a linear induction motor is roughly similar to that of the rotating induction motor but differs due to a number of factors. Induced currents in the secondary material entering the space under the primary tend to inhibit the establishment of air gap flux under the leading one or two poles. Also, secondary current tends to persist after the material has left the trailing pole of the primary. These end effects cause the linear motor to

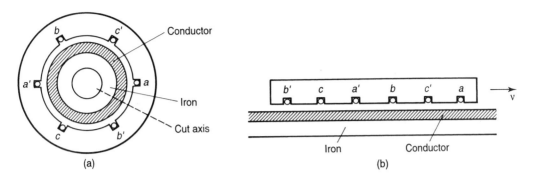

Figure 2-25. Evolution of a linear induction motor from a rotating induction motor.

which may require an air gap in excess of 10 mm. Thus, the power factor of a linear motor tends to be low.

Large linear induction motors have been used in public transit vehicles, high-speed trains, materials handling, extrusion processes, and the pumping of liquid metals. Small linear motors are used in a variety of applications such as curtain pullers and sliding door closers.

Linear synchronous motors have been proposed and tested for a number of low-speed transit and high-speed train applications [48, 49]. These motors have a primary similar to the three-phase stator shown in Figure 2-25b. The secondary may be a set of either permanent magnets or superconducting magnets. The magnets may be installed on the underside of the vehicle interacting with energized sections of primary on the track. The performance principles are similar to those of rotating synchronous motors.

2.12. CONCLUSION

A wide variety of choices of electrical motors is now available for use with variable speed, variable frequency drives. No single type is ideal for all applications. The optimum choice for a particular drive depends on a detailed assessment of the system design criteria that are considered to be important, criteria such as initial cost, life-cycle cost, dynamic performance, ease of maintenance, robustness, weight and environmental acceptability . It is hoped that the material included in this chapter will be useful in providing a rational overview of the motor choices.

Nomenclature

List of Symbols

A	area	p	number of poles
B_D	knee flux density	q	quadrature axis
B_r	residual flux density	R_R	equivalent rotor resistance
B_g	air gap flux density	R_s	stator winding resistance
d	direct axis	r	rotor radius
d_e	equivalent depth of conductor	T	torque
E, e	induced voltage	V, v	voltage
I, i	current	α	half angular width of magnet
k	dimensional factor	β	angle of stator field re magnet axis
k_r	factor for switched reluctance motor	γ	fraction covered by poles
k_φ	flux factor for switched PM motor	Δ	ratio of direct to quadrature inductance
K_s, K_r	linear current density	δ	angle of stator voltage re magnet axis
L_m	magnetizing inductance		
L_L	leakage inductance	Λ, λ	flux linkage
ℓ_g	stator to iron core length	μ_r	relative permeability
ℓ_m	magnet thickness	ν	linear velocity
ℓ	rotor length	ρ	resistivity
N	number of turns	σ	constant power speed ratio
P	power	ω	angular velocity

References

[1] Slemon, G. R., *Electric Machines and Drives*, Addison-Wesley, Reading, MA, 1992.

[2] Dewan, S. B., G. R. Slemon, and A. Straughen, *Power Semiconductor Drives*, Wiley Interscience, New York, 1984.

[3] Say, M. G., *The Performance and Design of Alternating Current Machines*, Pitman, London, U.K., 1948.

[4] Kusko, A., *Solid-State DC Motor Drives*, MIT Press, Cambridge, MA, 1969.

[5] Say, M. G., *Alternating Current Machines*, Pitman, London, U.K., 1976.

[6] Fitzgerald, A. E., C. Kingsley, and S. D. Umans, *Electric Machinery*, 5th ed., McGraw-Hill, New York, 1990.

[7] Slemon, G. R., "Modelling of induction machines for electric drives," *IEEE Trans. Ind. Appl.*, Vol. IA-25, No. 6, pp. 1126–1131, November/December 1989.

[8] de Jong, H. C. J., *AC Motor Design with Conventional and Converter Supplies*, Clarendon, Oxford, 1976.

[9] Levi, E., *Polyphase Motors*, John Wiley and Sons Inc., New York, 1984.

[10] Gallant, T. A., "The selection of an efficient electrical machine," *IEE Conf. Electric Machines and Drives*, London, pp. 136–140, 1991.

[11] Electric Power Research Institute, *Electric Motors: Markets, Trends and Applications*, EPRI, Palo Alto, CA, 1992.

[12] Richter, E., and T. J. E. Miller, "Technology of high efficiency motors," *Int. Conf. Electric Machines*, Manchester, U.K., pp. 190–194, September 1992.

[13] Werner, F. E., "Electrical Steels: 1970–1990," *TMS-AIME Conf. Energy Efficient Steels*, Pittsburgh, pp. 1–32, 1980.

[14] Slemon, G. R., "Scale Factors for physical modelling of magnetic devices," *Electric Machines and Electromechanics*, Vol. 1, No. 1, pp. 1–9, October 1976.

[15] Dubey, G. K., *Power Semiconductor Controlled Drives*, Prentice Hall, Englewood Cliffs, NJ, 1989.

[16] Leonard, W., *Control of Electric Drives*, Springer-Verlag, New York, 1985.

[17] Whiting, J. M. W., and J. A. Taufiq, "Recent development and future trends in traction propulsion drives," *Int. Conf. Electric Machines*, Manchester, U.K., pp. 195–199, September 1992.

[18] Vas, P., *Vector Control of AC Machines*, Oxford University Press, Oxford, 1993.

[19] Jansen, P. L., R. D. Lorenz, and D. W. Novotny, "Observer-based field orientation: Analysis and comparison of alternative methods, IEEE Industry Applications Society, *Annual Meeting Conf. Rec.* Part 1, pp. 536–543, October 1993.

[20] Wood, S. E., and D. Greenwood, "Forced ventilated motors—Advantages in fixed and variable speed applications," *IEE Conf. Electric Machines and Drives*, London, pp. 276–279, 1991.

[21] Jackson, W. J., "Mill motors for adjustable speed AC drives," *IEEE Trans. Ind. Appl.*, Vol. 29, No. 3, pp. 566–572, May/June 1990.

[22] Heck, C., *Magnetic Materials and Their Applications*, Crane, Russak, New York, 1974.

[23] Sagawa, M., S. Fugimura, N. Togawa, H. Yamamoto, and Y. Maysuura, "New material for permanent magnets on a base of Nd and Fe," *J. Appl. Phys.*, Vol. 55, No. 6, pp. 2083–2087, 1984.

[24] Slemon, G. R., and Xian Liu, "Modelling and design optimization of permanent magnet machines," *Electric Machines and Power Systems*, Vol. 20, pp. 71–92, 1992.

[25] Sebastian, T., and G. R. Slemon, "Operating limits of inverter driven permanent magnet motor drives," *IEEE Trans. Ind. Appl.*, Vol. IA–23, No. 2, pp. 327–333, March/April 1987.

[26] Sebastian, T., and G. R. Slemon, "Transient torque and short circuit capabilities of variable speed permanent magnet motors," *IEEE Trans. Magn.*, Vol. MAG-23, No. 5, pp. 3619–3621, September 1987.

[27] Slemon, G. R., and Liu Xian, "Core losses in permanent magnet motors," *IEEE Trans. Magn.*, Vol. MAG-26, No. 5, pp. 1653–1655, September 1990.

[28] Slemon, G. R., "High-efficiency drives using permanent-magnet motors," *Proc. IEEE Int. Conf. Industrial Electronics*, Maui, Hawaii, pp. 725–730, November 1993.

[29] Slemon, G. R., "Achieving a constant power speed range for PM drives," IEEE Industry Applications Society, *Annual Meeting Conf. Rec.*, Part 1, Toronto, pp. 43–50, October 1993.

[30] Slemon, G. R., "On the design of high performance PM motors," *IEEE Industry Applications Society Conf. Rec.* pp. 279–287, October 1992.

[31] Miller, T. J. E., *Brushless Permanent-Magnet and Reluctance Motor Drives*, Oxford Science, Oxford, 1989.

[32] Li, T., and G. R. Slemon, "Reduction of cogging torque in permanent magnet motors," *IEEE Trans. Magn.*, Vol. MAG-24, No. 6, pp. 2901–2903, November, 1988.

[33] Boldea, I., and S. A. Nasar, "Emerging electric machines with axially laminated anisotropic rotors," *Electric Machines and Power Systems*, Vol. 19, No. 6, pp. 673–703, 1991.

[34] Staton, D. A., T. J. E. Miller, and S. E. Wood, "Optimization of the synchronous reluctance motor geometry," *Electric Machines and Drives Conf.*, London, pp. 156–160, September 1991.

[35] Lipo, T. A., "Synchronous reluctance machines—A viable alternative for AC drives," *Electric Machines and Power Systems*, Vol. 19, No. 6, pp. 659–669, November/December 1991.

[36] Sebastian, T., G. R. Slemon, and M. A. Rahman, "Modelling of Permanent Magnet Synchronous Motors," *IEEE Trans. Magn.*, Vol. MAG-22, No. 5, pp. 1069–1071, September 1986.

[37] Schofield, N., P. H. Mellor, and D. Howe, "Field weakening control of brushless permanent magnet motors," *Int. Conf. Electrical Machines*, Manchester, U.K., Part 1, pp. 269–272, September 1992.

[38] Jack, A. G., and A. J. Mitcham, "Design and initial test results from a permanent magnet synchronous motor for a vehicle drive," *Int. Conf. Electric Machines*, Manchester, UK, pp. 751–757, September 1992.

[39] Scholfield, N., P. H. Mellor, and D. Howe, "Field weakening of brushless permanent magnet motors," *Int. Conf. on Electric Machines*, Manchester, UK, pp. 269–272, September 1992.

[40] Miller, T. J. E., *Switched Reluctance Motors and Their Control*, Oxford Science, Oxford, 1993.

[41] Lovatt, H. C., and J. M. Stephenson, "Influence of the number of poles per phase in switched reluctance motors," *IEE Proc.*, Vol. 139, No. 4, Part B, pp. 307–314, July 1992.

[42] Miller, T. J. E., D. A. Staton, and S. E. Wood, "Optimization of synchronous reluctance motor Geometry," *IEE Int. Conf. Electric Machines and Drives*, pp. 156–160, September 13–15, 1991.

[43] Lawrenson, P. J., "Switched reluctance drives: A perspective," *Int. Conf. Electric Machines*, Manchester, U.K., pp. 12–21, 1992.

[44] Lang, J., et al., "Origin of acoustic noise in variable reluctance motors," *IEEE Industry Applications Society Annual Meeting*, pp. 108–115, October 1989.

[45] Walker, J. H., *Large Synchronous Machines—Design, Manufacture and Operation*, Oxford Science, Clarendon Press, Oxford, 1981.

[46] Jackson, W. J., "Mill motors for adjustable speed AC drives," *IEEE Trans. Ind. Appl.*, Vol. 29, No. 3, pp. 566–573, May/June 1993.

[47] Nasar, S. A., and I. Bodea, *Linear Motion Electric Machines*, Wiley Interscience, New York, 1976.

[48] Slemon, G. R., P. E. Burke, and N. Terzis, "A linear synchronous motor for urban transit using rare earth magnets," *IEEE Trans. Magn.*, Vol. MAG-14, No. 5, pp. 921–923, 1978.

[49] Atherton, D. L., et al., "The Canadian high speed magnetically levitated vehicle system," *Canadian Electr. Eng. J.*, Vol. 3, No. 2, pp. 3–26, 1978.

J. D. van Wyk

Power Electronic Converters for Drives

3.1. INTRODUCTION

This chapter traces the development of power electronic converters to control AC machines up to the present state of the technology and gives a functional overview. The fundamental possibilities to control average power in a switching mode are considered, and the concepts of switching function, converter topology, and converter structure are introduced. A systematic approach to developing more complicated topologies and structures for singular and composite converters is then discussed and a functional classification of motion control converters is related to the switching nature of the power processing, illustrating the fundamental dilemma of this kind of highly efficient power control.

The different possibilities of using power electronic converters in variable frequency drives are reviewed systematically. We do not attempt to compile a comprehensive list of all possible converter circuits and variations but, rather, focus on generic issues and technologies. The switching technology in converters and its limitations are related to snubberless, snubbed, and resonant transition-type operation, with reference also to switch drive technology. Finally, some future generic possibilities related to electromagnetics of switching converters and other issues are discussed.

3.2. DEVELOPMENT OF POWER ELECTRONIC CONVERTERS AND ITS APPLICATION TO DRIVE TECHNOLOGY

This part reviews the development in the field of power converters for AC machines briefly, since extensive reviews can be found elsewhere [1–5]. After tracing the origins of using power electronics to control variable frequency drives we will outline systematics of converter circuits.

3.2.1. A Systematic Overview of Applied Power Electronic Converters

From the history of the development of power electronic technology [1, 5], it has become evident that a vast number of switching converter topologies and composite switching converter structures are possible. These possibilities are expanded in terms of the realization of all the switching functions (Figures 3-1 and 3-5) possible with the semiconductor switching devices available at present. This all adds up to a bewildering array of possible solutions in applying switching converters to AC drive problems. An attempt at producing an overview in commencing this chapter is therefore in order.

Figure 3-1 presents a functional classification of switching converter topologies and structures. Motion control applications with DC commutator machines use either pulse width modulated (PWM) choppers [6–9] in various configurations or high-frequency-link DC-to-DC converters in more specialized mobile applications [10, 11] where transformer isolation is required and high power density is a necessity. It can, therefore, be concluded that in the DC-to-DC conversion field for motion control, both DC coupled converter topologies and high-frequency link transformer coupled converter structures can be found in practice.

Converters in the AC-to-DC conversion field are the most widespread and are found as input converter topologies for the AC variable speed drives, whether that be as diode rectifier inputs to voltage fed inverter drives [12], controlled rectifier inputs to current-fed inverters [13], or PWM AC-to-DC converters for traction drives on locomotives [14, 15].

When evaluating the switching converter families with AC output, a much wider variation is found. The family of DC-to-AC converters include inverter topologies as current-fed or voltage-fed PWM inverters [16, 17], while in specialized applications where transformer isolation and high-power density is required, high-frequency link converters are again the solution [18, 19]. In the case of AC-to-AC converters, the converter topologies found are the amplitude controllers in the form of either AC phase controllers [20–24] for voltage amplitude, usually on induction machines as loads, or as AC chopper converters [25–28]. As proven adequately in the past, AC chopper controllers remove many of the harmonic generation and input power factor problems of phase controllers, at the expense of more complicated power electronics [26, 27]. As indicated in Figure 3-1, all the other AC-to-AC converters belong to more complicated converter structures, being either of parallel or series configuration.

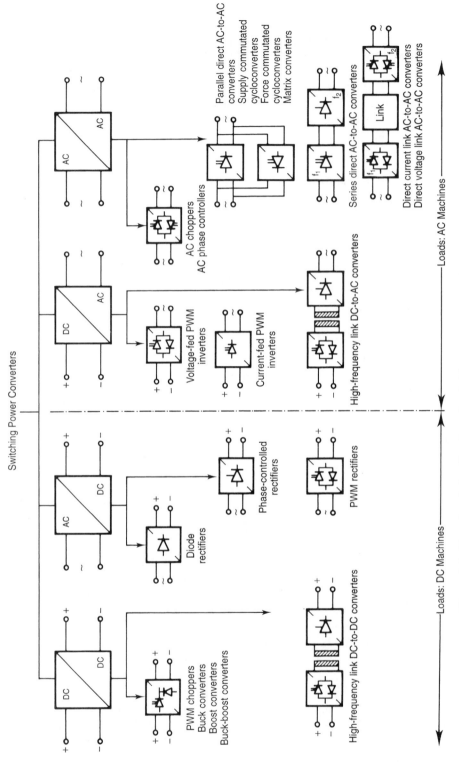

Figure 3-1. Functional classification of switching converter topologies and structures according to field of application.

82

The supply-commutated cycloconverters [29] have been around for a long time, while the advent of modern power semiconductor switching devices gave rise to the concept of force-commutated direct converters or matrix converters [13, 29–32]. The appearance of practically viable solutions have recently been reported in this field [32], but switch technology and commutation complication remain limiting factors to the envisaged general performance [32]. Series direct cycloconverters [33] simplify some of the commutation problems in these converter structures, leading to interesting possibilities and applications [19] in the case of high-frequency link compound converters as AC-to-AC converter structures for motion control.

The last group of the AC-to-AC converter structures—direct current and direct voltage link—have probably found the widest application of all switching converters in motion control at present [34, 35, 36]. In this regard the optimization of the direct current or voltage link quantities presents one of the most difficult problems in the practical application of converter families [37, 38].

3.2.2. Historical Development of Power Electronic Converters for Motion Control

A review on the history of converter development [1, 5] indicates that the technology started accelerated development only when solid-state switches became available, although the oldest known switches used in the control of electrical machine drives are mechanical. The mechanical commutator in DC machines was followed by a mechanical vibrating switching arrangement to control the field of generators, whereafter gaseous valves showed promise for generating the switching functions necessary to control machines [5]. Before the era of semiconductor switches dawned with the introduction of selenium power rectifiers, the development of switching technology still went through the stages of mechanical metallic rectifiers and transductors [5, 39]. Today we are still concerned with this development of solid-state switching converters, while silicon power device technology has developed to an advanced degree of refinement [2, 40]. It seems that solid-state power conversion devices still offer all the necessary capabilities for the foreseeable future [3, 4].

3.3. SOME FUNCTIONAL CONSIDERATIONS REGARDING SWITCHING CONVERTERS AND THEIR APPLICATIONS TO VARIABLE FREQUENCY DRIVES

The structure of a motion control system from which the different concepts to be discussed subsequently are derived is given schematically in Figure 3-2. The characteristics imparted to the drive train by the power electronic converter depends on its topology, the switching function impressed upon the switches, and the type of control applied.

Control of the energy flow between the source and the eventual mechanical motional output is normally achieved by the combination of the switched power electronic converter and the electromechanical converter as drive mechanism. The

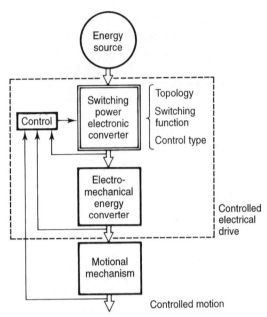

Figure 3-2. The structure of a drive system.

switching power converter is the controllable element, and as such the possibilities of topology and switching functions are discussed in this part. It has to be kept in mind that switching power converters for motion control applications are subjected to operational and parametric conditions fixed by the electrical machines that are connected to these converters as loads. These conditions differ appreciably in the following respects that are important considerations for power electronic converter topology, switching functions, and control:

1. The loads are inductive in character.
2. The loads are active; that is, they contain sources of induced electromotive force (EMF).
3. The loads require two directions of power flow, and often two directions of current flow through the converter.
4. The loads are dynamic; that is, they require large variations in power and direction of power flow, depending on the specific motion control application considered (from ultrafast servos to large and slowly varying mill and traction drives at multimegawatt levels).

This discussion includes both the fundamental thoughts on using switches to control average flow of energy, as well as the concepts of topology and structure for power electronic converters. Although this should be complemented by a consideration of the influence of control philosophy and implementation on the overall system transfer characteristics, this is not the subject of this chapter.

3.3.1. Controlling Average Energy Flow by Switching Converters

Efficiency and loss requirements dictate the control of power in a switching mode in power electronics. For a systematic approach, the technology of the specific types of power switches, as well as particular peculiarities of behavior, should be omitted. These characteristics—such as switching speed, switching losses, conduction losses, surge capacity, overvoltage capacity, temperature dependence, and power gain—are extremely important when considering a specific application and type of switching device in practice. For the present systematic approach, however, the power switch will be considered to change instantaneously from the nonconducting state to the conducting state or back at any chosen instant. This leads to the fundamental possibility to control power flow as shown in Figure 3-3. By operation of the switches S_1 and S_2, the switching function $s(t)$ is generated, so that the load voltage becomes

$$u_L(t) = s(t)\, u_s(t) \tag{3.1}$$

As the load is an electrical machine, it may contain inductance and/or sources of EMF as already stated. The function $s(t)$ is consequently generated by operating S_2 inverse to S_1, allowing circulating or "freewheeling" load current if necessary. The average power flow between source and load $P_s(t)$, therefore, becomes a function of $s(t)$. Ideally, this power transfer is now being controlled by S_1 and S_2 with no losses

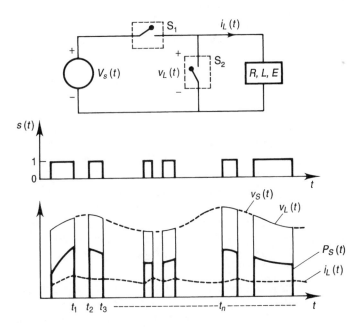

Figure 3-3. Fundamental possibilities for control of power flow by means of switching functions.

involved in S_1 and S_2, as either the voltage across S_1 or S_2 or the current conducted by S_1 and S_2 is always zero.

Switching Function. The generation of the switching function as pulse width modulation is treated in other contributions [41, 42] and will not be discussed here. It is normally required that the applied load voltage obey some externally prescribed modulation function through which control is effected, but that the average load voltage should be affected as little as possible by the switching function. By switching at a sufficiently high frequency with respect to the modulation function, the ripple due to the switching frequency in the load current is reduced. At the same time the wide separation between modulation frequency and switching frequency facilitates elimination of the latter in the output voltage by filtering—an effect that is normally achieved by the machine inductance. From this it would seem that the only require-ment for the switching function would be that it should be many times the frequency of the modulation function. However, limitations in practical switching devices severely limit the upper switching frequencies in power electronic converters in many instances, so that the switching function is in these cases usually related to the modulation function in such a way that the elimination of certain harmonic frequencies in the load voltage is obtained [41, 42]. The previous considerations therefore result in two large families of techniques for switching control, that is, modulation synchronous techniques and modulation asynchronous techniques, some of which are indicated in Figure 3-4. Although fundamentally the source voltage $v_s(t)$ could be any function of time, the two desired types almost exclusively encountered in power electronic converters for motion control are direct voltage and sinusoidal voltage, so that the latter has been selected for the illustrations in Figures 3-4a–d. Figure 3-4a represents the possibilities of pulse width modulation (PWM, the switching frequency f_s being constant) where the synchronous relationship holds:

$$f_s = nt \qquad n = 2, 3, 4, \ldots \tag{3.2}$$

The special case where $n = 2$, as illustrated in Figure 3-4c constitutes a very important class of modulation synchronous power electronic switching functions. For all the other values of n, the relationship obeyed by t_p and n is determined by the desired harmonic component elimination in v_L, with n normally larger than unity.

Pulse frequency modulation, as in Figure 3-4b, is an asynchronous technique, although the range of frequency sweep for f_s is known. Harmonic elimination in this case could only be effective if the switching frequency f_s is always much larger than the modulation frequency, implying a very short pulse width t_p. The high maximum switching frequency, short t_p, and extremely large sweep in f_s impart many draw-backs to this technique, so it is not often found in practical motion control con-verters.

In the examples discussed for Figures 3-4a, b, and c, the nature of the function $s(t)$ is determined in advance and is dependent on the value $v_r(t)$ to which it is referenced. This value may be derived from a load feedback, but will only change f_s and t_p in a previously determined fashion, this being one of the system character-istics. When the load-related bang-bang mechanism of Figure 3-4d is used to derive

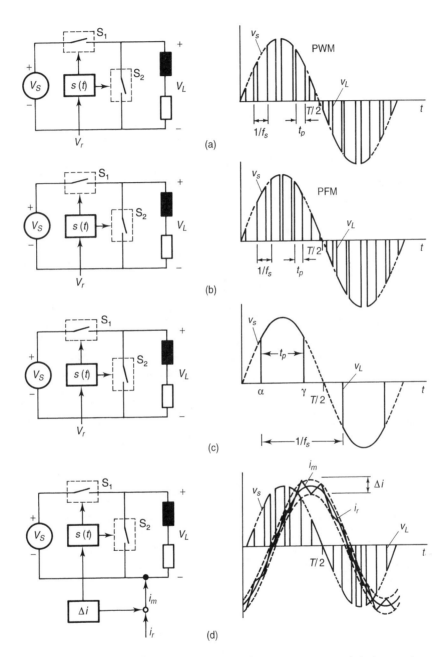

Figure 3-4. Fundamental relationships between some modulation and switching functions.

the switching function, a fully arbitrary switching function results, as shown. Under given load conditions the value of the current swing Δi will determine the shortest time between two consecutive switching cycles of S_1 and S_2. The technique is modulation asynchronous and of variable frequency. Owing to the current source characteristics of the output, and the good correspondence between i_r and i_m, the technique has found widespread acceptance in the practice of motion control converters. The characteristics of these systems have been discussed extensively [43–45] in relation to their control and switching characteristics.

It will be obvious that the characteristics of equipment using modulation asynchronous or synchronous techniques by derivation of $s(t)$ are very different regarding high-frequency electromagnetic radiation, conducted interference, supply reaction, and harmonics in the entire frequency spectrum and output filtering. In practice, it is often found that these boundary conditions dictate the technique selected for obtaining $s(t)$.

It should be noted that the frequency of the modulation function is not always synchronized with that of the source voltage as shown in Figure 3-4. If the modulation frequency and source frequency are not synchronized, the output contains components resulting from all these frequencies and beats between them, that is, that of the source, the modulation, and $s(t)$. For the case of a direct voltage source, only the modulation and the switching function affect the output components, with only $s(t)$ appearing when the modulation function also becomes a constant, as in the many instances when a direct voltage load is driven, such as DC machines. When this type of load is supplied by a system with specific AC source voltage, the output again contains these components, those of $s(t)$, and their cross-modulation.

The foregoing discussion of the switching function used voltage-fed systems and the resulting output voltage as an illustration. Perfect duality exists regarding all the techniques discussed with respect to current-fed systems. The asynchronous technique of Figure 3-4d could, for instance, also be applied by sensing the load voltage and deriving a switching function for the current into the load, as drawn from a current source $i_s(t)$.

Realization of Switching Functions. The switching functions discussed previously have to be realized by using practical power electronic devices. These devices allow different types of functions to be realized by being used alone or in combination with each other. The possible configurations of, and the symbols for, the power switches S_1, S_2 are shown systematically in Figure 3-5, while the possible device realizations are also shown. In the order that the switching functions are shown, they are

1. *Unidirectional conduction function or diode function.* As the diode conducts when the voltage across the device is positive, the current has to be turned off by an external voltage across the device, developed by some other part of the system. This voltage could be derived from either the supply or the load or both.

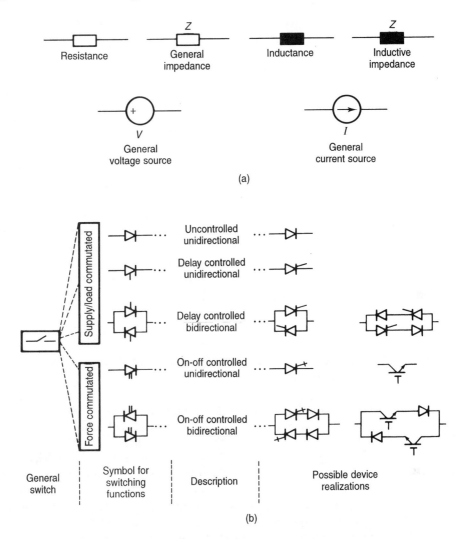

Figure 3-5. (a) Definitions for circuit elements; (b) symbols for different realisable switching functions.

2. *Unidirectional delayable conduction function.* While the current turn-off through the switch is still subject to an external voltage developed by the supply or load, control via a switching function may delay the conduction process in the switch. The switching function $s(t)$ as in Figure 3-4c, with $t_\alpha \leq T/2$ may be realized in this case, for unilateral conduction and this represents the large family of naturally commutated (load or supply) systems with in the past, grid-controlled mercury arc devices, ignitrons, magnetic amplifiers, and so on, and at present with thyristors.

3. *Bidirectional delayable conduction function or antiparallel function.* This adds the generation of symmetric AC outputs, to the previous possibilities, but it

is subject to the same limitations with $t_\alpha \leq T/2$. Apart from the circuits using antiparallel grid-controlled, mercury arc rectifiers, ignitrons, and thyristors, the triac has been added to this range of devices.

4. *Unidirectional on-off conduction function or force commutated switch.* The realization of pulse width modulation, pulse frequency modulation, and bang-bang control for unidirectional conduction by using this function in S_1 and S_2 as in Figures 3-4a–d is possible. Both the current turn-on and turn-off are under control, and the switch turns off current independently of load or supply voltage. The turn-off is achieved either by inherent device characteristics or by using an externally derived voltage for turning the delayable conduction function off, unrelated to the natural behavior of the circuit (forced commutation). The former refers to switches based upon devices such as bipolar power transistors, field effect transistors, insulated gate bipolar transistors, and gate turn-off thyristors, while the latter refers to a large number of force commutated thyristor circuits used previously.

5. *Bidirectional on-off conduction function or bidirectional force commutated switch.* This is the most general power electronic switch realizable and may be obtained by arranging the function or devices discussed in the preceding entry in antiparallel.

3.3.2. Topology and Structure of Power Electronic Converters

Up to now the discussion has concerned topologically the simplest converter configuration: that of a single source, single load, and single or double switch arrangement, as in Figure 3-3. The following discussion will formulate a systematic approach to the large number of configurations in which power electronic converters are found in practice, since the simplest circuit is only useful in a small number of motion control applications. As shown in Figure 3-6, when converters with different or the same topologies are cascaded in series, the configuration will be termed a "composite" converter, whereas the converters used as building blocks for these composite converters will be called "singular" converters.

Singular converters will be said to have "topology," determined by the way in which the switches are connected to each other and between load and supply. Composite converters have "structure" determined by the way in which the singular converters are connected together to form the composite converter. Finally, when composite converters of similar or different structure are cascaded, the structure will be said to be "compound."

Topology of Singular Power Electronic Converters. Those topologies where a source and a load are connected by a single power switch or a set of power switches will be considered as singular power electronic converters. Figure 3-7a represents the simple singular converter, while two of these topologies may be combined to give either the split-source double converter or the split-load double converter as shown

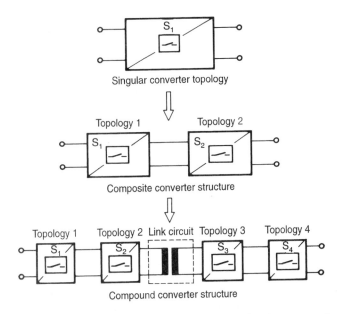

Figure 3-6. Illustrating topology and structure for singular, composite, and compound converters.

in Figures 3-7b and c, respectively. Although the first topology is invariant to the position of the source and the load, the last two are derived by an interchange of load and source positions. Extending these combinations leads to the bridge topologies shown in Figures 3-7d and e. A combination of two split-source double converter topologies leads to a bridge topology with the source in the bridge, whereas a combination of two split-load double converter topologies leads to a bridge converter with the load in the bridge. Upon close inspection, these topologies are seen to be identical. A systematic study of the many possible characteristics obtainable from this so-called single-phase bridge topology is available in [46].

The topologies of Figures 3-7d and e may be termed the generic topologies for switching converters, as all known variations may be derived from them. This process is illustrated in Figure 3-8. A combination of n of the simple topologies can result in two families of n-phase topologies, with, for example, $n = 3$ as in Figures 3-8a and b. For any one of these two families of multiphase topologies, the combination of two will lead to two families of multiphase bridge topologies, again either having n sources and a single load or n loads and a single source. Figure 3-9 illustrates this again for $n = 3$. Further consideration of the "twin" topologies of Figures 3-7d and e and Figures 3-8a and b indicate that they may be represented as in Figure 3-10. An interchange of load and source positions from state 1 to state 2 yields the well-known single-phase bridge, three-phase three-pulse and three-phase six-pulse topologies for rectification or inversion, well known from the literature. From the previous considerations, it will also be clear that the procedure for obtaining the topologies of Figure 3-9 may be extended to any multiphase converter. The topology of Figure 3-10b shows a "star" configuration and the bridge of Figure 3-10c shows

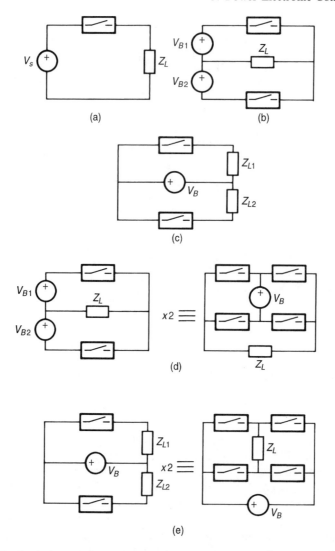

Figure 3-7. Evolution of the simplest bridge converter: (a) simple single-switch singular converter; (b) split-source double converter; (c) split-load double converter; (d) combination of two split-source converters into a bridge converter with source in the
bridge; (e) combination of two split-load converters into a bridge converter with
load in the bridge.

a "delta" connection. In fact, for any multiphase topology, either the center-con-
nected ("star") or ring ("delta") configuration may be used.

Examples of Singular Converters. In the topologies of Figures 3-7 through
3-10, the type of switching function to be inserted into each switch position in the
topology had not been given any consideration. In singular converters, the types of

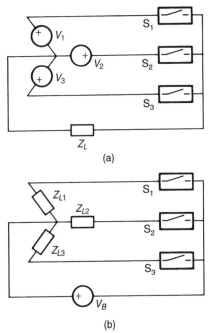

Figure 3-8. Multiphase topologies generated from the simple singular converter of Figure 3-7: (a) triple source combination of simple singular converter; (b) triple load combination of simple singular converter.

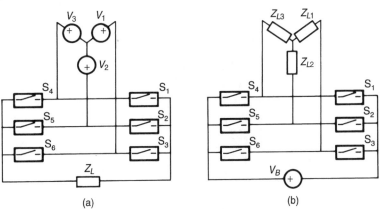

Figure 3-9. Generation of multiphase topologies: (a) combination of two of the converters of Figure 3-8a into a three-phase bridge; (b) combination of two of the converters of Figure 3-8b into a three-phase bridge.

functions introduced in Figure 3-5 are found as they are shown there, or they are found in combination with each other, as shown in Figure 3-11a. Figure 3-11b represents the well-known split-source single-phase inverter, while Figure 3-11c illustrates a transformer-fed three-phase cycloconverter for single-phase output at a lower frequency, as had been used in the past for a 50 Hz three-phase input to feed traction drives on locomotives with $16\frac{2}{3}$ Hz single phase. Further examples may be found in Figure 3-13, where these switch combinations have been used in

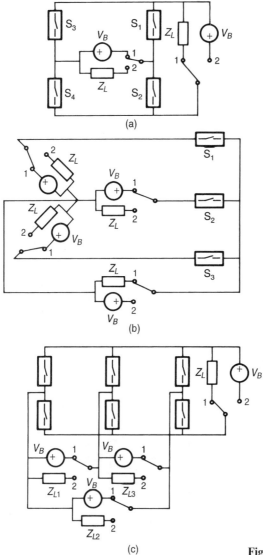

(a)

(b)

(c)   **Figure 3-10.** Unification of topologies.

the topologies discussed in the preceding section to obtain the singular converters that are joined together in the structures for composite converters. The switches of Figure 3-11a have been extended a step nearer their practical realization in Figure 3-13, by indicating power field effect transistors and bipolar transistors for the forced commutation function of Figure 3-5 for instance. It will be evident from the discussion in Section 3.2.1 that a large number of topologies may be synthesized by implementing all the different combinations of switches. In practice, all the different switch combinations of Figure 3-11a may also be used in different variations within each topology, so that a vast collection of these converters exist, as is evident from the review literature [1].

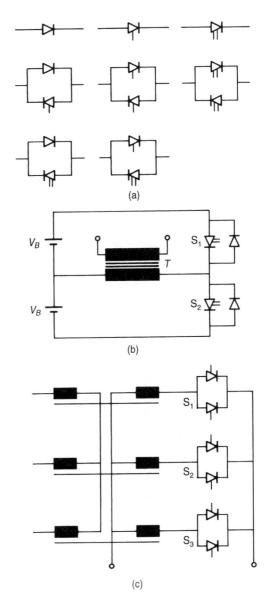

Figure 3-11. Examples of singular converters: (a) switches and combinations of switches that may be inserted in each position of a topology; (b) split-source single-phase, according to Figure 3-7b; (c) star connected three-phase cycloconverter, according to Figure 3-8b.

Structure of Composite Power Electronic Converters. Different topologies of singular switching converters may be used together in a structure comprising a composite switching converter. As shown in Figure 3-12, several fundamentally different structures are possible. To avoid discussing the detailed topology of the singular converters, each has been indicated symbolically as a p-pulse converter using the bidirectional on-off switching function from Figure 3-5. By definition, the left-hand singular converter is the primary converter and the right-hand converter is the secondary converter of each composite converter. The use of the

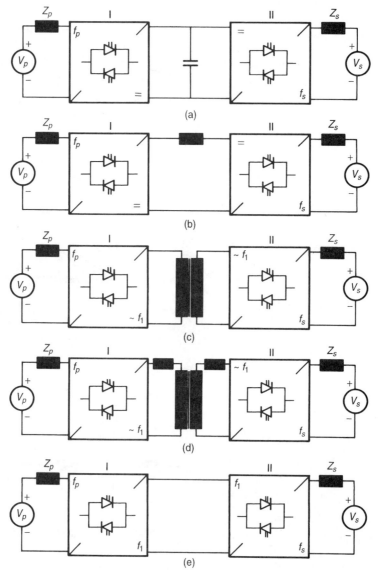

Figure 3-12. Structure of composite switching converters: (a) direct voltage
link converter; (b) direct current link converter; (c) alternating
voltage link converter; (d) alternating current link converter;
(e) directly linked composite converter.

bidirectional on-off switching function in the converters has the following consequences:

1. No restrictions are placed on the primary and secondary sources (loads).
2. All converters can pass current in two directions at both input and output.
3. All converters can pass power in both directions.

Fundamentally, the primary and secondary converters may be joined by a link circuit (Figures 3-12a–d), or the two converters may be coupled directly as in Figure 3-12e. A direct voltage link or a direct current link as in Figures 3-12a and b may be found, and an alternating voltage link and alternating current link via a coupling transformer as in c and d are also possible. In composite converters with a transformer link, it is usually found that the link frequency $f_l \gg f_p$, f_s, to use the advantages of a smaller transformer. In principle, the direct voltage link composite converter of Figure 3-12a has a steady link voltage, but alternating components of link current. For the direct current link converter of Figure 3-12b, the converse is true. The alternating link of Figure 3-12c is an alternating voltage link and that of Figure 3-12d an alternating current link, so that, for both these types of composite converters, only alternating voltages and currents are found in the (high-frequency) link circuit. Any directly linked composite structure as in Figure 3-12 will exhibit steady and alternating components of current and voltage in the link circuit as a matter of principle.

When taking into account that the singular converters shown, when used as elements in the composite converters of Figure 3-12, may have any of the hundreds of topologies for singular converters, as well as use any of the device switching functions indicated in Figure 3-5, it becomes evident that the number of detailed structures is so vast that an encyclopedic systematic discussion is not possible in the confines of the present review. This number is even further increased when considering the much larger family of compound converters. In these compound converters the primary and secondary converters may consist of direct composite converters or link composite converters, again with the numerous device switching functions determined by each specific application.

Finally, it should be noted that although the schematic representations in Figure 3-12 are given for single-phase structures, the number of input and output phases form another variable, with the pulse number p of singular converters as used in the topologies and structures being a further variable.

Examples of Composite Power Electronic Converters. Examples selected at random from among the structures of composite converters known from the literature to be useful for variable frequency drives are shown in Figure 3-13, each corresponding to the respective part of Figure 3-12. Figure 3-13a shows a voltage-fed inverter supplied by a so-called four-quadrant controller from a single-phase supply as has been applied for traction drives in locomotives [14, 15], and Figure 3-13b represents a machine commutated variable frequency three-phase inverter system supplied from the three-phase network, as has been applied for the acceleration of large gas turbines used for peak demand generation. When the switching function in singular converter II of this example is replaced by a unidirectional on-off conduction function (force-commutated thyristor or GTO, for instance), the well-known current-fed inverter is found. AC from a DC input via a high-frequency AC link is shown in Figure 3-13c. This leads to ultra lightweight and small dimensions [18] for mobile drives. Figure 3-13d shows a normal down chopper or buck converter for DC drives or as input converter to AC inverters as a composite converter.

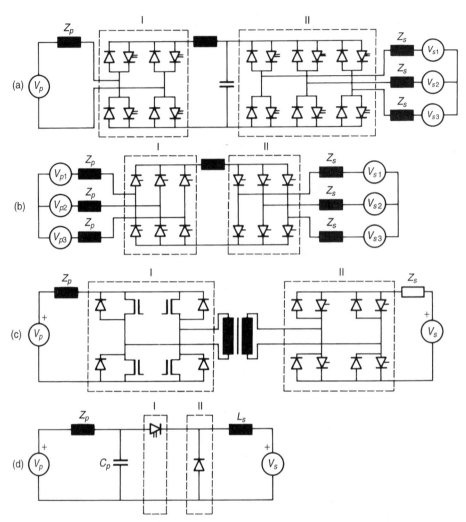

Figure 3-13. Examples of composite switching converters corresponding to
the different classes defined in Figure 3-12. The converters
numbered I and II refer to the primary and secondary converters as in Figure 3-12.

Many more examples of composite converter structures can be chosen from
those known in the subject literature. Again, as was the case with examples of
singular converters, no attempt can be made to be comprehensive in the examples
given: those given should each be seen as one from a very large set available in
practice.

3.3.3. The Fundamental Dilemma of Switching Converters

In Section 3.3.1 it was pointed out that efficiency and loss requirements in power electronic converters dictate switching operation as a fundamental requisite. The discussion then passed on to controlling average energy flow by this switching operation. However, the input supply system to the switching power network and the motion control load connected to this converter is subjected to the instantaneous values of voltage and current (and power) as generated by the converter. Sometimes differing drastically from the average values, these instantaneous values of voltage and current have a profound influence on the input network as well as the load. The general problem may be formulated as follows (refer to Figure 3-14).

The input amplitude A_i at desired input frequency f_i is always larger than the output amplitude A_u at desired output frequency f_u, due to the losses in C, or

$$A_i > A_u \tag{3.3}$$

Furthermore additional frequency components that did not exist in the supply network before operation appear at the converter input as

$$\sum_{n \neq i} A_n(f_n) \tag{3.4}$$

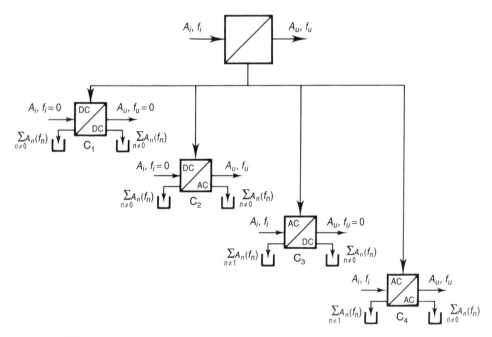

Figure 3-14. Converter families according to function, frequency, and amplitude transformation.

and at the output as

$$\sum_{n \neq u} A_n(f_n) \tag{3.5}$$

The relationships between these components on both input and output, between input and output components, and between these components and desired input and output frequencies depend on the topology and structure, switching functions, devices, and control methods. Much of the past literature on switching converters has been involved with the issues of determining these components, eliminating them [42, 47] or compensating their effects [48]. These undesired components are of critical importance for effects such as losses, stability, electromagnetic compatability/electromagnetic interference (EMC/EMI), acoustic noise, torque pulsations, and other detrimental aspects of switching power converters. Figure 3-14 gives a schematic representation for AC-to-DC converters (C1), DC-to-AC converters (C2), AC-to-DC converters (C3), and AC-to-AC converters (C4).

3.3.4. Converter Structures for Variable Frequency Drives

Variation of stator frequency on a synchronous machine or an induction machine changes the speed of rotation of the magnetic field in the air gap and, therefore, also the output speed of the mechanical drive shaft. Since the magnetic flux density in the machine must not increase with lowering of the frequency, the converter has to adapt output voltage amplitude applied to the AC machine directly proportional to frequency. This dual function of the converter (that is, adapting amplitude and changing frequency) in principle leads to a composite converter structure. However, in some cases these two functions are integrated into a single converter by the use of the appropriate switching function (such as pulse width modulation).

In many variable frequency drive systems the power is obtained from the constant frequency AC supply grid. This necessitates the use of another converter to effect transformation of the electrical power in such a way that it can be applied to the inverter input, again leading to a composite converter structure. Several possible converter structures for drives with synchronous machines (SMs) and induction machines (IMs) are given in Figures 3-15 and 3-16. Although other configurations are possible, the families given represent the configurations most commonly found. In all cases the direction of rotation is reversed by simply changing the phase rotation of the inverter through the sequence of driving the switches.

Converter Structures for Synchronous Machine Drives. Figure 3-15a represents the composite converter structure for a DC-fed system. Since the machine commutated converter can operate only when SM is already rotating and presenting a back EMF, provision has to be made for starting—usually by on-off operation of the input chopper. As shown by the dotted lines, a bidirectional chopper is necessary for regenerative operation—complicating the structure and the control.

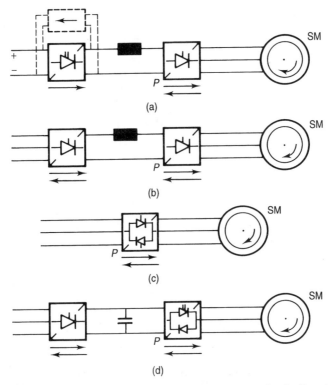

Figure 3-15. Some generic power electronics converter structures for feeding drives with synchronous machines (SM); arrows indicate power flow direction: (a) synchronous machine drive supplied from a DC bus via a chopper and a machine-commutated inverter (for a regenerative drive a bidirectional chopper is necessary, as shown dotted); (b) synchronous machine supplied from an AC bus via a controlled rectifier and a machine-commutated inverter; (c) synchronous machine supplied from an AC bus via a naturally commutated cycloconverter; (d) synchronous machine supplied from an AC bus via a controlled rectifier and a PWM voltage-fed converter.

A machine-commutated inverter supplied from the AC bus via a controlled rectifier is a fully regenerative system (Figure 3-15b). Starting presents the same problem as previously discussed, so that normally pulsed operation of the controlled rectifiers is used to start the drive. The naturally commutated cycloconverter of Figure 3-15c has no starting problems and is fully regenerative, yet has a limitation on the maximum frequency imposed by the supply frequency of the AC bus. By using forced commutation in the inverter and installing a direct voltage link, the quasi–square wave voltage-fed synchronous machine drive of Figure 3-15d is obtained. This has become attractive with the availability of turn-off switches (FETs, IGBTs, bipolar transistors, GTOs). It is also possible to use the voltage-fed converter in the PWM mode, and implement this as a field-orientation controlled drive [53].

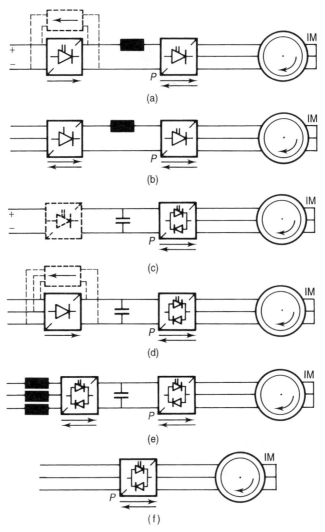

Figure 3-16. Some generic power electronic converter structures for feeding drives with induction machines (IM): (a) Induction machine drive supplied from a DC bus via a chopper and a current-fed inverter (for a regenerative drive, a bidirectional chopper is necessary, shown dotted). (b) Induction machine drive supplied from an AC bus via a controlled rectifier and current-fed inverter (PWM possible). (c) Induction machine drive supplied from a DC bus via a voltage-fed inverter. If the inverter uses square wave operation, a series chopper (shown dotted) is necessary. For PWM operation of the inverter, no chopper is used. (d) Induction machine drive supplied from an AC bus via a PWM voltage-fed inverter. For regeneration, a converter (shown dotted) is necessary. When the inverter is a quasi-square wave inverter, a controlled rectifier input is necessary. (e) Induction machine supplied from an AC bus via a PWM rectifier and PWM inverter. Structure is fully regenerative. (f) Induction machine supplied from an AC bus via a force-commutated cycloconverter (matrix converter). Structure is fully regenerative.

Converter Structures for Induction Machine Drives. Although the range of structures given in Figure 3-16 does not represent all possibilities, it represents most of the systems made possible by current power semiconductor technology. When driving an induction machine from a DC bus via a current-fed inverter, a buck chopper is necessary as shown in Figure 3-16a. For regenerative applications (road vehicles fed from batteries), the system is complicated by the necessity of a bidirectional chopper (shown dotted). Options of PWM or quasi–square wave operation of the inverter are possible.

The availability of switches with turn-off capability (FETs, BJTs, IGBTs) and GTOs have currently favored drives with voltage-fed PWM converters on induction machines as shown in Figure 3-16c and d. By using a PWM rectifier as primary converter in this composite structure (Figure 3-16e), both the problems of regeneration and line current distortion are succesfully solved—with the penalty of having a much more complicated converter structure and control system.

The force-commutated cycloconverter (matrix converter), Figure 3-16f, represents possibly the most advanced state of the art at present, enabling a good input and output current waveform [31], as well as eliminating the DC link components with very little limitation in input to output frequency ratio. This will again be discussed in one of the next parts of this chapter.

3.4. POWER ELECTRONIC CONVERTERS FOR CONTROL OF AMPLITUDE

In the control of DC machines in drives, it is necessary to adapt the voltage amplitude applied to the machine as a function of the operating conditions. Since—in terms of a functional view—inverters in AC drives perform the same function as mechanical commutators in DC machines [5], converters for amplitude control find application in various forms in variable frequency drives fed from inverters. Invariably, where these converters are applied in systems fed from a DC or an AC bus, a composite or a compound converter structure results (refer to Figure 3-6 for the definitions), since in AC variable speed drives the machines operate from a multiphase voltage/current system, supplied from a converter fed in many instances from one of the DC-to-DC or AC-to-DC converters [4].

3.4.1. DC-to-DC Converters

Although the buck, boost, and buck-boost DC-to-DC converters as shown in Figures 3-17a, b, and c have been used since the first chopper circuits were proposed and tried experimentally, low-loss turn-on and turn-off have been a continuous problem [49]. The original thyristor forced commutation circuits [50–52] limited the switching repetition frequency to low values. This was extended by the presently available power semiconductor switches [53] but still limited by the losses during the turn-off and turn-on transition of S. As will be discussed in Section 3.6, however, zero voltage switching (ZVS) and zero current switching (ZCS) realizations of the

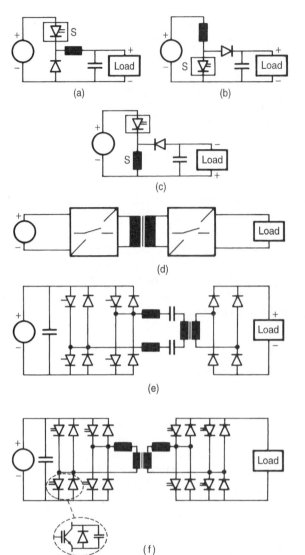

Figure 3-17. DC-to-DC converter topologies: (a) buck chopper; (b) boost chopper; (c) buck-boost chopper; (d) high-frequency link composite converter as DC-to-DC converter; (e) series resonant high-frequency link converter as DC-to-DC converter; (f) phase shifted high-frequency link converter as DC-to-DC converter.

switch S are possible in resonant, quasi-resonant, and resonant transition technology [54, 55] leading to a large family of these converters, but at lower power level due to the increased device stress. The DC-to-DC converters of this type used at high power levels [7] still employ hard switching techniques, with snubbers added to reduce switch stress to acceptable levels.

In applications requiring high-power and high-power density DC-to-DC converters (aerospace applications), soft switching is a prerequisite. Among the many variants suggested, the load-commutated series resonant converter with high-frequency link [11] and the phase-shifted high-frequency link converter [10] are among the few that have been successfully developed for this specialized application. As is evident from Figure 3-17d, this high-frequency link structure also enables transformer isolation between input and output. The series resonant system of Figure 3-17e allows the use of thyristors only—but this limits the frequency to some 10–20 kHz [11]. For regenerative operations, the diodes in the secondary converter have to be replaced by thyristors.

The phase-shifted high-frequency link converter structure of Figure 3-17f uses the resonant pole concept to achieve soft turn-on and uses capacitive snubbing for low turn-off loss. The power density of this system can be enhanced by using a three-phase topology for both primary and secondary converters [10]. For industrial drive applications requiring DC-to-DC converters, these types of converters are not preferred to the conventional structures of Figure 3-17a–c, unless transformer isolation and small mass are a requirement.

3.4.2. AC-to-DC Converters

In all AC variable frequency drives—with the exception of drives fed by direct converters—the AC-to-DC converter forms the important link between the drive and the AC bus. Traditionally uncontrolled and controlled rectifier bridges are used for this application [53], and their characteristics have been discussed extensively in the textbook literature [56–58]. The present accent on power quality and improved power factor have shifted the accent to more supply friendly AC-to-DC converters, some of which are shown in Figure 3-18a–c. The application of the single

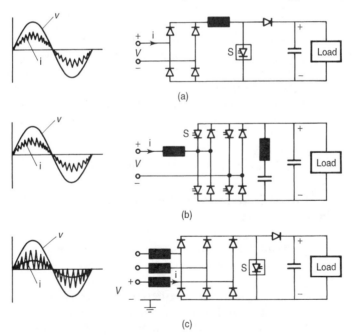

Figure 3-18. AC-to-DC converters for improved input current waveform: (a) diode rectifier with a boost converter; (b) four-quadrant converter for single-phase operation; (c) three-phase diode rectifier with active line current shaping.

phase diode rectifier with boost converter [59] is limited to some kilowatts due to being fed from the single-phase bus. In the three-phase diode rectifier circuit with output chopper [60] of Figure 3-18c, the input current pulses (chopper switch S operated at constant switching frequency) are discontinuous, leading to the consequence that S has to be operated at very high frequency to allow the filtering of the harmonics by a small input filter (not shown). This high required frequency puts a limitation on the maximum power level to which this technology is applicable, since the devices to be applied in S may be FETs or possibly IGBTs.

For application at power levels exceeding one megawatt (1MW), the four-quadrant PWM converter of Figure 3-18b was developed [1, 14, 15] first in the 1970s using thyristor forced commutation technology, later being adapted to the use of GTOs. As this singular converter is fed from a single phase (traction) supply, particular care is necessary in the layout of the DC link to absorb the second harmonic and supply a stable voltage to the follow-on PWM inverter. By using phase-shifted converters of this type in multiple operation [14], the ripple content is decreased—even at a switching frequency of a few hundred hertz (Hz)—to low levels, even in multimegawatt applications. The three-phase PWM rectifier (Figure 3-19d) is a follow-on from this, although the maximum applicable power level has yet to be established.

3.5. POWER ELECTRONIC CONVERTERS FOR AC VARIABLE FREQUENCY DRIVES

In Section 3.3.4 some overall converter structures for AC variable frequency drives were discussed. The machine-commutated inverters are described well in the literature [4, 53] and will not be discussed here. However, an overview of the many composite converters for variable frequency PWM operation—with chiefly induction machines as loads, is still deemed necessary—particularly in view of the variations in topology and structure brought about by the quest for supply friendly drives and soft switching converter systems at higher power levels. In this section the discussion concerns the topologies and structures, while the considerations regarding switching technology will be examined in Section 3.6.

3.5.1. AC-DC-AC Converters for Current-Fed Inverter Drives

Force-commutated inverter technology started shortly after controlled mercury arc rectifiers came into practice [5], both voltage-fed and current-fed topologies appearing. After thyristors appeared, attention focused initially on the voltage-fed topology, and subsequently current-fed inverters received attention [61], leading to the development of the well-known autosequentially commutated current-fed inverter [62], which has found widespread practical application [53] and is described adequately in textbooks [56]. Figure 3-19a indicates the generic structure of the current-fed AC-DC-AC converters. In industrial and traction applications these current-fed AC-DC-AC converters are robust in operation and reliable due to the

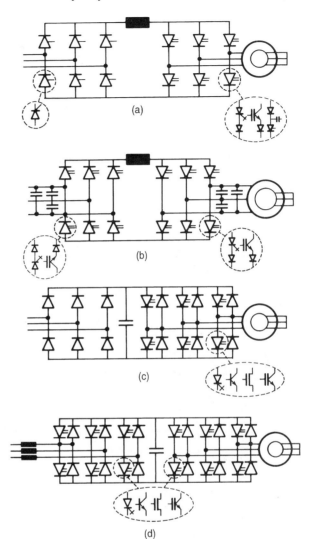

Figure 3-19. Composite AC-DC-AC converters: (a) current-fed inverter fed by a controlled three-phase rectifier; (b) current-fed PWM inverter fed by a current mode PWM rectifier; (c) voltage-fed PWM inverter fed by a three-phase diode rectifier; (d) voltage-fed PWM inverter fed by a three-phase PWM rectifier.

insensitivity to short circuits and a noisy environment. However, whether applied to synchronous or induction machines, the pulsating torque (at six times the inverter frequency) delivered by the drive can lead to mechanical resonances, additional losses, and other problems. In subsequent designs it was shown that PWM operation of the current-fed inverter can lead to elimation of harmonic current components, causing torque oscillations in the machine [47], even using thyristors with auto-sequential forced commutation. The availability of BJTs, IGBTs, and GTOs first led to replacement of the force commutated-thyristors with the addition of a small capacitor filter on the induction machine input terminals [17, 63], as shown on the inverter of Figure 3-19b, greatly improving the waveforms of voltage and current applied to the machine, as well as easing turning off of the devices in the inverter, since quasi-sinusoidal PWM could now be applied to the inverter output current.

This also solved the pulsating torque problem of these drives effectively. However, current-fed inverters need symmetrical devices as far as blocking voltage is concerned, so that for IGBTs and reverse conducting GTOs, a series diode is always necessary. A natural follow-on to the GTO current-fed inverter with capacitor filters was to add a similar topology to the input, leading to the double PWM current-fed structure of Figure 3-19b [63], representing the state-of-the-art development in this technology.

3.5.2. AC-DC-AC Converters for Voltage-Fed Inverter Drives

Voltage-fed inverters, as discussed in relation to Figure 3-16, have had a line of development closely coupled to the availability of the appropriate power semiconductor devices. Although in principle the voltage-fed inverter is more prone to reliability problems in a harsh environment than the current-fed inverter (with simultaneous conduction of two devices in one inverter leg being potentially the most dangerous), protection and control schemes were developed that enabled reliable application. Voltage-fed inverters do not need reverse blocking devices, favoring the modern devices such as IGBTs and GTOs.

However, voltage-fed inverters in forced commutation thyristor technology developed to a stage where it could be applied up to the highest power levels in main line electric locomotives. Since these converter structures remained complicated due to the additional commutation circuits, it was the availability of FETs, BJTs, and IGBTs that saw to the successful penetration of the entire power range by the three-phase diode rectifier-fed, PWM inverter drive for induction machines, shown in Figure 3-19c [2]. PWM technology enabled elimination of harmonics from the inverter output voltage, allowing quasi-sinusoidal machine waveforms and eliminating torque pulsations. The current-controlled, voltage-fed PWM inverter has gradually been superseding current-fed technology over the past years, especially since the availability of GTOs extended the applicability of turn-off devices in voltage-fed inverters up to the highest power levels.

3.5.3. Supply Interaction of AC-DC-AC Converters

The supply interaction has remained a problem over the entire power range, since the input converter is customarily an uncontrolled or a controlled rectifier bridge (see Figures 3-15, 3-16, and 3-19). This supply interaction could be solved for traction systems by the four-quadrant converter as discussed previously in relation to Figure 3-18b, while in the lower power range of AC-DC-AC converters fed from a three-phase bus, the solution of Figure 3-18c provides the appropriate input converter. The double PWM voltage-fed structure shown in Figure 3-19d [64] represents the ultimate power electronic solution in terms of PWM voltage-fed converter technology regarding a supply-friendly and machine-friendly converter system. Although this type of system can be implemented using any of the presently available turn-off devices (FETs, BJTs, IGBTs), general application still has to develop.

The supply interaction of controlled and uncontrolled rectifiers has been the subject of much work in the past and has already been treated extensively at textbook level [57, 65, 66] since awareness of the problem had been formulated in the times of mercury arc rectifiers [67]. It is only recently, however, that proliferation of AC-DC-AC converters for drives and other power electronics equipment has caused serious concern on the general problem of nonlinear loads on power networks, giving rise to problems regarding modeling, measurement, and compensation of supply interaction [48].

Of particular interest are the standards that have come into being internationally to control harmonic proliferation from converters in power networks. Notably, IEEE-519 regulates the harmonic injection at the point of common coupling (PCC) and is targeted toward larger plant and equipment [68, 69, 70]. IEC-555-2 regulates the harmonic emissions from each piece of equipment connected to an AC supply system and targets the smaller range (below 15 A, chiefly) [71]. It is therefore already evident that supply interaction of all AC-DC-AC converter-fed drives is going to be a central consideration in future, placing special emphasis on the development of solutions such as those of Figures 3-18 and 3-19d.

3.5.4. More Extended Converter Families

The preceding discussions have been concentrating on the classical topologies for singular converters that have evolved over time. In the context of converters for drives, this inevitably evolves toward six-pulse topologies. The field of power converters for drives remains ever interesting, since many possible variants of converter topology and structure continue to exist—especially for specialized applications. Some of these variants are discussed in this section.

Cycloconverters. Although these converters have been known since controlled switching devices became available [5], power semiconductors increased the application possibilities. Figure 3-20a represents the converter structure comprising the combination of nine AC switches [29]. With the AC switch consisting of two anti-parallel switches with delayable control (such as thyristors), the well-known naturally commutated cycloconverter—as widely applied—is obtained. However, since the advent of simpler turn-off technology in converters, there has been a quest for a "general converter" to couple two multiphase sets of AC voltages by having the nine AC switches capable of turning on and off at any instant [29–31]. Application possibilities have recently been proven [32], and the power level might be extended in the future.

Composite Converters with Specialized AC Links. When considering voltage-fed inverters, the link has generally been considered, first, to be a direct voltage or a direct current link and, second, to consist of passive elements only. During the earlier stages of the development of thyristor forced commutation technology, several schemes to actively reduce the DC link voltage or current to zero by employing a

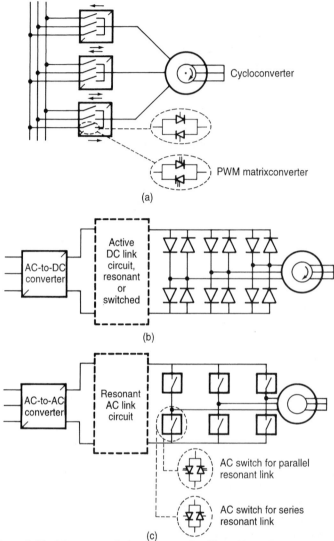

Figure 3-20. More extended converter families: (a) cycloconverter/matrix converter; (b) AC-DC-AC converter with an active DC link circuit; (c) AC-AC-AC converter with a resonant AC link and two AC-to-AC converters.

switching element in the link were proposed and tried [51, 52]. Turning off all the elements in the inverter bridge was consequently achieved. Hybrid BJT-thyristor inverters were developed, exhibiting the characteristic that all switching transitions could be undertaken only when the link was activated (and, for instance, the link voltage reversed or reduced to zero [72]).

The new generation of soft switched topologies, such as the resonant DC link inverters [73] and the resonant AC link converter [74], develops the idea of actively using the voltage variations in the link to style the turn-off of the devices in the converters further. Even further measures and active devices can be introduced to impart the necessary characteristics to the converter behavior [86]. Since these measures affect the switching characteristics of the converter, more attention will be devoted to a simplified discussion thereof in Section 3.6.

Multilevel Topologies. Extending the number of power switches (or individual power semiconductors) in the converter topology, as discussed in the first part of this section (for instance, cycloconverters), may enable more complicated switching functions to be realized. However, extension of the number of power switches may also be used to execute the same switching function in each—and by addition achieved desired results [75]. In the extrapolation of converters to higher powers, the fundamental question of increasing the converter current rating (devices in parallel) or increasing the converter voltage rating (devices in series) always has to be answered. The conduction losses in converters always favor the increased voltage model, and this is additionally accentuated by the fact that system input voltages for AC-DC-AC converters naturally rise as power level increases. All these factors contribute toward the necessity for a technology for multilevel converters, where series connection of devices is used to implement different subconverters in series, adding their switching functions at the converter output [76, 77]. As indicated in Figure 3-21a, multilevel converters can be configured in a generic way, not limiting the number of switching cells [78]. When utilizing *n*-switching cells in series, it has to be realized and explored that a commensurate large collection of switching functions may be used, starting with each switching cell in the phase arm using the same switching function—whether time shifted or not.

The present technology of multilevel inverters was initiated by a quest for better PWM output waveforms, using identical switching functions for each switch cell, originally resulting in the neutral point clamped inverter (Figure 3-21b [75]). With the advent of higher powers for PWM inverters, matched against the requirements for increased powers in systems using IGBTs and GTOs with limited voltage capability, this type of topology opened the possibility of increasing the power level without the troublesome direct series connection of devices [76], as shown in Figure 3-21c for a three-phase, three-level converter applicable in the MW range. These structures can again be doubled in power by using two in a twin arrangement on a motor with open windings [77].

3.5.5. Minimum Converter Topologies

The relatively large number of switching components needed for either a single-phase or a three-phase AC-to-AC converter has been a motivation for devising AC-to-DC and DC-to-AC converter topologies with a minimum number of components since the earliest times of power electronics. Especially when the volume (and cost) of power switches was an appreciable part of the system cost (originally mercury arc rectifiers, force-commutated thyristors, bipolar power devices), this approach would make sense. At the present state of the art, when the power semiconductor device

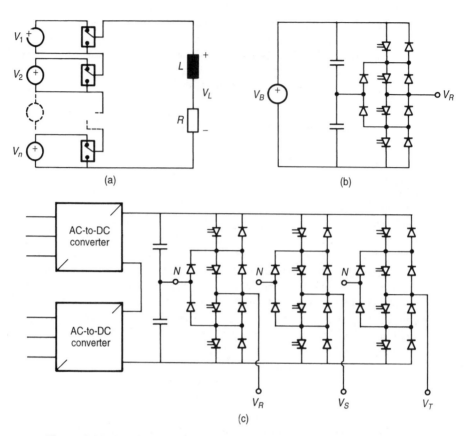

Figure 3-21. Multi-level converters: (a) generic topology for a multilevel
converter (load voltage of only one polarity); (b) topology of
phase arm for neutral point clamped inverter; (c) structure for
a three-phase three-level inverter.

and drive cost is becoming very small, this type of approach has to be evaluated
more for its disadvantages than for its advantages. These types of converter structures are still of present interest [79, 80].

Composite converter structures utilizing a minimum number of components [81]
can be devised for direct voltage link, direct current link, or high-frequency link
structures. The ideas proposed in the era of thyristors or bipolar transistors as
switches [81–85] are still valid. Examples of generic structures for direct voltage
link structures or high-frequency link structures, without specifying the switching
functions nearer, are given in Figure 3-22. The different direct voltage link circuit
variants and the different high-frequency link current variants have been investigated
fully in the past [81, 85], indicating what the generic problems are with these composite converters. The comments that follow concern the generic DC-to-AC topology and the high-frequency to low-frequency AC topology as the output converter
(Figure 3-22), but since the same generic topologies are applicable to the reverse
process at the input, these are not shown. Also note that the type of switching

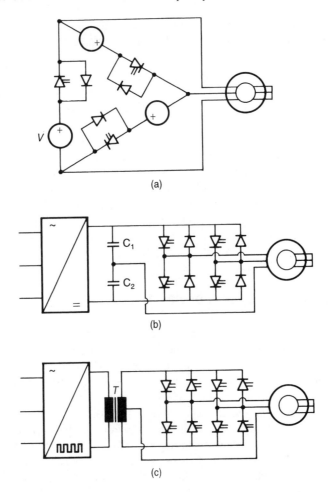

Figure 3-22. Converter topologies for minimum components: (a) delta
inverter for three-phase load; (b) four-switch voltage-fed
inverter for three-phase load; (c) four-switch high-frequency
link converter for three-phase load.

function [5] is not specified—general switches are indicated. Soft switching techniques can also be incorporated as switching technology. The further extension of the
composite high-frequency link structure by for instance introducing an AC-to-DC
stage before the output converter [80] is not discussed—this merely aggravates the
disadvantages.

The cited investigations have shown that in these families of converter structures
the reduction of the number of switches are paid for by

- Always an increase in the switch current ratings [81] and in some topologies
  also concurrently the voltage ratings [85].

- Severe increase in switching frequency to observe modulation functions that yield the same THD (total harmonic distortion) as in six-pulse PWM conversion [81].
- Increase in transformer requirements and/or in number of transformers and windings and excessive increase in reactive power transmission in high-frequency link systems [85], again increasing also switch ratings in the primary converters.

An evaluation of these topologies therefore shows that not only is increased device ratings the price to be paid, but also a substitution of capacitive and inductive components for silicon, with the capacitive and inductive components larger and subjected to more stresses than in comparable six-pulse topologies. In terms of the present trends in electronic power converter technology, this is not a good trade-off, so that application is reserved for very specialized situations.

3.6. SWITCH APPLICATIONS TECHNOLOGY

In previous sections a systematic discussion of converter topology and structure (with some examples) was followed by the different important possibilities where converters are being applied. From this it followed that a common limitation applicable to motion control converters (and also to converters in general) is due to the switching nature of the power processing within the converter. Similarly applications technology common to all types of converter can be identified, with specific emphasis on the requirements for variable frequency converters. These will be discussed in the paragraphs that follow, without first compiling an exhaustive review of all types of converter topologies and structures that can be used, since an excellent contribution of this nature can be found in [86], augmented well by some others [2, 4, 13, 36].

Important considerations for converter application are

- Turning power electronic switches on and off (forced commutation and/or driving the switch on and off).
- Reduction of switching losses in practical switches (snubbing and resonant circuits).
- Protection of practical switches (transient voltage suppression, clamping, overvoltage, and overcurrent detection).
- Cooling of practical switches.

A simplified summary of a typical application for these technologies to a motion control application on a main line all electric locomotive [87] is given in Figure 3-23. Whereas the complete composite converter structure is given in this figure, it should be noted that for all the other discussions (Figures 3-28 to 3-31), either an equivalent circuit reduced to a single switch or a phase arm is used as basis for discussion.

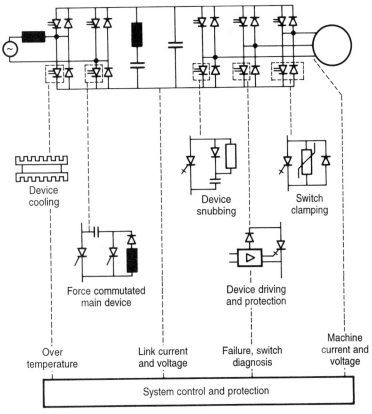

Figure 3-23. Important technologies in converter applications, illustrated by a motion control example on an AC-fed main line loco-motive.

In practical switching converters it must further be kept in mind that the following considerations—all related to the devices used and the switching action of the converter—apply:

- Power semiconductor switching devices are sensitive to the magnitude of dv/dt and di/dt transients regarding their stability, behavior, and reliability/ destruction [88, 89].
- The high-frequency harmonics of voltages and currents injected back into the supply system [48] and coupled electromagnetically, as well as conducted through the machine to earth, are all functions of the switching transition times and switching formats in the converter.
- The dv/dt and di/dt transients applied to the load are determined by the switching transitions in the converter, and in many cases the load is sensitive to these (insulation stress in electrical machines is an example).
- The switching transition behavior determines the converter switching losses for a given frequency. These losses can be reduced only by decreasing the

internal frequency of operation of the converter, once the format of the transitions is fixed by the accepted converter design and parameters. A reduction in internal frequency, however, carries with it the heavy penalty of increasing input and output ripple and losses in the electrical machine.

3.6.1. Turning Power Electronic Switches On and Off

Turning power switches on and off requires intricate techniques, both from the circuit and from the device gating point of view. With the original thyristors, gating requirements were modest, but turning the device off required complicated circuit arrangements. As turn-off capabilities became available in the power devices themselves, the advanced technology requirements shifted to the gating systems.

Thyristor Forced Commutation Technology. With the appearance of thyristors with shorter turn-off times, converter technology with higher switching repetition frequency enabled application to many new drive concepts. Much effort was expended in the development of many variations of parallel and series force commutated circuits for inverters [50–52], eventually developing into PWM inverter technology that proved to be applicable up to the highest powers on main line locomotives [87]. Current-fed inverter technology eased the commutation and protection problem considerably, and especially in the phase sequence commutated topology received much attention, even with pulsed operation for the elimination of torque pulsations in drives [47]. However, the availability of the present gate turn-off technology has led to the gradual obsolescence of forced commutation—certainly for any new developments [86]. It therefore serves little purpose to discuss these intricacies in this chapter. The same fate may befall current-fed inverter variations in motion control applications, since the development of current-controlled voltage-fed PWM inverters has advanced so rapidly.

Switch Driving Technology in Switching Converters. Soon after the advent of thyristors as the first of the modern high-power semiconductor devices, it was realized that special attention needed to be devoted to driving these switches. The turn-on process was studied thoroughly [90–92], with circuit techniques being developed for convenient triggering [93, 94]. With very small external circuit parasitics, the device becomes the limiting element for current transients in the converter circuit, so that exceptionally hard driving can increase the turn-on performance remarkably [95, 96]. For power switching converters this method of improving switching performance in the end becomes a question of economy (i.e., whether it is more cost effective in terms of system cost, performance, reliability, and losses to invest in more elaborate drive circuitry or in additional snubber elements in the converter). These considerations also apply to extreme switching of GTOs [97].

Large bipolar power transistors and Darlingtons require much care in proper driving to achieve reliable operation and high performance [98, 100]. Many driving technologies have been proposed in the past—for a comprehensive review, see [98,

101]. However, expertise and techniques developed for large bipolars aided in establishing the proper driving technology for gate turn-off thyristors. In GTOs the correct switching transition waveforms—as determined by the switch driving and snubber interaction—are essential for reliable operation [102, 103], with many interesting technical possibilities having been demonstrated [104–106].

MOS-gated switches present a different range of challenges regarding driving [107], as it is very soon found that the switching speed of these devices excite the circuit parasitics excessively. For IGBT devices the dynamic device input capacitance interaction with the drive circuit [108] is of primary importance in developing proper driving technology [109]. As device tail current behavior at turn-off and device tail voltage at turn-on are such important factors for switching losses in GTOs, IGBTs, bipolars, and Darlingtons [110], their drive dependence has to be optimized carefully in all switching converter applications in conjunction with snubbers and resonant circuits or with circuit parasitics in hard switching.

3.6.2. Reduction of Switching Losses in Practical Switches

Much attention has been given to the development of technology for switching loss reduction, and certainly in large variable frequency drives this is still essential.

The character of the switching transitions in converters is determined by an interactive relationship between the gating or driving of the power semiconductor devices and the external electromagnetic circuit elements. It has become customary, however, to discuss these two issues separately. This practice will also be followed here.

Since modern power semiconductor devices have switching transition times (turn-on and turn-off times) in the microsecond to nanosecond region, "parasitics" inductive and capacitive elements in the converter circuit always come into play during switching transitions. When only the electromagnetic circuit parasitics of the converter layout are influencing the switching transition, it has become accepted practice to term this "hard switching." With additional inductance and capacitance being added to slow down the turn-on and turn-off processes, respectively, snubbing is said to be added. As the size of these electromagnetic energy stores is increased in an attempt to reduce the stresses during transitions, the energy they store during each switching transition can no longer economically be dissipated during the next cycle (resetting action), so that the technology of regenerative snubbers was developed as discussed subsequently. When the energy stored in the electromagnetic elements increases even further, clear resonance is observed, leading to the technology of resonant converters (see Figure 3-24).

These three stages of influencing the voltage and current waveforms during a switching transition actually represent a continuous change in type of response, but will be discussed separately in the following paragraphs.

Snubbers in Converters: The Dissipative Approach. As discussed earlier, there is always some influence of "stray" inductance L_σ when turning a device on

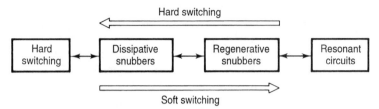

Figure 3-24. The development from hard switching through soft switching
to resonance in switching converters.

and "stray" capacitance C_σ in a circuit when turning a device off into an inductive
load (Figure 3-25a), such as experienced in motion control applications.

In the conscious use of electromagnetic storage of energy to slow down turn-on
transitions (magnetic energy) [111] or turn-off transients (capacitive energy) [112],
the inductance L_σ and the capacitance C_σ are enlarged artificially by at least an order
of magnitude. When the device S becomes conducting (approximately zero resis-
tance), the supply voltage is sustained by L_s during current build-up. When S is
turned off (approximately infinite resistance), the voltage build-up is slowed down
by C_s taking over the current. The time taken is now determined by the size of C and
the load current (Figure 3-25b). The energy stored during turn-on and turn-off
transitions, respectively, is dissipated in R_s [113, 114]. It must also be noted that
the turn-on and turn-off snubbers are not independent of each other, although they
are mostly analyzed as such [114]. From the linear fall time approximations for

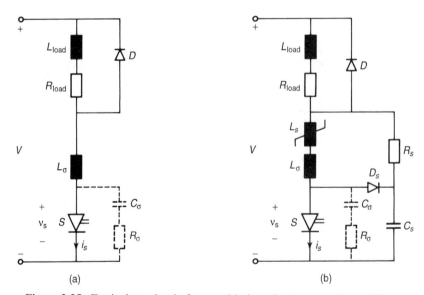

Figure 3-25. Equivalent circuit for considering electromagnetic snubbing
elements in a switching converter: (a) residual "stray" inductance and capacitance in circuit only; (b) inductive turn-on
and capacitive turn-off snubbers added.

voltage and current used in Figures 3-26a and b, it should be noted that the trade-offs are

1. The increase of overvoltage and overcurrent stresses on the switch with increase in L_s and C_s (Figure 3-26b), with decreasing of turn-on and turn-off losses.
2. The increase in energy to be dissipated in R_s. As will be discussed, this then subsequently leads to the technology of regenerative snubbers by replacing the resistor with a voltage source.

However, the original careful study of the problem [113] indicated clearly that by using a snubber the *total* switching losses can be minimized. As turn-on and turn-off times were continually reduced by advances in power semiconductor switch technology, the absolute value of the turn-on and turn-off energy losses per switching cycle reduced, creating the perception that the total switching losses need not be minimized. As the switching losses per cycle decreased, the switching repetition frequencies *increased*, however, so that the situation remains almost the same, with total loss minimization advantages being possible by using dissipative snubbers [114]—but for the complication of adding the additional circuit elements to the converter. This philosophy of striving for minimum components in the converter has led to the present position where snubbers are only used at high power levels when it is impossible to operate the converter without them. At lower power levels they are omitted, regardless of the increase of total converter switching losses because of this omission.

Snubbers in Converters: The Regenerative Approach. As the switching frequency and power level in converters increased, it became clear that important advantages in efficiency could be obtained in regenerating the snubber energy. Technology was developed to recover the energy stored in the inductive and capacitive snubbers and return it to the converter supply system [115–120] at high power levels. As an example of this type of technology, the generic study of turn-off snubbers can be used [114]. As a first approximation, linear rise and fall times for voltages and currents during turn-on and turn-off can be assumed, and the regenerative action be approximated by replacing the resistor R_s in Figure 3-25b by an energy recovery circuit of which many topologies can be found in the literature [121]. Since the added circuit elements comprising the regenerative part of the snubber—in principle a smaller, secondary converter—also introduce losses into the circuit, it is not easy to achieve a regeneration in excess of 80%, so that the 75% example of Figure 3-27b is a realistic illustration of what can be achieved. In comparison to dissipative snubbers (regeneration efficiency $\eta_r = 0$) shown in Figures 3-27a, it can be seen that the operational minimum is not as sharp, while the point of minimum total switch losses has moved toward a much larger value of the turn-off snubber (Figure 3-27b). From the point of view of minimizing thermal shock loads into the power semiconductor switch [114], this is an enormous advantage, since the power peak and

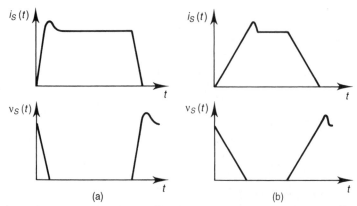

Figure 3-26. Switch voltage and current as influenced by electromagnetic circuit elements: (a) only residual parasitic inductance and capacitance in circuit (Figure 3-25a); (b) added inductive and capacitive snubbing (Figure 3-25b), increased stresses.

power dissipated in the semiconductor switching device are reduced by an order of magnitude.

The original assumptions of linear rise and fall times underlying the snubber analysis have to be adapted to modern power semiconductor switches—such as GTOs, IGBTs and MCTs—to arrive at a useful generic snubber design. These devices have fast fall times in voltage and current at turn-on and turn-off, but appreciable tails [110], changing the loss characteristics at turn-on and turn-off dramatically—as can be seen by comparing Figures 3-27a and c. In these analyses, the voltage or current waveforms during fall and tail time have been assumed to be representable by two linear segments. With the complication of the tail voltages/currents the number of parameters in the problem understandably increases—but a representative practical choice has been made to compute the graphs of Figure 3-27c and d. This is represented by the parameters A and B, which are used to define the waveform. These have also been calibrated against practical measurements [110]. It is clear from a comparison of Figure 3-27a and c that when large snubbers are used (such as in GTO technology), the design point is far above the size giving minimum losses—explaining the high losses incurred in these snubbers. The excellent improvement that can be obtained by a regenerative snubber with $\eta_r = 75\%$ follows from Figure 3-27d. However, the added complexity of the converter has unfortunately been a serious deterrent for industry to accept this up to the present, so that actual industrial practice in high power motion control converters at present is either no snubbers (just clamps) or dissipative snubbers.

Another method to reduce turn-on and turn-off losses in snubber circuits presents itself in the use of nonlinear energy storage elements, that is, nonlinear inductance for turn-on snubbers and nonlinear capacitance for turn-off snubbers. It has indeed been an accepted concept in large converters that when (linear) structural inductance is not sufficient for turn-on snubbing of devices, that nonlinear (saturable) inductance is added [111] as the only tolerable solution from a loss point of view. Saturable capacitance has been illustrated to be as effective in the

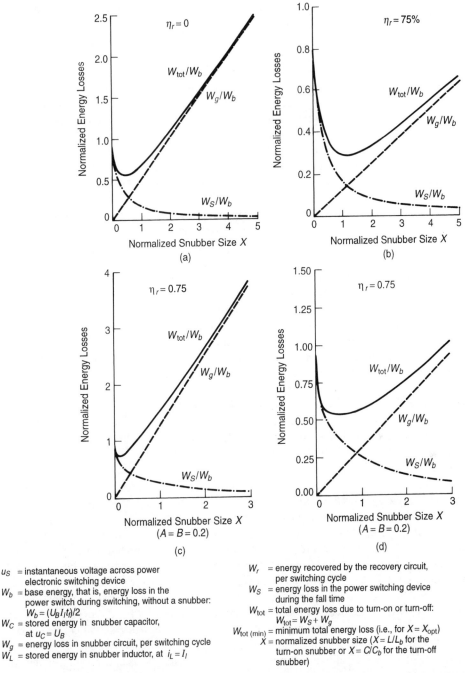

u_S = instantaneous voltage across power electronic switching device

W_b = base energy, that is, energy loss in the power switch during switching, without a snubber: $W_b = (U_B I_l t_l)/2$

W_C = stored energy in snubber capacitor, at $u_C = U_B$

W_g = energy loss in snubber circuit, per switching cycle

W_L = stored energy in snubber inductor, at $i_L = I_l$

W_r = energy recovered by the recovery circuit, per switching cycle

W_S = energy loss in the power switching device during the fall time

W_{tot} = total energy loss due to turn-on or turn-off: $W_{tot} = W_S + W_g$

$W_{tot\,(min)}$ = minimum total energy loss (i.e., for $X = X_{opt}$)

X = normalized snubber size ($X = L/L_b$ for the turn-on snubber or $X = C/C_b$ for the turn-off snubber)

Figure 3-27. Normalized energy loss as function of snubber size in converter switches. Regeneration efficiency of the snubber is given by η. The parameters A and B define the waveform of the turn-on or turn-off with tail. For the calculations the values of A and B have been chosen to coincide with modern IGBT waveforms: (a) dissipative snubber, linear fall time; (b) regenerative snubber, linear fall time; (c) dissipative snubber, tail time included; (d) regenerative snubber, tail time included.

case of turn-off [122], but due to various considerations, this has not found widespread practical application [5].

Converters Utilizing Resonant Transitions. Snubbers may be considered as auxiliary circuits included to reduce switching stresses, while in resonant converters the main circuit serves to reduce switching stresses. A number of classifications of the various resonant circuit families have been reported, for example [54, 55]. However, this type of converter has had limited application to variable frequency drives. Continuous resonance—as the name implies—involves the resonance being present during an entire cycle and switches are turned on and off at appropriate instances during the cycle, neccessitating that the machine be part of the resonant circuit.

For applications where switching frequency variations over a wide range are required, such as in motion control applications, resonant transition converters that limit resonance to a part of the switching period have been developed. The class of resonant transition converters that limit stresses on the semiconductors are suitable for high-power applications and examples are the actively clamped resonant DC link converter [73, 123] and the family of resonant pole inverters [124].

During the past decade much effort has been expended investigating and inventing a large number of topologies. Industrial acceptance of resonant technology has been slow because sufficient improvement in cost and performance lacked in the case of many topologies. Now that most resonant principles and topologies have been established, the challenge for the next decade will be to reduce the number of components and simplify the construction of resonant converters.

Operation of the DC resonant link (RL) and resonant pole (RP) inverters are complementary in many respects, and both introduce more than double stresses on the semiconductor switches. In case of the RL circuit the voltage stresses are in the order of 2–2.5 per unit (p.u.). Under steady-state conditions the average voltage across the inductor needs to be zero and the link voltage (V_{link}) has to reach zero to ensure soft switching. As indicated in Figure 3-28, the steady-state condition requires area 1 = area 2 which implies a voltage stress larger than 2 p.u. Similarly, in the case of the RP inverter, steady state requires that the average of voltage at the output should not change and charge balance can only be achieved if area 1 = area 2 for the inductor current (I_L), Figure 3-29, which implies larger than 2 p.u. current stresses on the switches. The 2–2.5 p.u. stresses on the switches in the RL and RP inverters demand that the installed rating of semiconductors needs to be doubled. The stresses can be reduced by modifying the resonant circuit. Two possibilities exist: the one involves the use of active switches to alter the properties of the resonant tank, and the other method makes use of nonlinear passive components.

The best known nonlinear resonant inverter is the actively clamped resonant link (ACRL) inverter shown in Figure 3-30a. The switch S_a is controlled at a certain link voltage, connecting capacitor C_2 in parallel to C_1, which represents a step change in the capacitance of the resonant link as indicated in Figure 3-30. By introducing a two-step capacitance, it is possible to clamp the voltage V_{link} to a value that is typically in the range 1.4–1.8 p.u.

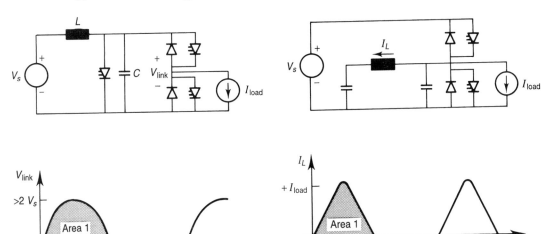

Figure 3-28. Resonant DC link inverter.

Figure 3-29. Resonant pole inverter.

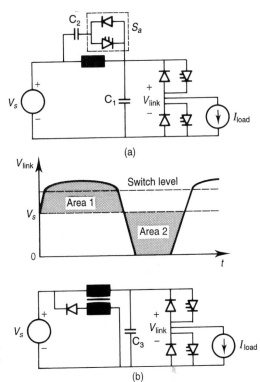

Figure 3-30. (a) Actively clamped resonant DC link inverter; (b) passively clamped resonant DC link inverter.

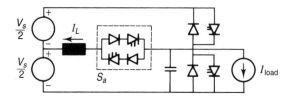

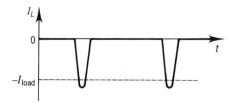

Figure 3-31. Auxiliary commutated resonant pole inverter.

The resonant pole inverter family achieves zero voltage turn-on by commutating the conducting diode in the phase arm with a current from the resonant circuit. This implies that the resonant circuit needs only be active at the end of the half cycle when the bottom diode conducts, which can be achieved by placing switch S_a in series with the inductor and connecting the resonant circuit to the phase arm only at the end of the diode conducting cycle: the so-called auxiliary commutated resonant pole (ACRP) inverter [125], as indicated in Figure 3-31. Switch S_a is turned on before the end of the diode cycle, and the diode is turned off by permitting the inductor current I_L to become larger than the load current.

Similar benefits can be obtained by using nonlinear inductors or capacitors instead of switched reactive components [126]. Such an approach has the advantage of decreasing the complexity of the circuit by eliminating switches. On the other hand, it brings another aspect into the design process, namely, achieving certain characteristics in nonlinear saturating reactive components as well as making the system essentially self-oscillating—with all the attendant advantages and disadvantages.

3.6.3. Converter Protection and Cooling

Although the fields of protection and cooling are vast, going far beyond the traditional boundaries of power electronics, they are essential technologies to review briefly in the context of the requirements for variable frequency drive converters. While fuse technology is an accepted part of low-voltage distribution, the availability of small, high-power density semiconductor devices such as thyristors after 1960 posed a very special problem, due to the obviously low thermal capacity in relation to their power control rating. The development of ranges of special semiconductor fuses [127–130] provided some protection for especially the larger devices. However, the steadily decreasing ratio between thermal capacity and power control rating, as well as the increasing switching speed, have necessitated the use of high-speed detection schemes with turn-off device capability, such as for GTOs and IGBTs [109, 131,

132], following on with sophisticated methods to design converters with appropriate protection [133–136]. Since protection against overvoltage remains much more of a problem than overcurrent, due to the external origin and uncontrollable nature of the former [133, 137], studies have led to special devices and methods to protect against this [138–141]. Finally, with converter system architecture becoming ever more complicated, diagnostic design with intelligent processing is becoming an essential feature [142] to utilize protection properly.

In contrast to application of converters elsewhere, cooling is often a limiting technology in many drive applications, due to both the space and mobility requirements [143, 144]. The review of [143] provides a good background in this regard. In possible future electromagnetically integrated high-power converters, cooling becomes even more critical [145, 146], since the spatially concentrated power processing leads to concentrated heat generation. In conventionally constructed converters, the heat generation is spread out spatially.

3.6.4. Further Converter Applications Technology

From the possibilities discussed for practical realization of converters for variable frequency drives, it follows that there are many other aspects that also present important considerations, dictating the specific choice of converter in each case, as summarized here:

1. Efficiency of switching converters [147].
2. Reliability of switching converters [148].
3. Stability of converters [149–152].
4. Power density of converters.
5. Acoustic converter-environment interaction [153–155].
6. Converter-supply interaction [48].
7. Converter-drive interaction, that is, pulsating torques, resonances [156–159].

In any specific motion control application, the foregoing can be as important as all the considerations on topology, structure, and switching technology. However, even the literature cited is only a preliminary start on these subjects, so that space limitations preclude a discussion on these aspects in this chapter.

3.7. FUTURE CONVERTER DEVELOPMENT IN RELATION TO ELECTROMAGNETICS

In many of the previous reviews cited in the bibliography, considerations for future growth and expansion of power electronics have been formulated, such as

1. New device possibilities, extending the present range of possible frequencies and power levels.

2. New materials for existing device structures, extending ranges of operation (temperature, frequency).
3. New topologies and structures for switching converters.
4. New signal processing and control possibilities through advanced microprocessor technology and new control methods.

In the context of future development it is perhaps worthwhile looking at a new dimension that up to the present has not received this kind of concentrated attention. None of the growth stimulants (1) to (4) can realize their full potential without proper attention to what may be termed "converter electromagnetics."

3.7.1. Switching Converter Electromagnetics

In any motion control application, it is taken for granted that the electrical machine as output actuator is an electromagnetic device. The power electronic converter is traditionally seen from a circuit topological/structural point of view. However, from all the previous discussion it is clear that "parasitic" as well as designed-in electromagnetic energy storage is essential for the proper functioning of the converter. These elements are traditionally treated by lumped element approximations.

Historically the limitations of the switching devices that started the power electronics development restricted the first systems to the millisecond region. Each subsequent advance in power converter technology has been an extension of the value of the power-frequency product. This has led to the present situation where the internal power-frequency product of converters is pushing the technology into a range where the electromagnetic implications will become a limitation, unless properly approached in future.

Recently, magnetic components have received specific attention at the higher converter frequencies—see, for example, [160]. It is necessary to take into account the entire electromagnetic layout and design of the converter [160]. A specific difficulty with this type of approach at present, however, is that the conventional discrete construction methods of converters lead to extremely complicated three-dimensional electromagnetic field problems, hampering the progress in total electromagnetic design of converters [161].

3.7.2. Electromagnetics and EMI/ EMC

Pushing the power-frequency product of converters to the present value has created an awareness that electromagnetic interference/electromagnetic compatibility (EMI/EMC) problems are fast becoming a limitation in some areas of converter technology. Although the knowledge has been existing that the electromagnetic layout of converters relates in a specific way to the EMI performance [162, 163], the problem has mostly been treated by countermeasures afterwards (installation of filters and EMI traps). There is growing evidence that—especially at higher power levels—this approach is becoming self-defeating in terms of cost, bulk, and complication. The source of the EMI should be traced to the topology and layout during design and development [164, 165] and the switching transitions tailored appropriately by converter design and snubbing. For larger converters this problem also overlaps with the power quality issue, especially where these converters are on-board electric locomotives and signaling via the traction circuits are involved.

3.8. CONCLUSION

A functional overview of generic converter topologies and structures useful for application to drives has been developed, followed with fundamental considerations on controlling average energy flow by switching converters. The realization of the switching function takes into account the different possibilities of executing the conduction function: that is, unidirectional conduction function, undirectional delayable conduction function, bidirectional delayable conduction function, unidirectional on-off conduction function, and bidirectional on-off conduction function. The questions of topology and structure of power electronic converters were discussed, leading up to a systematic synthesis of single-pulse, two-pulse, three-pulse, four-pulse and six-pulse converter topologies.

Converter structures for variable frequency drives were covered on a block diagram level, including those for synchronous machines and induction machines. The discussion of converter structures covers DC-to-DC converters, AC-to-DC converters, AC-DC-AC converters (current-fed and voltage fed), supply interaction of AC-to-DC converters, more extended converter families (such as AC link converters, three-level converters, and cycloconverters), and converter topologies with a minimum number of components.

Attention was given to switch applications technology with regard to turning power switches on and off and reduction of switching losses in converters. The discussion of regenerative snubbers led in to a discussion on resonant converters and generic topologies for resonant transition converters, covering the resonant DC link, resonant pole, and auxiliary resonant pole concepts. Converter switch applications technology is concluded with a short discussion on converter protection and cooling and a summary of aspects not discussed in this chapter. This chapter concludes with a look at future converter development in relation to electromagnetics.

References

[1] van Wyk, J. D., H.-Ch. Skudelny, and A. Müller Hellman, "Power electronics, control of the electromechanical energy conversion process and some applications," *Proc. Inst. Elec. Eng.,* Vol. 133, Part B, pp. 369–399, November 1986. Also in B. K. Bose (ed.), *Modern Power Electron.,* IEEE Press, New York, NY, pp. 43–73, 1992.

[2] Bose, B. K., "Motion control technology—present and future," *IEEE Trans. Ind. Appl.,* Vol. 1A-21, No. 6, pp. 1337–1342, November 1985.

[3] Bose, B. K., "Power electronics—an emerging technology," *IEEE Trans. Ind. Electron.,* Vol. 36, No. 3, pp. 403–411, August 1989.

[4] Bose, B. K., "Variable frequency drives—technology and applications," *Proc. IEEE Int. Symp. on Industrial Electron. (ISIE),* Budapest, pp. 1–18, June 1993.

[5] van Wyk, J. D., "Power electronic converters for motion control," *Proc. IEEE,* Vol. 82, No. 8, pp. 1164–1193, August 1994.

[6] Heintze, K. and R. Wagner, "Elektronischer Gleichstromsteller zur Geschwindigkeitssteuerung von aus Fahrleitungen gespeisten Gleichstrom—Triebfahrzeugen," *Elektrotech Z-A,* Vol. 87, pp. 165–170, 1966.

[7] Ohno, E., and M. Akamatsu, "High-voltage multiple phase thyristor chopper for motor control," *IEEE Trans. Magn.,* Vol. MAG-3, pp. 232–236, 1967.

[8] Bose, B. K., and R. L. Steigerwald, "A DC motor control system for electric vehicle drive," *IEEE Trans. Ind. Appl.,* Vol. IA-14, pp. 565–572, 1978.

[9] Allan, J., B. Melitt, and W. Y. Wu, "A gate-turn-off thyristor chopper for traction drives," *Proc. Inst. Elec. Eng.,* Vol. 132, Part B, pp. 245–250, 1985.

[10] de Doncker, R. W., D. M. Divan, and M. H. Kheraluwala, "A three-phase soft switched high power density DC-to-DC converter for high-power applications," *IEEE Trans. Ind. Appl.,* Vol. 27, pp. 63–73, 1991.

[11] Schwarz, F. C., and J. B. Klaassens, "A controllable 45kW current source for DC machines," *IEEE Trans. Ind. Appl.,* Vol. 24, pp. 437–444, July–August 1979.

[12] Heumann, K., "Power electronics—present status and future trends," *Proc. 6th Int. Conf. Power Electron. and Motion Control (PEMC),* Budapest, Vol. 3, pp. 786–796, 1990.

[13] Lipo, T. A., "Recent progress in the development of solid state AC motor drives," *IEEE Trans. Power Electron.,* Vol. 3, pp. 105–117, April 1988.

[14] Kehrmann, H., W. Lienau, and R. Nill, "Vierquadrantensteller—eine netzfreundliche Einspeisung für Triebfahrzeuge mit Drehstromantrieb," *Elektr. Bahnen,* Vol. 45, pp. 135–142, 1974.

[15] Appun, P., and W. Lienau, "Der Vierquadrantensteller bei induktivem und kapazitivem Betrieb," *ETZ Arch.,* Vol. 6, pp. 3–8, 1984.

[16] Phillips, K. P., "Current source converter for AC motor drives," *IEEE Trans. Ind. and Gen. Appl.,* Vol. 8, pp. 679–683, 1972.

[17] Hombu, M., S. Ueda, A. Ueda, and Y. Matsuda, "A new current source GTO—inverter with sinusoidal voltage and current," *IEEE Trans. Ind. Appl.,* Vol. IA-21, pp. 1192–1198, September/October 1985.

[18] van Wyk, J. D., and D. B. Snyman, "High frequency link systems for specialized power control applications," *Conf. Rec., IEEE-IAS Annual Meeting*, pp. 793–801, 1982.

[19] Stielau, O. H., and J. D. van Wyk, "A high frequency link system with self-oscillating inverter and direct converter using ZTO's," *Conf. Rec., IEEE-IAS Annual Meeting*, pp. 735–741, 1988.

[20] Erlicki, M. S., J. Ben-Uri, and Y. Wallach, "Switching drive of induction motors," *Proc. Inst. Elec. Eng.*, Vol. 110, pp. 1441–1450, 1963.

[21] Lipo, T. A., "The analysis of induction motor voltage control by symmetrically triggered thyristors," *IEEE Trans. Power App. Syst.*, Vol. PAS-90, pp. 515–525, 1971.

[22] Rowan, T. R., and T. A. Lipo, "A quantitative analysis of induction motor performance improvement by SCR-voltage-control," *IEEE Trans. Ind. Appl.*, Vol. IA-19, pp. 545–553, 1983.

[23] Al-Kababjie, M. F., and W. Shepherd, "A speed and power factor controller for small three phase induction motors," *IEEE Trans. Ind. Appl.*, Vol. IA-20, pp. 1260–1266, 1984.

[24] Sobczyk, T. J., and B. L. Sapinski, "Analysis of phase-controlled converters for induction motors," *IEEE Trans. Power Electron.*, Vol. 5, pp. 172–181, 1990.

[25] Emanuel-Eigeles, A., and J. Appelbaum, "AC-voltage regulator by means of chopper circuits," *Electron. Letters*, Vol. 4, pp. 26–28, 1968.

[26] Emanuel-Eigeles, A., and J. Appelbaum, "Analyse einer Regelungsschaltung für Wechselstromspannungen mittels Steller," *Arch. Elektrotech.*, Vol. 53, pp. 326–336, 1970.

[27] Addoweesh, K. E., and A. L. Mohamadein, "Microprocessor based harmonic elimination in chopper type AC voltage regulators," *IEEE Trans. Power Electron.*, Vol. 5, pp. 191–200, 1990.

[28] Hamed, S. A., "Steady-state modelling, analysis and performance of transistor controlled AC power conditioning systems," *IEEE Trans. Power Electron.*, Vol. 5, pp. 305–313, 1990.

[29] Gyugi, L., and B. R. Pelly, *Static Power Frequency Changers*, Wiley Interscience, New York, 1976.

[30] Venturini, M., "A new sinewave in sinewave out conversion technique eliminates reactive components," *Proc. Power Con.*, Vol. 7, pp. E3-1–E3-15, 1980.

[31] Venturini, M., and A. Alesina, "The generalized transformer: A new bidirectional sinusoidal waveform frequency converter with continuously adjustable input power factor," *Conf. Rec., IEEE PESC*, pp. 242–252, 1980.

[32] Neft, C. L. and C. D. Schauder, "Theory and design of 30-hp matrix converter," *IEEE Trans. Ind. Appl.*, Vol. 28, pp. 546–551, 1992.

[33] Emanuel, A. E., and R. A. Dziura, "A simplified dual cycloconverter," *IEEE Trans. Ind. Electron. Contr. Instrum.*, Vol. IECI-28, pp. 249–252, 1981.

[34] Leonhard, W., *Control of Electrical Drives*, 346 pp., Springer, Berlin, 1985.

[35] Bose, B. K., "Power electronics—a technology review," *Proc. IEEE*, Vol. 80, pp. 1303–1334, August 1992.

[36] Bose, B. K., "Recent advances in power electronics," *IEEE Trans. Power Electron.*, Vol. 7, pp. 2–16, January 1992.

[37] Rajashekara, K. S., V. Rajagopalan, A. Sevigny, and J. Vithayathil, "DC link filter design considerations in three phase voltage source inverter fed induction motor drive system," *IEEE Trans. Ind. Appl.*, Vol. IA-23, pp. 673–680, 1987.

[38] Evans, P. D., and R. J. Hill-Cottingham, "DC-link current in PWM inverters," *Proc. Inst. Elec. Eng.*, Part B, Vol. 133, pp. 217–234, 1986.

[39] Milnes, A. G., *Transductors and Magnetic Amplifiers*, 286 pp., London, Macmillan, 1957.

[40] Baliga, B. J., "Power semiconductor devices for variable frequency drives," *Proc. IEEE*, Vol. 2, No. 8, pp. 1112–1122, August 1994.

[41] Murphy, J. M. D., and M. G. Egan, "A comparison of PWM-strategies for inverter-fed induction motors," *IEEE Trans. Ind. Appl.*, Vol. IA-19, No. 3, pp. 363–369, 1983.

[42] Bowes, S. R., "Advanced regular sampled PWM-control techniques for drives and static power converters," *Proc. IEEE, Int. Conf. on Industrial Electronics, Control and Instrumentation (IECON)*, Maui, Hawaii, pp. 662–669, November 15–19, 1993.

[43] Brod, D. M., and D. W. Novotny, "Current control of VSI-PWM inverters," *IEEE Trans. Ind. Appl.*, Vol. IA-21, pp. 562–570, May/June 1985.

[44] Gaio, E., R. Piovan, and L. Malesani, "Evaluation of current control methods for voltage source inverters," *Proc. Int. Conference on Electrical Machines (ICEM)*, Pisa, pp. 345–350, 1988.

[45] McMurray, W., "Modulation of the chopping frequency in DC choppers and PWM inverters having current-hysteresis controllers," *IEEE Trans. Ind. Appl.*, Vol. IA-20, pp. 763–768, July/August 1984.

[46] Müller-Hellmann, A., and H.-Ch Skudelny, "Beitrag zur Systematik der Einphasen Brückenschaltungen," *Elektrotech, Z. ETZ-A*, Vol. 98, No. 12, pp. 803–807, 1977.

[47] Lienau, W., and A. Müller-Hellman, "Möglichkeiten zum Betrieb von Stromeinprägenden Wechselrichtern ohne niederfrequente Oberschwingungen," *Elektrotech, Z. ETZ-A*, Vol. 97, pp. 97–105, 1976.

[48] van Wyk, J. D., "Power quality, power electronics and control," *Proc. 5th European Conf. on Power Electron. and Applications (EPE)*, Vol. 1, pp. 17–32, 1993.

[49] Satpati, H., G. K. Dubey, and L. P. Singh, "A comparative study of some chopper commutation circuits," *Int. J. Electron*, Vol. 53, No. 1, pp. 47–56, 1982.

[50] Mapham, N. W., "The classification of SCR-inverter circuits," *IEEE Int. Conf. Rec.*, Part 4, pp. 99–105, 1964.

[51] Abraham, L., and F. Koppelmann, "Die Zwangskommutierung, ein neuer Zweig der Stromrichtertechnik," *Elektrotech, Z. ETZ-A*, Vol. 87, No. 18, pp. 649–658, 1966.

[52] Heumann, K., "Elektrotechnische Grundlagen der Zwangskommutierung—neue Möglichkeiten der Stromrichtertechnik," *Elektrotech. Maschinenbau*, Vol. 84, No. 3, pp. 99–112, 1967.

[53] Bose, B. K., "Introduction to power electronics," *Modern Power Electron.*, B. K. Bose (ed.), pp. 3–40, IEEE Press, New York, 1992.

[54] Lee, F. C., "High frequency quasi-resonant converter topologies," *Proc. IEEE*, Vol. 76, pp. 377–390, 1988.

[55] Bhat, A. K. S., "A unified approach to characterization of PWM and quasi-PWM switching converters: topological constraints, classification and sythesis," *IEEE Trans. Power Electron.*, Vol. 6, pp. 719–725, 1991.

[56] Mohan, N., T. M. Undeland, and W. P. Robbins, *Power Electronics*, 2nd Ed., 802 pp., Wiley, New York, 1995.

[57] Wasserrab, Th., *Schaltungslehre der Stromrichtertechnik*, 466 pp., Springer, Berlin, 1962.

[58] Dewan, S. B., and A. Straughen, *Power Semiconductor Circuits*, 526 pp., Wiley, New York, 1975.

[59] Morimoto, M., K. Oshitani, K. Sumito, S. Sato, M. Ishida, and S. Okuma, "New single-phase unity power factor PWM converter—inverter system," *Conf. Rec., IEEE Power Electron. Specialists Conf. PESC*, pp. 585–589, 1989.

[60] Prasad, A. R., P. D. Ziogas, and S. Manias, "An active power factor correction technique for three phase diode rectifiers," *Conf. Rec., IEEE PESC*, pp. 58–66, 1989.

[61] Ward, E. E., "Inverter suitable for operation over a range of frequency," *Proc. Inst. Elec. Eng.*, Vol. 111, pp. 1423–1434, August 1964.

[62] Kazumo, H., "Commutation of a three phase thyristor bridge with commutation capacitors and series diodes," *Electrical Engineering in Japan*, Vol. 90, pp. 91–100, 1970.

[63] Hombu, M., S. Ueda, and A. Ueda, "A current source GTO inverter with sinusoidal input and output," *IEEE Trans. Ind. Appl.*, Vol. IA-23, pp. 247–255, March/April 1987.

[64] Kohlmeier, H., and D. Schröder, "Control of a double voltage fed inverter system coupling a three phase mains with an AC drive," *Conf. Rec. IEEE-IAS Annual Meeting*, pp. 593–599, 1987.

[65] Arillaga, J., D. A. Bradley, and P. S. Bodger, *Power System Harmonics*, 336 pp., Wiley, New York, 1985.

[66] Heydt, C. T., *Electric Power Quality*, 541 pp., Stars in a Circle Publications, Indianapolis, 1991.

[67] Rissik, H., "The influence of mercury-arc rectifiers upon the power factor of the supply system," *Inst. Elec. Eng. J.*, Vol. 35, No. 5, pp. 435–455, 1935.

[68] Groetzbach, M., W. Dirnberger, and R. Redmann, "Simplified predetermination of line current harmonics," *IEEE Ind. Appl. Mag.*, Vol. 1, No. 2, pp. 17–27, March/April 1995.

[69] Duffey, C. K., and R. P. Stratford, "Update of harmonic standard IEEE-519: IEEE recommended practice and requirements for harmonic control in electric power systems," *IEEE Trans. Ind. Appl*, Vol. 25, No. 6, pp. 1025, 1989.

[70] Stratford, R. P., "Harmonic pollution on power systems—a change in philosophy," *IEEE Trans. Ind. Appl.*, Vol. 16, No. 5, pp. 617–623, 1980.

[71] Redl, R., "Power factor correction in single phase switching mode power supplies—an overview," *Int. J. Electron.*, Vol. 77, No. 5, pp. 555–582, 1995.

[72] Junge, G., J. Nestler, and H. Wrede, "Loeschung von Thyristoren mittels Leistungstransistoren in selbstgefuehrten Stromrichtern," *Elektrotech, Z. ETZ-A*, Vol. 99, pp. 678–681, 1987.

[73] Divan, D. M., "Resonant DC link inverter—a new concept in static power conversion," *IEEE Trans. Ind. Appl.*, Vol. 25, pp. 317–325, 1989.

[74] Sul, S. K., and T. A. Lipo, "Design and performance of a high-frequency link induction motor drive operating at unity power factor," *IEEE Trans. Ind. Appl.*, Vol. 26, pp. 434–440, May/June 1990.

[75] Nabae, A, I. Takahashi, and H. Akagi, "A new neutral-point-clamped PWM inverter," *IEEE Trans. Ind. Appl.*, Vol. IA-17, No. 5, pp. 518–523, September/October 1981.

[76] Stemmler, H., "High-power industrial drives," *Proc. IEEE*, Vol. 82, No. 8, pp. 1266–1286, August 1994.

[77] Stemmler, H., and P. Guggenbach, "Configurations of high-power voltage source inverter drives," *Proc. 5th European Conf. Power Electron. and Appl. (EPE)*, Brighton, UK, Vol. 5 pp. 7–14, September 1993.

[78] Rufer, A. C., "An aid in the teaching of multilevel inverters for high power applications," *Conf. Rec., IEEE, PESC*, Atlanta, GA, pp. 347–352, June 1995.

[79] Enjeti, P. N., and A. Rahman, "A new single-phase to three-phase converter with active input current shaping for low cost AC motor drives," *IEEE Trans. Ind. Appl.*, Vol. 29, No. 4, pp. 806–813, July/August 1993.

[80] Da Silva, E. R. C., S. B. De Souza Filho, and F. A. Coelho, "A single to three phase soft-switched converter, isolated and with input current shaping," *Conference Rec., IEEE Power Electron. Specialists Conference, PESC*, Atlanta, GA, pp. 1252–1257, June 1995.

[81] Van der Broeck, H. W., and J. D. Van Wyk, "A comparative investigation of a three-phase induction machine drive with a component minimized voltage-fed inverter under different control options," *IEEE Trans. Ind. Appl.*, Vol. IA-20, No. 2, pp. 309–320, March/April 1984.

[82] Yair, A., and J. Ben-Uri, "New three phase inverters with three thyristors," *Proc. Inst. Elec. Eng.*, Vol. 127 Part B, pp. 333–340, 1980.

[83] Eastham, J. F., A. R. Daniels, and R. T. Lipcynski, "A novel power inverter configuration," *Conf. Rec., IEEE IAS Ind. Appl. Society Annual Meeting*, pp. 748–751, 1980.

[84] Evans, P. D., R. C. Dodson, and J. F. Eastham, "Delta inverter," *Proc. Inst. Elec. Eng.*, Vol 127, Part B, pp. 333–340, November 1980.

[85] Stielau, O. H., J. D. Van Wyk, and J. J. Schoeman, "A high density three phase high frequency link system for variable frequency output," *Conf. Rec., IEEE-IAS Ind. Appl. Society Annual Meeting*, San Diego, CA, pp. 1031–1036, October 1989.

[86] Bose, B. K., "Introduction to AC drives," *Adjustable Speed AC Drive Systems*, B. K. Bose, (ed.), IEEE Press, New York, pp. 1–21, 1981.

[87] Appun, P., and J. Koerber, "Von der Lokomotive Baureihe 120 zum ICE," *Elektr. Bahnen*, Vol. 84, pp. 257–264, 1986.

[88] Pelly, B. R., "Power semiconductor devices—a status review," *Proc. IEEE-IAS, Int. Semiconductor Power Converter Conf.*, Orlando, FL, pp. 1–19, May 1982.

[89] Adler, M. S., and S. R. Westbrook, "Power semiconductor switching devices—a comparison based on inductive switching," *IEEE Trans. Electron. Devices*, Vol. ED-29, No. 6, pp. 947–952, 1982.

[90] Bergman, G. D., "The gate triggered turn-on process in thryistors," *Solid-State Electron.*, Vol. 8, pp. 757–765, 1965.

[91] Dodson, W. H., and R. L. Longini, "Probed determination of turn-on spread in thyristors," *IEEE Trans. Electron Devices*, Vol. ED-13, pp. 478–484, 1966.

[92] Adler, M. S., "Details of the plasma spreading process in thyristors," *IEEE Trans.*, Vol. ED-27, pp. 495–502, 1980.

[93] Turnbull, F. G., "A carrier frequency gating circuit for static inverters, converters and cycloconverters," *IEEE Trans. Magn.*, Vol. MAG-2, pp. 14–17, 1966.

[94] van Wyk, J. D., "On carrier frequency gating systems for static switching circuits," *IEEE Trans. Magn.*, Vol. MAG-5, pp. 140–142, 1969.

[95] Hudgins, J. L., and W. M. Portnoy, "Gating effects on thyristor anode current di/dt," *IEEE Trans. Power Electron.*, Vol. PE-2, pp. 149–153, 1987.

[96] Hudgins, J. L., and W. M. Portnoy, "High di/dt pulse switching of thyristors," *IEEE Trans. Power Electron.*, Vol. PE-2, pp. 143–148, 1987.

[97] Wirth, W. F., "High speed snubberless operation of GTOs using a new gate drive technique," *IEEE Trans. Ind. Appl.*, Vol. 24, pp. 127–131, 1988.

[98] Swanepoel, P. H., J. D. van Wyk, and J. J. Schoeman, "Transformer coupled direct base drive technology for high voltage high power bipolar transistor PWM converters," *IEEE Trans. Ind. Appl.*, Vol. 25, pp. 1158–1166, 1989.

[99] Antic, D., and J. Holtz, "High efficiency dual transistor base drive circuit based on the Cuk converter topology," *IEEE Trans. Ind. Electron.*, Vol. 38, pp. 161–165, 1991.

[100] Biswas, S. K., and B. Basak, "Some aspects in the design of discrete MOS-bipolar Darlington power switches," *IEEE Trans. Ind. Appl.*, Vol. 27, pp. 340–345, 1991.

[101] Prest, R. B., and J. D. van Wyk, "Pulsed transformer base drives for high efficiency, high current low voltage switches," *IEEE Trans. Power Electron.*, Vol. 3, pp. 137–146, 1988.

[102] Johnson, C. M., and P. R. Palmer, "Influence of gate drive and anode circuit conditions on the turn-off performance of GTO-thyristors," *Proc. Inst. Elec. Eng.-B*, Vol. 139, pp. 62–70, 1992.

[103] Ho, E. Y. Y., and P. C. Sen, "Effect of gate drive circuits on GTO-thyristor characteristics," *IEEE Trans. Ind. Electron.*, Vol. IE-33, pp. 325–331, 1986.

[104] Harada, K., H. Sakamoto, and M. Shoyama, "On the effective turn-off of GTO by a small saturable core," *IEEE Trans. Power Electron.*, Vol. PE-2, pp. 20–27, 1987.

[105] Stielau, O. H., J. J. Schoeman, and J. D. van Wyk, "A high performance gate/base drive using a current source," *IEEE Trans. Ind. Appl.*, Vol. 29, pp. 933–939, 1993.

[106] Matsuo, H., F. Kurokawa, K. Iida, T. Koga, and T. Kishimoto, "A new gate amplifier for high power GTO thyristors," *Conf. Rec. IEEE PESC*, pp. 557–561, 1991.

[107] Hinchcliffe, S., and L. Hobson, "High frequency switching of power MOSfets," *Int. J. Electron.*, Vol. 65, pp. 127–138, 1988.

[108] Hefner, A. R., "An investigation of the drive requirements for the power insulated gate bipolar transistor (IGBT)," *IEEE Trans. Power Electron.*, Vol. 6, pp. 208–219, 1991.

[109] Ackva, A., T. Reckhorn, and M. Poluszny, "Kurzschlussfestigkeit und system-eigene Sicherheit durch optimale Ansteurerung: Eine Treiberstufe fur Bipolartransistoren und IGBT im Vergleich," *ETZ-Archiv.*, Vol. 12, pp. 291–295, 1990.

[110] Swanepoel, P. H., and J. D. van Wyk, "Analysis and optimization of regen-erative linear snubbers applied to switches with voltage and current tails," *IEEE Trans. Power Electron.*, Vol. 9, No. 4, pp. 433–442, 1994.

[111] Paice, D. A., and P. Wood, "Nonlinear reactors as protective elements for thyristor circuits," *IEEE Trans. Magn.*, Vol. MAG-3, pp. 228–232, 1967.

[112] Calkin, E. T., and B. H. Hamilton, "Circuit techniques for improving the switching loci of transistor switches in switching regulators," *IEEE Trans. Ind. Appl.*, Vol. IA-12, pp. 364–369, 1976.

[113] McMurray, W., "Selection of snubbers and clamps to optimize the design of transistor switching converters," *IEEE Trans. Ind. Appl.*, Vol. IA-16, pp. 513–523, 1980.

[114] Steyn, C. G., "Analysis and optimization of regenerative linear snubbers," *IEEE Trans. Power Electron.*, Vol. 4, pp. 362–370, 1986.

[115] McMurray, W., "Efficient snubbers for voltage-source GTO inverters," *IEEE Trans. Power Electron.*, Vol. PE-2, pp. 264–272, 1987.

[116] Bendien, J. C., H. W. van der Broeck, and G. Fregien, "Recovery circuit for snubber energy in power electronic applications with high switching frequen-cies," *IEEE Trans. Power Electron.*, Vol. PE-3, pp. 26–30, 1988.

[117] Undeland, T. M., et al., "A snubber configuration for both power transistors and GTO PWM-inverters," *Conf. Rec. IEEE PESC*, pp. 42–53, 1984.

[118] Zach, F. C., et al., "New lossless turn-on and turn-off (snubber) networks for inverters, including circuits for blocking voltage limitation," *Conf. Rec. IEEE PESC*, pp. 34–41, 1984.

[119] Holtz, J., and S. F. Salama, "High power transistor PWM inverter drive with complete switching energy recovery," *Proc. Int. Conf. Electrical Machines (ICEM)*, pp. 995–998, 1986.

[120] Boehringer, A., and H. Knöll, "Transistorschalter im Bereich hoher Leistungen and Frequenzen," *Elektrotech. Z.*, Vol. 100, pp. 664–670, 1979.

[121] Ferraro, A., "An overview of low loss snubber technology for transistor inver-ters," *Conf. Rec. IEEE PESC*, pp. 466–477, 1982.

[122] Steyn, C. G., and J. D. van Wyk, "Optimum nonlinear turn-off snubbers: design and applications," *IEEE Trans. Ind. Appl.*, Vol. 25, pp. 298–306, 1989.

[123] Divan, D. M., and G. Skibinski, "Zero switching loss inverters for high power applications," *Conf. Rec., IEEE IAS Annual Meeting*, pp. 626–639, 1987.

[124] Patterson, O. D., and D. M. Divan, "A pseudo-resonant full bridge DC-to-DC converter," *Conf. Rec. IEEE PESC*, pp. 424–430, 1987.

[125] De Doncker, R. W., and J. P. Lyons, "The auxiliary resonant commutated pole converter," *Conf. Rec., IEEE-IAS Annual Meeting*, pp. 1228–1235, 1990.

[126] Ferreira, J. A., A. van Ross, and J. D. van Wyk, "A hybrid phase arm module with nonlinear resonant tank," *Conf. Rec., IEEE-IAS Annual Meeting*, pp. 1679–1685, 1990.

[127] Gutzwiller, F. W., "The current limiting fuse as fault protection for semiconductor rectifiers," *AIEE Trans.*, Vol. 77, pp. 751–754, November 1958.

[128] Jacobs, P. C., "Fuse application to solid-state motor and power control," *IEEE Trans. Ind. Gen. Appl.*, Vol. IGA-2, pp. 281–285, 1966.

[129] Crnko, T. M., "Current limiting fuse update—a new style fuse for protection of semiconductor devices," *IEEE Trans. Ind. Appl.*, Vol. IA-15, pp. 308–312, May/June 1979.

[130] Howe, A. F., and P. G. Newbury, "Semiconductor fuses and their applications," *Proc. Inst. Elec. Eng.*, Vol. 127, Part B, pp. 155–168, May 1980.

[131] Dubhashi, A. U., H. J. Rathod, and T. K. Sundararajaran, "Power transistor protection schemes," *Int. J. Electron.*, Vol. 59, pp. 397–403, 1985.

[132] Fregien, G., "Overcurrent protection for GTO-thyristors," *Proc. European Conf. Power Electron. and Appl. (EPE)*, pp. 431–436, September 1987.

[133] Berger, T., "Schutzmassnahmen gegen Schaltueberspannungen in Thyristoranlagen," *Elektrie*, Vol. 21, pp. 207–210, June 1967.

[134] Tunia, H., and A. Wojciak, "Surge suppressor design by means of nomograms," *Proc. 2nd IFAC Symp. on Control in Power Electron. and Electrical Drives*, Duesseldorf, Germany, pp. 345–353, 1977.

[135] Barili, A., A. Brambilla, E. Dallago, and R. Romano, "Computer aided design of protection networks for power SCR choppers," *Proc. European Conf. Power Electron. (EPE)*, pp. 27–30, September 1987.

[136] Kawabata, T., T. Asaeda, M. Sigenobu, and T. Nakamura, "Protection of voltage source inverters," *Proc. Int. Power Electron. Conf. (IPEC)*, Tokyo, pp. 882–893, 1983.

[137] Skibinski, G. L., J. D. Thunes, and W. Mehlhorn, "Effective utilization of surge protection devices," *IEEE Trans. Ind. Electron*, Vol. IA-22, pp. 641–652, July/August 1986.

[138] Wetzel, P., "Ueberspannungsschutzelemente fuer die Leistungselektronik," *Elektrotech. Z. ETZ*, Vol. 100, pp. 431–433, 549–551, 1979.

[139] Wetzel, P., "Metalloxid-Varistoren schuetzen Leistungshalbleiterbauelemente," *BBC Nachrichten*, pp. 97–105, 1981.

[140] Jinzenji, T., T. Kudor, H. Nishikawa, and T. Sakuma, "Application of zinc oxide varistors to DC low-voltage thyristor circuit breakers," *Proc. Int. Power Electron. Conf. (IPEC)*, Tokyo, pp. 904–915, 1983.

[141] Lawatsch, H. M., and J. Vitins, "Protection of thyristors against overvoltage with breakover diodes," *IEEE Trans Ind. Appl.*, Vol. 24, pp. 444–448, May/June 1988.

[142] Berendsen, C. S., G. Champanois, J. Davoine, and G. Rostang, "How to detect and to localize a fault in a DC-to-DC converter," *Proc. IEEE Int.*

Conf. Ind. Electron., Control and Instrumentation (IECON), San Diego, pp. 536–541, 1992.

[143] Crees, D. E., G. Humpston, D. M. Jacobson, and D. Newcombe, "Silicon/ heatsink assemblies for high power device applications: present technology and its applications," *GEC J. of Res.*, Vol. 6, pp. 71–79, 1988.

[144] Noren, D., "Heatpipe/sink combination cools sealed power modules," *Power Conversion and Intelligent Motion*, pp. 15–18, June 1988.

[145] Smit, M. C., J. A. Ferreira, J. D. van Wyk, and M. Ehsani, "Technology for manufacture of integrated planar LC structures for power electronic applications," *Proc. European Conf. Power Electron. and Appl. (EPE)*, Vol. 2, pp. 173–178, 1993.

[146] Hofsajer, I. W., J. D. van Wyk, and J. A. Ferreira, "Functional component integration in a multikilowatt resonant converter," *Proc. European Conf. Power Electron. and Appl. (EPE)*, Vol. 2, pp. 125–130, 1993.

[147] Reimers, E., "Power transfer in DC chopper motor drive," *IEEE Trans. Ind. Appl.*, Vol. IA-17, pp. 302–314, May/June 1981.

[148] Ludbrook, A., and M. Ehsani, "Burndown prevention in static power converter equipment," *Conf. Rec., IEEE-IAS*, Toronto, pp. 899–904, 1985.

[149] Ishida, M., and M. Ueda, "Stability of current-source induction motor drive system," *Elec. Eng. in Japan,* Vol. 98, No. 6, pp. 54–64, 1978.

[150] Ahmed, M. M., J. A. Taufiq, C. J, Goodman, and M. Lockwood, "Electrical instability in a voltage source inverter-fed induction motor drive at constant speed," *Proc. Inst. Elec. Eng.*, Vol. 133, Part B, pp. 299–307, 1986.

[151] Itoh, R., "Stability of a PWM current source rectifier/inverter fed induction motor drive," *Proc. Inst. Elec. Eng.*, Vol. 137, Part B, pp. 348–354, 1990.

[152] Mutoh, N., A. Ueda, K. Sakai, M. Hattori, and K. Nandoh, "Stabilizing control method for suppressing oscillations of induction motors by PWM inverters," *IEEE Trans. Ind. Electron.*, Vol. 37, pp. 48–56, February 1990.

[153] Takahashi, I., and H. Mochikawa, "Optimum PWM waveforms of an inverter for decreasing acoustic noise of an induction motor," *IEEE Trans. Ind. Electron*, Vol. IA-22, pp. 828–834, September/October 1986.

[154] Belmans, R. J. M., L. d'Hondt, A. J. Vandenput, and W. Geysen, "Analysis of the audible noise of three-phase squirrel-cage induction motors supplied by inverters," *IEEE Trans. Ind. Electron.*, Vol. IA-23, pp. 842–847, September/ October 1987.

[155] Habetler, T. G., and D. M. Divan, "Acoustic noise reduction in sinusoidal PWM drives using a randomly modulated carrier," *Conf. Rec. IEEE PESC*, pp. 665–671, 1989.

[156] Sattler, P. K., "Parasitaere Drehmomente von Stromrichtermotoren," *Elektrotech, Z. ETZ-A*, Vol. 88, pp. 89–93, February 1967.

[157] Harders, H., and B. Weidemann, "Pendelmomententwicklung bei der stromrichtergespeisten Asynchronmschine mit Beruecksichtigung des welligen Zwischenkreisstromes," *Arch. f. Elektrotech.*, Vol. 64, pp. 297–305, 1982.

[158] Andresen, E., and K. Bieniek, "On the torques and losses of voltage- and current-source inverter drives," *IEEE Trans. Ind. Appl.*, Vol. IA-20, pp. 321–327, March/April 1984.

[159] Wahsh, S., "Torque pulsation harmonics in PWM inverter induction motor drives," *ETZ Arch.*, Vol. 11, pp. 267–271, 1989.

[160] Ferreira, J. A., *Electromagnetic Modelling of Power Electronic Converters*, 192 pp., Kluwer, Boston, 1989.

[161] Van Wyk, J. D., and J. A. Ferreira, "Some present and future trends in power electronic converters," *Proc. IEEE Int. Conf. on Industrial Electronics, Control and Instrumentation (IECON)*, San Diego, Vol. 1, pp. 9–18, 1992.

[162] Ninomiya, T., and K. Harada, "Common mode noise generation in a DC-to-DC converter," *IEEE Trans. Aerosp. Electron Syst.*, Vol. AES-16, pp. 130–137, 1980.

[163] Nakahara, M., T. Ninomiya, and K. Harada, "Surge and noise generation in a forward DC-to-DC converter," *IEEE Trans.*, Vol. AES-21, pp. 619–630, 1985.

[164] Sinclair, A. J., J. A. Ferreira, and J. D. van Wyk, "A systematic study of EMI-reduction by physical converter layout and suppressive circuits," *Proc. IEEE Industrial Electron. Control and Instrumentation (IECON)*, Maui, Hawaii, Vol. 2, pp. 1059–1065, 1993.

[165] Dos Reis, F. S., J. Sebastian, and J. Uceda, "Characterization of conducted noise generation for Sepic, Cuk, and Boost converters working as power factor preregulators," *Proc. IEEE Industrial Electron. Control and Instrumentation (IECON)*, Maui, Hawaii, Vol. 2, pp. 965–970, 1993.

[166] Chew, W. M., P. D. Evans, and W. J. Heffernan, "High frequency inductor design concepts," *Conf. Rec., IEEE PESC*, pp. 673–678, 1991.

J. Holtz

Chapter 4

Pulse Width Modulation for Electronic Power Conversion

4.1. INTRODUCTION

The efficient and fast control of electric power constitutes a key technology of modern automated production. It is performed using electronic power converters. The converters transfer energy from a source to a controlled process in a quantized fashion, using semiconductor switches that are turned on and off at fast repetition rates. The algorithms that generate the switching functions—pulse width modulation techniques—are manifold. They range from simple averaging schemes to involved methods of real-time optimization. This chapter gives an overview.

Predominant applications are in variable speed AC drives. The machine loads require a three-phase supply of variable voltage and variable frequency. While the rotor speed is controlled through the supply frequency, the machine flux is determined by the supply voltage. The power requirements range from a few kilowatts to several megawatts. It is generally preferred to take the power from a DC source and convert it to three-phase AC using electronic DC-to-AC converters. The input DC voltage, mostly of constant magnitude, is obtained from a public utility through rectification. Alternatively, the power is taken from a storage battery. This applies for uninterruptable power supplies and electric vehicle drives.

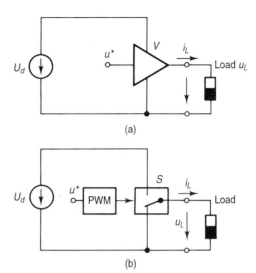

Figure 4-1. DC-to-AC power conversion: (a) linear amplifier; (b) switched mode amplifier.

4.2. DC-TO-AC POWER CONVERSION

4.2.1. Principles of Power Amplification

Two basic principles can be employed for controlled DC-to-AC power conversion. Figure 4-1a shows a linear power amplifier V, the output voltage of which is proportional to the controlling input signal $u^*(t)$, provided the maximum DC source voltage U_d is not exceeded. Since the load current i_L flows from the source through the amplifier, the idealized amplifier losses

$$P_L = (U_d - u_L) \cdot i_L > 0 \tag{4.1}$$

The amplifier losses can reach the same magnitude as the power consumption of the load, which makes linear amplifiers uneconomic at power levels exceeding a few 100 W.

The switched mode amplifier in Figure 4-1b is an alternative solution. There is a pulse width modulator inserted between the input signal and the electronic power switch S. The modulator converts the continuous input signal $u^*(t)$ into a sequence of switching instants t_i. The modulation process is illustrated in Figure 4-2a. The switching instants are determined as the intersections between the reference signal $u^*(t)$ and a triangular carrier signal $u_{cr}(t)$ having the constant frequency $f_s = 1/T_s$. The linear slopes of $u_{cr}(t)$ ensure that the duty cycle

$$d = \frac{T_{on}}{T_s} \tag{4.2}$$

of the switched output voltage u_L in Figure 4-2b varies in proportion to the reference signal $u^*(t)$. This is always true as long as f_s is sufficiently high such that the value of $u^*(t)$ can be considered constant during a time interval T_s.

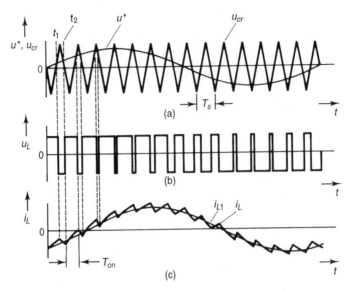

Figure 4-2. Waveforms at pulse width modulation: (a) reference signal and carrier signal; (b) load voltage; (c) load current. The waveforms refer to the topology in Figure 4-5.

In most cases the load circuit comprises an inductive component, which provides low-pass characteristics between the load voltage and the load current. If the time constant $T_L = R_L/L_L$ of the load circuit is large against the modulator switching cycle, $T_L \gg T_s$, the switching harmonics contained in the load voltage appear largely suppressed in the load current. The fundamental current becomes then predominant as compared to the harmonic components. The fundamental load voltage is proportional to the reference signal, $u_{L1}(t) \propto u^*(t)$, which significates linear amplification of the fundamental signal.

The losses of the idealized switched mode amplifier are always zero,

$$P_L = (U_d = u_L) \cdot i_L \equiv 0 \qquad (4.3)$$

since the voltage $U_d - u_L$ across the switch is zero when the switch is closed, while the current i_L through the switch is zero when the switch is open. Hence the losses do not constitute a basic limitation for the amount of power that can be handled by a switched mode amplifier. In fact, switched three-phase converters are in practical use up to power levels of several tens of megawatts.

4.2.2. Semiconductor Switches

Modern semiconductor technology has provided a wide variety of electronic switches that can be used for switched power conversion. Figure 4-3 shows the most important semiconductor devices in symbolic form and gives their typical ratings.

Power MOSFET devices permit the highest switching frequency, although their power-handling capability is limited. Their on-state resistance increases approximately with the power of 2.6 as the blocking voltage increases, rendering the efficiency of high-voltage MOSFET devices low.

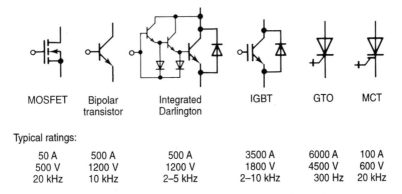

MOSFET	Bipolar transistor	Integrated Darlington	IGBT	GTO	MCT

Typical ratings:

50 A	500 A	500 A	3500 A	6000 A	100 A
500 V	1200 V	1200 V	1800 V	4500 V	600 V
20 kHz	10 kHz	2–5 kHz	2–10 kHz	300 Hz	20 kHz

Figure 4-3. Power semiconductor switches and ratings.

The insulated gate bipolar transistor (IGBT) has emerged as a new and attractive device, replacing the integrated Darlington transistor as the preferred power semiconductor switch. IGBTs combine the advantages of integrated Darlington transistors with those of MOSFETs. They have the merit of low switching losses and require very little drive power at the gate. Their fast switching process produces high dv/dt rates. This requires special care in the converter design to reduce electromagnetic radiation and conducted emissions along the cables that connect the power converter to the load. Special isolation materials that withstand high dv/dt stress must be used for the windings of the load. IGBT converters cover a power range up to about one megawatt and higher using paralleled devices. The switching frequencies range from a few kilohertz up to 10 kHz at derated power capability.

The power range beyond 2–3 mW is the domain of the gate turn-off thyristor (GTO). Owing to its internal four-layer structure, a GTO maintains a low on-state voltage as there are inherently charge carriers generated in proportion to the external load current. Having transferred to the state of conduction by a short gate current impulse, a GTO can absorb high-surge currents independently from the gate current magnitude. GTOs have large transition times at switching. They are generally sensitive against fast load current changes at turn-on and fast voltage changes at turn-off. This makes the use of protective circuits mandatory in order to limit the respective di/dt and dv/dt values. High dynamic losses are incurred during the switching transitions, limiting the switching frequency to a few 100 Hz.

The MOS controlled thyristor (MCT) is a promising high switching frequency device that has not yet reached a mature state of development to compete with the existing electronic switches at higher power levels.

The electrical properties of power semiconductor devices are characterized in Figure 4-4 in terms of the maximum on-state current I_{max}, the maximum blocking voltage U_{max}, and the maximum switching frequency $f_{s\,max}$. The general limiting quantity, as in every power system, is the maximum thermal dissipation. The semiconductor losses are dumped into heat sinks to maintain the interior device temperature at a safe level. The absolute temperature limit is around 125°C for most power semiconductors. It is anticipated that this temperature limit can be raised in future by resorting to semiconductor materials other than silicon.

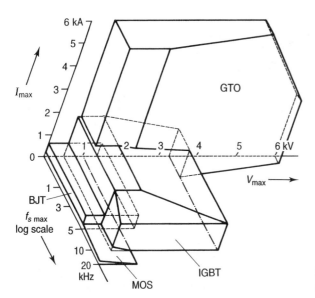

Figure 4-4. Properties of power semi-conductor switches.

The losses of power semiconductors subdivide into two major portions: the on-state losses

$$P_{\mathrm{on}} = f(u_{\mathrm{on}}, i_L) \qquad (4.4\mathrm{a})$$

and the dynamic losses

$$P_{\mathrm{dyn}} = f_s \cdot g(U_d, i_L) \qquad (4.4\mathrm{b})$$

where u_{on} is the forward voltage drop of the device.

It is apparent from equations 4.4a and 4.4b that, once the power level has been fixed by the DC supply voltage U_d and the maximum load current $i_{L\,\mathrm{max}}$, the switching frequency f_s is an important design parameter. It is discussed in Section 4.4.6 that f_s must be high enough to ensure that the switching harmonics do not impair the performance of the load, for example, an AC motor. On the other hand, the dynamic losses (4.4b) increase in proportion to the switching frequency f_s. To make things worse, semiconductor devices tend to have increased dynamic losses as their power ratings increase. This is obvious from the graph in Figure 4-4, where either the switching frequency or the product of voltage and current is high.

This conflicting situation can be alleviated by choosing sophisticated pulse width modulation methods with a view to reducing the switching harmonics and the dynamic device losses. The higher complexity and cost are offset, especially at high power ratings, by a reduction of harmonic losses and by performance improvements.

4.2.3. The Half-Bridge Topology

The existing power semiconductor switches are unipolar. This means that the quality of a fully controllable switch exists only for one direction of current flow. To

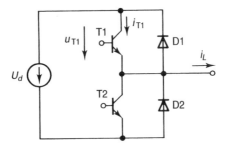

Figure 4-5. Half-bridge configuration.

control the bidirectional AC currents in a voltage source configuration requires two antiparallel semiconductor switches, one for each current direction. The half-bridge circuit Figure 4-5, a basic building block of voltage source DC-to-AC converters, satisfies this requirement. It consists of two bidirectional switches, forming the equivalent of switch S in Figure 4-1b. Each bidirectional switch is formed by a fully controllable power semiconductor arrangement, represented in Figure 4-5 by a bipolar transistor T1 (or T2) and an antiparallel diode D1 (or D2, respectively).

Although the respective diodes are not fully controllable, the half-bridge arrangement Figure 4-5 is. Assuming that the load current i_L is positive, it flows through the upper switch T1 when T1 is on or through the lower switch D2 when T1 is off. The load current commutates from T1 to D2 when T1 is turned off. During this process, a positive voltage u_{T1} builds up across T1, finally reaching a value $u_{T1} > U_d$. Consequently, the voltage $L_\sigma di_{T1}/dt$ gets negative, and i_{T1} reduces. L_σ is the commutating inductance, which is either a lumped circuit element or formed by the distributed stray inductances of the interconnecting conductors in the power circuit topology.

When T1 is turned on again, the source voltage U_d forces a circulating current i_c in the loop U_d - T1 - D2 which transfers i_L back to T1. The diode D2 assumes the blocking state when $i_c > i_L$. A similar process takes place at negative load current with T2 replacing T1 and D1 replacing D2.

4.2.4. Three-Phase Power Conversion

The generation of a single-phase AC waveform using pulse width modulated (PWM) switched mode amplifiers is a fairly straightforward task. Most single-phase loads have a moderate power consumption, which permits operating the feeding power amplifier at high switching frequency. Any of the methods discussed in Sections 4.5 and 4.6 can be easily adapted for single-phase operation.

The problem becomes more involved when pulse width modulation for a three-phase voltage system is considered. The preferred power circuit is the basic voltage source topology, consisting of three half-bridge circuits as shown in Figure 4-6. Depending on the respective state of switching, each load terminal can assume one of the two voltage potentials, $+U_d/2$ or $-U_d/2$, at a given time.

The resulting waveforms are shown in Figure 4-7a. The fundamental period gets subdivided into six equal time intervals by the switching instants of the three half-

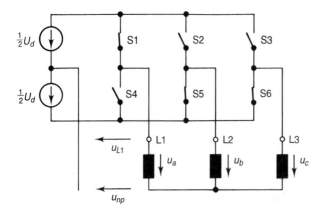

Figure 4-6. Three-phase power converter; the switch pairs S1–S4 (and S2–S5 and S3–S6) form half-bridges; one and only one switch in a half-bridge is closed at a time.

bridges. The operation is referred to as the six-step mode, in which the switching frequency of the power switches equals the fundamental frequency.

The neutral point potential u_{np} of the load is positive when more than one upper half-bridge switch is closed, Figure 4-7b; it is negative when more than one lower half-bridge switch is closed. The respective voltage levels given in Figure 4-7b hold for symmetrical load impedances.

The waveform of the phase voltage $u_a = u_{L1} - u_{np}$ is shown in the upper trace of Figure 4-7c. It forms a symmetrical, nonsinusoidal three-phase voltage system along with the other phase voltages u_b and u_c. Since the waveform u_{np} has three times the frequency of u_{Li}, $i = 1, 2, 3$, while its amplitude equals exactly one-third of the amplitudes of u_{Li}, this waveform contains exactly all triplen of the harmonic components of u_{Li}. Because of $u_a = u_{L1} - u_{np}$, there are no triplen harmonics left in the phase voltage. This is also true for the general case of balanced three-phase pulse width modulated waveforms. As all triplen harmonics form zero-sequence systems, they produce no currents in the machine windings, provided there is no electrical connection to the star point of the load, for example, u_{np} in Figure 4-6 must not be shorted.

The example Figure 4-7 demonstrates also that a change of any half-bridge potential invariably influences upon the other two-phase voltages. It is therefore expedient for the design of PWM strategies and for the analysis of PWM waveforms to consider the three-phase voltages as a whole, instead of looking at the individual phase voltages separately.

The space vector approach complies exactly with this requirement. Since the predominant application of three-phase PWM is in controlled AC drive systems, the stator winding of an AC machine will be subsequently considered as a representative AC load.

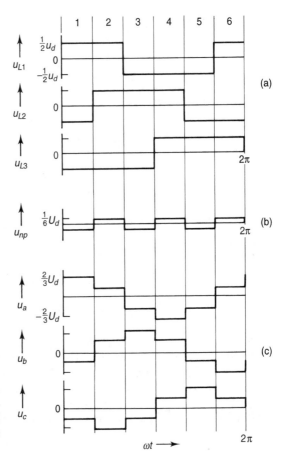

Figure 4-7. Switched three-phase waveforms: (a) voltage potentials at the inverter terminals; (b) neutral point potential at the load; (c) phase voltages of the load.

4.3. AN INTRODUCTION TO SPACE VECTORS

4.3.1. Definitions

Consider a symmetrical three-phase winding located in the stator of an electric machine, located for instance in the stator, Figure 4-8a. The three phase axes are defined by the respective unity vector $\mathbf{1}$, $\mathbf{a}$, and $\mathbf{a}^2$, where $\mathbf{a} = \exp(j^2\pi/3)$. Figure 4-8b is a symbolic representation of the winding. Neglecting space harmonics, the phase currents i_{sa}, i_{sb}, and i_{sc} generate a sinusoidal current density wave (MMF wave) around the air gap as symbolized in Figure 4-8a. The MMF wave rotates at the angular frequency of the phase currents. Like any sinusoidal distribution in time or space, it can be represented by a complex MMF phasor $\mathbf{A}_s$ as shown in Fig 8a. It is preferred, however, to describe the MMF wave by the equivalent current phasor $\mathbf{i}_s$, because this quantity is directly linked to the three stator currents i_{sa}, i_{sb}, i_{sc} that can be measured at the machine terminals:

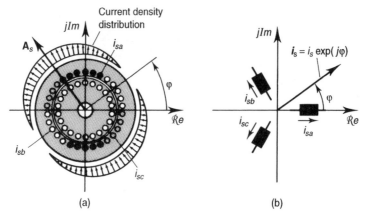

Figure 4-8. Definition of a current space vector: (a) cross section of an induction motor; (b) stator windings and stator current space vector in the complex plane.

$$i_s = \tfrac{2}{3}\left(i_{sa} + \boldsymbol{a}\,i_{sb} + \boldsymbol{a}^2 i_{sc}\right) \tag{4.5}$$

The subscript s refers to the stator of the machine.

The complex phasor in equation 4.5, more frequently referred to in the literature as a current space vector [1] has the same direction in space as the magnetic flux density wave produced by the MMF distribution A_s.

In a similar way, a sinusoidal flux density wave can be described by a space vector. It is preferred, however, to choose the corresponding distribution of the flux linkage with a particular three-phase winding as the characterizing quantity. For example, we write the flux linkage space vector of the stator winding in Figure 4-8 as

$$\boldsymbol{\Psi}_s = l_s\,\boldsymbol{i}_s \tag{4.6}$$

In the general case, when the machine develops nonzero torque, the two space vectors $\boldsymbol{i}_s$ of the stator current and $\boldsymbol{i}_r$ of the rotor current are nonzero, yielding the stator flux linkage vector as

$$\boldsymbol{\Psi}_s = l_s\,\boldsymbol{i}_s + l_h\,\boldsymbol{i}_r \tag{4.7}$$

where $l_s = l_h + l_{s\sigma}$ is the three-phase stator winding inductance, l_h is the three-phase mutual inductance between the stator and rotor windings, and $l_{s\sigma}$ is the leakage inductance of the stator. The expression three-phase inductance relates to an inductance value that results from the flux linkage generated by all three-phase currents. For example, the three-phase stator winding inductance is $l_s = \Psi_{sa}/i_{sa}$, where Ψ_{sa} and i_{sa} are the phase a components of the stator flux linkage and the stator current, respectively. However, the magnetic field that contributes to Ψ_{sa} is excited all three phase currents, i_{sa}, i_{sb}, and i_{sc}. Hence $l_s = 3/2L_{sh} + L_{s\sigma}$, where L_{sh} is the main inductance per phase of the stator winding and $L_{s\sigma}$ is the stator leakage inductance per phase. Note that $l_{s\sigma} = L_{s\sigma}$, [1].

Furthermore,

$$i_r = \tfrac{2}{3}\left(i_{ra} + \boldsymbol{a}\,i_{rb} + \boldsymbol{a}^2\,i_{rc}\right) \tag{4.8}$$

is the rotor current space vector, and i_{ra}, i_{rb} and i_{rc} are the three rotor currents. Note that flux linkage vectors like $\boldsymbol{\Psi}_s$ also represent sinusoidal distributions in space, which can be seen from an inspection of equations 4.6 or 4.7.

The rotating stator flux linkage wave $\boldsymbol{\Psi}_s$ generates induced voltages in the stator windings which are described by

$$\boldsymbol{u}_s = \frac{d\boldsymbol{\Psi}_s}{dt} \tag{4.9}$$

where

$$\boldsymbol{u}_s = \tfrac{2}{3}\left(u_{sa} + \boldsymbol{a}\,u_{sb} + \boldsymbol{a}^2\,u_{sc}\right) \tag{4.10}$$

is the stator voltage space vector defined by the three stator phase voltages u_{sa}, u_{sb}, u_{sc}.

The individual phase quantities associated to any space vector are obtained as the projections of this space vector on the respective phase axes. Given the space vector $\boldsymbol{u}_s$, for example, we obtain the phase voltages as

$$\begin{aligned}
u_{sa} &= \mathcal{R}e\{\boldsymbol{u}_s\} \\
u_{sb} &= \mathcal{R}e\{\boldsymbol{a}^2 \cdot \boldsymbol{u}_s\} \\
u_{sc} &= \mathcal{R}e\{\boldsymbol{a} \cdot \boldsymbol{u}_s\}
\end{aligned} \tag{4.11}$$

4.3.2. Normalization

Normalized quantities are used throughout this chapter. Space vectors are normalized with reference to the nominal values of the connected AC machine. The respective base quantities are

- The rated peak phase voltage, $\sqrt{2}U_{phR}$
- The rated peak phase current, $\sqrt{2}I_{phR}$ $\qquad\qquad$ (4.12)
- The rated stator frequency, ω_{sR}

Using the definition of the maximum modulation index in Section 4.4.4, the normalized DC bus voltage becomes $u_d = \pi/2$

4.3.3. Switching State Vectors

The space vector resulting from a balanced sinusoidal voltage system u_{sa}, u_{sb}, u_{sc} of frequency ω_s is

$$\boldsymbol{u}_s = u_s \cdot \exp(\mathrm{j}\omega_s t) \tag{4.13}$$

which can be shown by inserting the phase voltages into equation 4.10.

A three-phase machine being fed from a switched power converter Figure 4-6 receives the symmetrical rectangular three-phase voltages shown in Figure 4-7. The three phase potentials are constant over every sixth of the fundamental period.

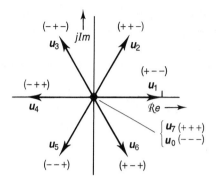

Figure 4-9. Switching state vectors in the complex plane; in brackets: switching polarities of the three half-bridges.

Inserting these phase voltages into equation 4.10 yields the typical set of six active switching-state vectors $u_1 \ldots u_6$ shown in Figure 4-9.

When operating with pulse width modulated waveforms, two zero vectors u_0 and u_7 add to the pattern in Figure 4-9. The zero vectors are associated to those inverter states with all upper half-bridge switches closed, or all lower switches closed, respectively. The three machine terminals are then short-circuited, and the voltage vector assumes zero magnitude. The existence of two zero vectors introduces an additional degree of freedom for the design of PWM strategies.

Using equation 4.11, the three-phase voltages of Figure 4-7c can be reconstructed from the switching-state pattern Figure 4-9.

4.3.4. Generalization

Considering the case of a three-phase DC-to-AC power supplies, an *LC* output filter and the connected load replace the motor at the inverter output terminals. Although not distributed in space, such load circuit behaves exactly the same way as a motor load. It is permitted and common practice therefore to extend the space vector approach to the analysis of equivalent lumped parameter circuits.

4.4. PERFORMANCE CRITERIA

Considering an AC machine drive, it is the leakage inductances of the machine and the inertia of the mechanical system which account for low-pass filtering of the harmonic components that are contained in the switched voltage waveforms. Remaining distortions of the current waveforms, harmonic losses in the power converter and the load, and oscillations of the electromagnetic machine torque are due to the operation in the switched mode. Such undesired effects can be valued by performance criteria [2–7]. These provide the means of comparing the qualities of different PWM methods and support the selection of a pulse width modulator for a particular application.

4.4.1. Current Harmonics

The harmonic currents primarily determine the copper losses of the machine, which account for a major portion of the machine losses. The rms harmonic current

$$I_{h\,\text{rms}} = \sqrt{\tfrac{1}{T}\textstyle\int_T [i(t) - i_1(t)]^2\,dt} \tag{4.14}$$

does not depend only on the performance of the pulse width modulator, but also on the internal impedance of the machine. This influence is eliminated when the distortion factor is used as a figure of merit. This quantity is derived from the normalized rms harmonic content

$$\frac{I_{h\,\text{rms}}}{I_1} = \frac{1}{I_1}\sqrt{\sum_{\nu=2}^{\infty} I_\nu^2} = \frac{\omega_1 l_\sigma}{U_1}\sqrt{\sum_{\nu=2}^{\infty}\left(\frac{U_\nu}{\nu\omega_1 l_\sigma}\right)^2} \tag{4.15}$$

of a periodic current waveform, where I_1 is the rms fundamental current and I_ν are the rms Fourier current components. The right-hand side of equation 4.15 is obtained with reference to the simplified equivalent circuit of an AC load in Figure 4-10. Assuming that the load is supplied by an inverter operated in the six-step mode, the Fourier voltage components U_ν can be easily determined from the rectangular voltage waveform, and we have from equation 4.15

$$\frac{I_{h\,\text{rms six-step}}}{I_1} = 0.0464 \tag{4.16}$$

Dividing equation 4-15 by this well-defined normalized current yields the distortion factor

$$d = \frac{I_{h\,\text{rms}}}{I_{h\,\text{rms six-step}}} \tag{4.17}$$

In this definition, the distortion current $I_{h\,\text{rms}}$ in equation (4.14) of a given switching sequence is referred to the distortion current $I_{h\,\text{rms six-step}}$ of same AC load operated in the six-step mode, that is, with the unpulsed rectangular voltage waveforms Figure 4-7c. The definition equation 4.17 values the AC-side current distortion of a PWM method independently from the properties of the load. We have $d = 1$ at six-step operation by definition. Note that the distortion factor d of a pulsed waveform can be much higher than that of a rectangular wave. This is demonstrated by Figure 4-26.

The harmonic content of a current space vector trajectory is computed as

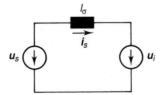

Figure 4-10. Simplified equivalent circuit of an AC load, for example, an induction motor.

$$I_{h\,\text{rms}} = \sqrt{\tfrac{1}{T}\textstyle\int_T [\boldsymbol{i}(t) - \boldsymbol{i}_1(t)] \cdot [\boldsymbol{i}(t) - \boldsymbol{i}_1(t)]^* \, dt} \tag{4.18}$$

from which d can be determined by equation 4.17. The asterisk in equation 4.18 marks the complex conjugate.

The harmonic copper losses in the load circuit are proportional to the square of the harmonic current: $P_{L\,\text{Cu}} \propto d^2$, where d^2 has the significance of a loss factor.

4.4.2. Harmonic Spectrum

The contributions of individual frequency components to a nonsinusiodal current wave are expressed in a harmonic current spectrum, which offers a more detailed description than the global distortion factor d. We obtain discrete current spectra $h_i(k \cdot f_1)$ in the case of synchronized PWM, where the switching frequency $f_s = N \cdot f_1$ is an integral multiple of the fundamental frequency f_1. N is the pulse number, or gear ratio, and k is the order of the harmonic component. Note that all harmonic spectra in this chapter are normalized as per the definition 4.17:

$$h_i(k \cdot f_1) = \frac{I_{h\,\text{rms}}(k \cdot f_1)}{I_{h\,\text{rms six-step}}} \tag{4.19}$$

They describe the properties of a pulse modulation scheme independently from the parameters of the connected load.

Nonsynchronized pulse sequences produce harmonic amplitude density spectra $h_d(f)$ of the currents, which are continuous functions of frequency. A measured spectrum generally contains periodic components $h_i(k \cdot f_1)$ as well as nonperiodic components $h_d(f)$. Since these quantities have different physical dimensions, they must be displayed with reference to two different scale factors on the ordinate axis, as for example, Figure 4-46. While the normalized discrete spectra $h_i(k \cdot f_1)$ do not have a physical dimension, the amplitude density spectra $h_d(f)$ are measured in $\text{Hz}^{-1/2}$.

The normalized harmonic current equation 4.17 is computed from the discrete spectrum equation 4.19 as

$$d = \sqrt{\sum_{k \neq 1} h_i^2(k \cdot f_1)} \tag{4.20}$$

and from the amplitude density spectrum as

$$d = \sqrt{\int\limits_{0,\, f \neq f_1}^{\infty} h_d^2(f) \, df} \tag{4.21}$$

Another figure of merit for a given PWM scheme is the product $d \cdot f_s$ of the distortion factor and the switching frequency of the inverter. This value can be used to compare different PWM schemes operated at different switching frequencies. The criterion $d \cdot f_s$ holds provided that the pulse number $N > 15$, since the relation becomes nonlinear at lower values of N.

4.4.3. Space Vector Trajectories

PWM schemes can be also evaluated by visual inspection of the trajectories of the current and the stator flux linkage vector in the complex plane. Such trajectories can be recorded in real time during the operation of the modulator and inverter. The current space vector is then computed from the measured phase currents using equation 4.5, while the flux linkage vector is derived from the measured phase voltages by subsequent integration. The flux linkage trajectory is erroneous at lower fundamental frequency since the voltage drop across the winding resistance is normally neglected.

The harmonic content of a steady-state trajectory is observed as the deviation from its fundamental component, with the fundamental trajectory describing a circle. Although deviations in amplitude are easily discernible, differences in phase angles have equal significance. The example in Figure 4-11 demonstrates that a nonoptimal modulation method, Figures 4-11a and 4-11b, produces both higher deviations in amplitude and in phase angle as compared with the respective optimized patterns shown in Figures 4-11c and 4-11d.

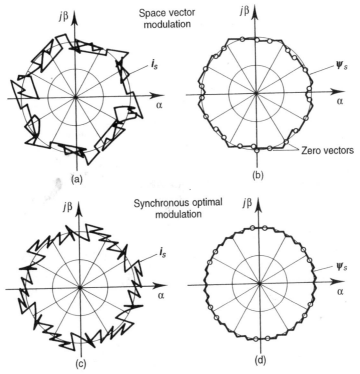

Figure 4-11. Space vector trajectories of nonoptimal (upper patterns) and optimized modulation techniques (lower patterns): (a) and (c) current trajectories; (b) and (d) stator flux trajectories.

4.4.4. Maximum Modulation Index

The modulation index is the normalized fundamental voltage, defined as

$$m = \frac{u_1}{u_{1\,\text{six-step}}} \qquad (4.22)$$

where u_1 is the fundamental voltage of the modulated switching sequence and $u_{1\,\text{six-step}} = 2u_d/\pi$ the fundamental voltage at six-step operation. We have $0 \le m \le 1$, and hence the unity modulation index, by definition, can be attained only in the six-step mode.

The maximum value m_{max} of the modulation index may differ in a range of about 25%, depending on the respective pulse width modulation method. As the maximum power of a PWM converter is proportional to the maximum voltage at the AC side, the maximum modulation index m_{max} constitutes an important utilization factor of the equipment.

4.4.5. Torque Harmonics

The torque ripple produced by a given switching sequence in a connected AC machine can be expressed as

$$\Delta T = \frac{T_{\text{max}} - T_{av}}{T_R} \qquad (4.23)$$

where $T_{\text{max}} = $ maximum air gap torque

$T_{av} = $ average air gap torque

$T_R = $ rated machine torque

Although torque harmonics are produced by the harmonic currents, there is no stringent relationship between both of them.

4.4.6. Switching Frequency and Switching Losses

The losses of power semiconductors subdivide into two major portions: the on-state losses

$$P_{\text{on}} = g_1(u_{\text{on}}, i_L) \qquad (4.24a)$$

and the dynamic losses

$$P_{\text{dyn}} = f_s \cdot g_2(U_d, i_L) \qquad (4.24b)$$

where g_1 and g_2 are characteristic functions.

The power rating of a particular application is given by the DC supply voltage U_d and the maximum load current $i_{L\,\text{max}}$. While the dynamic losses (4.24b) increase as the switching frequency increases, the harmonic distortion of the AC-side currents reduces in the same proportion. Yet the switching frequency cannot be deliberately increased for the following reasons:

- The switching losses of semiconductor devices increase proportional to the switching frequency.
- Semiconductor switches of higher power rating generally produce higher switching losses. The switching frequency must be reduced accordingly. Megawatt switched power converters using GTOs are switched at only a few 100 Hertz.
- The regulations regarding electromagnetic compatibility (EMC) are stricter for power conversion equipment operating at switching frequencies higher than 9 kHz [8].

Another important aspect related to switching frequency is the radiation of acoustic noise. The switched currents produce fast-changing electromagnetic fields that exert mechanical Lorentz forces on current-carrying conductors. They also produce magnetostrictive mechanical deformations in ferromagnetic materials. It is especially the magnetic circuits of the AC loads that are subject to mechanical excitation in the audible frequency range. Resonant amplification may take place in the active stator iron, being a hollow cylindrical elastic structure, or in the cooling fins at the outer structure of an electric machine.

The dominating frequency components of acoustic radiation are strongly related to the spectral distribution of the harmonic currents, and to the switching frequency of the feeding power converter. The sensitivity of the human ear makes switching frequencies below 500 Hz and above 10 kHz less critical, while the maximum sensitivity is around 1–2 kHz.

4.4.7. Polarity Consistency Rule

The switching of one half-bridge of a three-phase power converter influences also the two other phase voltages. The result of such interaction can be seen when looking at the individual phase voltages, or at the α, β-components of the switching-state vector. These waveforms are composed of up to five different voltage levels, while the phase potentials assume only two levels.

In a well-designed modulation scheme, the phase voltage at a given time instant should not differ much from the value of the sinusoidal reference wave. Generally speaking, the switched voltage should assume at least the same polarity as the reference voltage. The polarity consistency rule is a means to identify ill-designed PWM schemes from the waveforms they produce.

4.4.8. Dynamic Performance

Usually a current control loop is designed around a switched mode power converter, the response time of which essentially determines the dynamic performance of the overall system. The dynamics are influenced by the switching frequency and/or the PWM method used. Some schemes require feedback signals that are free from current harmonics. Filtering of feedback signals increases the response time of the loop [9].

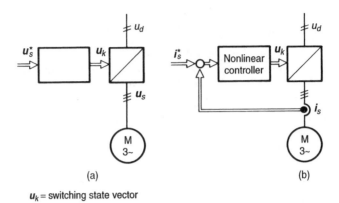

u_k = switching state vector

Figure 4-12. Basic PWM structures: (a) open loop scheme; (b) feedback scheme.

PWM methods for the most commonly used voltage source inverters impress either the voltages, or the currents into the AC load circuit. The respective approach determines the dynamic performance and, in addition, influences upon the structure of the superimposed control system: the methods of the first category operate in an open loop fashion, Figure 4-12a. Closed loop PWM schemes, in contrast, inject the currents into the load and require different structures of the control system, Figure 4-12b.

4.5. OPEN LOOP SCHEMES

Open loop schemes refer to a reference space vector $u^*(t)$ as an input signal, from which the switched three-phase voltage waveforms are generated such that the time average of the associated normalized fundamental space vector $u_{s1}(t)$ equals the time average of the reference vector. The general open loop structure is represented in Figure 4-12a.

4.5.1. Carrier-Based PWM

The most frequently used methods of pulse width modulation are carrier based. They have as a common characteristic subcycles of constant time duration, a subcycle being defined as the time duration $T_0 = 1/2 f_s$ during which any of the inverter half-bridges, as formed for instance by S1 and S2 in Figure 4-6, assumes two consecutive switching-states of opposite voltage polarity. Operation at subcycles of constant time duration is reflected in the harmonic spectrum by pairs of salient sidebands, centered around the carrier frequency f_s, and additional frequency bands located on either side around integral multiples of the carrier frequency. An example is shown in Figure 4-24.

There are various ways to implement carrier based PWM; these will be discussed next.

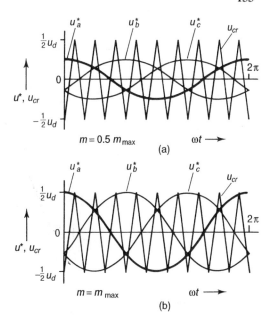

Figure 4-14. Reference signals and carrier signal: modulation index (a) $m = 0.5\,m_{max}$, (b) $m = m_{max}$.

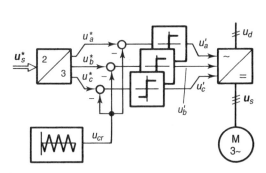

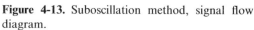

Figure 4-13. Suboscillation method, signal flow diagram.

Suboscillation Method. This method employs individual carrier modulators in each of the three phases [10]. A signal flow diagram of a suboscillation modulator is shown in Figure 4-13. The reference signals u_a^*, u_b^*, u_c^* of the phase voltages are sinusoidal in the steady state, forming a symmetrical three-phase system as in Figure 4-14. They are obtained from the reference vector u^*, which is split into its three phase components u_a^*, u_b^*, u_c^* on the basis of equation 4.11. Three comparators and a triangular carrier signal u_{cr}, which is common to all three-phase signals, generate the logic signals u_a', u_b', and u_c' that control the half-bridges of the power converter.

Figure 4-15 shows the modulation process in detail, expanded over a time interval of two subcycles. T_0 is the subcycle duration. Note that the three-phase potentials u_a', u_b', u_c' are of equal magnitude at the beginning and at the end of each subcycle. The three line-to-line voltages are then zero, and hence u_s results as the zero vector.

A closer inspection of Figure 4-14 reveals that the suboscillation method does not fully utilize the available DC bus voltage. The maximum value of the modulation index $m_{max\,1} = \pi/4 = 0.785$ is reached at a point where the amplitudes of the reference signal and the carrier become equal. This situation is shown in Figure 4-14b. Computing the maximum line-to-line voltage amplitude in this operating point yields $u_a^*(t_1) - u_b^*(t_1) = \sqrt{3} \cdot u_d/2 = 0.866\,u_d$. This is less than what is obviously possible when the two half-bridges that correspond to phases a and b are switched to $u_a = u_d/2$ and $u_b = -u_d/2$, respectively. In this case, the maximum line-to-line voltage would equal u_d.

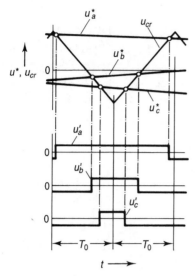

T_0 = subcycle duration

Figure 4-15. Determination of the switching instants.

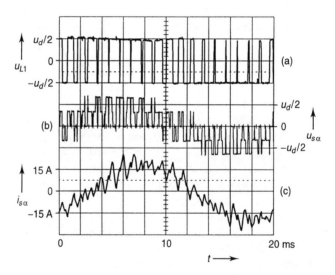

Figure 4-16. Recorded three-phase PWM waveforms (suboscillation method): (a) voltage potential u_{L1} at one inverter terminal: (b) phase voltage $u_{s\alpha}$; (c) load current $i_{s\alpha}$.

Measured waveforms obtained with the suboscillation method are displayed in Figure 4-16. This oscillogram was taken at 1 kHz switching frequency and $m \approx 0.75$.

Modified Suboscillation Method. The deficiency of a limited modulation index, inherent to the suboscillation method, is cured when distorted reference waveforms are used. Such waveforms must not contain other components than zero-

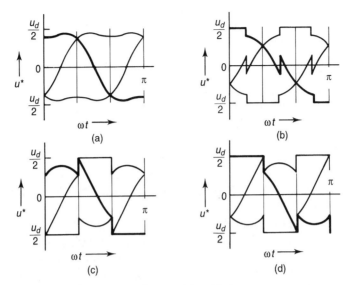

Figure 4-17. Reference waveforms with added zero-sequence systems: (a) with added third harmonic; (b), (c), (d) with added rectangular signals of triple fundamental frequency.

sequence systems in addition to the fundamental. The reference waveforms shown in Figure 4-17 exhibit this quality. They have a higher fundamental content than sinewaves of the same peak value. As explained in Section 4.2.4, such distortions are not transferred to the load currents.

There is an infinity of possible additions to the fundamental waveform that constitute zero-sequence systems. The waveform in Figure 4-17a has a third harmonic content of 25% of the fundamental; the maximum modulation index is increased here to $m_{max} = 0.882$ [11]. The addition of rectangular waveforms of triple fundamental frequency leads to reference signals as shown in Figure 4-17b through Figure 4-17d; a limit value of $m_{max2} = \sqrt{3}\pi/6 = 0.907$ is reached in these cases. This is the maximum value of modulation index that can be obtained with the technique of adding zero-sequence components to the reference signal [12, 13].

Sampling Techniques. The suboscillation method is simple to implement in hardware, using analog integrators and comparators for the generation of the triangular carrier and the switching instants. Analog electronic components are very fast, and inverter switching frequencies up to several tens of kilohertz are easily obtained.

When digital signal processing methods based on microprocessors are preferred, the integrators are replaced by digital timers, and the digitized reference signals are compared with the actual timer counts at high repetition rates to obtain the required time resolution. Figure 4-18 illustrates this process, which is referred to as natural sampling [14].

To relieve the microprocessor from the time-consuming task of comparing two time variable signals at a high repetition rate, the corresponding signal processing

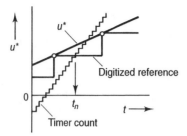

Figure 4-18. Waveforms at natural sampling.

functions are implemented in on-chip hardware. Modern microcontrollers comprise of capture/compare units or waveform generators that generate digital control signals for three-phase PWM when loaded from the CPU with the corresponding timing data [15].

If the capture/compare function is not available in hardware, other sampling PWM methods can be employed [16]. In the case of symmetrical regular sampling, Figure 4-19a, the reference waveforms are sampled at the very low repetition rate f_s which is determined by the switching frequency. The sampling interval $1/f_s = 2T_0$ extends over two subcycles. The individual sampling instants are t_{sn}. The triangular carrier shown as a dotted line in Figure 4-19a is not really existent as a signal. The

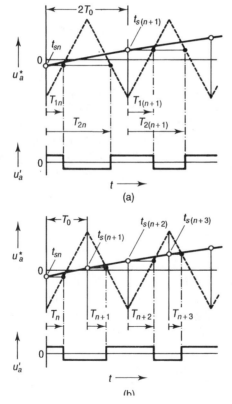

Figure 4-19. Sampling techniques: (a) symmetrical regular sampling; (b) asymmetric regular sampling.

time intervals T_1 and T_2, which define the switching instants, are simply computed in real time from the respective sampled value $u_a^*(t_s)$ using the geometrical relationships

$$T_1 = \tfrac{1}{2}T_0 \cdot [1 + u_a^*(t_s)] \tag{4.25a}$$

$$T_1 = T_0 + \tfrac{1}{2}T_0 \cdot [1 - u_a^*(t_s)] \tag{4.25b}$$

which can be established with reference to the dotted triangular line.

Another method, referred to as asymmetric regular sampling [17], operates at double sampling frequency $2f_s$. Figure 4-19b shows that samples are taken once in every subcycle. This improves the dynamic response and produces somewhat less harmonic distortion of the load currents.

Space Vector Modulation. The space vector modulation technique differs from the aforementioned methods in that there are not separate modulators used for each of the three phases. Instead, the complex reference voltage vector is processed as a whole [18, 19]. Figure 4-20a shows the principle. The reference vector u^* is sampled at the fixed clock frequency $2f_s$. The sampled value $u^*(t_s)$ is then used to solve the equations

$$2f_s \cdot (t_a u_a + t_b u_b) = u^*(t_s) \tag{4.26a}$$

$$t_0 = \frac{1}{2f_s} - t_a - t_b \tag{4.26b}$$

where u_a and u_b are the two switching-state vectors adjacent in space to the reference vector u^*, Figure 4-20b. The solutions of equation 4.26 are the respective on-durations t_a, t_b, and t_0 of the switching-state vectors u_a, u_b, u_0:

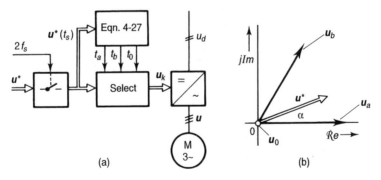

(a)   (b)

Figure 4-20. Space vector modulation: (a) signal flow diagram; (b) switching-state vectors, shown in the first 60°-sector.

$$t_a = \frac{1}{2f_s} \cdot u^*(t_s) \frac{3}{\pi} \left(\cos \alpha - \frac{1}{\sqrt{3}} \sin \alpha \right) \qquad (4.27a)$$

$$t_b = \frac{1}{2f_s} \cdot u^*(t_s) \frac{2\sqrt{3}}{\pi} \sin \alpha \qquad (4.27b)$$

$$t_0 = \frac{1}{2f_s} - t_a - t_b \qquad (4.27c)$$

The angle α in these equations is the phase angle between the reference vector and u_a.

This technique in effect averages the three switching-state vectors over a sub-cycle interval $T_0 = 1/2f_s$ to equal the reference vector $u^*(t_s)$ as sampled at the beginning of the subcycle. It is assumed in Figure 4-20b that the reference vector is located in the first 60° sector of the complex plane. The switching-state vectors adjacent to the reference vector are then $u_a = u_1$ and $u_b = u_2$, Figure 4-9. As the reference vector enters the next sector, $u_a = u_2$ and $u_b = u_3$, and so on. When programming a microprocessor, the reference vector is first rotated back by $n \cdot 60°$ until it resides in the first sector, and then equation 4.27 is evaluated. Finally, the switching states to replace the provisional vectors u_a and u_b are identified by rotating u_a and u_b forward by $n \cdot 60°$ [20].

Having computed the on-durations of the three switching-state vectors that form one subcycle, an adequate sequence in time of these vectors must be determined next. Associated to each switching-state vector in Figure 4-9 are the switching polarities of the three half-bridges, given in brackets. The zero vector is redundant. It can be either formed as u_0 $(- - -)$, or u_7 $(+ + +)$. u_0 is preferred when the previous switching-state vector is u_1, u_3, or u_5; u_7 will be chosen following u_2, u_4, or u_6. This ensures that only one half-bridge in Figure 4-6 needs to commutate at a transition between an active switching-state vector and the zero vector. Hence, the minimum number of commutations is obtained by the switching sequence

$$u_0 \langle t_0/2 \rangle \cdot \cdot u_1 \langle t_1 \rangle \cdot \cdot u_2 \langle t_2 \rangle \cdot \cdot u_7 \langle t_0/2 \rangle \qquad (4.28a)$$

in any first, or generally in all odd subcycles, and

$$u_7 \langle t_0/2 \rangle \cdot \cdot u_2 \langle t_2 \rangle \cdot \cdot u_1 \langle t_1 \rangle \cdot \cdot u_0 \langle t_0/2 \rangle \qquad (4.28b)$$

for the next, or all even subcycles. The notation in equation 4.28 associates to each switching-state vector its on-duration in brackets.

Modified Space Vector Modulation. The modified space vector modulation [21, 22, 23] uses the switching sequences

$$u_0 \langle t_0/3 \rangle \cdot \cdot u_1 \langle 2t_1/3 \rangle \cdot \cdot u_2 \langle t_2/3 \rangle \qquad (4.29a)$$

$$u_2 \langle t_2/3 \rangle \cdot \cdot u_1 \langle 2t_1/3 \rangle \cdot \cdot u_0 \langle t_0/3 \rangle \qquad (4.29b)$$

or a combination of equations 4.28 and 4.29. Note that a subcycle of the sequences equation 4-29 consists of two switching states, since the last state in equation 4.29a is the same as the first state in equation 4.29b. Similarly, a subcycle of the sequences (4.28) comprises three switching states. The on-durations of the switching-state vec-

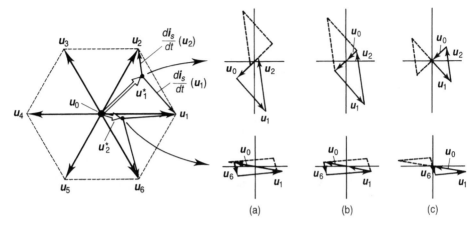

Figure 4-21. Linearized trajectories of the harmonic current for two voltage
references u_1^* and u_2^*: (a) suboscillation method; (b) space vector modulation; (c) modified space vector modulation.

tors in equation 4.29 are consequently reduced to 2/3 of those in 4.28 in order to
maintain the switching frequency f_s at a given value.

The choice between the two switching sequences (4.28) and (4.29) should depend
on the value of the reference vector. The decision is based on the analysis of the
resulting harmonic current. Considering the equivalent circuit Figure 4-10, the differential equation

$$\frac{di_s}{dt} = \frac{1}{l\sigma}(u_s - u_i) \tag{4.30}$$

can be used to compute the trajectory in space of the current space vector i_s. In
equation 4.30, u_s is the actual switching-state vector. If the trajectories $di_s(u_s)/dt$ are
approximated as linear, the closed patterns of Figure 4-21 will result. The patterns
are shown in this graph for the switching-state sequences in equations 4.28 and 4.29,
and two different magnitude values, u_1^* and u_2^*, of the reference vector are considered. The harmonic content of the trajectories is determined using 4.18. The result
can be confirmed just by a visual inspection of the patterns in Figure 4-21: the
harmonic content is lower at high-modulation index with the modified switching
sequence 4.29; it is lower at low-modulation index when the sequence 4.28 is applied.

Figure 4-23 shows the corresponding characteristics of the loss factor d^2: curve
svm corresponds to the sequence 4.28, and curve c to sequence 4.29. The maximum
modulation index extends in either case up to $m_{\max 2} = 0.907$.

Synchronized Carrier Modulation. The aforementioned methods operate at
constant carrier frequency, while the fundamental frequency is permitted to vary.
The switching sequence is then nonperiodic in principle, and the corresponding
Fourier spectra are continuous. They contain also frequencies lower than the lowest
carrier sideband, Figure 4-24. These subharmonic components are undesired as they

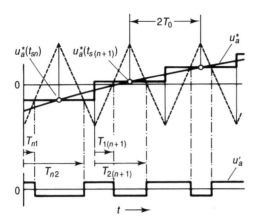

Figure 4-22. Synchronized regular sampling.

produce low-frequency torque harmonics that may stimulate resonances in the mechanical transmission train of the drive system. Resonant excitation leads to high mechanical stresses and entails fatigue problems. A synchronization between the carrier frequency and the controlling fundamental avoids these drawbacks, which are especially prominent if the frequency ratio, or pulse number

$$N = \frac{f_s}{f_1} \tag{4.31}$$

is low. In synchronized PWM, the pulse number N assumes only integral values [24].

When sampling techniques are employed for synchronized carrier modulation, an advantage can be drawn from the fact that the sampling instants $t_{sn} = n/(f_1 \cdot N)$, $n = 1 \ldots N$ in a fundamental period are a priori known. The reference signal is $u_a^*(t) = m/m_{max} \cdot \sin 2\pi f_1 t$, and the sampled values $u_a^*(t_s)$ in Figure 4-22 form a discretized sinefunction that can be stored in the processor memory. Based on these values, the switching instants are computed on-line using equation 4.25.

Performance of Carrier-Based PWM. The loss factor d^2 of suboscillation PWM depends on the zero-sequence components added to the reference signal. A comparison is made in Figure 4-23 at 2 kHz switching frequency. Letters "a" through "d" refer to the respective reference waveforms in Figure 4-17.

The space vector modulation exhibits a better loss factor characteristic at $m > 0.4$ as the suboscillation method with sinusoidal reference waveforms. The reason becomes obvious when comparing the harmonic trajectories in Figure 4-21. The zero vector appears twice during two subsequent subcycles, and there is a shorter and a subsequent larger portion of it in a complete harmonic pattern of the suboscillation method. Figure 4-15 shows how the two different on-durations of the zero vector are generated. Against that, the on-durations of two subsequent zero vectors Figure 4-21b are almost equal in the case of space vector modulation. The contours of the harmonic pattern come closer to the origin in this case, which reduces the harmonic content.

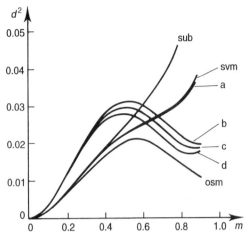

Figure 4-23. Performance of carrier modulation at $f_s = 2$ kHz; for (a) through (d), refer to Figure 4-17.

osm = optimal subcycle method
sub = suboscillation method
svm = space vector modulation

The modified space vector modulation, curve d in Figure 4-23, performs better at higher modulation index, and worse at $m < 0.62$.

A measured harmonic spectrum produced by the space vector modulation method is shown in Figure 4-24. The carrier frequency, and integral multiples thereof, determine the predominant harmonic components. The respective amplitudes vary with the modulation index. This tendency is exemplified in Figure 4-25 for the case of the suboscillation method.

The loss factor curves of synchronized carrier PWM are shown in Figure 4-26 for the suboscillation technique and the space vector modulation. The latter appears superior at low pulse numbers, the difference becoming less significant as N increases. The curves exhibit no differences at lower modulation index. Operating in this range is of little practical use for constant v/f_1 loads where higher values of N are permitted, and, above all, d^2 decreases if m is reduced (Figure 4-23).

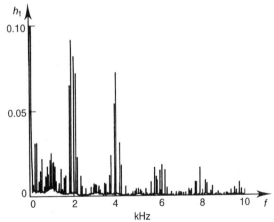

Figure 4-24. Space vector modulation, harmonic spectrum, 2 kHz switching frequency.

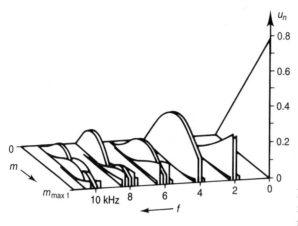

u_n = harmonic voltage component

Figure 4-25. Spectral amplitudes of carrier based pulse width modulation versus modulation index, 2 kHz switching frequency.

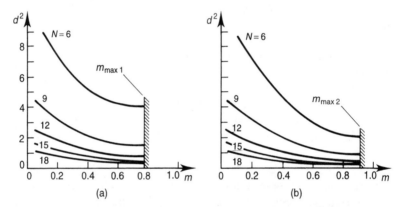

Figure 4-26. Synchronized carrier modulation, loss factor d^2 versus modulation index: (a) suboscillation method, (b) space vector modulation.

The performance of a pulse width modulator based on sampling techniques is slightly inferior to that of the suboscillation method, but only at low pulse numbers.

Because of the synchronism between f_1 and f_s, the pulse number must necessarily change as the modulation index varies over a broader range. Such changes introduce discontinuities to the modulation process. They generally originate current transients, especially when the pulse number is low [25]. This effect is discussed in Section 4.5.4.

4.5.2. Carrierless PWM

The typical harmonic spectrum of carrier-based pulse width modulation exhibits prominent harmonic amplitudes around the carrier frequency and its harmonics, Figure 4-24. Increased acoustic noise is generated by the machine at these frequen-

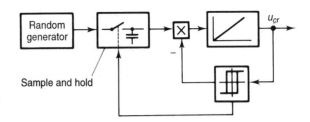

Figure 4-27. Random frequency carrier signal generator.

cies through the effects of magnetostriction. The vibrations can be amplified by mechanical resonances. To reduce the mechanical excitation at particular frequencies, it may be preferable to have the harmonic energy distributed over a larger frequency range instead of being concentrated around the carrier frequency.

Such concept is realized by varying the carrier frequency in a random manner. When applying this to the suboscillation technique, the slopes of the triangular carrier signal must be maintained linear to conserve the linear input-output relationship of the modulator. Figure 4-27 shows how a random frequency carrier signal can be generated. Whenever the carrier signal reaches one of its peak values, its slope is reversed by a hysteresis element, and a sample is taken from a random signal generator which imposes an additional small variation on the slope. This varies the durations of the subcycles in a random manner [26]. The average switching frequency is maintained constant such that the power devices are not exposed to changes in temperature.

The optimal subcycle method (Section 4.5.4) classifies also as carrierless. Another approach to carrierless PWM is explained in Figure 4-28; it is based on the space vector modulation principle. Instead of operating at constant sampling frequency $2f_s$ as in Figure 4-20a, samples of the reference vector are taken whenever the duration t_{act} of the actual switching-state vector $\boldsymbol{u}_{act}$ terminates. t_{act} is determined from the solution of

$$t_{act}\boldsymbol{u}_{act} + t_1\boldsymbol{u}_1 + \left(\frac{1}{2f_s} - t_{act} = t_1 \right)\boldsymbol{u}_2 = \frac{1}{2f_s} \cdot \boldsymbol{u}^*(t) \qquad (4.32)$$

where $\boldsymbol{u}^*(t)$ is the reference vector. This quantity is different from its time discretized value $\boldsymbol{u}^*(t_s)$ used in equation 4.26a. As $\boldsymbol{u}^*(t)$ is considered a continuously time-variable signal in equation 4.32, the on-durations t_{act}, t_1, and $(1/2f_s - t_{act} - t_1)$ of the respective switching-state vectors $\boldsymbol{u}_a, \boldsymbol{u}_b$, and $\boldsymbol{u}_0$ are different from the values in equation 4.27, which introduces the desired variations of subcycle lengths. Note that t_1 is another solution of equation 4.32, which is disregarded. The switching-state vectors of a subcycle are shown in Figure 4-28b. For the solution of equation 4.32 $\boldsymbol{u}_1$ is chosen as $\boldsymbol{u}_b$, and $\boldsymbol{u}_2$ as $\boldsymbol{u}_0$ in a first step. Once the on-time t_{act} of $\boldsymbol{u}_{act}$ has elapsed, $\boldsymbol{u}_b$ is chosen as $\boldsymbol{u}_{act}$ for the next switching interval, $\boldsymbol{u}_1$ becomes $\boldsymbol{u}_0$, $\boldsymbol{u}_2$ becomes $\boldsymbol{u}_a$, and the cyclic process starts again [27].

Figure 4-28c gives an example of measured subcycle durations in a fundamental period. The comparison of the harmonic spectra Figure 4-28d and Figure 4-24 demonstrates the absence of pronounced spectral components in the harmonic current.

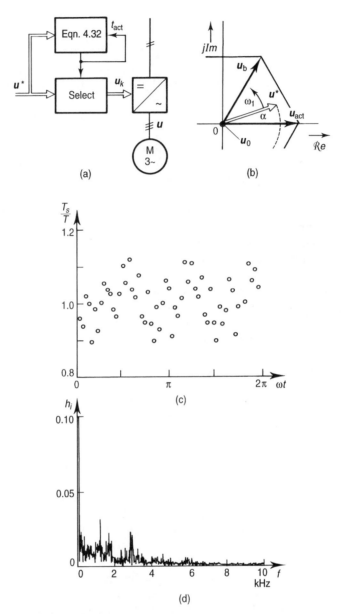

Figure 4-28. Carrierless pulse width modulation: (a) signal flow diagram;
(b) switching-state vectors of the first 60°-sector; (c) measured
subcycle durations; (d) harmonic spectrum.

Carrierless PWM equalizes the spectral distribution of the harmonic energy. The energy level is not reduced. To lower the audible excitation of mechanical resonances is a promising aspect. It remains difficult to decide, though, whether a clear, single tone is better tolerable in its annoying effect than the radiation of white noise.

4.5.3. Overmodulation

It is apparent from the averaging approach of the space vector modulation technique that the on-duration t_0 of the zero vector $\boldsymbol{u}_0$ (or $\boldsymbol{u}_7$) decreases as the modulation index m increases. The value $t_0 = 0$ in equation 4.27 is first reached at $m = m_{\max 2}$, which means that the circular path of the reference vector $\boldsymbol{u}^*$ touches the outer hexagon, which is opened up by the six active switching-state vectors Figure 4-29a. The controllable range of linear modulation methods terminates at this point.

An additional singular operating point exists in the six-step mode. It is characterized by the switching sequence $\boldsymbol{u}_1 - \boldsymbol{u}_2 - \boldsymbol{u}_3 - \ldots - \boldsymbol{u}_6$ and yields the highest possible fundamental output voltage corresponding to $m = 1$.

Control in the intermediate range $m_{\max 2} < m < 1$ can be achieved by overmodulation [28]. It is expedient to consider a sequence of output voltage vectors $\boldsymbol{u}_k$, averaged over a subcycle to become a single quantity $\boldsymbol{u}_{av}$, as the characteristic variable. Overmodulation techniques subdivide into two different modes. In mode I, the trajectory of the average voltage vector $\boldsymbol{u}_{av}$ follows a circle of radius $m > m_{\max 2}$ as long as the circle arc is located within the hexagon; $\boldsymbol{u}_{av}$ tracks the hexagon sides in the remaining portions (Figure 4-29b). Equations 4.27a–c are used to derive the switching durations while $\boldsymbol{u}_{av}$ is on the arc. A value $t_0 < 0$ as a solution of equation 4.27a–c

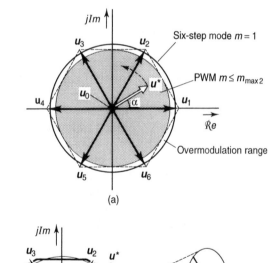

(a)

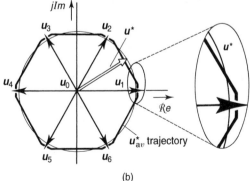

Figure 4-29. Overmodulation: (a) definition of the overmodulation range; (b) trajectory of $\boldsymbol{u}_{av}$ in overmodulation mode I.

(b)

indicates that u_{av} is on the hexagon sides. The switching durations are then $t_0 = 0$, and

$$t_a = T_0 \frac{\sqrt{3} \cos \alpha - \sin \alpha}{\sqrt{3} \cos \alpha + \sin \alpha} \tag{4.33a}$$

$$t_b = T_0 - t_a \tag{4.33b}$$

Overmodulation mode I reaches its upper limit at $m > m_{\max 3} = 0.952$ when the length of the arcs reduces to zero and the trajectory of u_{av} becomes purely hexagonal.

The output voltage can be further increased using the technique of overmodulation mode II. In this mode, the average voltage moves along the linear trajectories that form the outer hexagon. Its velocity is controlled by varying the duty cycle of the two switching-state vectors adjacent to u_{av}. As m increases beyond $m_{\max 3}$, the velocity becomes gradually higher in the center portion of the hexagon side, and lower near the corners. Eventually the velocity in the corners reduces to zero. The average voltage vector remains then fixed to the respective hexagon corner for a time duration that increases as the modulation index m increases. Such operation is illustrated in Figure 4-30 showing the locations of the reference vector u^* and the average voltage vector u_{av} in an equidistant time sequence that covers one-sixth of a fundamental period. As m gradually approaches unity, u_{av} tends to get locked at the corners for an increasing time duration. The lock-in time finally reaches one-sixth of the fundamental period. Overmodulation mode II has then smoothly converged into six-step operation, and the velocity along the edges has become infinite.

Throughout mode II, and partially in mode I, a subcycle is made up by only two switching-state vectors. These are the two vectors that define the hexagon side on which u_{av} is traveling. Since the switching frequency is normally maintained at constant value, the subcycle duration T_0 must reduce due to the reduced number of switching-state vectors. This explains why the distortion factor reduces at the beginning of the overmodulation range (Figure 4-31).

The average voltage waveforms of one phase, Figure 4-32, demonstrate that the modulation index is increased by the addition of harmonic components beyond the limit that exists at linear modulation. The added harmonics do not form zero-sequence components as those discussed in Section 4.5.1. Hence they are fully

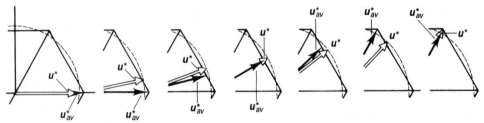

Figure 4-30. Overmodulation mode II. Sequence at equidistant time intervals showing the locations of the reference voltage vector and the average voltage vector.

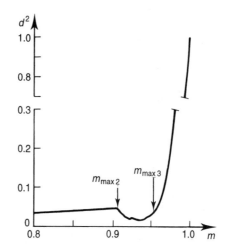

Figure 4-31. Loss factor d^2 at overmodulation (different d^2 scales).

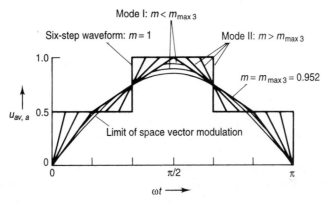

Figure 4-32. Voltage waveform of phase a at overmodulation mode I and mode II, including six-step operation.

reflected in the current waveforms Figure 4-33, which classifies overmodulation as a nonlinear technique.

4.5.4. Optimized Open Loop PWM

PWM inverters of higher power rating are operated at very low switching frequency to reduce the switching losses. Values of a few 100 Hertz are customary in the upper megawatt range. If the choice is an open loop technique, only synchronized pulse schemes should be employed for modulation in order to avoid the generation of excessive subharmonic components. The same applies for drive systems operating at very high fundamental frequency while the switching frequency is in the lower kilohertz range. The pulse number in equation 4.31 is low in both cases. There are only a few switching instants t_k per fundamental period, and small variations of

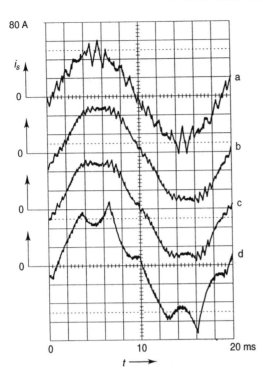

Figure 4-33. Current waveforms at overmodulation: (a) space vector modulation at $m_{\mathrm{max}\,2}$; (b) at the transition between range I and range II; (c) overmodulation range II, (d) operation close to the six-step mode.

the respective switching angles $\alpha_k = 2\pi f_1 \cdot t_k$ have considerable influence on the harmonic distortion of the machine currents.

It is advantageous in this situation to determine the finite number of switching angles per fundamental period by optimization procedures. Necessarily the fundamental frequency must be considered constant for the purpose of defining a sensible optimization problem. A numerical solution can be then obtained off line. The precalculated optimal switching patterns are stored in the drive control system to be retrieved during operation in real time [29].

The application of this method is restricted to quasi steady-state operating conditions. Operation in the transient mode produces waveform distortions worse than with nonoptimal methods; see Section 5.6.3.

The best optimization results are achieved with switching sequences having odd pulse numbers and quarter-wave symmetry. Off-line schemes can be classified with respect to the optimization objective [30].

Harmonic Elimination. This technique aims at the elimination of a well-defined number $n_1 = (N-1)/2$ of lower order harmonics from the discrete Fourier spectrum. It eliminates all torque harmonics having 6 times the fundamental frequency at $N = 5$, or 6 and 12 times the fundamental frequency at $N = 7$, and so on [31]. The method can be applied when specific harmonic frequencies in the machine torque must be avoided to prevent resonant excitation of the driven mechanical system (motor shaft, couplings, gears, load). The approach is suboptimal as regards other performance criteria.

Objective Functions. An accepted approach is the minimization of the loss factor d^2 [32], where d is defined by equation 4.17 or 4.20. Alternatively, the highest peak value of the phase current can be considered a quantity to be minimized at very low pulse numbers [33]. The maximum efficiency of the inverter/machine system is another optimization objective [34].

The objective function that defines a particular optimization problem tends to exhibit a very large number of local minimums. This makes the numerical solution extremely time consuming, even on today's modern computers. A set of switching angles that minimize the harmonic current ($d \rightarrow$ min) is shown in Figure 4-34. Figure 4-35 compares the performance of a $d \rightarrow$ min scheme at 300 Hz switching frequency with the suboscillation method and the space vector modulation method.

The improvement of the optimal method is due to a basic difference in the organization of the switching sequences. Such sequences can be extracted from the stator flux vector trajectories Figure 4-11b and Figure 4-11d, respectively. Figure 4-36 compares the switching sequences that generate the respective trajectories in Figure 4-11 over the interval of a quarter cycle. While the volt-second balance over a subcycle is always maintained in space vector modulation, the optimal method does not strictly obey this rule. Repeated switching between only two switching-state vectors prevails instead, which indicates that smaller volt-second errors, which needed correction by an added third switching state, are left to persist throughout a larger number of consecutive switchings. This method is superior in that the error component that builds up in a different spatial direction eventually reduces without an added correction.

Synchronous optimal pulse width modulation is inherently restricted to steady-state operation since it is hardly possible to predefine dynamic conditions. An optimal switching pattern generates a well-defined steady-state current trajectory $i_{ss}(t)$ of

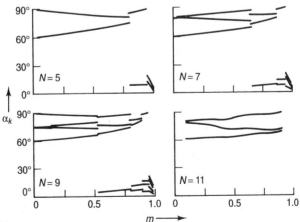

Figure 4-34. Optimal switching angles versus modulation index.

N = pulse number

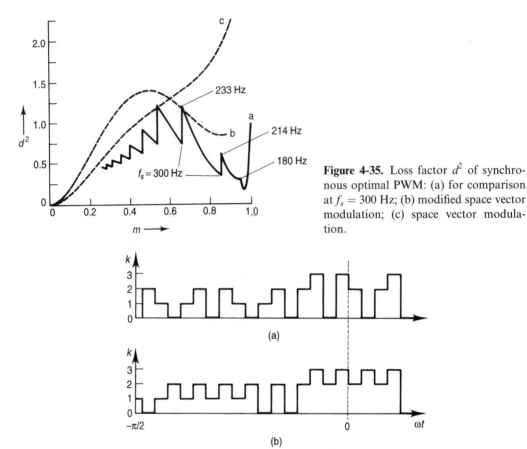

Figure 4-35. Loss factor d^2 of synchronous optimal PWM: (a) for comparison at $f_s = 300$ Hz; (b) modified space vector modulation; (c) space vector modulation.

Figure 4-36. Switching state sequences of a quarter cycle: (a) space vector modulation; (b) synchronous optimal modulation (k refers to the switching-state vectors u_k as defined in Figure 4-9, time axis not to scale).

minimum harmonic distortion, as the one shown in the left half of Figure 4-37a. An assumed change of the operating point at $t = t_1$ commands a different pulse pattern, to which a different steady-state current trajectory is associated. Since the current must be continuous, the actual trajectory resulting at $t > t_1$ exhibits an offset in space, Figure 4-37b. This is likely to occur at any transition between pulse numbers since two optimized steady-state trajectories rarely have the same instantaneous current values at any given point of time. The offset in space is called the dynamic modulation error $\delta(t)$ [53]. This error appears instantaneously at a deviation from the preoptimized steady state.

The dynamic modulation error tends to be large at low switching frequency. It is therefore almost impossible to use a synchronous optimal pulse width modulator as part of a fast current control system. The high harmonic current components are then fed back to the modulator input, heavily disorganizing the preoptimized switching sequences with a tendency of creating a harmonic instability where steady-state

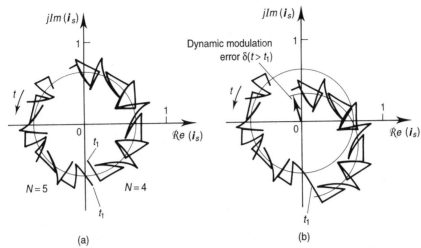

Figure 4-37. (a) Two sections of optimized current trajectories having different pulse numbers N; (b) resulting current trajectory with the pulse number changing from $N = 5$ to $N = 4$ at $t = t_1$.

operation is intended. The situation turns worse at transient operation. The reference vector of the modulator then changes its magnitude and phase angle very rapidly. Sections of different optimal pulse patterns are pieced together to form a real-time pulse sequence in which the preoptimized balance of voltage-time area is lost. The dynamic modulation error accumulates, and overcurrents occur which may cause the inverter to trip. Figure 4-62 gives an example.

Optimal Subcycle Method. This method considers the durations of switching subcycles as optimization variables, a subcycle being the time sequence of three consecutive switching-state vectors. The sequence is arranged such that the instantaneous distortion current equals zero at the beginning and at the end of the subcycle. This enables the composition of the switched waveforms from a precalculated set of optimal subcycles in any desired sequence without causing undesired current transients under dynamic operating conditions. The approach eliminates a basic deficiency of the optimal pulse width modulation techniques that are based on precalculated switching angles.

A signal flow diagram of an optimal subcycle modulator is shown in Figure 4-38a. Samples of the reference vector $u^*(t_s)$ are taken at $t = t_s$, whenever the previous subcycle terminates. The time duration $T_s(u^*(t_s))$ of the next subcycle is then read from a table which contains off-line optimized data as displayed in Figure 4-38b. The curves show that the subcycles enlarge as the reference vector comes closer to one of the active switching-state vectors, both in magnitude as in phase angle. This implies that the optimization is particularly worthwhile in the upper modulation range.

The modulation process itself is based on the space vector approach, taking into account that the subcycle length is variable. Hence T_s replaces $T_0 = 1/2f_s$ in equation 4-27. A predicted value $u^*(t_s + 1/2T_s(u^*(t_s)))$ is used to determine the on-times.

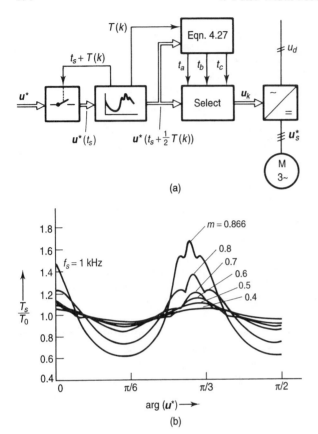

(a)

(b)

Figure 4-38. Optimal subcycle PWM: (a) signal flow diagram; (b) subcycle duration versus fundamental phase angle with the switching frequency as parameter.

The prediction assumes that the fundamental frequency does not change during a subcycle. It eliminates the perturbations of the fundamental phase angle that would result from sampling at variable time intervals [35].

The performance of the optimal subcycle method is compared with the space vector modulation technique in Figure 4-39. The Fourier spectrum is similar to that of Figure 4-28. It lacks dominant carrier frequencies, which reduces the radiation of acoustic noise from connected loads.

4.5.5. Switching Conditions

It was assumed until now that the inverter switches behave ideally. This is not true for almost all types of semiconductor switches. The devices react delayed to their control signals at turn-on and turn-off. The delay times depend on the type of semiconductor, on its current and voltage rating, on the controlling waveforms at the gate electrode, on the device temperature, and on the actual current to be switched.

Minimum Duration of Switching States. To avoid unnecessary switching losses of the devices, allowance must be made by the control logic for minimum

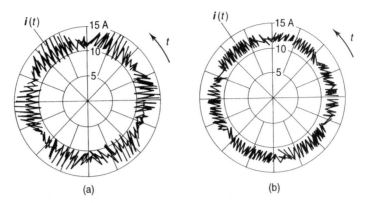

Figure 4-39 Current trajectories: (a) space vector modulation; (b) optimal subcycle modulation.

time durations in the on-state and the off-state, respectively. An additional time margin must be included so as to allow the snubber circuits to energize or deenergize. The resulting minimum on-duration of a switching-state vector is of the order 1–100 μs, depending on the respective type of semiconductor switch. If the commanded value in an open loop modulator is less than the required minimum, the respective switching state must be either extended in time or skipped (pulse dropping [36]). This causes additional current waveform distortions and also constitutes a limitation of the maximum modulation index. The overmodulation techniques described in Section 4.5.3 avoid such limitations.

Dead-Time Effect. Minority carrier devices in particular have their turn-off delayed owing to the storage effect. The storage time T_{st} varies with the current and the device temperature. To avoid short circuits of the inverter half-bridges, a lock-out time T_d must be introduced by the inverter control. The lock-out time counts from the time instant at which one semiconductor switch in a half-bridge turns off and terminates when the opposite switch is turned on. The lock-out time T_d is determined as the maximum value of storage time T_{st} plus an additional safety time interval.

We have now two different situations, displayed in Figure 4-40a for positive load current in a bridge leg. When the modulator output signal k goes high, the base drive signal k_1 of T1 gets delayed by T_d, and so does the reversal of the phase voltage u_{ph}. If the modulator output signal k goes low, the base drive signal k_1 is immediately made zero, but the actual turn-off of T1 is delayed by the device storage time $T_{st} < T_d$. Consequently, the on-time of the upper bridge arm does not last as long as commanded by the controlling signal k. It is decreased by the time difference $T_d < T_{st}$, [37].

A similar effect occurs at negative current polarity. Figure 4-40b shows that the on-time of the upper bridge arm is now increased by $T_d < T_{st}$. Hence, the actual duty cycle of the half-bridge is always different from that of the controlling signal k. It is

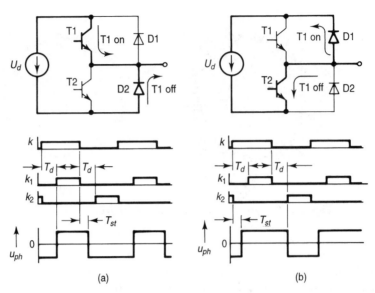

k_1, k_2 = base drive signals

Figure 4-40. Inverter switching delay: (a) positive load current; (b) negative
load current.

either increased or decreased, depending on the load current polarity. The effect is
described by an error voltage vector

$$\Delta u = \frac{T_d - T_{st}}{T_s} \; \mathbf{sig}(i_s), \tag{4.34}$$

which changes the inverter output from its intended value $u_{av} = u^*$ to

$$u_{av} = u^* - \Delta u \tag{4.35}$$

where u_{av} is the inverter output voltage vector averaged over a subcycle. The error
magnitude Δu is proportional to the actual safety time margin $T_d - T_{st}$; its direction
changes in discrete steps, depending on the respective polarities of the three phase
currents. This is expressed in equation 4.34 by a polarity vector of constant magnitude

$$\mathbf{sig}(i_s) = \tfrac{2}{3} \left[\text{sign}(i_a) + a \; \text{sign}(i_b) + a^2 \cdot \text{sign}(i_c) \right] \tag{4.36}$$

where $a = \exp(j2\pi/3)$, and i_s is the current vector. The notation $\mathbf{sig}(i_s)$ was chosen to
indicate that this complex nonlinear function exhibits properties of a sign function.
The graph $\mathbf{sig}(i_s)$ is shown in Figure 4-41a for all possible values of the current vector
i_s. The space vector $\mathbf{sig}(i_s)$ is of constant magnitude; it always resides in the center of
that $60°$ sector in which the current space vector is located. The three phase currents
are denoted as i_a, i_b, and i_c.

The dead-time effect described by equations 4.34 through 4.36 produces a nonlinear distortion of the average voltage vector trajectory u_{av}. Figure 4-41b shows an
example. The distortion does not depend on the magnitude u^* of the fundamental
voltage and hence its relative influence is very strong in the lower-speed range where

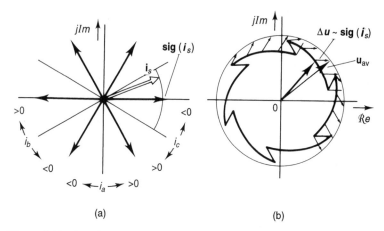

Figure 4-41. Dead-time effect: (a) location of the polarity vector **sig**(*i*); (b) trajectory of the distorted average voltage $\boldsymbol{u}_{av}$.

u^* is small. Since the fundamental frequency is low in this range, the smoothing action of the load circuit inductance has little effect on the current waveforms, and the sudden voltage changes become clearly visible, Figure 4-42a. As a reduction of the average voltage occurs according to equation 4.36 when one of the phase currents changes its sign, these currents have a tendency to maintain their values after a zero crossing, Figure 4-43a. The situation is different in the generator mode of the machine. The average voltage then suddenly increases, causing a steeper rise of the respective phase current after a zero crossing.

The machine torque is influenced in any case, exhibiting pulsations in magnitude at six times the fundamental frequency in the steady state. Electromechanical stability problems may result if this frequency is sufficiently low. Such a case is illustrated in Figure 4-43b, showing one phase current and the speed signal in permanent instability.

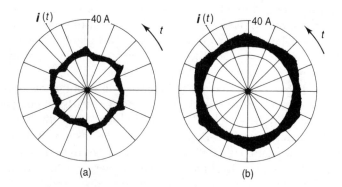

Figure 4-42. Dead-time effect: (a) measured current trajectory with sixth harmonics and reduced fundamental; (b) as in (a), with dead-time compensation ($f_s = 200$ Hz).

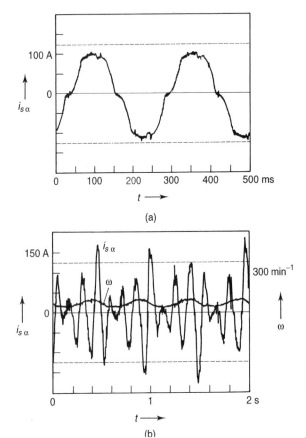

(a)

(b)

Figure 4-43. Dead-time effect measured at two different operating frequencies: (a) phase current distortion at $f_1 = 3.8$ Hz, (b) electromechanical instability at $f_1 = 7.4$ Hz.

Dead-Time Compensation. If the pulse width modulator and the inverter form part of a superimposed high-bandwidth current control loop, the current waveform distortions caused by the dead-time effect are compensated to a certain extent. This may eliminate the need for a separate dead-time compensator. A compensator is required when fast current control is not available, or when the machine torque must be very smooth. Dead-time compensators can be implemented in hardware or in software.

The hardware compensator Figure 4-44a operates by closed loop control [38]. Identical circuits are provided for each bridge leg. Each compensator forces a constant time delay between the logic output signal k of the pulse modulator and the actual switching instant. To achieve this, the instant at which the phase voltage changes is measured at the inverter output. A logic signal $\text{sign}(u_{\text{ph}})$ is obtained that is fed back to control an up-down counter, which, in turn, controls the bridge: a positive count controls a negative phase voltage, and vice versa.

Figure 4-44b shows the signals at positive load current. The half-bridge output is negative at the beginning, and $\text{sign}(u_{\text{ph}}) = 0$. The counter holds the measured storage time T_{st} of the previous commutation. It starts downcounting at fixed

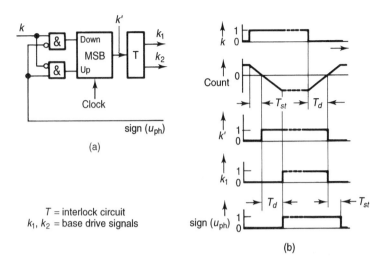

T = interlock circuit
k_1, k_2 = base drive signals

Figure 4-44. Dead-time compensation: (a) compensation and delay circuit per phase; (b) signal waveforms.

clock rate when the modulator output k turns high. The inverter control logic in block T receives the on-signal k' after T_{st}, and then inserts the lockout time T_d before k_1 turns the bridge on. The total time delay of the turn-on process amounts to $T_d + T_{st}$.

Turn-off is initiated when the modulator output k turns low. The counter starts upcounting and turns the signal k' low after T_d. k_1 is reduced to zero immediately, but the associated switch T_2 turns off later when its storage delay time T_{st} has elapsed. The total time delay of this process is also $T_d + T_{st}$. Hence the switching sequence gets delayed in time, but its duty cycle is conserved.

When T_{st} changes following a change of the current polarity, the initial count of T_{st} is wrongly set, and the next commutation gets displaced. Thereafter, the duty cycle is again maintained as the counter starts with a revised value of T_{st}.

Software compensators are mostly designed in the feedforward mode. This eliminates the need for potential-free measurement of the inverter output voltages. Depending on the sign of the respective phase current, a fixed delay time T_{st0} is either added, or not added, to the control signal of the half-bridge. As the actual storage delay T_{st} is not known, a complete compensation of the dead-time effect may not be achieved.

The changes of the error voltage vector Δu act as sudden disturbances on the current control loop. They are compensated only at the next switching of the phase leg. The remaining transient error is mostly tolerable in induction motor drive systems; synchronous machines having sinusoidal back-EMF behave more sensitively to these effects as they tend to operate partly in the discontinuous current mode at light loads. The reason for this adverse effect is the absence of a magnetizing component in the stator current. Such machines require more elaborate switching delay compensation schemes when applied to high-performance motion control systems.

Alternatively, a *d*-axis current component can be injected into the machine to shorten the discontinuous current time intervals at light loads [39].

4.6. CLOSED LOOP PWM CONTROL

Closed loop PWM schemes generate the switching sequences inherently in a closed control loop, Figure 4-12b. The feedback loop is established either for the stator current vector or for the stator flux vector. These are machine state variables that can be measured or observed and that reflect a switching action of the inverter. Closed loop control is mostly fast enough to compensate the nonlinear effects of pulse dropping and variable switching delay.

4.6.1. Nonoptimal Methods

Hysteresis Current Control. The signal flow diagram in Figure 4-45a shows three hysteresis controllers, one for each phase. Each controller determines the switching-state of one inverter leg such that the error of the corresponding phase current is maintained within the hysteresis band. The width of the hysteresis band is $\pm \Delta i$.

The control method is simple to implement, and its dynamic performance is excellent. There are some inherent drawbacks, though [40].

- There is no intercommunication between the individual hysteresis controllers of the three phases and hence no strategy to generate zero-voltage vectors. This increases the switching frequency at lower modulation index.

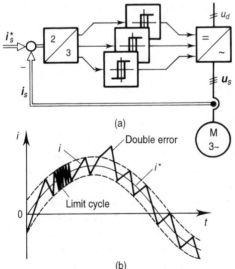

Figure 4-45. Hysteresis current control: (a) signal flow diagram; (b) basic current waveform.

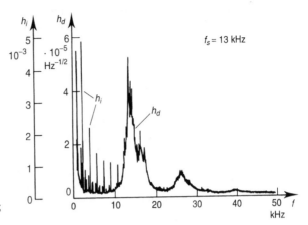

Figure 4-46. Hysteresis current control; measured harmonic spectrum.

- There is a tendency at lower speed to lock into limit cycles of high-frequency switching which comprise only nonzero-voltage vectors (Figure 4-45b).
- The current error is not strictly limited. The signal will leave the hysteresis band whenever the zero vector is turned on while the back-EMF vector has a component that opposes the previous active switching-state vector. The maximum overshoot is $2\Delta i$ (Figure 4-45b).
- The modulation process generates subharmonic components.

The amplitude density spectrum $h_d(f)$, shown in the Figure 4-46, includes also discrete components $h_i(k \cdot f_1)$ at subharmonic frequencies; the spectrum is almost independent of the modulation index. The switching frequency of a hysteresis current controller is strongly dependent on the modulation index, having a similar tendency as curve a in Figure 4-51 if the aforementioned special effects are not considered.

This effect can be explained with reference to Figure 4-21. It is illustrated there that the current distortions reduce when the reference vector u^* reaches proximity to one of the seven switching-state vectors. u^* is in permanent proximity to the zero vector at low modulation index and in temporary proximity to an active switching-state vector at high modulation index. Consequently, the constant harmonic current amplitude of a hysteresis current controller lets the switching frequency drop to near zero at $m \approx 0$, and toward a higher minimum value at $m \to m_{max}$. This results in a behavior as basically demonstrated in the graph Figure 4-51, showing that the switching capability of the inverter is not well utilized. Especially the very low switching frequency in the lower modulation range favors the generation of subharmonics.

Hysteresis controllers are preferred for operation at higher switching frequency, as this compensates for their inferior quality of modulation. The switching losses restrict the application to lower power levels. Improvements have aimed at eliminating the basic deficiencies of this attractive modulation technique. The current overshoot and limit cycles can be done away with at the expense of additional comparators and logic memory [41]. In an alternative approach, the current error signals are transformed to a synchronous reference frame before acting on the hysteresis controllers [42]. Operation at constant switching frequency can be achieved by adapting the hysteresis bandwidth. The width must be adjusted basically in proportion to the inverse function of Figure 4-51a, which implies further circuit complexity [43, 44].

Suboscillation Current Control. A carrier-based modulation scheme as part of a current control loop eliminates the basic shortcomings of the hysteresis controller. Figure 4-47 shows that a proportional integral (PI-type) controller is used to derive the reference voltage u^* for the pulse width modulator from the current error. The back-EMF of the machine acts as a disturbance in the current control loop. This voltage is free from harmonics and discontinuities in amplitude and phase angle. It is therefore possible to almost compensate the influence of the back-EMF through the I channel of the PI controller. However, a steady-state tracking error must persist so as to generate the time-varying signal u_s^* at the controller input [45].

The tracking error is kept low by choosing a high gain for the PI controller. The gain is limited, on the other hand, as it amplifies the harmonic currents. In order not to impair the proper operation of the pulse width modulator, the slope of the current error signal must be always less than the slope of the triangular carrier signal.

The scheme cannot be simply looked at as a pulse width modulator having a superimposed current control loop. This becomes obvious when comparing its harmonic spectrum Figure 4-48a with that of a space vector modulator, Figure 4-24. The difference is explained by Figure 4-48b, which shows that the current distortions exert an influence on the switching instants. This entails the advantage of a fast current response, provided that the modulator can react on instantaneous changes of its reference signal u^*. Hence, an analog circuit implementation of the suboscillation method is the adequate solution.

When the reference signal amplitude is driven beyond the carrier amplitude, the pulse width modulation is periodically interrupted and pulse dropping occurs. The beneficial effect is that the fundamental output voltage increases beyond the limit of proportional control; as compared with overmodulation techniques (Section 4.5.3), the harmonic distortion is higher since the switching frequency reduces.

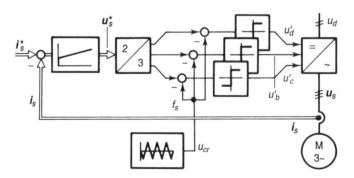

Figure 4-47. Suboscillation current control, signal flow graph.

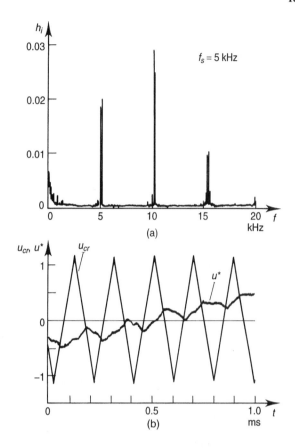

Figure 4-48. Suboscillation current control: (a) harmonic spectrum; (b) carrier signal and reference signal.

Space Vector Current Control. The nonzero current error in the steady state, inherent to the previous scheme, may be undesired in a high-performance vector-controlled drive. The error can be eliminated by using the back-EMF voltage u_i and the leakage voltage $\sigma l_s di_s/dt$ of the machine as compensating feedforward signals (Figure 4-49). The estimation is based on a machine model. The task of the current controller is then basically reduced to correct minor errors that originate from a mismatch of the model parameters or the model structure. The dynamic performance is improved by feedforward control based on the derivative of the current reference. The leakage inductance σl_s enters here as a parameter.

A space vector modulator is preferred in this scheme. Since its reference signal u^* is periodically sampled, Figure 4-49, this signal must be free from the harmonic components contained in the current feedback signal. These harmonics are eliminated by sampling the measured current signal in synchronism with the space vector modulator.

Figures 4-21b and 4-21c show that the vector trajectory of the harmonic current passes through zero while the zero vector is on. The zero crossing occurs in the center of the zero time interval. The respective switching sequences (4.28 and 4.29) begin with one-half of the zero vector time, such that each subcycle starts at zero harmonic current. The sampled current feedback signal $i_s(t_s)$ equals the fundamental current i_{s1}

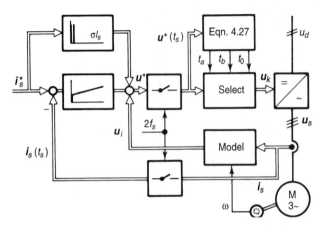

Figure 4-49. Space vector current control, signal flow diagram.

at this time instant. The instant is predetermined by the modulator. The method requires high-bandwidth A-to-D conversion. Furthermore, the pulse width modulator must operate in synchronism with the digital algorithm that executes the current controllers [19].

Note that the suboscillation method cannot provide the time instant of zero harmonic current; it is obvious from the harmonic patterns Figure 4-21a and from the waveforms in Figure 4-15 that this time instant depends on the respective operating point and cannot be determined by a simple algorithm.

Look-Up Table Methods. In a closed loop control scheme for a suitable space vector signal (stator current [46] or stator flux vector [47, 48]), the error is also a space vector quantity, for instance, $\Delta i(t) = i^*(t) - i(t)$. Limiting the magnitude $| \Delta i |$ of this error vector, or the respective magnitudes of suitable error components, either Δi_a and Δi_β, or $\Delta \Psi_s$ and $\Delta \arg(\Psi_s)$, by predetermined boundary values is a means to terminate an actual switching-state at time t_s. The next switching-state vector is then read from a look-up table. The table is adressed by the error vector and other state variables, like the back-EMF vector and/or the actual switching-state vector (Fig. 4-50).

These schemes generate asynchronous pulse sequences. Since the loss factor d^2 is mostly fixed by predefined boundary conditions, the performance at varying modulation index is reflected in the switching frequency. This is demonstrated in Figure 4-51a for a flux look-up table scheme [50].

4.6.2. Closed Loop PWM with Real-Time Optimization

Optimal pulse width modulation techniques are attractive for industrial and traction drives at very high power ratings, especially when GTOs are used as power switches. The switching frequency is then limited to values well below 1 kHz, Figure 4-4, and the minimization of the switching harmonics becomes worth-

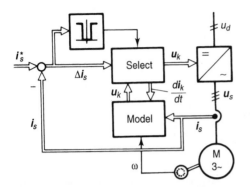

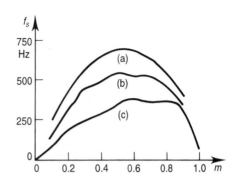

Figure 4-50. Current look-up table method, signal flow graph.

Figure 4-51. Performance of nonlinear controllers: (a) flux look-up table; (b) predictive control, circular error boundary; (c) current control with field orientation.

while. Another important issue for such applications is fast control of the fundamental current as part of the modulation process.

Predictive Current Control. Pulsewidth modulation by predictive current control, Figure 4-52, has common elements with the look-up table methods discussed earlier. In both methods, the switching instants are determined by suitable error boundaries. As an example, Figure 4-53 shows a circular boundary, the location of which is controlled by the current reference vector i_s^*. When the current vector i_s touches the boundary line, the next switching-state vector is determined by prediction and optimization.

The trajectories of the current vector for each possible switching-state are then computed, and predictions are made of the respective time intervals required to reach the error boundary again. These events depend also on the location of the error boundary, which is considered moving in the complex plane as commanded by the

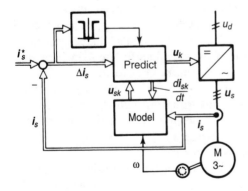

Figure 4-52. Predictive current control, signal flow diagram.

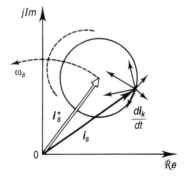

Figure 4-53. Predictive current control, boundary circle and space vectors; dashed circle: boundary at the next instant of switching.

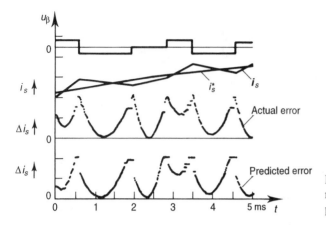

Figure 4-54. Predictive current control, measured waveforms with the double prediction method.

predicted current reference. The movement is indicated by the dotted circle in Figure 4-53. The predictions of the switching instants are based on a simplified mathematical model of the machine, Figure 4-10. The switching-state vector that produces the maximum on-time is finally selected. This corresponds to minimizing the switching frequency [49]. The optimization can be extended to include the next two switching-state vectors [50].

The algorithms which determine the optimal switching-state vector take about 20 μs on a DSP. Such delay is tolerable at lower switching frequency. Higher frequencies are handled by employing the double prediction method: well before the boundary is reached, the actual current trajectory is predicted in order to identify the future time instant at which the boundary transition is likely to occur. The back-EMF vector at this time instant is also predicted. It is used for the optimal selection of the switching-state vector for that future time instant, based on the procedure earlier described. The corresponding signals are shown in Figure 4-54 [51].

The performance of a predictive current control scheme which maximizes the on-durations of the next two switching-state vectors is illustrated in Figure 4-51b.

Pulse Width Control with Field Orientation. A further reduction of the switching frequency, which may be needed in very-high-power applications, can be achieved by defining a current error boundary of rectangular shape, having the rectangle aligned with the rotor flux vector of the machine, Figure 4-55. This transfers a major portion of the unavoidable current harmonics to the rotor field axis where they have no direct influence on the machine torque; the large rotor time constant eliminates their indirect influence on torque through the rotor flux. The selection of the switching-state vectors is based on prediction, satisfying the objectives that the switching frequency is minimized and that switching at d-current boundaries is avoided to the extent possible. This can be seen in the oscillogram of Figure 4-55. Using a rectangular boundary area in field coordinates leads to a reduction of switching frequency over what can be achieved with a circular boundary area (Figure 4-51c) [52]. The torque harmonics are reduced at the expense of increased current harmonics, since the d-axis current increases.

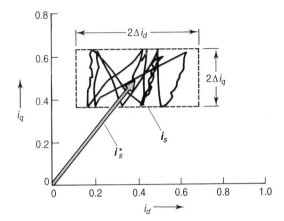

Figure 4-55. Pulse width modulation with field orientation; current trajectory and rectangular error boundary.

4.6.3. Real-Time Adaptation of Preoptimized Pulse Patterns

Pulse width modulation methods with on-line optimization target at the minimization of an objective function within a restricted time interval. They rely only on the next, at maximum on the next two switching instants as the basis of optimization. Only a conditional optimum can be obtained, therefore. Every solution depends on the actual initial conditions, which are the outcome of the optimization in the previous time interval. A small sacrifice on the optimum criterion in one time interval might entail larger benefits in a following interval. Such benefits cannot materialize, however, when the respective portion of time is not included in the optimization. This explains why a global optimum cannot be reached.

On-line methods also fail to provide synchronous switching, which is important if the pulse number is low. Higher subharmonic currents are generated in consequence, from which low-frequency torque pulsations result. A definite advantage is the excellent dynamic performance of on-line optimization methods.

Against this, the off-line optimization approach provides a global optimum as long as the restricting condition of steady-state operation is satisfied. No restriction exists with respect to the time interval of optimization, and hence the complete set of switching instants in a fundamental period can be optimized in a closed algorithmic procedure. This enforces synchronism between the switching frequency and the fundamental. It eliminates subharmonic currents and undesired torque pulsations in the eigenfrequency range of the mechanical subsystem of the drive. However, the steady-state restriction makes the dynamic performance poor, which renders such scheme nearly impracticable. More explanations are given in Section 4.5.4.

Trajectory Tracking Control. The combination of off-line optimization for the steady-state and on-line optimization for transient operation exploits the advantages of both methods [53]. The synchronous optimal pulse width modulator is shown in the upper portion of Figure 4-56. The pulse pattern $P(m, N)$ to control

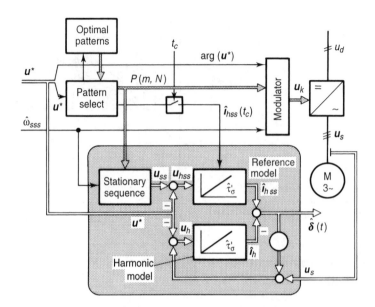

Figure 4-56. Estimation of the dynamic modulation error.

the inverter in a given operating point is selected by the reference magnitude u^* from the stored set of optimal patterns, where $u^* \propto m$, and the pulse number $N = f_s/f_1$.

It is important to use each switching sequence exactly as preoptimized whenever the system is in a steady state. Only then are the same steady-state conditions established in the drive system that had been the basis of the off-line pattern optimization. To achieve this, the controlling signal u^* of the modulator must be free from harmonics. Its dominating portion $u^{*\prime}$ is derived as a smooth signal from an accurate machine model which is shown in Figure 4-57. The model represents the interaction between the fundamental components of the stator current i_s and the rotor flux linkage Ψ_r. The large rotor time constant τ_r in Figure 4-57 serves to eliminate the harmonics of the measured stator current i_s. The model also generates a smoothed stator frequency signal $\hat{\omega}_{s\,ss}$ which does not contain the fast torque-dependent changes of ω_s. The signal is used to convert the optimal switching angles α_k into switching instants $t_k = \alpha_k/\hat{\omega}_{s\,ss}$, Figure 4-56.

The fundamental machine model is adapted to the state of magnetic leakage saturation by the normalized signal $l_\sigma(i_1)/\hat{l}_\sigma$. Residual errors of the reference voltage $u^{*\prime}$ are compensated using a superimposed control loop for the fundamental current component i_{s1}. This loop forms part of a parallel channel current control system, which will be described in the next chapter. The fundamental current loop is shown there in the upper portion of Figure 4-60. It operates very slowly as it need not contribute to the dynamic performance.

Fast current changes subdivide in three categories: the unavoidable harmonics $i_{h\,ss}(t)$, commanded transients $i_{tr}(t)$, which include transients generated by the load, and the dynamic modulation error $\delta(t)$; hence,

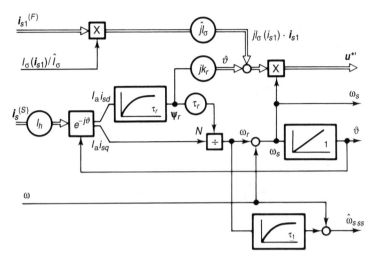

Figure 4-57. Saturation-dependent fundamental machine model.

$$i_s(t) = i_{s1}(t) + i_{tr}(t) + i_{h\,ss}(t) - \delta(t) \tag{4.37}$$

The dynamic modulation error was introduced with reference to Figure 4-37; it is an inherent imperfection of any voltage-controlled pulse width modulator [53].

Representing the commanded transients $i_{tr}(t)$ as a separate component in equation 4.37 classifies the harmonic current $i_{h\,ss}(t)$ as a steady-state quantity, which is indicated by the additional subscript ss. Therefore, $i_{h\,ss}(t)$ can be reconstructed from the optimal pulse pattern in actual use. The optimal patterns, by definition, also refer to the steady state. Using equation 4.36, the dynamic modulation error can be then determined from the measured machine current $i_s(t)$ and the commanded values $i_1(t)$ and $i_{tr}(t)$. The commanded values are indirectly available in the reference voltage u^*.

The lower portion of Figure 4-56 shows the signal flow for the estimation of the dynamic modulation error. The reference voltage vector u^* of the optimal pulse width modulator represents the fundamental machine voltage. It is substracted from the measured stator voltage vector u_s so as to form the vector u_h of the instantaneous harmonic stator voltage. This voltage is passed through a harmonic machine model to yield an estimated harmonic machine current $\hat{i}_h$; the model parameter $\hat{\tau}_\sigma'$ approximates the transient machine time constant; it will be shown that the exact value is not required.

The harmonic current vector $i_{h\,ss}$ moves on a particular steady-state trajectory which is associated to the selected optimal pulse pattern $P(m, N)$. Each pulse pattern has a different steady-state trajectory, and the transitions between them, when the pattern changes, are discontinuous. Figure 4-58 shows examples of steady-state harmonic trajectories for different values of m and N. To generate the steady-state trajectory in real-time, the actual switching sequence u_{ss} is reconstructed from the selected switching pattern $P(m, N)$ in a first step. The reference voltage u^* is then substracted, yielding the steady-state voltage harmonics $u_{h\,ss}$. A reference machine model serves to generate the steady-state harmonic current $i_{h\,ss}$ using a well deter-

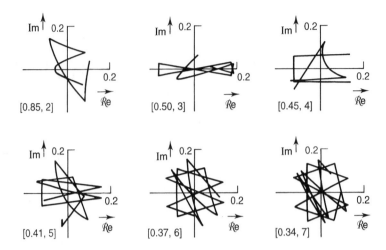

Figure 4-58. Steady-state harmonic trajectories $i_{h\,ss}(\omega_s\,t,\;m,\;N)$ at different values of the parameters $[m, N]$, stator coordinates; only one-sixth of a fundamental period is shown $(0 \leq \omega_1 t \leq \pi/3)$.

mined initial current value. The model receives a new initial setting $i_{h\,ss}(t_c)$ at every time instant t_c at which another optimal switching pattern becomes active.

The steady-state harmonic current $i_{h\,ss}$ is a synthesized waveform containing only AC components. Contrasting to that, the actual harmonic current $\hat{i}_h$ includes the DC offsets that occur at transient operation of the machine. The transient machine currents decay as commanded by the transient machine time constant τ_σ'. The same behavior is given to the estimated harmonic current $\hat{i}_h$ in Figure 4-56 by a feedback term $\hat{i}_h r_{sr}$ in the harmonic model, where $r_{sr} = r_s + l_s/l_r \cdot r_r$. The feedback term is not required for the reference model as its input signal comprises only steady-state harmonics which do not contain DC offsets. It is particularly important to note that by processing only the harmonic components in the error estimator, the estimation itself is independent of the varying load conditions of the machine.

The dynamic modulation error $\hat{\delta}(t)$ is now estimated as the difference between the harmonic reference trajectory $\hat{i}_{h\,ss}$ and the harmonic current $\hat{i}_h$. According to Figure 4-59, the error $\hat{\delta}(t)$ is fed to a trajectory controller, which, at nonzero error, modifies the selected pulse pattern using a dead-beat algorithm for fast dynamic response. The modification is represented by the signal ΔP. It forces the dynamic modulation error to zero, which means that the space vector of the machine current is made to track a time-moving reference point on the preoptimized steady-state trajectory. This effectively bridges the transition gap between trajectories of different pulse numbers, Figure 4-34. It also avoids undesired transients when the reference vector u^* executes rapid changes in magnitude and phase angle during dynamic operation of the drive system.

The modification of the pulse pattern in actual use must depend on the transient machine time constant. Here, $\hat{\tau}_\sigma'$ controls the conversion of the dynamic modulation error into a complex volt-second difference; in a reversed way, the modulation error itself is computed from a volt-second unbalance using the same time constant. An

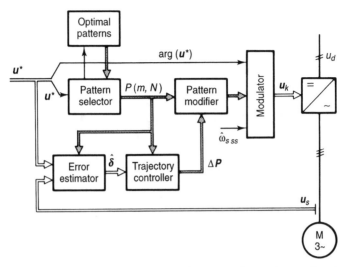

Figure 4-59. Optimal pulse width modulator and superimposed trajectory control system.

inaccuracy in the selected value of $\hat{\tau}_\sigma{}'$ is therefore ineffective, and the control of the harmonic current is parameter independent. This underlines the fact that the dynamic modulation error is an imperfection of the modulation process alone, on which the load has no influence. The trajectory tracking control system Figure 4-59 requires the generated switching sequence u_k at the modulator output as a feedback signal; using the machine voltage u_s instead includes the inverter in the control loop. This provides the additional benefit that the nonlinear switching delay of the inverter gets compensated by the trajectory controller, Section 4.5.5.

Parallel Channel Current Control. Since the voltage reference u^* of the optimal pulse width modulator is derived from the machine model as a smooth signal without harmonic components, it is the machine itself in Figure 4-60 that generates its feeding stator voltage depending on the actual load condition. It becomes a self-controlling machine that tends to perpetuate its operation at any speed, if once established. The required dynamic performance is achieved by the added action of a second current control channel which is shown as the lower channel in Figure 4-60. The current error ε_i is adapted to the actual state of leakage saturation by the factor $l_\sigma(i_1)/\hat{l}_\sigma$. It then acts on the tracking controller, making use of the available fast dead-beat control algorithm. The output is a voltage pattern modification ΔP.

Advantage is taken from the fact that the estimated harmonic current $\hat{i}_h$ is available in the tracking controller. An on-line correlation between this waveform and the measured stator current yields the ratio between the estimated and the actual harmonic amplitudes. This time-varying factor, which is identical to $l_\sigma/\hat{l}_\sigma$, expresses the magnetic saturation of the leakage inductance l_σ of the machine. The i_{s1}-extractor in the lower portion of Figure 4-60 computes the instantaneous fundamental current as is $i_{s1}(t) = i_s(t) - l_\sigma(i_1)/\hat{l}_\sigma \cdot [\hat{i}_h(t) + \hat{\delta}(t)]$. This signal is fed back to the parallel

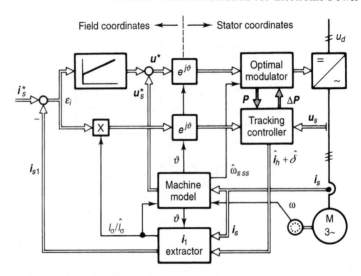

Figure 4-60. Trajectory-based parallel channel control scheme.

channel control system. Here, the upper channel provides accuracy and undisturbed synchronous optimal pulse width modulation in the steady state. The high dynamic performance is contributed by the lower channel [54].

Figure 4-61 shows the dynamic response and the extracted fundamental current i_{s1}. The shape of the oscillographed harmonic pattern A in Figure 4-61b gives proof of achieved synchronous optimal modulation before the step. The pattern B is optimal, although not steady state, since the machine starts accelerating after the step. With the trajectory control disengaged, shown in Figure 4-62 at C, the modulator fails to achieve optimal modulation even in the steady state. A high dynamic modulation error builds up after a step change, causing subsequently an overcurent trip of the inverter at D.

4.7. MULTILEVEL CONVERTERS

4.7.1. Twelve-Step Operation

Multilevel converters are fed from more than one source voltage on the DC side. This permits generating three or more different voltage levels on the AC side. The voltage potentials at the output of the inverter legs are no longer restricted to pure rectangular waveforms. Staircased waveshapes are possible instead, permitting better approximation of the desired continuous, mostly sinusoidal fundamental waveform. The harmonic content of the AC-side voltage is favorably reduced.

The penalty of multilevel converters is an increased number of power semiconductor switches. Figure 4-63 shows one inverter leg of a three-level topology, requiring four power switches and six diodes. The inverter leg corresponds to the half-bridge configuration Figure 4-5 of a two-level inverter. A total of three inverter legs

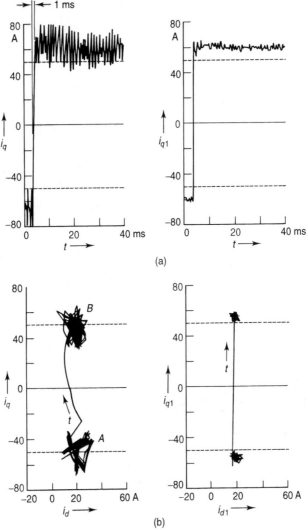

Figure 4-61. Torque reversal: (a) current waveforms; (b) current trajectories (left: torque producing current i_q, right: extracted fundamental component i_{q1}).

is used in a six-pulse bridge, schematically shown in Figure 4-64. A systematic overview on circuit topologies of multilevel converter is given in [55].

The output voltages of a three-phase, three-level power converter [56, 57] are shown in Figure 4-65a at unpulsed operation. There are 12 switching transitions in a fundamental period. Each power switch is operated exactly once per fundamental period, from which the denomination 12-step operation is derived. The same way as in Figure 4-7, the neutral point potential u_{np} contains all zero-sequence voltages. Since zero-sequence components do not appear in the phase voltages of the load, their harmonic content is favorably reduced.

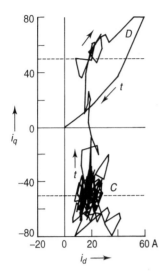

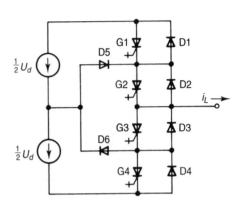

Figure 4-62. Torque reversal as in Figure 4-61, tracking control disengaged.

Figure 4-63. Inverter leg of a three-level topology

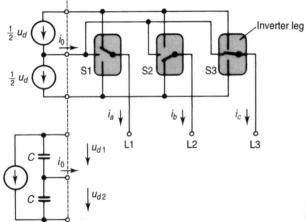

Figure 4-64. Three-level six-pulse bridge, schematic representation; lower portion: supply from a single DC source.

The fundamental content of a 12-step waveform is controllable by varying the control angle α. Figure 4-66 shows how the fundamental voltage and the distortion factor vary with the control angle. The fundamental voltage is represented by the modulation index m in Figure 4-66 as per the definition (4.22). The graph shows that the distortion factor d rises steeply at $m < 0.5$. Control of the fundamental voltage in this range requires pulse width modulation.

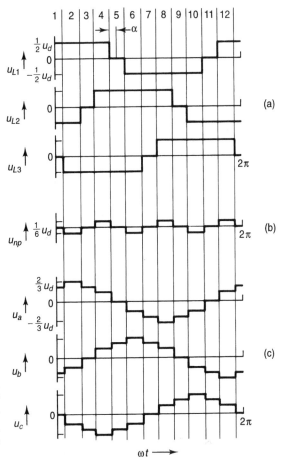

Figure 4-65. Voltage waveforms of a three-phase, three-level power converter at 12-step operation: (a) voltage potentials at the inverter terminals; (b) neutral point potential; (c) phase voltages of the load.

4.7.2. Switching-State Vectors

Inserting the voltage waveforms of Figure 4-65a or of Figure 4-65c into (4.10) defines the switching-state vectors of the 12-step mode. Generally, there is a total of $3^3 = 27$ different switching-states possible in a three-phase, three-level converter. The switching-state vectors are shown in Figure 4-67. Included in brackets are the respective polarities of the voltage potentials at the three output terminals. Twelve switching-state vectors are located on the periphery of an outer hexagon. Another six switching-state vectors point to the vertices of an inner hexagon. The polarity notations in brackets show that a redundancy exists here in that a particular switching-state vector can be formed by two different switching states. Note that the voltage differences between the pertaining potentials at the three output terminals define a zero sequence component over a fundamental period.

The zero voltage vector in the origin is represented by three different switching-states.

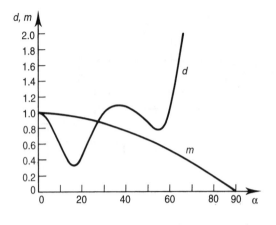

Figure 4-66. Control of the fundamental voltage at 12-step operation. Normalized fundamental voltage m and distortion factor d versus control angle α.

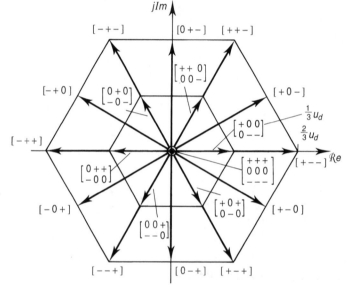

Figure 4-67. Switching state vectors of a three-phase three-level power converter; in brackets, voltage polarities at the three phase terminals.

The redundancy of switching-states offers additional degrees of freedom for the design of PWM patterns. These are generally used to satisfy the condition

$$\int_0^T i_0 \, d\tau = 0 \qquad (4.38)$$

where i_0 is the current from the center tap of the DC source, Figure 4–64. Condition 4.38 permits supplying the DC side of a three-level converter from a single-source voltage $U_d = u_{d1} + u_{d2}$, such that the center tap voltage is maintained at $u_{d1} = u_{d2}$ by a capacitive voltage divider as shown in the lower portion of Figure 4-64. This is a convenient solution as there are anyway two capacitors required in series in most applications, given the existing voltage a limit of electrolytic capacitors.

Condition 4.38 is satisfied as follows: whenever a switching-state pertaining to the inner hexagon is active, the three-phase load is supplied only by the upper DC source u_{d1} in Figure 4-64 or only by the lower source u_{d2}, depending on which of the two redundant switching-state vectors is selected. The respective choice determines the polarity of i_0 for the time duration of one switching state. The decision is made by a closed loop control system that receives the difference $u_{d1} - u_{d2}$ between the two DC link voltages as an error signal.

4.7.3. Three-Level Pulse Width Modulation

The design of pulse width modulators for multilevel topologies follows the same principles as discussed before for the special case of the three-level topology.

Suboscillation Method. In principle, any PWM method that is suited for controlling a two-level inverter can be adapted for three-level inverter control. Special care must be taken with carrier-based methods, for example, the suboscillation method. It is important to observe the polarity consistency rule (Section 4.7), which, in the case of a three-level inverter, implies that switching between low voltage levels must occur around a zero crossing of the phase voltages. An example of a well-designed pulse pattern is shown in Figure 4-72b.

When applying the suboscillation technique, a reduction of the harmonic distortion can be achieved by adding adequate zero-sequence waveforms to the sinusoidal reference signal. A viable example of a zero-sequence signal is the triangular wave in Figure 4-68b [58]. The amplitude of this signal varies in proportion to the modulation index m. It is added to each of the signals $u_a{}'$, $u_b{}'$, and $u_c{}'$ in Figure 4-68a. These signals represent a set of balanced, sinusoidal three-phase voltages having their amplitudes adjusted in proportion to the modulation index. The resulting signals form the set of reference voltages, two of which are shown in Figure 4-68c and e.

The carrier signals are synchronized with the reference waveforms to avoid the generation of subharmonic components. Note that each of the carrier signals has a defined phase relationship with respect to its particular reference signal to ensure minimum harmonic distortion of the modulation process. There is a positive carrier $u_{cr\,b+}$, the intersections of which with the reference signal u_b^* determine the switching instants of the positive voltage pulses at the phase terminal L2, Figure 4-64; similarly the negative carrier $u_{cr\,b-}$ generates the negative voltage pulses. Different carriers are used for the other phases, see Figure 4-68e. The resulting loss factor is displayed in Figure 4-69a as function of the modulation index m and the pulse number N.

The performance of an ill-designed three-level pulse width modulator shall be derived for comparison. Figure 4-70 characterizes a suboscillation modulator that uses sinusoidal reference signals and a common synchronized carriers for all three

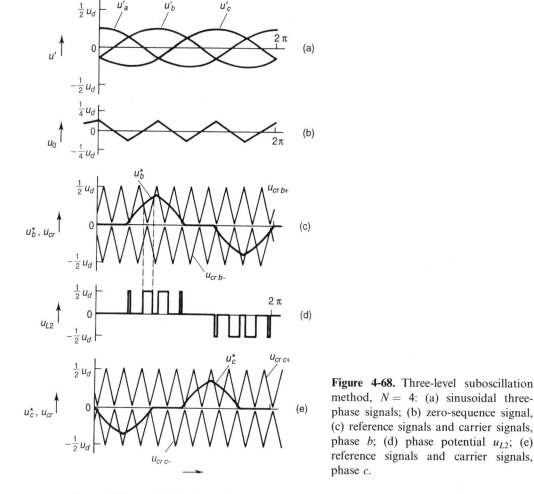

Figure 4-68. Three-level suboscillation method, $N = 4$: (a) sinusoidal three-phase signals; (b) zero-sequence signal, (c) reference signals and carrier signals, phase b; (d) phase potential u_{L2}; (e) reference signals and carrier signals, phase c.

phases. Figure 4-69b shows that the harmonic distortion is poor as compared with Figure 4-69a.

Space Vector Modulation. Also the space vector modulation method is well applicable for the control of three-level inverters. The location of the reference vector $u^*(t_s)$ at a given sampling instant t_s determines those three switching-state vectors from the pattern in Figure 4-67 that are in closest distance from the reference vector. The three switching-state vectors that form the modulation subcycle are named u_a, u_b, and u_c. The on-durations t_a, t_b, and t_c of the pertaining switching-state vectors u_a, u_b, and u_c are obtained from the solution of

$$2f_s \cdot (t_a u_a + t_b u_b + t_c u_c) = u^*(t_s) \tag{4.39a}$$

$$t_c = \frac{1}{2f_s} - t_a - t_b \tag{4.39b}$$

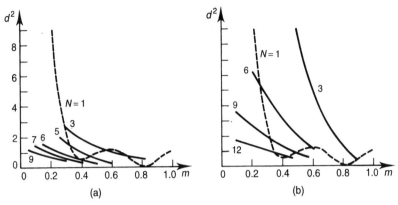

Figure 4-69. Loss factor d^2 as functions of the modulation index m and the pulse number N, suboscillation method: (a) with added zero-sequence signals and different carrier signals per phase; (b) sinusoidal reference signals and common carriers.

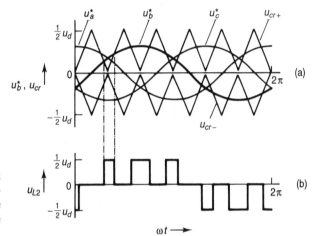

Figure 4-70. Three-level suboscillation method, $N = 3$; sinusoidal reference voltages and common carrier: (a) reference signals and carrier signals; (b) phase potential u_{L2}.

The sequence in time of the switching-state vectors is determined such that the number of commutations of the converter bridges is minimum. The selection exploits also the redundancy of switching-states.

Optimal Pulse Width Modulation. In high-power applications, the larger number of power switches in a three-level inverter is well utilized since the power output increases approximately in proportion to the number of switches. Since high-power switches generally require operation at low-switching frequency, an optimal PWM method is a good choice. The same optimization methods apply as described in Sections 4.5.4, 4.6.2, and 4.6.3.

The optimization using the $d \rightarrow$ min criterion yields the set of switching angles α_k that is shown in Figure 4-71a as a function of the modulation index. The resulting distortion factor d is displayed in Figure 4-71b in comparison with two other syn-

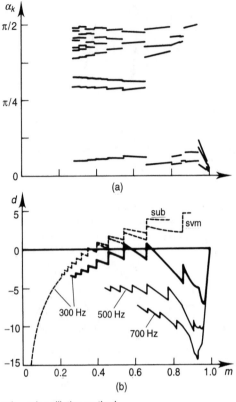

(a)

(b)

sub = suboscillation method
svm = space vector modulation

Figure 4-71. Three-level synchronous optimal PWM: (a) optimal switching angles α_k and (b) distortion factor d in decibels versus modulation index.

chronous PWM methods. Two decisive advantages of the optimal method can be observed:

- The distortion factor is much lower, particularly in the upper modulation range.
- The modulation range extends continuously up to the theoretical limit $m = 1$, which means that overmodulation performance is included.

Figure 4-72 shows the voltage and current waveforms pertaining to a synchronous optimal three-level pulse sequence. Notwithstanding the low switching frequency, the current distortion is low. The current waveform Figure 4-72c and the appertaining trajectory in Figure 4-73a demonstrate that this is owed to a density variation of the switching instants which carrier-based PWM methods cannot provide. The representation of the current trajectory in synchronous coordinates Figure 4-73b shrinks to the area of a small circle. The harmonic spectrum Figure 4-74 shows that the harmonic amplitudes are fairly well distributed. Thus the acoustic noise is minimum.

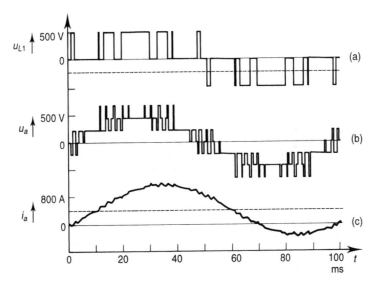

Figure 4-72. Three-level synchronous optimal PWM, steady-state waveforms: (a) voltage potential at the output of one inverter
leg; (b) phase voltage; (c) phase current.

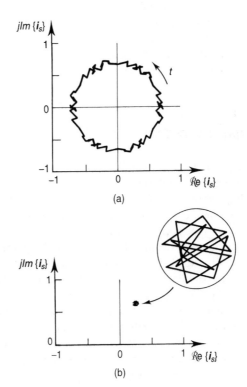

Figure 4-73. Three-level synchronous
optimal PWM, steady-state trajectories:
(a) stationary coordinates; (b) synchronous coordinates.

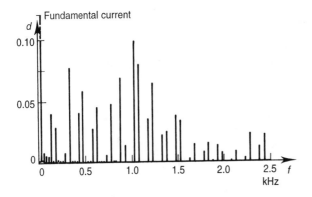

Figure 4-74. Three-level synchronous optimal PWM, harmonic spectrum of the waveform in Figure 4-72(c)

Superimposing a fast current control scheme around a synchronous optimal pulse width modulator requires trajectory tracking control. This method eliminates the dynamic modulation error. It is described in Section 4.6.3. Alternatively, the current controller must operate slowly enough so as to maintain quasi-steady-state conditions for pulse width modulation.

Extension. Multilevel converters having more than three voltage levels are less common. Their switching-state vectors are arranged similar to those in Figure 4-67, with each additional voltage level defining another hexagon. The switching-state vectors pointing to the most outward hexagon are unique. The redundancy of switching-state vectors increases stepwise by one for any next inner hexagon. The set of zero vectors is reached after the last step. For instance, there are five different zero vectors in a five-level converter.

4.8. CURRENT SOURCE INVERTER

The preceding discussions on pulse width modulation techniques made reference to a power circuit configuration as in Figure 4-6, in which the DC power is delivered by a voltage source U_d. Rectangular voltage waveforms are impressed on the load circuit, from which current waveforms result that depend on the actual load impedances.

Another approach, dual to the aforementioned principle of power conversion, is the current source inverter. A switched rectangular current waveform is injected into the load, and it is the voltage waveforms which develop under the influence of the load impedances. The fundamental frequency is determined by the switching sequence, exactly as in the case of a voltage source inverter.

There are two different principles employed to control the fundamental current amplitude in a current sourve inverter. Most frequently the DC current source is varied in magnitude, which eliminates the need for fundamental current control on the AC side, and pulse width modulation is not required. An alternative, although not very frequently applied, consists controlling the fundamental current by pulse width modulation [59]. There is a strong similarity to PWM techniques for voltage source inverters, although minor differences exist.

The majority of applications are based on the voltage source principle, which is owed in the first place to favoring properties of the available power semiconductor switches.

4.9. CONCLUSION

Pulse width modulation for the control of three-phase power converters can be performed using a large variety of different methods. Their respective properties are discussed and compared based on mathematical analyses and on measured results obtained from controlled drive systems in operation.

Performance criteria assist in the selection of a PWM scheme for a particular application. An important design parameter is the switching frequency since it determines the system losses. These are hardly a constraint at low power levels, permitting high-frequency switching combined with straightforward modulation methods. The important selection factors in this range are cost of implementation and dynamic performance. As the losses force the switching frequency to be low at higher power, elaborate techniques are preferred including off-line and on-line optimization. These permit that the contradicting requirements of slow switching and fast response can be satisfied.

Nomenclature

a unity vector
d distortion factor
d^2 loss factor
f_1 fundamental frequency
f_s switching frequency
h_d amplitude density spectrum
h_i discrete current spectrum
i_h harmonic current
$\boldsymbol{i}_s$ stator current vector
i_{tr} transient current
l_h mutual inductance
l_s stator inductance
$l_{s\sigma}$ stator leakage inductance
l_σ total leakage inductance
L inductance per phase
m modulation index
N pulse number
P set of pulse patterns P
$\mathbf{sig}(\boldsymbol{i}_s)$ polarity vector
T mechanical torque
T_d lock-out time

T_{on} on-time of a switch
t_0 constant subcycle duration
T_s variable subcycle duration
T_{st} storage time
T_0 constant subcycle duration
u_a voltage of phase a
u_d normalized DC link voltage
U_d DC link voltage
u_h harmonic voltage
$\boldsymbol{u}_i$ back-EMF voltage
$\boldsymbol{u}_k$ discrete sequence of switching state vectors
u_L phase to neutral voltage
u_{np} neutral point potential
u_{ph} phase voltage
$\boldsymbol{u}_s$ stator voltage vector
$\boldsymbol{u}_1 \dots \boldsymbol{u}_6$ switching state vectors
$\boldsymbol{u}^*$ reference voltage vector

Greek Symbols

α switching angle
δ dynamic modulation error
σ total leakage factor
τ_r rotor time constant
$\tau_\sigma{}'$ transient time constant
ω_s stator frequency
$\omega_{s\,ss}$ steady-state stator frequency
$\boldsymbol{\Psi}$ flux linkage vector

Indices

a, b, c phases
av average value
1 fundamental quantity
max maximum value
min minimum value
r rotor
R rated value
s stator
ss steady-state value
$\langle \dots \rangle$ bracket encloses on-duration
$\hat{}$ estimated value

References

[1] Kovács, P. K., *Transient Phenomena in Electrical Machines*, Elsevier Science Publishers, Amsterdam, 1994.

[2] Klingshirn, E. A., and H. E. Jordan, "A polyphase induction motor performance and losses on nonsinusoidal voltage sources," *IEEE Trans. Power App. Syst.*, Vol. PAS-87, pp. 624–631, March 1968.

[3] Murphy, J. M. D., and M. G. Egan, "A comparison of PWM strategies for inverter-fed induction motors," *IEEE Trans. Industry Appl.*, Vol. IA-19, no. 3, pp. 363–369, May/June 1983.

[4] Boys, J. T., and P. G. Handley, "Harmonic analysis of space vector modulated PWM waveforms," *IEE Proc.*, Vol. 137, Part B, no. 4, pp. 197–204, July 1990.

[5] Zach, F. C., R. Martinez, S. Keplinger, and A. Seiser, "Dynamically optimal switching patterns for PWM inverter drives," *IEEE Trans. Industry Appl.*, Vol. IA-21, no. 4, pp. 975–986, July/August, 1985.

[6] Lipo, T. A., P. C. Krause, and H. E. Jordan, "Harmonic torque and speed pulsations in a rectifier-inverter induction motor drive," *IEEE Trans. Power App. Syst.*, Vol. PAS-88, pp. 579–587, May 1969.

[7] Habetler, T. G., and D. M. Divan, "Performance characterization of new discrete pulse modulated current regulator," *IEEE Industry Appl. Soc. Ann. Meet.*, Pittsburgh, PA, 1988, pp. 395–405.

[8] P519, *IEEE Recommended Practices and Requirements for Harmonics Control in Electric Power Systems.*

[9] Leonhard, W., *Control of Electrical Drives*, Springer-Verlag, Berlin, Heidelberg, 1985.

[10] Schönung, A., and H. Stemmler, "Static frequency changers with subharmonic control in conjunction with reversible variable speed AC drives," *Brown Boveri Rev.*, pp. 555–577, 1964.

[11] Grant, D. A., and J. A. Houldsworth, "PWM AC motor drive employing ultrasonic carrier," *IEE Conf. Pow. Electronics Var. Speed Drives*, London, pp. 234–240, 1984.

[12] Kolar, J. W., H. Ertl, and F. C. Zach, "Influence of the modulation method on the conduction and switching losses of a PWM converter system," *IEEE Trans. Industry Appl.*, Vol. 27, no. 6, pp. 1063–1075, November/December 1991.

[13] Dependbrock, M., "Pulse width control of a 3-phase inverter with nonsinosoidal phase voltages," *IEEE/IAS Int. Semicond. Power Conv. Conf.*, Orlando, FL, pp. 399-398, 1975.

[14] Bowes, S. R., "New sinosoidal pulse width modulated inverter," *Proc. Inst. Elec. Eng.*, Vol. 122, Part B, pp. 1279–1285, November 1975.

[15] Intel User's Manual 8XC196MC, Intel Corporation, 1992.

[16] Bose, B. K., and H. A. Sutherland, "A high-performance pulse width modulator for an inverter-fed drive system using a microcomputer," *IEEE Trans. Industry Appl*, Vol. IA-19, no. 2, pp. 235–243, March/April 1983.

[17] Bowes, S. R., and R. R. Clements, "Computer-aided design of PWM inverter systems," *IEE Proc. B*, Vol. 129, no. 1, pp. 1–17, January 1982.

[18] Busse, A., and J. Holtz, "Multiloop control of a unity power factor fast-switching AC to DC converter," *IEEE Pow. Elec. Special. Conf., Cambridge,* pp. 171–179, 1982.

[19] Pfaff, G., A. Weschta, and A. Wick, "Design and experimental results of a brushless AC servo drive," *IEEE/IAS Annual Meeting*, San Francisco, pp. 692–697, 1982.

[20] Holtz, J., P. Lammert, and W. Lotzkat, "High-speed drive system with ultrasonic MOSFET PWM inverter and single-chip microprocessor control," *IEEE Trans. Industry Appl.*, Vol. IA-23, no. 6, pp. 1010–1015, November/December 1987.

[21] Holtz, J., and E. Bube, "Field oriented asynchronous pulse width modulation for high performance AC machine drives operating at low switching frequency," *IEEE Trans. Industry Appl.*, Vol. IA-27, no. 3, pp. 574–581, May/June 1991.

[22] Ogasawara, O., H. Akagi, and A. Nabae, "A novel PWM scheme of voltage source inverters based on space vector theory," *EPE Europ. Conf. Power Electronics and Appl.*, Aachen, pp. 1197–1202, 1989.

[23] Kolar, J. W., H. Ertl, and F. C. Zach, "Calculation of the passive and active component stress of three-phase PWM converter systems with high pulse rate," *EPE Europ. Conf. Power Electronics and Appl.*, Aachen, pp. 1303–1312, 1989.

[24] Heintze, K., et. al., "Pulse width modulating static inverters for the speed control of induction motors," *Siemens-Z*, Vol. 45, no. 3, pp. 154–161, 1971.

[25] Kliman, G. B., and A. B. Plunkett, "Development of a modulation strategy for a PWM inverter drive," *IEEE Trans. Industry Appl.*, Vol. IA-15, no. 1, pp. 702–709, January/February, 1979.

[26] Habetler, T. G., and D. Divan, "Acoustic noise reduction in sinusoidal PWM drives using a randomly modulated carrier," *IEEE Trans. Pow. Electr.*, Vol. 6, no. 3, pp. 356–363, July 1991.

[27] Holtz, J., and L. Springob, "Reduced harmonics PWM controlled line-side converter for electric drives," *IEEE Trans. Industry Appl*, Vol. 29, no. 4, pp. 814–819, November 1993.

[28] Holtz, J., W. Lotzkat, and A. Khambadkone, "On continuous control of PWM inverters in the overmodulation range with transition to the six-step mode," *IECON, 18th Ann. Conf. IEEE Industrial Elect. Soc.*, San Diego, pp. 307–312, 1992.

[29] Jackson, S. P., "Multiple pulse modulation in static inverters reduces selected output harmonics and provides smooth adjustment of fundamentals," *IEEE Trans. Indust. and General Appl.*, Vol. IGA-6, no. 4, pp. 357–360, July/August 1970.

[30] Holtz, J., "On the performance of optimal pulse width modulation techniques," *European Power Electronics*, Vol. 3, no. 1, pp. 17–26, March 1993.

[31] Patel, H. S., and R. G. Hoft, "Generalized techniques of harmonic elimination and voltage control in thyristor inverters," *IEEE Trans. Industry Appl.*, Vol. IA-9, no. 3, pp. 310–317, May/June 1973.

[32] Buja, G. S., and G. B. Indri, "Optimal pulse width modulation for feeding AC motors," *IEEE Trans. Industry Appl*, Vol. IA-13, no. 1, pp. 38–44, January/February 1977.

[33] Holtz, J., S. Stadtfeld, and H.-P. Wurm, "A novel PWM technique minimizing the peak inverter current at steady-state and transient operation," *Elektr. Bahnen*, Vol. 81, pp. 55–61, 1983.

[34] Zach, F. C., and H. Ertl, "Efficiency optimal control for AC drives with PWM inverters," *IEEE Trans. Industry Appl.*, Vol. IA-21, no. 4, pp. 987–1000, July/August 1985.

[35] Holtz, J., and B. Beyer, "Optimal pulse width modulation for AC servos and low-cost industrial drives," *IEEE Industry App. Soc. Ann. Meeting*, Houston, pp. 1010–1017, 1992.

[36] Mutch, M., K. Sakai, et al., "Stabilizing methods at high frequency for an induction motor drives driven by a PWM inverter," *EPE Europ. Conf. Power Electronics and Appl.*, Florence, pp. 2/352–358, 1991.

[37] Klug, R. D., "Effects and correction of switching dead-times in 3-phase PWM inverter drives," *EPE Europ. Conf. Power Electronics and Appl.*, Aachen, pp. 1261–1266, 1989.

[38] Murai, Y., T. Watanabe, and H. Iwasaki, "Waveform distortion and correction circuit for PWM inverters with switching lag-times," *IEEE Trans. Industry Appl.*, Vol. IA-23, no. 5, pp. 881–886, September/October 1987.

[39] Wang, Y., and H. Grotstollen, "Control strategies for the discontinuous current mode of AC drives with PWM inverters," *EPE Europ. Conf. Power Electronics and Appl.*, Florence, pp. 3/217–222, 1991.

[40] Brod, D. M., and D. W. Novotny, "Current control of VSI-PWM inverters," *IEEE Trans. Industry Appl.*, Vol. IA-21, no. 3, pp. 562–570, May/June 1985.

[41] Salama, S., and S. Lennon, "Overshoot and limit cycle free current control method for PWM inverter," *EPE Europ. Conf. Power Electronics and Appl.*, Florence, pp. 3/247–251, 1991.

[42] Rodriguez, J., and G. Kastner, "Nonlinear current control of an inverter-fed induction machine," *ETZ Archiv.*, pp. 245–250, 1987.

[43] Malesani, L., and P. Tenti, "A novel hysteresis control method for current-controlled VSI-PWM inverters with constant modulation frequency," *IEEE Industry Appl. Soc. Ann. Meet.*, Atlanta, GA, pp. 851–855, 1987.

[44] Bose, B. K., "An adaptive hysteresis-band current control technique of a voltage-fed PWM inverter for machine drive system," *IEEE Trans. Industrial Electron.*, Vol. 37, no. 5, pp. 402–408, October 1990.

[45] Stefanovic, V. R., "Present trends in variable speed AC drives," *Int. Pow. Elec. Conf. IPEC*, Tokyo, pp. 438–449, 1983.

[46] Nabae, A., S. Ogasawara, and H. Akagi, "A novel control scheme for PWM controlled inverters," *IEEE Industry Appl. Soc. Ann. Meet.*, Toronto, pp. 473–478, 1985.

[47] Török, V. G., "Near-optimum on-line modulation for PWM inverters," *IFAC Symp.*, Lausanne, pp. 247–254, 1983.

[48] Takahashi, I., and N. Toshihiko, "A new quick-response and high-efficiency control strategy of an induction motor," *IEEE Industry Appl. Soc. Ann. Meet.*, Toronto, pp. 820–827, 1985.

[49] Holtz, J., and S. Stadtfeld, "A predictive controller for the stator current vector of AC machines fed from a switched voltage source," *IPEC*, Tokyo, pp. 1665–1675, 1983.

[50] Boelkens, U., "Comparative study on trajectory-based pulse width modulation methods for three-phase converters feeding induction machines," Ph.D. Thesis (in German), Wuppertal University, 1989.

[51] Holtz, J., and S. Stadtfeld, "A PWM inverter drive system with on-line optimized pulse patterns," *EPE Europ. Conf. Power Electronics and Appl.*, Brussels, pp. 3.21–3.25, 1985.

[52] Khambadkone, A., and J. Holtz, "Low switching frequency high-power inverter drive based on field-oriented pulse width modulation," *EPE Europ. Conf. Power Electronics and Appl.*, Florence, pp. 4/672–677, 1991.

[53] Holtz, J., and B. Beyer, "The trajectory tracking approach—a new method for minimum distortion PWM in dynamic high-power drives, *IEEE Trans. Industry Appl.*, Vol. 30, no. 4, pp. 1048–1057, July/August 1994.

[54] Holtz, J., and B. Beyer, "Fast current trajectory tracking control based on synchronous optimal pulse width modulation," *IEEE Trans. Industry Appl.*, 1995.

[55] Meynard, T. A., and H. Foch, "Multi-level conversion: High voltage choppers and voltage source inverters," *IEEE Ann. Pow. Electr. Specialists Conf. PESC*, pp. 397–403, 1992.

[56] Holtz, J., "Self-commutated power converter with stair-cased output voltage for high power and high switching frequency," (in German, *Siemens Research and Development Reports*, Vol. 6, no. 3, pp. 164–171, 1977.

[57] Nabae, A., I. Takahashi, and H. Akagi, "A new neutral-point-clamped PWM inverter, *IEEE Trans. on Industry Appl.*, pp. 518-523, 1981.

[58] Wurm, H. P., "Increased utilization of high-power converters by the three-level approach," (in German), Ph.D. Thesis, Wuppertal University, 1983.

[59] Amler, G., "A PWM current-source inverter for high quality drives," *EPE Journal*, Vol. 1, no. 1, pp. 21-32, July 1991.

R. D. Lorenz
T. A. Lipo
D. W. Novotny

Chapter 5

Motion Control with Induction Motors

5.1. INTRODUCTION

An important factor in the worldwide industrial progress during the past several decades has been the increasing sophistication of factory automation. Manufacturing lines in an industrial plant typically involve one or more variable speed motor drives that serve to power conveyor belts, robot arms, overhead cranes, steel process lines, paper mills, and plastic and fiber processing lines to name a few. Prior to the 1950s all such applications required the use of a DC motor drive since AC motors were not capable of true adjustable or smoothly varying speed since they inherently operated synchronously or nearly synchronously with the frequency of electrical input. The inherent disadvantages of DC drives, however, have prompted continual attempts to find better solutions to the problem. To a large extent, applications that require only a gradual change in speed are now being replaced by what can be called *general-purpose* AC drives. In general, such AC drives often feature a cost advantage over their DC counterparts and, in addition, offer lower maintenance, smaller motor size, and improved reliability. However, the control flexibility available with these drives is very limited and their application is, in the main, restricted to fan, pump, and compressor types of applications where the speed need be regulated only roughly and where transient response and low-speed performance are not critical.

Drives used in machine tools, spindles, high-speed elevators, dynamometers, mine winders, rolling mills, glass float lines, and the like have much more sophisticated requirements and must afford the flexibility to allow for regulation of a number of variables, such as speed, position, acceleration, and torque. Such high-performance applications typically require a high-speed holding accuracy better than 0.5%, a wide speed range of at least 20:1, and fast transient response, typically better than 50 rad/s, for the speed loop. Until recently, such drives have almost exclusively been the domain of DC motors combined with various configurations of AC-to-DC converters depending upon the application. With suitable control, however, induction motor drives are more than a match for DC drives in a high-performance application. It will be shown in this chapter that control of the induction machine is considerably more complicated than its DC motor counterpart. However, with continual advancement of microelectronics, these control complexities are rapidly being overcome. The gradual replacement of DC drives has already begun, for example, in machine tool drives, and induction motors drives can be expected to continue to overtake them over the next decade. It is still too early to determine if DC drives will eventually be relegated to the history book as have nearly all other commutator machines such as rotary converters, amplidynes, and the like. However, the next decade will surely witness a marked increase in the use of induction motor drives.

5.2. INVERTERS FOR ADJUSTABLE SPEED

The process of converting DC to AC power is called *inversion*, and it is the *inverter* that creates the variable frequency from the DC source which is used to drive an induction motor at a variable speed [1]. In general, two basic types of inverters exist that are totally different in their behavior. The so-called voltage source inverter or VSI is more common and this type of inverter creates a relatively well-defined switched voltage waveform at the terminals of the motor. A stiff DC bus voltage is maintained by the use of a large capacitor in the DC link. The resulting motor current is then governed primarily by the motor load and the speed. The VSI is typically subdivided into two forms, the so-called *six-step inverter* and the *pulse width modulated inverter*.

The second type of inverter, the current source inverter (CSI), provides a switched current waveform at the motor terminals. A stiff DC bus current is maintained by use of a large inductor in the DC link. The voltage waveform is now governed primarily by the motor load and speed.

5.2.1. Basic Six-Step Voltage Source Inverter

A simplified diagram of a basic three-phase voltage source inverter (VSI) bridge is shown in Figure 5-1. The rectifier serves to establish a DC potential in much the same manner as in a DC motor drive. In addition, a relatively large electrolytic capacitor is inserted to "stiffen" the link voltage and provide a path for the rapidly changing currents drawn by the inverter. It is the rms value of this ripple current that

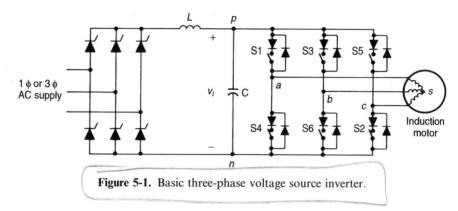

Figure 5-1. Basic three-phase voltage source inverter.

determines the value of the capacitor used. The capacitor is of appreciable size and cost, frequently 2000 to 20,000 microfarads, and is a major cost item in the system. It is also usual to interpose some inductive reactance between the rectifier and the AC supply to limit fault current and to reduce the severity of the commutation dips produced by the rectifier. This series impedance is also helpful in attenuating the voltage spikes that may enter through the rectifier bridge due to switching or lightning strikes out in the utility system.

The inverter acts somewhat as the equivalent of the commutator assembly in a DC motor and converts the DC voltage to a variable frequency AC voltage. The inverter bridge is similar to the rectifier used in a DC motor drive except, because of the lagging power factor presented by the induction motor, the thyristors must be replaced with devices which are capable of being turned "off" as well as "on." The choice at present is

- Thyristors plus external commutation network
- Bipolar junction transistors (BJTs)
- MOS field effect transistors (MOSFETs)
- Insulated Gate Bipolar Transistors (IGBTs)
- Gate turn-off thyristors (GTOs)
- MOS controlled thyristors (MCTs)

Since thyristors are not inherently capable of turn-off, they require external commutation circuitry not shown in Figure 5-1. The types of commutation circuits are numerous but typically require one additional "commutation" thyristor and one capacitor (or pair of capacitors) per phase. Since the capacitor needs to have its polarity reversed in preparation for a commutation, a technique known as "resonant reversal" is used that requires at least one and often two resonant reversal inductors per phase. In addition, extra "snubbing" circuits are placed across and in series with each thyristor to keep them within their specified tolerances with respect to their turn-on and turn-off capabilities. The result is a complex circuit having numerous components and requiring a precisely timed firing sequence to prevent accidental short circuits. Thyristors have one important advantage over transistors, which is their ability to withstand a substantial fault current for a brief time before protective devices (fuses or circuit breakers) operate. However, when used as an externally commutated switch, they are presently of more interest for historical than for practical reasons.

Transistors have nearly completely replaced thyristors in inverter circuits below 100 kW. They are available in ratings to 1200 V and have the great advantage of being able to be turned off as well as to turn on the current in the device. Hence, auxiliary components to accomplish turn-off are not needed. Snubbing is still necessary if the transistors are to be used near their maximum ratings, which is generally an economic necessity. Of the transistor family, the bipolar junction transistor or BJT is the lowest cost and thus most widely used. The circuitry required to drive the transistor base is, however, relatively complex and is usually fabricated in a Darlington configuration to reduce the demand on the base current driver. A small reverse voltage must typically be applied to the base of a transistor in the off-state, and at turn-off it is usually necessary to extract current from the base. IGBTs and MOSFETs are more recent additions to the transistor family, and IGBTs have effectively replaced BJTs in lower power applications. They do not suffer from second breakdown, a destructive loss mechanism in bipolar transistors that must be carefully controlled. Since both devices are turned off by field effect techniques, the current demand on the gate drive is minimal, and they can even be turned off with an integrated circuit chip. Overall losses, parts count, and driver cost are markedly reduced with these devices resulting in an increasingly competitive product even though the devices remain more expensive than a BJT.

Gate turn-off thyristors and MOS controlled thyristors (a relative newcomer to the scene) have the benefits of both thyristors and transistors. GTOs are both turned on and turned off by applying short gate current pulses. However, they require complex gate circuitry similar to BJTs. GTOs are presently available to more than 3000 volts and are becoming widely used in inverters operating at the utility medium voltage level (> 1000 V). Such voltages are a necessary operating condition when the rating of the motor exceeds a few hundred horsepower.

In addition to the "turn-off" element denoted as an ideal switch in Figure 5-1, each arm of the bridge normally contains an inverse parallel connected diode. These diodes, which are called the *return current* or *feedback* diodes, must be provided to allow for an alternate path for the inductive motor current that continues to flow when the main power device is turned off. When regeneration occurs, the roles of main power device and diode reverse. The diodes now return the regenerated power to the DC link while the thyristors carry the reactive current. The return of power to the DC link will raise the link voltage above its normal value, and steps must be taken to absorb this regenerated power to prevent a dangerous link voltage build-up. Typically, a resistor is switched in parallel with the DC link capacitor to absorb this energy or the input bridge is made bidirectional by adding a second inverse parallel bridge.

The basic operation of the six-step voltage inverter can be understood by considering the inverter to effectively consist of the six ideal switches shown in Figure 5-1. While it is possible to energize the motor by having only two switches closed in sequence at one time, it is now accepted that it is preferable to have three

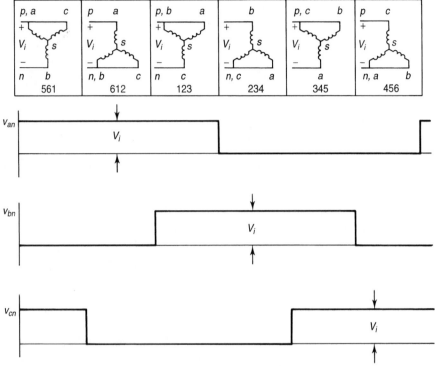

Figure 5-2. Illustrating the six connections of the induction motor when switched by a six-step voltage source inverter.

switches closed at any instant since this strategy produces a higher output voltage under any operating condition [3]. This second pattern produces the voltage waveforms of Figure 5-2 at the terminals a, b, and c referred to the negative DC potential n. The numbers written on the level parts of the waveforms indicate which switches are closed. This sequence of switching is in the order 561, 612, 123, 234, 345, 456, and back to 561. The line-to-line voltages V_{ab}, V_{bc}, V_{ca} and line-to-neutral voltages V_{as}, V_{bs}, V_{cs} then have the waveform shown in Figure 5-3. The line-to-line voltage contains an rms fundamental component of

$$V_{11(\text{rms fund})} = \frac{\sqrt{6}V_I}{\pi}$$
(5.1)

Thus a standard 460 V, 60 Hz induction motor would require 590 volts at the DC terminals of the motor to operate the motor at its rated voltage and speed. For this reason, a 600 volt DC bus is quite standard in the United States for inverter drives.

Although the induction motor functions as an active rather than a passive load, the effective impedances of each phase remain "balanced." That is, so far as voltage drops are concerned, the machine may be represented by three equivalent impedances as shown in Figure 5-2 for the six possible connections. Note that a specific

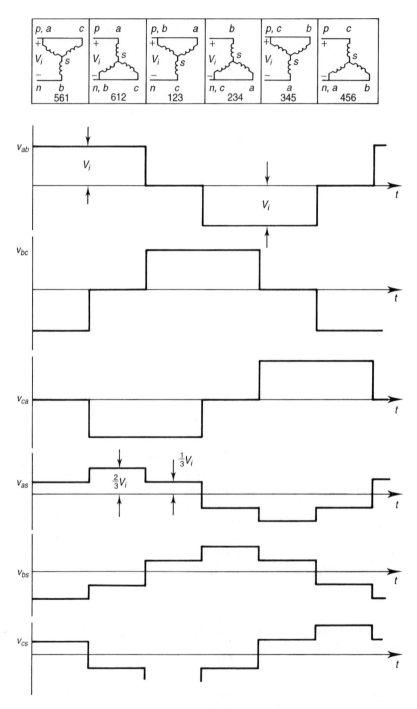

Figure 5-3. Line-to-line and line-to-neutral voltages across the induction motor phases when operating from a six-step voltage source inverter.

phase is alternately switched from positive pole to negative pole and that it is alternately in series with the remaining two phases connected in parallel or it is in parallel with one of the two phases and in series with the other. Hence, the voltage drop across the phase is always 1/3 or 2/3 of the DC bus voltage with the polarity of the voltage drop across the phase being determined by whether it is connected to the positive or negative pole. A plot of one of the three motor line-to-neutral voltages is also given in Figure 5-3. Note that harmonics of order three and multiples of three are absent from both the line-to-line and the line-to-neutral voltages (and consequently absent from the currents). The presence of six "steps" in the line-to-neutral voltage waveform is the reason for this type of inverter being called a *six-step* inverter. A Fourier analysis of these waveforms indicates a "square wave" type of geometric progression of the harmonics. That is, the line-to-line and line-to-neutral waveforms contain 1/5th of the 5th harmonic, 1/7th of the 7th harmonic, 1/11th of the 11th harmonic, and so forth.

5.2.2. The Pulse Width Modulated VSI Inverter

The PWM inverter maintains a nearly constant DC link voltage but combines both voltage control and frequency control within the inverter itself. It uses an uncontrolled diode rectifier or a battery (or perhaps both) to provide the nearly constant DC link voltage. In this case the power switches in the inverter are switched at a high-frequency thus operating, in effect, as choppers. In general, modulation techniques fall into two classes: those that operate at a fixed switching ratio to the fundamental switching frequency (block or "picket fence" modulation) and those in which the switching ratio is continuously changing, usually sinusoidally, to synthesize a more nearly sinusoidal motor current (called sinusoidal pulse width modulation or sinusoidal PWM).

Block Modulation. Block modulation is the simplest type of modulation and is closest to simple six-step operation. Instead of varying the amplitude of the motor voltage waveform by variation of the DC link voltage, it is varied by switching one or two of the inverter thyristors at a fixed switching ratio and adjusting the notches of the resulting pulses to control the motor voltage to suit the speed. The number of pulses is maintained constant over a predetermined speed range, although the number of pulses may change discretely at several prescribed speeds.

A simple form of block modulation is shown in Figure 5-4, where the chopping is limited to the middle 60° of each device conduction period, resulting in minimum switching duty on the semiconductor switches. In spite of the similarities between block modulation and the six-step mode, the torque pulsations at low speed are much less severe than for the six-step inverter. The harmonics typical of a six-step inverter are present also with block modulation, but there are also higher harmonics associated with the chopping frequency. The motor losses and motor noise, however, are significant compared to more elegant modulation algorithms, and this method is no longer widely used.

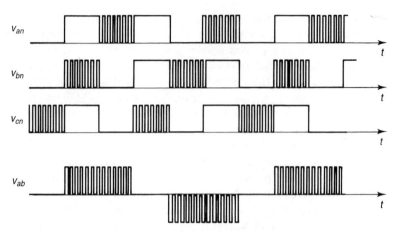

Figure 5-4. Typical block modulation waveforms showing voltages from line-to-negative DC bus and line-to-line voltage, V_{ab}.

✳ *Sinusoidal PWM Modulation.* The objective of sinusoidal PWM is to synthe-size the motor currents as near to a sinusoid as economically possible. The lower voltage harmonics can be greatly attenuated, leaving typically only two or four harmonics of substantial amplitude close to the chopping or carrier frequency. The motor now tends to rotate much more smoothly at low speed. Torque pulsations are virtually eliminated and the extra motor losses caused by the inverter are sub-stantially reduced. To counterbalance these advantages, the inverter control is com-plex, the chopping frequency is high (typically 500–2500 Hz for GTOs and up to 5,000 or more for BJT transistors), and inverter losses are higher than for the six-step mode of operation. To approximate a sine wave, a high-frequency triangular wave is compared with a fundamental frequency sine wave, as shown in Figure 5-5. When the low-frequency sine waves are used with 120° phase displacement, the switching pattern for the six inverter devices ensues.

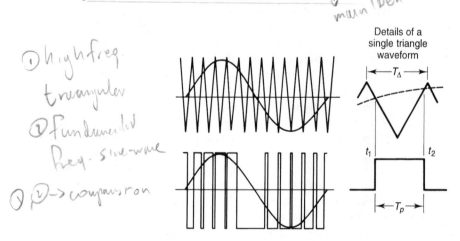

Figure 5-5. Illustration of a sinusoidal pulse width modulation technique.

Synchronous Versus Asynchronous Modulation. If the carrier and modulation are to be synchronized, there must be a fixed number of carrier cycles in each modulation cycle. If this number is chosen to give a "good" sinusoidal current waveform at low frequency, say, at 1 Hz, then the ratio of carrier frequency to fundamental frequency will be in the neighborhood of 100–200 kHz. If the same ratio is then used at a motor frequency of 100 Hz, the chopping frequency becomes 10–20 kHz. This switching frequency is far too high for normal thyristor inverters and somewhat too high even for bipolar transistor inverters. Hence, a technique of "gear changing" must be employed that reduces the switching ratio in steps as the fundamental frequency increases. A commercial system employing a custom chip designed for such a purpose is available [4]. Hysteresis must be included at each change in switching frequency ratio so that the system does not cycle continuously between two differing switching ratios at certain speeds as shown in Figure 5-6. Whereas the carrier frequency must be a fixed ratio of fundamental frequency at moderate and high speed to prevent undesirable "subharmonics" from appearing, asynchronous operation of the carrier frequency becomes acceptable at low speeds where the effect of the differing number of carrier cycles per modulation cycle is small. However, a change over to synchronous operation is required at some speed, and some type of "phase locking" technique must be introduced to achieve the synchronization.

While Figure 5-5 provides a good picture of the modulation process, once the possibility of using a microprocessor is introduced, the "analog" solution obtained by the intersection of a triangle wave and a sine wave becomes only one of many possibilities. This so-called "natural modulation" can be replaced to advantage by "regular modulation" in which the modulating waveform is piece wise constant, that is, sampled at the carrier frequency. An example of carrier and modulated waveforms for such a scheme is reproduced in Figure 5-7 for a switching ratio of 9 [5].

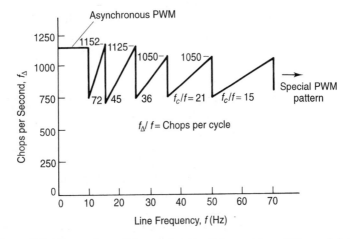

Figure 5-6. Illustration of "gear changing" in a pulse width modulation inverter.

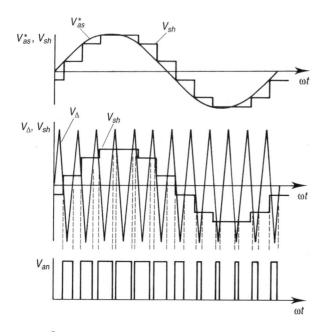

Figure 5-7. Illustration of the switching pattern using regular sampling modulation.

48Hz ➚

Pulse Dropping. Referring to Figure 5-7, it can be noted that the center pulses of the PWM modulation are narrower than the pulses where the fundamental voltage passes through zero. These center pulses typically become progressively more and more narrow as the speed (and hence voltage) increases. When the conduction pulses become very narrow, there is insufficient time to reverse bias the off-going power device in force-commutated systems. Hence, if the voltage across this device increases in a positive sense too soon, the device will again conduct and high losses will occur. Hence, if the modulation scheme calls for a pulse width below this minimum time, then this pulse must be omitted altogether and a canceling adjustment made to a nearby pulse to correct for the change in the fundamental component. Once again, pulse dropping must be coupled with a small hysteresis to avoid cycling across the threshold condition.

50Hz ➚

Changeover to Six-Step Mode. Without pulse dropping techniques, it is apparent that sinusoidal modulation prevents the motor terminal phase voltage from reaching its maximum value ($0.78\ V_I$) since "notches" of minimum width imposed by the switching limitations of the solid-state switching devices prevent full fundamental voltage from being reached. Operation at a reduced "maximum" fundamental voltage is costly since the attendant motor must be designed for a lower voltage, implying, in turn, higher currents for a given horsepower requirement. This increased current duty then impacts on the choice of switching device used in the inverter design and results in a more costly system. This disadvantage can be overcome if the modulation process is allowed to "saturate" at high speeds, so that the waveform takes on first several forms of block modulation and, finally, full six-step operation. During the intervening process when the transition is made from sinu-soidal PWM through the various block modes to square wave operation, additional harmonics will exist that cause extra heating in the motor. However, since the motor is running near its rated speed, cooling is good, and there is an overall benefit.

5.2.3. The Current Source Inverter Drive

Since induction motors have traditionally been designed to operate from a voltage source, the voltage source inverter was developed and used first since it is, in principle, an approximation of the waveform presented to the motor by the utility. The current source inverter, on the other hand, is very different in concept. Current-fed inverter drives have been in use slightly more than 20 years. They have, however, several properties which make them attractive as well as an inevitable number of undesirable effects. As the name implies, the inverter switches of a CSI are fed from a constant current source. While a true constant current source can never be a reality, it is reasonably approximated by a controlled rectifier or chopper with a current control loop as well as a large DC link inductor to smooth the current. The circuit is shown in Figure 5-8. In this case, the current is switched sequentially into one of the three motor phases by the top half of the inverter and returns from another of the phases to the DC link by the bottom half of the inverter. Since the current is con-stant, there will be zero voltage drop across the stator winding self-inductance and a constant voltage drop across the winding resistances. Hence, the motor terminal voltage is not set by the inverter but by the motor.

Since the motor is wound with sinusoidally distributed windings, the consequent voltages that appear on the terminals of the motor are nearly sinusoidal. Ideally, the current waveform is an exact replica of the voltage waveform of a six-step voltage stiff inverter. In practice, the motor currents cannot change instantaneously, and the transitions in the current waveform have a finite slope. During these transitions, the current transfers from one inverter thyristor to the next with the aid of one of six

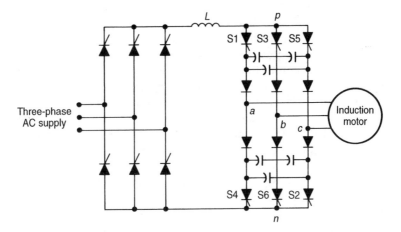

Figure 5-8. Schematic of a current source inverter drive.

commutating capacitors. A sketch of the motor voltage and currents is shown in Figure 5-9.

A very attractive feature of a CSI is its ability to ride through a commutation failure and to return naturally to normal operation. This feature is in marked contrast to most voltage-fed inverters in which costly preventive measures are necessary. Another benefit of the CSI is its ability to regenerate back into the utility supply by simply reversing the polarity of the rectifier DC output voltage. Again, this in contrast with the VSI where the current flow rather than DC voltage must be reversed, thereby requiring an additional inverse parallel connected six-pulse bridge. This again is a desirable advantage over a VSI. Operation of the motor at negative slip automatically causes a reversal in the DC link voltage since the link current rather than voltage is the controlled variable. Hence, power is automatically regenerated into the AC supply.

An important limitation in the application of a CSI drive is the fact that open loop operation in the manner of a VSI is not possible. Figure 5-10 shows the torque/speed curve of a typical induction motor fed from both a voltage source and a current source.

A marked "peaking" exists in the case of the current source inverter. Inspection of this torque/speed curve suggests two possible operating points: one on the negatively sloped region which is usually regarded as stable and one on the positively sloped region which occurs at a speed below the point where the maximum torque is reached (breakdown torque) and is generally unstable (depending, however, upon the load torque versus speed characteristics). Careful examination of the point on the stable side of the torque speed curve reveals that continuous operation is not feasible here because the working flux in the machine at this point is high, resulting in saturated operation and excessive magnetizing current and iron losses. On the other hand at the other point, the so-called unstable point, the flux in the machine is near its rated value and losses are not excessive. (Note that this point corresponds to the intersection of the torque/speed curve for rated voltage with the torque/speed curve for rated current. At this point, both voltage and current are thus at their rated values.) Unfortunately, being on the unstable side of the torque speed curve (for a current source) means that operation is impossible without some sort of feedback control that forces the machine to remain at this operating point. A system that is widely used is a motor voltage control loop that regulates the motor voltage by controlling the input phase control rectifier (or DC link chopper). A representative block diagram of a system of this type is shown in Figure 5-11.

Typically, an internal current loop is used as shown with the voltage error serving as a reference signal for the current regulator. Some IR drop compensation is often added, and since motor current and DC link current are proportional, this is easily accomplished. Additional compensating circuits are usually employed to improve system dynamics.

The size and cost of the AC commutating capacitors, the DC link inductor, and the reverse voltage device requirements are the major disadvantages of this inverter. The capacitors are large because they must absorb the total energy stored in the leakage field of the winding when the current is commutated. To keep this energy interchange at a minimum, motors of special design are advantageous. In particular,

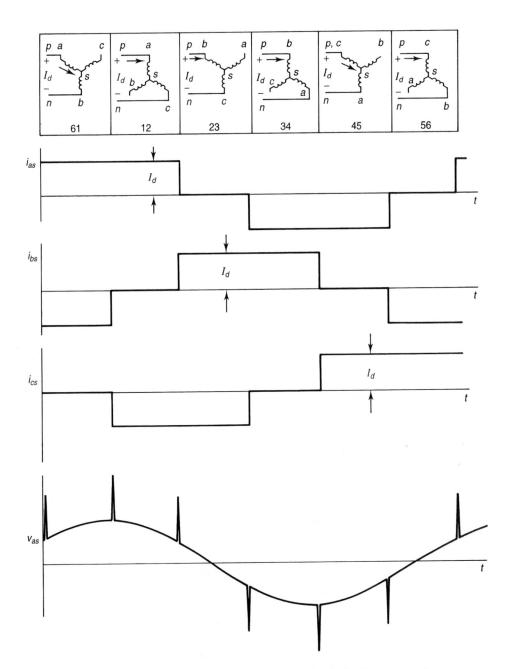

Figure 5-9. Illustration of the six connections of an induction motor when switched by a six-step current source inverter showing motor line currents and one phase voltage.

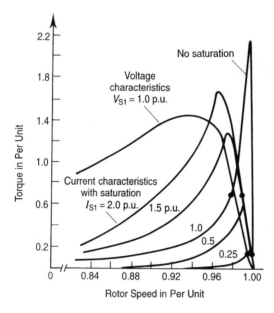

Figure 5-10. Induction motor torque-speed curves when operating from a VSI and from a CSI.

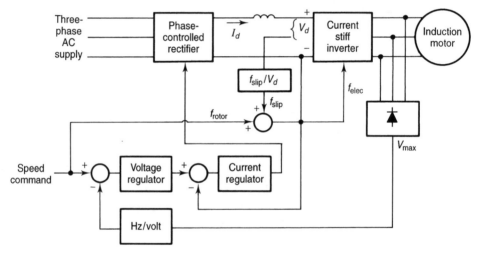

Figure 5-11. Block diagram of a CSI-induction motor drive with motor voltage loop.

these motors are designed to have as small a leakage reactance as possible. Unfortunately, this requirement is at odds with conventional motor design since small leakage reactances imply an overdesigned motor from the point of view of continuous rating. Finally, the use of thyristors rather than transistors or turn-off thyristors in this circuit can be noted. While the CSI could potentially employ switches with turn-off capability, they would not enhance the performance of the converter significantly since the rate of change of motor current and, thus, the voltage rise across the motor would remain fixed by the capacitors.

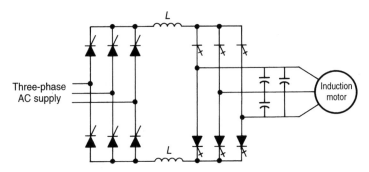

Figure 5.12. Current source inverter drive employing turn-off devices (GTO is shown as an illustration).

Turn-off devices can be used in a current source link configuration if the capacitors are freed from their commutating requirements and simply placed across the terminals of the machine. Figure 5-12 shows a motor drive configuration that is becoming increasingly popular in high-horsepower applications. In this case, the three top switches are sequentially switched, one at a time, to feed the link current into the motor/capacitor network and one of the bottom three switches returns the current to the link. The capacitors are now selected to absorb only the harmonic currents and need not be sized to absorb the total energy in the commutated phase during switching. The motor current is much more sinusoidal than the square wave current impressed on the motor in the conventional scheme. Pulse width modulation is again used to suppress the low frequency 5th and 7th harmonic torque pulsations, which are inherent in six-step operation. A major disadvantage of this scheme is the potential for resonance between the capacitors and the motor inductances. Care must be taken to avoid impressing current harmonics into the motor/capacitor network that will excite one of the system resonant frequencies. This possibility can be avoided by careful use of pulse width modulation. However, since the motor parameters must be known to implement such an approach, the drive is presently not popular for general-purpose applications.

5.3. MOTION CONTROL SYSTEMS

Motion control systems have traditionally been developed by engineers who focus on the mechanical system for which the motion must be controlled. This emphasis has, over the last 25 years, resulted in the dominant use of permanent magnet (PM), DC servomotors with high-bandwidth current regulators and has made high-bandwidth, electromagnetic torque control a de facto standard for motion control servo drives. It should be noted that this capability is largely a result of the high switching frequencies of modern power electronic converters and the small transient inductances of the PM machine which inherently allows the current to be slewed (in energy state) at very high rates. Because the induction motor typically has a very small transient inductance, a similar capability can exist with this class of machines.

However, serious inroads to the wide acceptance of the DC PM servomotor are only recently taking place.

The de facto torque control standard has fostered development of motion control algorithms that assume that torque control is nearly instantaneous and that no appreciable low-frequency torque distortion is produced, such that an electromagnetic torque proportional to the command ($T_{em} \approx T^*$) can be used to model the drive. This is a reasonably accurate assumption for many drives, but it is consistent only with an ideally tuned field-oriented induction motor (FO-IM) servo as will be demonstrated in the following section. Such requirements indeed are major issues for acceptance of induction motor motion control as an alternative to DC PM servos.

There are two basic measures of motion control system performance which will be used in this section: (1) command tracking and (2) disturbance rejection. These measures will be augmented by considerations of motion sensors and how they affect IM motion control.

5.3.1. Classical, Industry Standard, Digital Motion Control with FO-IM

Classical, industry standard, motion control with induction motors assumes an ideal field-oriented torque controller in a cascade control topology. The principle of field orientation is discussed in Section 5.4. This cascade topology assumes that the torque control loop is the innermost loop and receives its command from the velocity loop controller. The de facto industry standard for motion control is to use a PI velocity loop and a proportional position loop with a velocity command separately fed via what is generally described as a "velocity feedforward" path. Figure 5-13 shows the digital implementation of such a FO-IM motion controller.

This FO-IM motion controller generally uses an optical encoder as a position feedback device. This is because the commonly used field orientation algorithms presented in Section 5.4 also require rotor incremental position. Thus, the optical, incremental encoder and its inexpensive resettable digital counter interface is often preferred over resolver feedback. Since velocity feedback is required to close the velocity inner loop and velocity sensing is expensive, the industry standard approach is to estimate velocity, (deonoted $\hat{\omega}(z)$), by numerically differentiating position as shown in Figure 5-13.

Command Tracking Performance of Classical Motion Control. For such a classical motion control approach, the closed loop transfer function for position, assuming an ideal FO-IM torque controller is

$$\frac{\theta(z)}{\theta^*(z)} = \frac{\dfrac{\hat{\omega}(z)}{\omega^*(z)} + K_{pp} \dfrac{\hat{\omega}(z)}{\omega^*(z)} \left(\dfrac{T}{1 - z_{-1}}\right)}{1 + K_{pp} \dfrac{\hat{\omega}(z)}{\omega^*(z)} \left(\dfrac{T}{1 - z_{-1}}\right)} \tag{5.2}$$

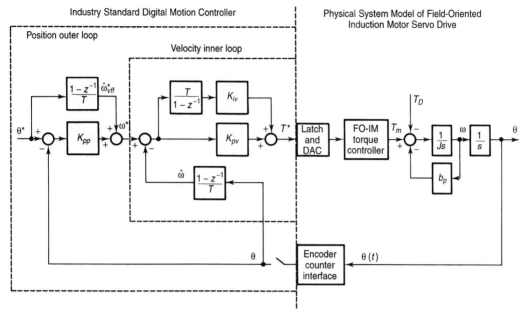

Figure 5-13. Classical industrial motion controller with field-oriented induction motor (FO-IM).

From this formulation of the transfer function, it can be seen that the system is error driven. Thus, the tracking accuracy of this system is limited to commands which have dynamic content which is well below the outermost loop bandwidth (typically from 0.5 to 10 Hz).

Disturbance Rejection Performance of Classical Motion Control. Utilizing this classical motion controller approach, it can be shown that the FO-IM motion controller has a dynamic stiffness property which is best illustrated by adoption of an equivalent state variable control topology.

5.3.2. State Variable, Digital Motion Control with FO-IM

The equivalent, state variable, digital motion controller with FO-IM is shown in Figure 5-14 [6]. This topology is mathematically identical to that of Figure 5-13 with the gain relationships:

$$b_a(\text{nm/rad/sec}) = K_{pv}$$
$$K_{sa}(\text{nm/rad}) = K_{pp}K_{pv} + K_{iv} \qquad (5.3)$$
$$K_{ia}(\text{nm/rad} \cdot \text{sec}) = K_{pp}K_{iv}$$

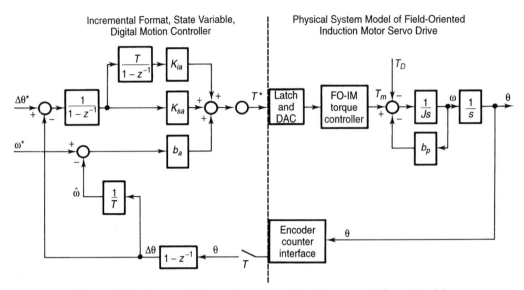

Figure 5-14. State variable motion controller (incremental format) with FO-IM.

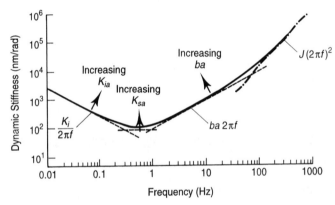

Figure 5-15. Dynamic stiffness of the state variable motion controller with FO-IM.

One direct benefit of this topology is physical units scaling, which relates directly to the disturbance rejection (dynamic stiffness) of the system as shown in Figure 5-15 [6]. From Figure 5-15, it can be seen that the disturbance rejection dynamics of the state variable motion control with FO-IM has predictable properties, including "infinite static stiffness" as provided by the integral-of-position-error state. This control topology provides a system with extremely high stiffness to disturbance inputs. It should be noted that the state variable system topology of Figure 5-14 is still error driven and thus will produce dynamic tracking errors. Figure 5-14 also shows the incremental format typically used so that the position command variables can be implemented with the smallest word size needed to allow for continuous operation at maximum speed.

5.3.3. Zero Tracking Error, State Variable, Digital Motion Control with FO-IM

State variable methods inherently facilitate the separation of command tracking issues from disturbance rejection issues. For example, to provide zero tracking errors, a state feedforward approach may be implemented with very little dependence on the state feedback controller as shown in Figure 5-16 [6]. It should be noted that such feedforward (computed torque) methods provide means for zero error tracking for nonlinear as well as linear systems. Such approaches also do not affect the stability nor the disturbance rejection dynamics provided by the state feedback controller. However, such computed torque feedforward methods cause a more intense focus on torque accuracy and torque dynamics. In effect, by enabling zero tracking error to be achieved, these approaches enable the motion control designer to verify if accurate, dynamically capable, low-distortion torque control has been achieved. Such performance requirements are indeed challenging for FO-IM drives as will be discussed in Section 5.3 concerning sensitivity to parameter errors.

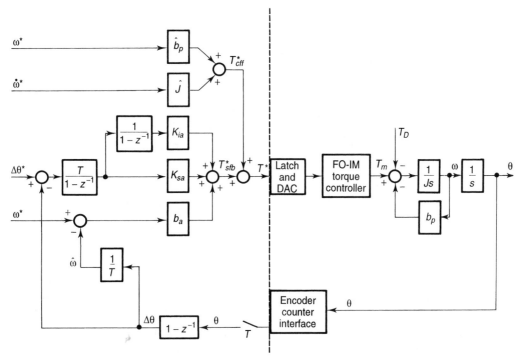

Figure 5-16. Zero tracking error state variable motion controller with FO-IM.

5.3.4. Feedback Sensor Issues for Motion Control with FO-IM

As was discussed earlier, the optical incremental encoder is generally selected for motion control with FO-IM systems. The normal interface for optical encoder feedback is to use one-bit A/D converters (Schmitt triggers) on the two channels of optical signals to form the industry standard "A quad B" outputs which produce 2 bits (four states) for each line space of the encoder. These pulse inputs are fed into an up/down counter with direction decoding to produce a digital word for the angular position. The position resolution for such systems is then

$$\theta_{res} = \frac{2\pi \text{ rad}}{4 \times \text{number of encoder lines}} \tag{5.4}$$

Typical line counts are from 512 to 4096, although both lower and higher resolutions are used for specialized applications. The typical incremental position resolution used for industrial servo drives is 1024 lines (alternatively 1000 lines).

Velocity Estimation via Numerical Differentiation (Incremental Encoder Pulse Counts). The difficulty in applying such feedback devices is related to their use in velocity estimation methods, especially when using numerical differentiation approaches as shown in Figures 5-13, 5-14, and 5-16. For these cases,

$$\hat{\omega}(k) = \frac{\theta(k) - \theta(k-1)}{T} \tag{5.5}$$

Such an estimator has several limitations that can prove to be important in motion control with FO-IM drives. The first limitation is the limited velocity resolution which can be obtained. That is,

$$\omega \text{ resolution} = \frac{\theta_{res}}{T} = \frac{\text{digital position resolution}}{\text{sample period}} \tag{5.6}$$

For example, one can calculate a typical case as follows

$$1024 \text{ lines} \Rightarrow 4096 \text{ ppr @ 1 kHz sample rate}$$

$$\frac{1}{4096} \text{ rev} \cdot \frac{1}{0.0001 \text{ sec}} \cdot \frac{60 \text{ sec}}{1 \text{ in}} = 14.6 \text{ RPM}$$

For a peak speed of $3 \times 1800 \text{ RPM} = 5400 \text{ RPM}$

$\Rightarrow$ velocity resolution 1 : 386, that is, ≈ 8 bit resolution

The result of such coarse velocity quantums (also referred to as quantization noise) is erroneous FO-IM torque commands via the wide bandwidth current regulators and an increase in rms currents and rms torque, resulting in motor heating as well as more pervasive voltage and current limiting problems with the FO-IM drive. One solution is to reduce the gain, b_a, on velocity. However, this approach reduces dynamic stiffness.

A second problem with this approach is the lagging nature of the velocity estimate. In fact, note that this estimator actually calculates the average velocity; that is,

$$\omega_{average} \left|_{\frac{k}{k-1}} \equiv \frac{\theta(k) - \theta(k-1)}{T}\right. \tag{5.7}$$

This results in a lagging characteristic (linear phase lag with frequency) which causes the digital velocity loop to have lower bandwidth than its analog counterpart. However, this difference is small if the sample rate is made much faster than the desired closed loop bandwidth.

Velocity Estimation via Clock Pulse Counting (Time) Between Encoder Pulses. For low-speed operation an alternative feedback velocity estimation solution is often implemented as shown in Figure 5-17 [6]. This approach uses the relatively high temporal resolution of the microprocessor clock (typically 25–100 nanosec) to measure the time period between the most recent pulses from the digital position transducer. If the basic digital position pulse resolution is Δq, and the computer clock period is T_{clk}, then the clock pulse feedback velocity, W_{cp}, calculation is given by

$$\omega_{cp}(k) = \frac{\Delta\theta}{[N(k) - N(k-1)] \cdot T_{clk}} = \frac{\Delta\theta}{\Delta N(k) T_{clk}} \tag{5.8}$$

or conversely the incremental clock pulse count for a given velocity, ω, is

$$\Delta N(k) = \text{integer} \left(\frac{\Delta\theta}{T_{clk}} \frac{1}{\omega} \right) \tag{5.9}$$

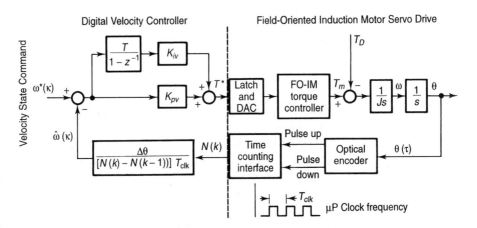

Figure 5-17. Digital velocity controller for FO-IM drive using clock pulse counting velocity feedback via optical encoder position pulses.

This clock pulse measurement approach substantially improves the velocity measurement resolution for low-speed operation. The incremental velocity resolution is given by

$$\omega_{\text{res-cp}} \equiv \frac{\Delta\theta}{\Delta N(k)T_{\text{clk}}} - \frac{\Delta\theta}{[\Delta N(k) + 1]T_{\text{clk}}} = \frac{\Delta\theta}{\Delta N(k)T_{\text{clk}}} \left[\frac{1}{\Delta N(k) + 1}\right]$$

$$\omega_{\text{res-cp}} \equiv \omega_{cp}(k)\left[\frac{1}{\Delta N(k) + 1}\right] \tag{5.10}$$

Note that the velocity resolution for this technique is not dependent on the computer control sample period, T, but is dependent on the current velocity. Thus, for low velocities, the resolution is vastly improved because of the large number of clock pulses which occur between pulses.

It is also important to note this velocity measurement technique is suitable for a relatively wide range of speed. The velocity above which one would have superior resolution from the position pulse method (5.6) than from the clock pulse method (5.10) is given by

$$\omega_{\text{res-pp}} \geq \omega_{\text{res-cp}}$$

which from (5.2), (5.6), and (5.7) may be determined as

$$\omega_{cp} \leq \omega_{\text{res-pp}} \left(\frac{1}{2} + \sqrt{\frac{1}{4} + \frac{T}{T_{\text{clk}}}}\right) \tag{5.11}$$

Because $T \gg T_{\text{clk}}$, the highest velocity for which clock pulse methods should be used (based on resolution) may be approximated by

$$\omega_{cp} \approx \omega_{\text{res-pp}} \sqrt{\frac{T}{T_{\text{clk}}}} \tag{5.12}$$

Example: For the prior example, and given a master clock frequency of 25 MHz, that is, $T_{\text{clk}} = 40$ nanosec, a sample period of 1 msec, and a position pulse velocity resolution of 14.6 RPM, the maximum velocity for which the clock pulse method would have superior resolution would be

$$\omega_{cp} \approx 14.6 \text{ RPM } \sqrt{\frac{1 \cdot 10^{-3}}{40 \cdot 10^{-9}}} = 14.6 \text{ RPM} \cdot 158.1 = 2308 \text{ RPM}$$

Thus for operation up to base speed, the clock pulse method has superior resolution as compared to the position pulse counting method.

The velocity measurement lag properties for the clock pulse method are inherently less severe than those of the position pulse method because the information being supplied always comes from the most recent position pulse (i.e. just prior to the real-time sample period interrupt). Thus, the delay is never any worse than that of the position pulse method.

The primary limitations of this clock pulse method are the inherently large word length needed for very low-speed counts and the accuracy problems (including jitter) caused by the lack of uniformity of the actual encoder line spacing. Since this approach measures the time between individual lines, the individual line spacing directly affects the accuracy of this method and any lack of uniformity of the spacing shows up as a speed jitter on the drive. This specification is quite problematic in current industrial encoders.

5.3.5. Observer-Based Feedback Issues for Motion Control with FO-IM

The problems with numerical differentiation to obtain velocity feedback can be greatly diminished by applying observer-based velocity estimation approaches. Such approaches can be implemented in hardware or software [9]. Figure 5-18 shows a software observer that achieves the objectives of avoiding numerical differentiation as well as eliminating estimator lag [7]. Such closed loop observers are not very sensitive to the mechanical parameters for inputs within the bandwidth, f_b, of the observer as shown in Figure 5-19. However, they are clearly sensitive to torque accuracy and torque dynamics. Torque errors will lead to velocity estimation errors since the observer model is being fed the torque command whereas the real motor experiences the actual torque control limitations of the FO-IM algorithms.

The use of these estimators for motion feedback with FO-IM is shown in Figure 5-20 [6]. The motion controller with FO-IM of Figure 5-20 displays greatly diminished problems due to quantization error and estimation lag. In addition, it retains the zero tracking error properties provided by computed torque command feedforward. It is assumed, however, that the observer bandwidth is sufficiently high that the dynamic stiffness will not be degraded. It should be noted that the observer methods described here can easily be extended to allow further improvement in dynamic stiffness (disturbance rejection) by adding acceleration feedback.

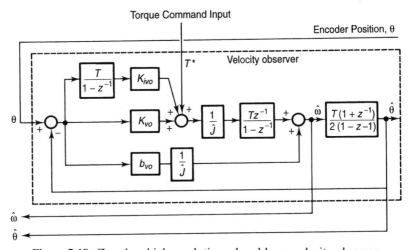

Figure 5-18. Zero lag, high-resolution, closed loop, velocity observer.

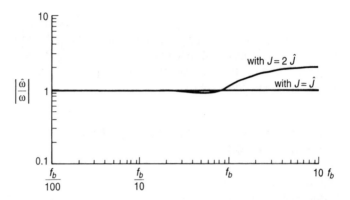

Figure 5-19. Observer parameter error sensitivity versus bandwidth, f_b.

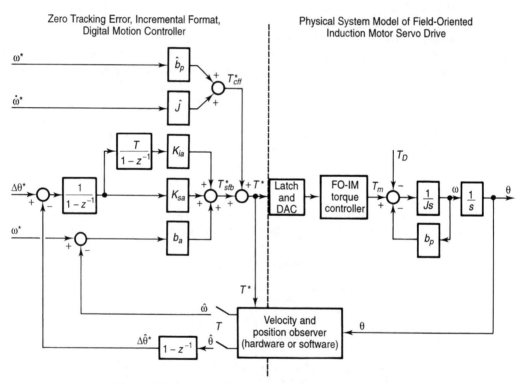

Figure 5-20. Observer-based motion control with FO-IM.

5.3.6. State Variable, FO-IM, Digital Motion Control with Acceleration Feedback

Dynamic stiffness can be greatly improved by adding acceleration sensing and acceleration feedback as shown in Figure 5-21 [8, 9]. The units of the gain, J_a, on the acceleration error are those of inertia, that is, nm per rad/sec^2. Thus, this feedback

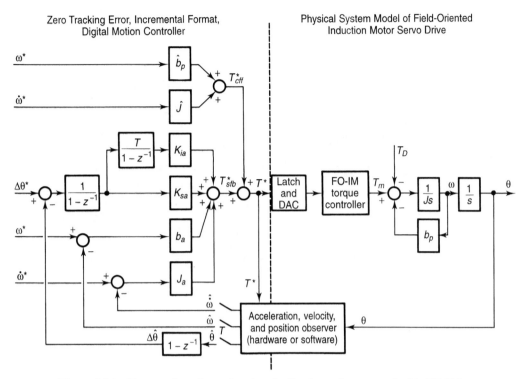

Figure 5-21. Observer-based, acceleration feedback motion control with FO-IM.

acts to add "active inertia" to the system, causing the disturbances to perceive a more "massive" system that responds less to disturbances. Furthermore, the use of acceleration feedback and active inertia, J_a, allows the entire dynamic stiffness characteristic to improve, assuming all other bandwidths are kept constant.

Figure 5-22 shows the improvement in dynamic stiffness that can be achieved with this approach. This approach is limited by the difficulty in obtaining suitable acceleration feedback signals. To obtain such signals, the velocity observer of Figure 5-18 can be extended to construct a cascaded topology, which produces corrected acceleration estimates as shown in Figure 5-23 [6, 8, 9].

5.3.7. Summary of Motion Control Requirements for FO-IM

From this discussion, it is clear that torque control dynamics and torque control accuracy are essential ingredients for motion control with FO-IM. Furthermore, because of the high torque bandwidth of FO-IM, the quality of sensor signals can have a dramatic impact on the rms currents, thermal heating, and accuracy of motion control with FO-IM. The next section will examine how such accurate, high-bandwidth torque dynamics are produced using field orientation.

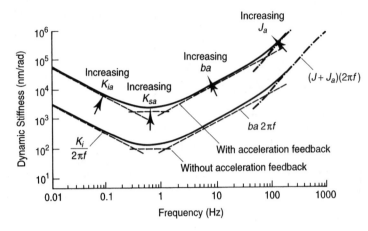

Figure 5-22. Dynamic stiffness for acceleration feedback motion control with FO-IM.

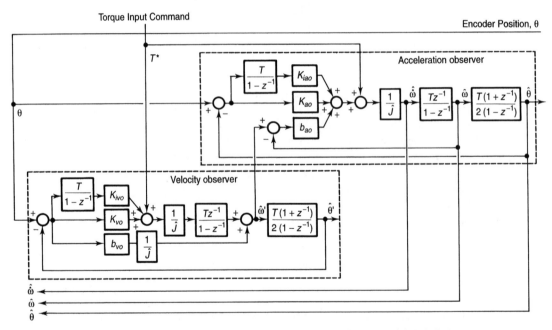

Figure 5-23. Zero lag, closed loop acceleration observer with FO-IM.

Furthermore, the accuracy and dynamic distortion which can be produced by FO-IM are presented as limiting factors in successful application of motion control with induction motors.

5.4. FIELD ORIENTATION (FO) CONTROL PRINCIPLES FOR INDUCTION MOTORS

The motion control system requirements outlined in the preceding section are typically fulfilled by employing torque control concepts in the induction machine which are patterned after DC machine torque control. The action of the commutator of a DC machine in holding a fixed, orthogonal spatial angle between the field flux and the armature MMF is emulated in induction machines by orienting the stator current with respect to the rotor flux so as to attain independently controlled flux and torque. Such controllers are called *field-oriented controllers* and require independent control of both magnitude and phase of the AC quantities and are, therefore, also referred to as vector controllers. "Field orientation" and "vector control" are used virtually interchangeably.

A basic understanding of the decoupled flux and torque control resulting from field orientation can be attained from the d-q–axis model of an induction machine with the reference axes rotating at synchronous speed ω_e [10, 11]

$$v_{qds}^e = r_s\, i_{qds}^e + p\lambda_{qds}^e + j\omega_e\, \lambda_{qds}^e \tag{5.13}$$

$$0 = r_r i_{qdr}^e + p\lambda_{qdr}^e + j(\omega_e - \omega_r)\lambda_{qdr}^e \tag{5.14}$$

$$T_m = \frac{3}{2}\frac{P}{2}\frac{L_m}{L_r}\,(\lambda_{dr}^e\, i_{qs}^e - \lambda_{qr}^e\, i_{ds}^e) \tag{5.15}$$

where

$$v_{qds}^e = v_{qs}^e - jv_{ds}^e, \quad i_{qds}^e = i_{qs}^e - ji_{ds}^e, \quad i_{qdr}^e = i_{qr}^e - ji_{dr}^e \tag{5.16}$$

$$\lambda_{qds}^e = \lambda_{qs}^e - j\lambda_{ds}^e = L_{ls}\,i_{qds}^e + L_m(i_{qds}^e + i_{qdr}^e) \tag{5.17}$$

$$\lambda_{qdr}^e = \lambda_{qr}^e - j\lambda_{dr}^e = L_{lr}\,i_{qdr}^e + L_m(i_{qds}^e + i_{qdr}^e) \tag{5.18}$$

The field orientation concept implies that the current components supplied to the machine should be oriented in phase (flux component) and in quadrature (torque component) to the rotor flux vector λ_{qdr}^e. This can be accomplished by choosing ω_e to be the instantaneous speed of λ_{qdr}^e and locking the phase of the reference system such that the rotor flux is entirely in the d-axis (flux axis), resulting in the mathematical constraint

$$\lambda_{qr}^e = 0 \tag{5.19}$$

Assuming the machine is supplied from a current regulated source so the stator equations can be omitted, the d-q equations in a rotor flux-oriented (field-oriented) frame become

$$0 = r_r i_{qr}^e - (\omega_e - \omega_r)\lambda_{dr}^e \tag{5.20}$$

$$0 = r_r i_{dr}^e - p\lambda_{dr}^e \tag{5.21}$$

$$\lambda_{qr}^e = L_m i_{qs}^e + L_r i_{qr}^e = 0 \tag{5.22}$$

$$T_m = \frac{3}{2} \frac{P}{2} \frac{L_m}{L_r} \lambda_{dr}^e i_{qs}^e \tag{5.23}$$

The torque equation (5.23) clearly shows the desired torque control property of providing a torque proportional to the torque command current i_{qs}^e. A direct (ampere-turn) equilibrium relation between the torque command current i_{qs}^e and the rotor current i_{qr}^e follows immediately from (5.22)[1]

$$i_{qr}^e = \frac{L_m}{L_r} i_{qs}^e \tag{5.24}$$

and combining (5.20) and (5.24) yields what is commonly called the slip relation

$$S\omega_e = -\frac{r_r}{L_r} \frac{L_m i_{qs}^e}{\lambda_{dr}^e} \tag{5.25}$$

Equation (5.21) shows that in the steady-state when $p\lambda_{dr}^e$ is zero, the rotor current component i_{dr}^e is zero. However, during flux changes, i_{dr}^e is not zero but is given by

$$i_{dr}^e = \frac{\lambda_{dr}^e - L_m i_{ds}^e}{L_r} \tag{5.26}$$

Combining (5.26) and (5.21) to eliminate i_{dr}^e yields the equation relating i_{ds}^e and λ_{dr}^e (flux and flux command)

$$(r_r + L_r p) \lambda_{dr}^e = r_r L_m i_{ds}^e \tag{5.27}$$

The close parallel to the DC machine is clear. Equation (5.23) emphasizes this correspondence in terms of torque production. The relation between the flux command current i_{ds}^e and the rotor flux λ_{dr}^e is a first-order linear transfer function with a time constant τ_r where

$$\tau_r = \frac{L_r}{r_r} \tag{5.28}$$

This corresponds to the field circuit of a DC machine, where the time constant τ_r is that associated with the field winding time constant resulting from damping currents in the field pole magnetic structure. The slip relation expresses the slip frequency which is inherently associated with the division of the input stator current into the desired flux and torque components.

Field orientation to fluxes other than the rotor flux is also possible [12] with the stator and air gap fluxes being the most important alternatives. Only the rotor flux yields complete decoupling, however, for some purposes (wide range field weakening operation for example) the advantages of choosing stator flux orientation can outweigh the lack of complete decoupling [13].

---

[1]The ampere turn balance expressed in equation 5.24 implies there is no "armature reaction" in a field-oriented controlled induction machine. The cross-magnetizing component i_{qs}^e that produces the torque is cancelled by i_{qr}^e, and thus there is no effect on rotor flux even under saturated conditions.

5.4.1. Direct Field Orientation

In direct field orientation the position of the flux to which orientation is desired is directly measured using sense coils or estimated from terminal measurements. Since it is not possible to directly sense rotor flux, a rotor flux-oriented system must employ some computation to obtain the desired information from a directly sensed signal. Figure 5-24 illustrates the nature of these computations for terminal voltage and current sensing.

In cases where flux amplitude information is available, a flux regulator can be employed to improve the flux response. A variety of flux observers can be employed to obtain improved response and less sensitivity to machine parameters. Some of these are discussed in a later section. A major problem with most direct orientation schemes is their inherent problems at very low speeds where the machine IR drops are dominant and/or the required integration of signals becomes problematic.

5.4.2. Indirect (Feedforward) Field Orientation

An alternative to direct sensing of flux position is to employ the slip relation to estimate the flux position relative to the rotor. Figure 5-25 illustrates this concept and shows how the rotor flux position can be obtained by adding the slip frequency position calculated from the flux and torque commands to the sensed rotor position. In the steady-state this corresponds to setting the slip to the specific value which correctly divides the input stator current into the desired magnetizing (flux producing) and secondary (torque producing) currents. Indirect field orientation does not have inherent low-speed problems and is thus preferred in most systems that must operate near zero speed.

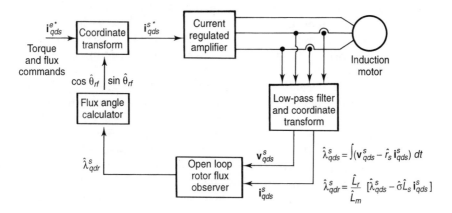

Figure 5-24. Field angle determination from terminal voltage and current.

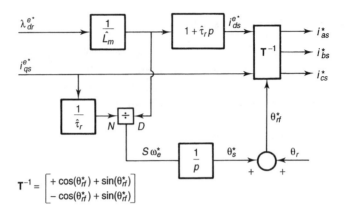

Figure 5-25. Indirect field orientation controller using flux and torque current commands (compensated flux response).

5.4.3. Influence of Parameter Errors

Both basic types of field orientation have some sensitivity to machine parameters and provide nonideal torque control characteristics when control parameters differ from the actual machine parameters. In general both steady-state torque control and dynamic response differ from the ideal instantaneous torque control achieved by a correctly tuned controller.

In indirect control the major problem is the rotor open circuit time constant τ_r, which is sensitive to both temperature and flux level [14]. When this parameter is incorrect in the controller the calculated slip frequency is incorrect and the flux angle is no longer appropriate for field orientation. This results in an instantaneous error in both flux and torque which can be shown to excite a second-order transient characterized by eigenvalues having a real part equal to $-1\tau_r$ and an oscillation frequency equal to the (incorrect) commanded slip frequency. Since τ_r is an open circuit time constant and therefore rather large, these oscillations can be poorly damped. There is also a steady-state torque amplitude error since the steady-state slip is also incorrect. Steady-state slip errors also cause additional motor heating and reduced efficiency.

Direct field orientation systems are generally sensitive to stator resistance and total leakage inductance, but the various systems have individual detuning properties. Typically, parameter sensitivity is less than in indirect control, especially when a flux regulator is employed. In all cases, both direct and indirect, parameter sensitivity depends on the L/R ratio of the machine (which in steady-state determines the location of the peak torque and thus the shape of the torque versus slip frequency characteristic), with larger values giving greater sensitivity. Thus, large, high-efficiency machines tend to have high sensitivity to parameter errors, and field-weakened operation further enhances this sensitivity. Typically, parameter sensitivity in small machines is low enough that its problems are less serious.

5.4.4. Selection of Flux Level

A major advantage of induction machines over field-oriented PM synchronous machines is the ability to adjust the flux to meet operating requirements. Reducing the flux to extend the high-speed range in the constant power mode is widely known and is a clear advantage of the induction machine. Less appreciated is the ability to operate at substantially above the nominal flux at low speed to enhance the torque per ampere and thus better utilize the available power supply current [15]. This is feasible since at low speed the core losses are not important and thus operating in a more saturated magnetic state is acceptable and desirable if the torque per ampere is improved. It is also possible to utilize trapped rotor flux to produce large torque pulses at low speed by first utilizing all of the available current to build up flux and then switching to all torque producing current to obtain the torque pulse [16].

5.5. CURRENT REGULATORS FOR MOTION CONTROL WITH FO-IM

The power converter in a high-performance induction motor drive used in motion control essentially functions as a power amplifier, reproducing the low power level control signals generated in the field orientation controller at power levels appropriate for the driven machine. As noted in Section 5.2, high-performance drives utilize control strategies which develop command signals for the AC machine currents. The basic reason for the selection of current as the controlled variable is the same as for the DC machine; the stator dynamics (effects of stator resistance, stator inductance, and induced EMF) are eliminated. Thus, to the extent that the current regulator functions as an ideal current supply, the order of the system under control is reduced and the complexity of the controller can be significantly simplified.

Current regulators for AC drives are more complex than for DC drives because an AC current regulator must control both the amplitude and phase of the stator current. In addition, the steady-state currents are AC currents not DC currents, so that a straightforward application of conventional proportional integral control (PI control) as applied in DC drives cannot be expected to provide performance comparable to that of a DC drive current regulator. As in the DC drive, the AC drive current regulator forms the inner loop of the overall motion controller. As such, it must have the widest bandwidth in the system and must, of necessity, have zero or nearly zero steady-state error. Achieving these goals in AC current regulators has proven to be a challenging task, and it is only recently that satisfactory, fundamentally sound techniques have been developed.

Both CSI and PWM inverters can be operated in controlled current modes. The current source inverter is a "natural" current supply and can readily be adapted to controlled current operation. The PWM inverter requires more complexity in the current regulator but offers much higher bandwidth and elimination of current harmonics as compared to the CSI and is almost exclusively used for motion control applications. Figure 5-26 shows a PWM system.

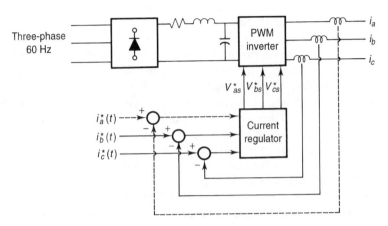

Figure 5-26. PWM system with current regulation acting as a three-phase current source.

Usually only two current sensors are employed, since in the absence of a neutral connection the three currents must add to zero. It is, in fact, advantageous to use only two current sensors since this avoids error signals containing a zero sequence component resulting from sensor errors. The nature of the controlled output current consists of a reproduction of the reference current with high-frequency PWM ripple superimposed.

From Figure 5-26, it should be noted that the role of the AC current regulator is to select the correct voltage vector such that motor current will follow the desired current commands. The criteria for comparison of current regulators can be based upon how well this basic objective is met, as well as the method's robustness to parameter variations and cost of implementation.

While the sensor requirement and the development of good regulators has required considerable developmental effort, the excellent response and low harmonic content of the PWM inverter make it the best currently available regulated current supply. The subject of current regulators remains a subject of intense activity. In general, current controllers can be classified into three groups: hysteresis and bang-bang, PI with ramp-comparison PWM, and predictive (optimal) voltage vector selection controllers.

5.5.1. Hysteresis and Bang-Bang Current Regulators

A low-cost current regulator using bang-bang control with hysteresis is shown in Figure 5-27. In this controller the desired current of a given phase, say, i_a^*, is summed with the negative of the measured current, i_a. The error is fed to a comparator having a prescribed hysteresis band $2\Delta I$. Switching of the leg of the inverter (T_{A+} off, T_{A-} on) occurs when the current attempts to exceed a value set corresponding to the desired current $i_a + \Delta I$. The reverse switching (T_{A+} on, T_{A-} off) occurs when the current attempts to become less than $i_a + \Delta I$. A "lock-out circuit" is normally incorporated to allow for inverter switch recovery time and thus avoid

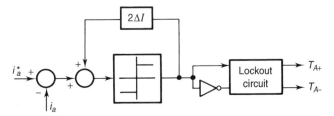

Figure 5-27. Hysteresis current controller for one phase.

short circuits across the DC link. Unfortunately, with this type of control, the switching frequency does not remain constant but varies along different portions of the desired current waveform due to the effect of the counter EMF.

While this system is very simple and provides good current amplitude control, hysteresis regulation has the major disadvantage of producing a highly variable PWM switching rate. Figure 5-28 shows a typical waveform illustrating the variable nature of the PWM switching frequency. Unfortunately, the variation of the switching rate is also opposite to the needs for good current control with the highest switching rates associated with the lowest reference frequencies. The hysteresis controller also has the somewhat unexpected property of limiting the current error to twice the hysteresis band rather than to the a value within the band itself.

The behavior of the hysteresis controller can be explained in terms of a complex plane switching diagram [17, 18], as shown in Figure 5-29. Figure 5-29a shows the reference current vector i_{qds}^{s*}, the actual current vector i_{qds}^{s}, and the current error vector Δi_s in the complex plane along with the a-axis of a three-phase reference system. The line current error Δi_a is the projection of Δi_s on this axis. The hysteresis controller will switch the a-phase inverter leg when Δi_a exceeds the hysteresis band as represented by the two switching lines drawn perpendicular to the a-axis. The switch-

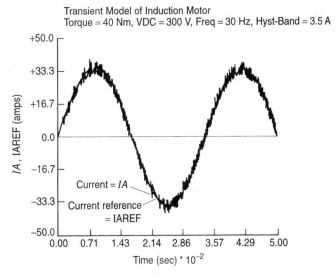

Figure 5-28. Current waveform for simple hysteresis regulator.

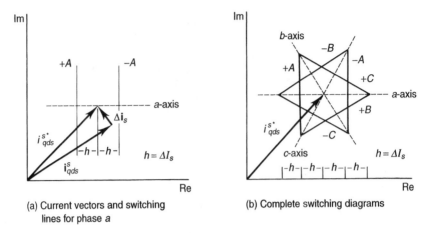

(a) Current vectors and switching   (b) Complete switching diagrams
    lines for phase a

Figure 5-29. Hysteresis controller switching diagrams using three independent controllers.

ing lines are located a distance h, equal to the hysteresis band, from the tip of the
current reference vector. Similar switching lines can be drawn for phases b and c; the
resulting complete switching diagram is shown in Figure 5-29b. The entire diagram
moves with the current reference vector with its center remaining fixed at the tip of
the vector.

The typical or expected behavior of the controller is to confine operation to the
interior, hexagonal region of the switching diagram. Thus, whenever the current
error touches one of the switching lines, that inverter leg is switched driving the
current error in that leg in the opposite direction. Note, however, that the current
error can be carried to one of the switching lines by motion of the current vector i^s_{qds},
or by motion of the current reference vector i^{s*}_{qds} *and* the attached switching diagram.
Thus, the situation occurs, where, for example, the $(-a)$ switching line is encountered with the inverter in the state $(a+, b-, c-)$. The system is then effectively short
circuited and can "coast" out to the tip of the switching diagram (in the region of the
tips of the "star") before a new switching of the inverter occurs. During this period
the motor is effectively "out of control" since the exact trajectory is determined only
by the motor parameters and the internal EMF at this condition and not by the
applied voltages. Hence, an error of $2\Delta I_s$, twice the expected value equal to the
hysteresis band, can occur on an almost random basis. This problem can be overcome by incorporating a second threshold that senses when the boundary is being
reached with the inverter in a state that would produce a short circuited condition
and avoiding this mode [20].

Figure 5-30 shows a current trajectory, indicated by the solid line, which can
occur and represents a high-frequency limit cycle. The initial voltage vector $\mathbf{v}_1$ forces
the tip of the current vector to travel in the same direction as the voltage vector since
the resistance and counter EMF are assumed to be small. The current vector hits the
$+c$ switching line causing inverter leg c to switch and produce the new voltage vector
$\mathbf{v}_2$ $(a+, b-, c+)$. Next the current vector will hit the $-a$ switching line producing the
voltage vector $\mathbf{v}_3$ $(a-, b-, c+)$. Continuing this reasoning, the six possible nonzero

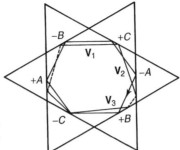

Figure 5-30. Hysteresis controller trajector illustrating possible high-frequency limit cycles.

voltage vectors are applied repeatedly and a high switching frequency results if the inductance is low and the hysteresis band small. Note that the magnitude of the current error vector is not reduced to zero during the limit cycle.

The problems associated with rapid limit cycle oscillations can be remedied by use of a bang-bang controller in which the switching of the inverter is clocked or controlled. Such an algorithm is commonly termed a delta modulator. Its simplest form as a part of a current regulator is illustrated in Figure 5-31. The delta modulator encodes continuous signals as pulse width modulated digital signals via a comparator output that is clocked at a frequency $f_s = 1/T$. The current regulated delta modulator (CRΔM) is very parameter insensitive, is very low cost, and has excellent transient performance. The major limitation of the CRΔM is that the current ripple and rms harmonic content can be large for normal values of the induction motor transient inductance unless a high sampling (switching) frequency is used, typically in excess of 20.

5.5.2. PI Current Control with Ramp Comparison, Constant Frequency PWM

At present, most systems employ current regulators in which the switching frequency is either nearly constant or at least bounded and known. Constant switching frequency PWM can be easily obtained by a so-called ramp comparison modulator. The ramp comparison modulator is a direct carry-over from DC machine current regulators employing the four-transistor bidirectional chopper or H-bridge. Figure 5-32 illustrates the basic concept.

A triangular waveform signal at the desired switching frequency is used to provide the "ramp" and the output of the PI controller forms the "ramp comparison". The comparator output directly determines the inverter switching. As shown in Figure 5-32, a PI controller is commonly used to provide a high DC gain, which

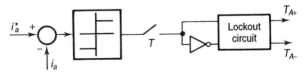

Figure 5-31. Current regulated delta modulator for one inverter phase; T is sampling period.

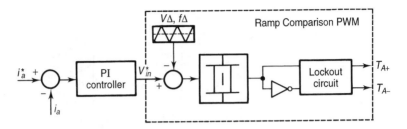

Figure 5-32. Basic PI current controller, ramp comparison PWM for one inverter phase.

eliminates steady-state errors and provides a controlled roll-off of the high-frequency response. This approach is also a direct carry-over from DC systems where the PI controller effectively decouples the steady-state effect of back EMF voltage (i.e. speed).

A more detailed description is as follows. The PI controller generates a voltage command, v_{in}^*, which is compared to the triangular "ramp" waveform. As long as the voltage command is greater than the triangle waveform, the inverter leg is held switched to the positive polarity. When the voltage command is less than the triangle waveform, the inverter leg is switched to the negative polarity. The inverter leg is forced to switch at the frequency of the triangle wave and produces an output voltage proportional to the voltage command signal from the PI controller (assuming the output voltage is well below the inverter DC bus voltage) [19]. If the switching frequency is high compared to the frequency content of the voltage command, the proportional relationship between the voltage command input, v_{in}^*, and the line-to-neutral fundamental component of the output voltage can be expressed by

$$v_{out} = \frac{1}{2} V_{dc} \frac{v_{in}^*}{V_\Delta} = K_\Delta v_{in}^* \tag{5.29}$$

where $K_\Delta = (1/2)(V_{dc}/V_\Delta)$, V_{dc} is the inverter DC bus voltage, and V_Δ is the zero-to-peak value of the triangle signal. The input voltage v_{in}^* is, effectively, the voltage at the input of the summer leading to the hysteresis block. The ratio of the input voltage command to the peak of the triangle wave voltage is typically termed the amplitude modulation index, m_i. Hence, in terms of the modulation index equation 5.29 can be rewritten as

$$v_{out} = \frac{1}{2} V_{dc} m_i, \qquad m_i \le 1 \tag{5.30}$$

so that the output voltage is linearly proportional to the command. When the instantaneous amplitude of the input voltage exceeds the triangle wave, intersections of the input voltage and triangle wave are eliminated, and consequently PWM pulses are dropped. In this case, it can be shown that the output voltage can be expressed by the describing function [19]

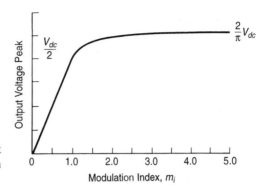

Figure 5-33. Fundamental component of phase voltage versus modulation index m_i.

$$v_{\text{out}} = m_i \frac{V_{dc}}{\pi} \left[\sin^{-1}\left(\frac{1}{m_i}\right) + \left(\frac{1}{m_i}\right) \sqrt{1 - \left(\frac{1}{m_i}\right)^2} \right], \qquad m_i > 1 \qquad (5.31)$$

A sketch of the variation of the fundamental component of the output voltage as a function of m_i is given in Figure 5-33. Note that, depending on the frequency, the amplitude of the input voltage command v_{in} must exceed a factor of roughly 5 times the amplitude of the triangle wave for full output voltage to be reached, that is, square wave voltage operation. In effect, the gain falls off rapidly as pulses are eliminated from the PWM pattern, indicating a loss of control of the stator current as the PWM inverter approaches square wave operation. The hysteresis element shown in Figure 5-32 is included to prevent multiple switching of the inverter leg if the time rate of change of the input voltage exceeds that of the ramp. It is usually omitted when the input voltage rate is controlled.

In the typical DC current controller, a single ramp comparison controller is employed with the output used to control the four transistors of the H-bridge in a complimentary fashion, that is, either $a+$, $b-$ (positive output) or $a-$, $b+$ (negative output). The current error signal, as noted previously, is usually conditioned using a simple PI controller and very satisfactory static and dynamic performance can be attained.

An attempt to extend this basic concept to three-phase current controllers by using three ramp comparison controllers, one for each phase, immediately poses a problem since the three PI controllers result in an attempt to regulate three independent states when only two exist (because the three-phase currents must add to zero). This immediate problem can be overcome by using only two PI controllers and slaving the third phase, an approach which is used in some cases. Alternatively, a zero-sequence current can be synthesized and fed back to the three regulators to decouple them. An intellectually pleasing as well as very effective approach is to view the current control problem in d-q coordinates, which immediately indicates the necessity of using only two regulators and algebraically establishes the three-phase inverter gating signals.

It is easily shown that this simple system also has inherent regulation problems compared to its DC machine counterpart by considering the steady-state response.

In the DC machine case, the steady-state response is characterized by zero current error because of the integration in the forward path. However, for the AC controller, the steady-state condition requires sinusoidal output at the reference frequency and clearly the forward path integration in the PI controller does not produce zero current error.

The inherent problems of regulating AC signals in the stationary frame were first recognized by Schauder and Caddy [21]. It was shown that regulators like that of Figure 5-32 can be implemented in other "reference frames" which exhibit quite different characteristics. That this is so is quite clear if one considers that the frequency of the current is different in different reference frames and hence the regulator performance (if it is frequency dependent) will also be different. In machine analysis it is convenient to view the stator current, voltage, and flux linkages as vectors as has already been discussed. Since the current is viewed as a rotating vector in this representation, it is logical to consider the possibility of regulating this variable from a set of orthogonal axes which rotate together with the current vector itself. In this case the axes are said to be "synchronously rotating." From a control perspective, a synchronous frame seems especially appropriate since the steady-state currents represented in these rotating axes are DC currents and a simple PI controller will result in zero steady-state error. A system diagram for a synchronous d-q frame regulator is shown in Figure 5-34.

The block involving the exponential $e^{-j\omega_e t}$ represents a coordinate transformation from a stationary to a rotating coordinate system given by

$$i_{qs}^e = i_{qs}^s \cos \omega_e t + i_{ds}^s \sin \omega_e t \qquad (5.32)$$
$$i_{ds}^e = i_{qs}^s \sin \omega_e t + i_{ds}^s \cos \omega_e t \qquad (5.33)$$

The inverse quantity $e^{-j\omega_e t}$ represents the transformation from "synchronous" to stationary variables; that is,

$$i_{qs}^s = i_{qs}^e \cos \omega_e t + i_{ds}^e \sin \omega_e t \qquad (5.34)$$
$$i_{ds}^s = i_{qs}^e \sin \omega_e t + i_{ds}^e \cos \omega_e t \qquad (5.35)$$

The superscripts in these expressions are used to distinguish between the stationary "s" and synchronous "e" reference frames. This regulator is clearly more complex and requires more hardware for implementation because of the requirement to transform the measured currents to a synchronous frame and subsequently to transform the error amplifier outputs back to a stationary frame to drive the ramp comparison controller. These transformations require explicit knowledge of the frequency ω_e, although precision is not required since the PI regulator can handle low-beat-frequency components.

A much simpler hardware implementation has been suggested [22] in which the synchronous regulator of Figure 5-34 is "transformed" to the stationary frame. It can be shown that in vector form, the stationary frame version of the equations of the synchronous regulator are

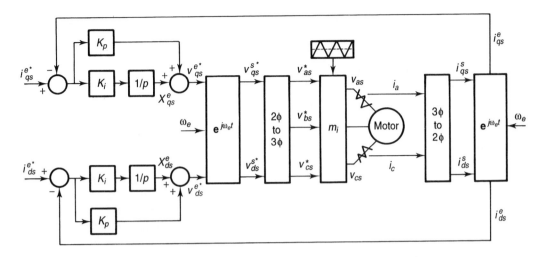

Figure 5-34. System diagram of synchronous d-q frame PI current regulator with constant frequency ramp comparison PWM.

$$\mathbf{v}_{qds}^{s^*} = \mathbf{K}_p \left(\mathbf{i}_{qds}^{s^*} - \mathbf{i}_{qds}^s \right) + \mathbf{x}_s^s \tag{5.36}$$

$$\mathbf{v}_{qds}^s = \frac{1}{p} \left[K_i (\mathbf{i}_{qds}^{s*} - \mathbf{i}_{qds}^s) + jwe\, \mathbf{x}_s^s \right] \tag{5.37}$$

where $1/p$ denotes integration with respect to time. The corresponding stationary frame equivalent of the synchronous regulator is shown in Figure 5-35. In this form, cross-coupling terms are introduced. However, the cross-coupling is very useful and significant since the local oscillator formed in the cross-coupling provides the

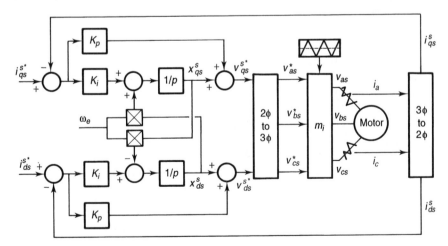

Figure 5-35. The stationary frame equivalent of a synchronous frame PI current regulator with constant frequency ramp comparison PWM.

required AC signal to allow the current error to go to zero in the steady-state. The performance of the regulator is identical to the synchronous frame regulator of Figure 5-34. Note, however, that only two multipliers are required for implementation beyond that needed for the ordinary stationary regulator. Reference 26 provides an in-depth study of the synchronous regulator and a detailed comparison with the stationary regulator.

5.5.3. Predictive (Optimal) Current Controllers

Predictive (optimal) current controllers are a relatively recent development. The control of stator current is seen as an optimization problem that can be described as choosing an optimal voltage vector to best move the current in the vector plane as shown in Figure 5-36. In this case, the future path of the current vector is predicted for all possible inverter output voltage vector states (inverter switch connections). Six active inverter output voltage vector states can be identified together with 2 null or zero voltage states in which the currents of the inverter "coast" in a similar manner as described previously. At each clock cycle the inverter switching state is chosen that best keeps the current vector within a prescribed circle on this plane, the radius of which denotes the tolerable current error band [23] while minimizing some cost function, such as rms current ripple. While the principle is very attractive, it is difficult to implement in practice due to (1) the need for calculating possible future current trajectories for all switching states at every clock cycle and (2) the necessity of knowing the instantaneous value of the motor EMF, a quantity which is not easily measured.

Means for minimizing the computing effort of predictive (optimal) controllers has been a productive area of research recently. One attractive approach is the method of Nabae and colleagues [24]. This method is based on the principle of predicting the direction of the derivative of the current error vector for all possible inverter states, and subsequently choosing the state that provides a small di/dt in the opposite direction to the current error vector for steady-state operating conditions. Alternatively, the state that opposes the current error vector most is chosen during large dynamic transients. Such prediction rules can be summarized by means of simple tables (a 6×6 table for steady-state and 6×1 table for the transient operat-

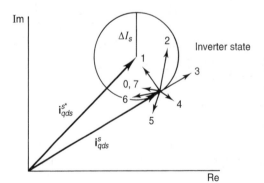

Figure 5-36. Possible trajectories of current vector for the eight inverter states, and $\mathbf{i}_{qds}^{s^*}$ (current command), $\mathbf{i}_{qds}^{s}$ (stator current vector location) when the error bound $\Delta \mathbf{i}_s$ is reached.

ing conditions) where the angle of the current error vector along with the angle of the equivalent voltage vector are the inputs and the optimal inverter state is the output. The equivalent voltage vector is defined as the sum of the unity voltage vector and the derivative of the reference current vector multiplied by the filter inductance, all of which can be readily measured. The switching action is executed whenever the current error vector exceeds defined boundaries in a similar manner to that illustrated in Figure 5-36.

Another algorithm which approximates the optimal voltage vector solution is the method by Habetler, and colleagues [25]. This method switches nearly constant-width output voltage pulses resulting from a high-frequency resonant link to produce a pulse density modulation of the inverter output voltage. This controller can be implemented in a very inexpensive set of hybrid hardware, which is shown in Figure 5-37. For this current regulator the decision of which voltage vector to select is simple digital logic. A voltage vector is preselected by a 3ϕ current-regulated delta modulator (CRΔM). Based on the present output vector,

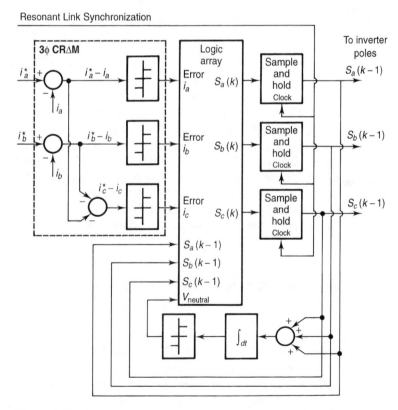

Figure 5-37. Approximate optimal current controller using adjacent state algorithm with high-frequency resonant link converter for discrete pulse modulation.

TABLE 5-1   ADJACENT STATE SELECTION LOGIC FOR THE NEXT ALLOWABLE OUTPUT
STATE

3ϕ CR$\varDelta$M State	Current State						
	0	1	2	3	4	5	6
1	1	1	1	0	0	0	0
2	2	2	2	2	0	0	0
3	3	0	3	3	3	0	0
4	4	0	0	4	4	4	0
5	5	0	0	0	5	5	5
6	6	6	0	0	0	6	6

$S_a(k-1)$, $S_b(k-1)$, $S_c(k-1)$, no direct jumps between nonzero vectors are allowed, except for a jump to an adjacent voltage vector (such as from vector 2 to vector 3 or 1 in Figure 5-36). If the vector is not adjacent, the inverter is forced to select the zero vector for one pulse width before selecting the new nonzero voltage vector (such as from vector 2 to zero to vector 5 in Figure 5-36). The logic for this algorithm is contained in Table 5-1. In addition to this simple switching logic, this "adjacent state" controller has a simple bang-bang loop on the neutral voltage. This allows it to maintain a zero average neutral voltage.

This controller is an approximation to the optimal controller in that the CR$\varDelta$M does not select "optimal" states. However, most of its decisions have been shown to be consistent with the optimal voltage vector selection [24]. Furthermore, the adjacent state algorithm does effectively reduce the rms ripple currents by never allowing the worst case switching to occur. This is a heuristic approximation to a cost function optimal controller, but again, experimental results have shown this to be a reasonable approximation. Given the low cost of this controller and its approximately optimal performance, it can be viewed as a viable alternative.

5.5.4. Summary of Current Regulators for Motion Control with FO-IM

Current regulators are essential for instantaneous torque control for FO-IM. The alternatives for such regulators include a number of viable solutions, for example, the current regulated delta modulator, the synchronous frame PI (especially its stationary frame implementation), and the nearly optimal, predictive, voltage vector selection controllers such as the di/dt controller and the adjacent state controller. The various solutions differ in implementation costs, in robustness with respect to parameter variations, and in their ability to track current commands with high fidelity and low distortion.

5.6. ADVANCED FLUX AND TORQUE REGULATION METHODS FOR MOTION CONTROL WITH FO-IM

Section 5.2 identified high-performance torque dynamics, accurate torque gains (calibration), and low amplitudes of distortion as primary requirements for motion control. Section 5.3 presented the principles of field orientation, which enable such properties to be achieved with induction servo motors, but it also pointed out the inherent parameter sensitivity that affects indirect as well as direct field-oriented control when it is based on open loop flux observers. This section will present some advanced methods which substantially ameliorate the problems associated with parameter sensitivity in FO-IMs. These alternatives are based on the decreased parameter sensitivity and stability of certain topologies of closed loop flux observers [26–40]. Their use and their natural extension to closed loop flux and closed loop torque control for motion control with induction motors will be presented.

As discussed in Section 5.2, position sensing will generally be required for motion control. Thus for motion control with induction motors, it will be assumed that angular position is a measured variable and is available for flux estimation purposes.

5.6.1. Flux Accuracy Issues

As described in Section 5.3, direct field orientation uses the spatial location of the field to control current such that torque and flux are independently manipulated. This approach will produce nonideal, oscillatory torque dynamics depending on the accuracy of the flux angle measurement (or estimation). It will furthermore produce incorrect torque levels in direct relation to the accuracy of the flux amplitude. Thus, for motion control, the accuracy of the flux magnitude and the accuracy of the flux spatial angle are both important issues. Furthermore, the dynamics of the flux estimation can limit the ability to regulate flux.

5.6.2. Open Loop Flux Observers for Direct Field Orientation at Zero Speed

Direct field orientation techniques as discussed in Section 5.3 used an open loop flux observer described by the basic flux equations in Figure 5-38. This open loop observer is shown in block diagram form in Figure 5-38 (estimated parameters and states are again denoted by the ˆ). This voltage model is limited by the open loop integrator, which is acting on the stator voltage terms. For low-frequency operation the free integrator exhibits both noise and stability problems and is also sensitive to stator resistance. For high frequencies (above 5 to 10 Hz), the voltage model is less parameter sensitive because the back-EMF voltage tends to dominate at high speeds [33]. However, the low-frequency limitations preclude the use of open loop, voltage model, flux observers for direct field orientation in motion control applications.

Zero speed direct field orientation can be achieved via an open loop flux observer if rotor position (as available in motion control systems) is used. Figure 5-39

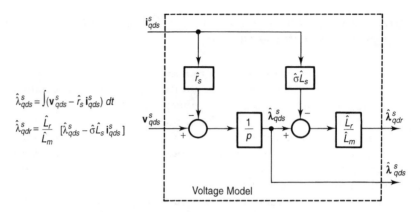

$$\hat{\lambda}_{qds}^s = \int (v_{qds}^s - \hat{r}_s \, i_{qds}^s) \, dt$$

$$\hat{\lambda}_{qdr}^s = \frac{\hat{L}_r}{\hat{L}_m} \, [\hat{\lambda}_{qds}^s - \hat{\sigma}\hat{L}_s \, i_{qds}^s]$$

Figure 5-38. Open loop observer for stationary frame stator and rotor flux based on motor voltage and current (hereafter described as "the voltage model").

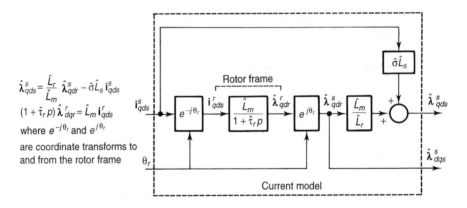

$$\hat{\lambda}_{qds}^s = \frac{\hat{L}_r}{\hat{L}_m} \, \hat{\lambda}_{qdr}^s - \hat{\sigma}\hat{L}_s \, i_{qds}^s$$

$$(1 + \hat{\tau}_r \, p) \, \hat{\lambda}_{dqr}^r = \hat{L}_m \, i_{qds}^r$$

where $e^{-j\theta_r}$ and $e^{j\theta_r}$
are coordinate transforms to
and from the rotor frame

Figure 5-39. Open loop observer for stationary frame stator and rotor flux based on motor current and rotor position (hereafter described as "the current model").

shows such an open loop, stationary frame, observer, which is hereafter described as the "current model" [32–35].

This current model open loop observer is globally stable in the stationary frame and does work for direct field orientation even at zero speed. It is still parameter sensitive, however and has been demonstrated to be less accurate than the voltage model at high speeds [35].

5.6.3. Open Loop Flux Observers in Indirect Field Orientation

As described in Section 5.3, indirect field orientation offers a feedforward solution which is stable but parameter sensitive at all frequencies. For this technique, the field angle is determined by integration of the slip frequency, which is derived from the slip relation, equation 5.24. The slip frequency relation is formed as a feedforward controller by using commanded values of flux and current (denoted by the *) and estimated parameters (denoted by ^)

$$S\omega_e^* = -\frac{1}{\hat{\tau}_r}\frac{\hat{L}_m i_{qs}^{e^*}}{\lambda_{dr}^{e^*}} \tag{5.38}$$

The field angle produced by integration of this slip frequency is the relative angle of the rotor flux with respect to the excitation. The excitation angle for rotor flux orientation is determined by summing this relative angle with the instantaneous rotor position measurement, θ_r.

$$\theta_{rf}^* = \theta_r + \int S\omega_e^* \, dt = \theta_r + \int \frac{\hat{r}_r}{\hat{L}_r}\frac{\hat{L}_m i_{qs}^{e^*}}{\lambda_{dr}^{e^*}} \, dt \tag{5.39}$$

This form is consistent with the indirect field orientation scheme shown in Figure 5-39. The flux amplitude used for torque scaling and current input calculation is based on the ideal model of rotor flux for field orientation and can be formed as an implicit, open loop, flux observer as shown in Figure 5-40 below (based on the field orientation flux equation 5.27).

Although this form of the rotor flux observer functions at all speeds, it was shown in Section 5.3 to be parameter sensitive and can result in poor torque dynamics as well as incorrect torque amplitudes. It is particularly problematic under field weakening conditions where the magnetizing inductance may vary substantially. To minimize these effects, certain closed loop observer topologies can be used.

5.6.4. Closed Loop Flux Observers and Direct Field Orientation—Rotor Flux

It is possible to apply closed loop observer techniques to form flux observers. Many of the possible configurations have significant limitations, most especially at low speeds. One topology has evolved which seems to have particularly good attributes since it is derived from using the best features of both the current model and the voltage model open loop observers. This topology is shown in Figure 5-41 [33].

This observer tracks the current model at zero and low speeds and the voltage model at high speeds. The transition occurs seamlessly since the closed loop observer

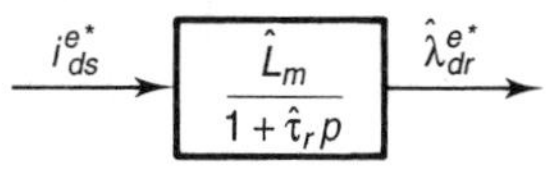

Figure 5-40. Implicit, open loop observer for rotor flux from motor current command in the synchronous frame, assuming field orientation is accurately implemented.

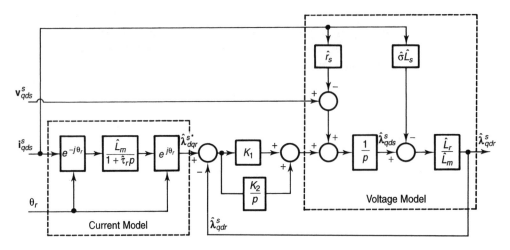

Figure 5-41. Closed loop observer for rotor flux from motor current and voltage, and rotor position measurements, based on current and voltage models.

tracks the current model reference within its bandwidth and the feedforward voltage model beyond its bandwidth. The transition between models is governed by the bandwidth of the closed loop observer which is user selectable by gains K_1 and K_2. Typical bandwidths are in the range of 1 to 10 Hz. A lower-frequency bandwidth is generally selected when parameter estimates for the rotor time constant and magnetizing inductance are poor. The lower bandwidth forces the transition to the voltage model at lower frequencies because the voltage model is not very sensitive to these terms. This closed loop observer is then formed into a direct field-oriented controller as shown in Figure 5-42 [33–34].

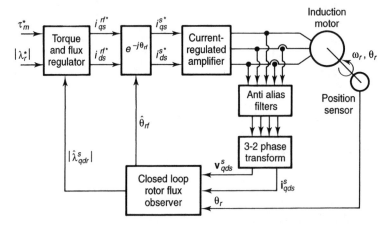

Figure 5-42. Direct field orientation and rotor flux regulation using closed loop observer for rotor flux based on motor current and voltage and rotor position measurements.

Because the flux observer produces both a magnitude and spatial angle estimate, it is possible to perform both field orientation (using the spatial angle) and field flux control (using the flux magnitude). This separation is important for motion control trajectories in which flux must be altered dynamically. Special cases of considerable concern include motion trajectories using a wide dynamic range that require field weakening and efficiency-driven, field weakening trajectories such as for electric vehicles that also require rapid acceleration response and correspondingly high-response field flux control. For such cases, a second form of control may be desired, that of stator flux.

5.6.5. Closed Loop Flux Observers and Direct Field Orientation—Stator Flux

A closed loop stator flux observer may be formed which allows direct field orientation including zero speed operation. Figure 5-43 shows such an observer which is formed in a manner analogous to the rotor flux observer [35].

This approach uses a form of both the current model and the voltage model, after the leakage terms have been moved from the voltage model to the current model. This approach is very parameter insensitive at high speeds, since there are no leakage parameters remaining in the voltage model. It still retains the zero speed robustness of the current model and thus allows true zero speed operation based on stator flux. Such performance is not possible with the open loop voltage model type of stator flux-oriented systems. The primary limitation of this technique lies in the approach needed to obtain decoupled control of torque and flux as identified in Figure 5-44 [36].

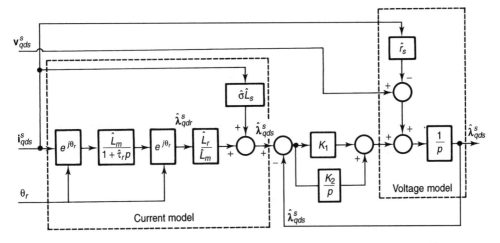

Figure 5-43. Closed loop observer for stator flux from motor current and voltage, and rotor position measurements, based on current and voltage models.

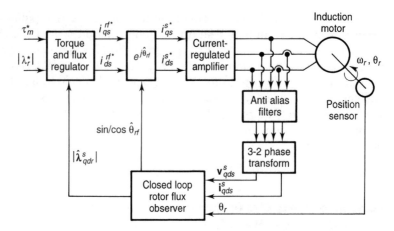

Figure 5-44. Direct field orientation and stator flux regulation using closed loop observer for stator flux based on motor current and voltage and rotor position measurement.

Unlike rotor flux decoupling with its very simple, one-step decoupling structure, stator flux decoupling of flux and torque by forcing the q-axis stator flux to zero is considerably more complicated as shown in Figs. 5-45 and 5-46 [41]. Figure 5-46 shows the combination of terms that would cause complete decoupling. Because of complexity and parameter sensitivity this method is less practical to implement than rotor flux field orientation.

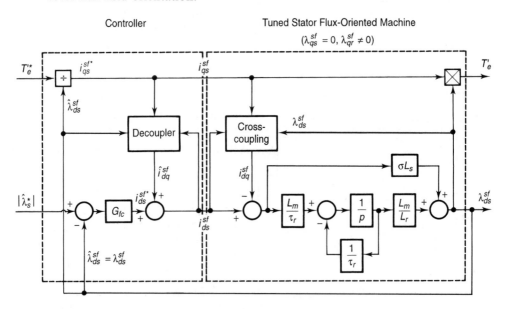

Figure 5-45. Direct field orientation and stator flux regulation showing decoupling controller used to separate torque and stator flux control.

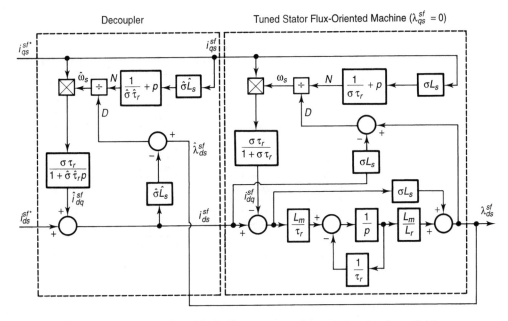

Figure 5-46. Decoupling details for torque and stator flux in direct field orientation.

5.6.6. Direct Rotor Flux Orientation, Stator Flux Regulation, and Closed Loop Flux Observers

The best features of all these approaches can be combined to form a rotor flux-oriented, stator flux-regulated, direct field-oriented induction motor servo drive suitable for a complete range of motion control applications. This configuration is shown in Figure 5-47 [36].

This composite system retains the simplicity of rotor flux decoupling but has wider dynamic range and decreased parameter sensitivity compared to indirect field orientation. It also allows optimum utilization of the DC bus voltage and the converter current ratings by controlling the stator flux linkage, thereby dynamically modulating the back-EMF voltage of the motor.

5.6.7. Summary of Advanced Flux and Torque Regulation Methods for Motion Control using FO-IM

This section has shown how performance of indirect field orientation can be surpassed by using direct field orientation and flux and torque control based on flux observers. It was shown how the use of a position sensor as required for motion control along with current sensors will facilitate a simple open loop observer for rotor flux. This observer will allow direct field orientation at zero speed, and it can be formed for either rotor flux or for stator flux.

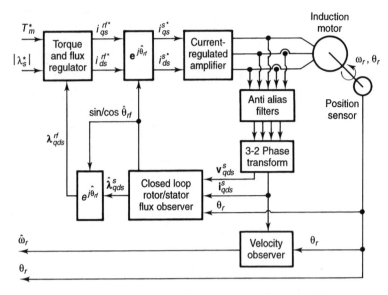

Figure 5-47. Direct rotor flux orientation with stator flux regulation using closed loop flux observers and motion control velocity observers.

The formulation of a closed loop observer will allow even better performance to be achieved by seamlessly incorporating two models, one that performs best at zero and low speed and the other that performs best at high speeds, including field weakening.

The use of closed loop observers allows for high-performance flux regulation as well as flux orientation. Thus, it is possible to combine approaches to yield the best compromises in performance, parameter insensitivity, and converter utilization for FO-IM used in motion control.

5.7. SELF-COMMISSIONING AND CONTINUOUS SELF-TUNING FOR FO-IM

There are two basic problems in tuning all forms of AC drives: initial tuning at start-up of a drive (or of the application) and tuning during operation to reflect changes in the system. Automated forms of initial tuning are generally called *self-commissioning* and automated tuning during operation is generally called *continuous self-tuning*.

Current technologies for self-commissioning and self-tuning of motion controllers with FO-IM are primarily focused on parameters needed for accurate indirect field orientation. This is because it is the most widely used approach for induction motor servo drives. As previously discussed in Section 5.3, indirect field orientation is sensitive to

- The rotor time constant, $\tau_r = L_r/r_r$, for correct torque dynamics.
- The magnetizing inductance, L_m, for steady-state flux and thus torque scaling.

Both parameters are unique in the challenge they offer for self-commissioning and self-tuning of FO-IM motion controllers.

In addition to the FO-IM parameter needs, most motion control systems require some form of load parameter estimation. From the perspective discussed in Section 5.2, the motion control system is sensitive to

- The total connected load inertia, J.
- The total connected process drag, b_p, for correct feedforward command tracking.

Both parameters offer similar challenges for self- commissioning and self-tuning of motion controllers with FO-IM.

In general, self-commissioning and self-tuning routines are based upon the model shown in Figure 5-48 where stator voltage, stator current, and rotor position are measured and available. There are two basic approaches to parameter estimation or adaptation that will be discussed in this section:

- Statistical fitting of parameters from measured data under appropriate conditions.
- Adaptive (recursive) control to vary parameters and force convergence.

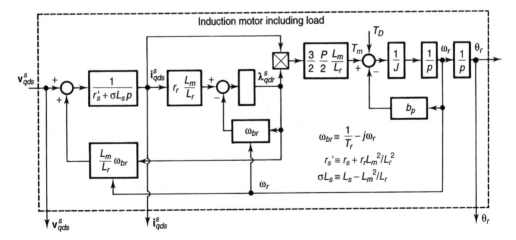

Figure 5-48. Induction motor and load model showing variables available for self-commissioning and self-tuning of motion controllers with FO-IM.

In general, statistical approaches are the most widely used approach for self-commissioning at standstill, while adaptive control approaches are used for continuous self-tuning during use.

5.7.1. Statistical Approaches to Parameter Estimation

The statistical approaches sample the inputs, that is, $\mathbf{v}_{qds}^{e}$, and the outputs, that is, $\mathbf{i}_{qds}^{e}$ and θ_r, developing a data record sufficient to assure accuracy tolerances. The data record is used in estimating transfer function coefficients which minimize squared deviation between the actual response and the estimated model's response to the input data. The physical parameter estimates needed for the field orientation and motion control algorithms are then extracted from the transfer function coefficients as shown conceptually in Figure 5-49.

The least squared error models are based on algorithms commonly known as *multivariable regression analysis* [44, 45]. Unfortunately, all such approaches are based on linear models. Thus, to be applied to nonlinear models such as induction motors, operating point models (also called perturbation or small signal analysis models) must be formed. Since the induction motor is a physical system with continuous nonlinearities, such models may be readily formed through several techniques. One common technique used to form a linear, operating point model is to

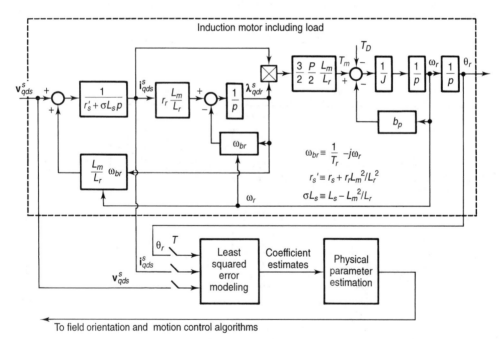

Figure 5-49. Induction motor and load model showing statistical methods for parameter estimation suitable for motion control with FO-IM.

apply partial differential analysis to the induction motor differential equations. While mathematically correct, that approach is relatively tedious. A simpler technique can be used if output velocity is changing slowly and if state derivatives can be determined by observer-based state estimators [47, 48] This simplified technique will now be described.

5.7.2. Statistical Regression Model Formulation— Induction Motor Estimation at Constant Speed

For the stationary frame induction motor model of Figure 5-49, the vector format state equations are

$$p\mathbf{i}_{qds}^{s} = \frac{1}{\sigma L_s}\left(\mathbf{v}_{qds}^{s} - r_s{}'\,\mathbf{i}_{qds}^{s} + \frac{L_m}{L_r}\omega_{br}\,\lambda_{qdr}^{s}\right) \tag{5.40}$$

$$p\lambda_{qdr}^{s} = \frac{1}{\tau_r}\,L_m\mathbf{i}_{qds}^{s} - \omega_{br}\lambda_{qdr}^{s} \tag{5.41}$$

The complex coefficient electrical transfer function (for constant average velocity) may be written as

$$\frac{\mathbf{i}_{qds}^{s}}{\mathbf{v}_{qds}^{s}} = \frac{L_r/L_sL_r - L_m^2\left[p + (1/\tau_r - j\omega_r)\right]}{p^2 + (r_sL_r + r_rL_s/L_sL_r - L_m^2 - j\omega_r)\,p + r_sL_r/L_sL_r - L_{m^2}\,(1/\tau_r - j\omega_r)}$$
$$= \frac{\mathbf{B}_1\,p + \mathbf{B}_0}{p^2 + \mathbf{A}_1\,p + \mathbf{A}_0} \tag{5.42}$$

This continuous time model can be formed as a linear equation suitable for multivariable regression analysis as follows:

$$\frac{d^2\mathbf{i}_{qds}^{s}}{dt^2} = -\mathbf{A}_1\frac{\mathbf{i}_{qds}^{s}}{dt} - \mathbf{A}_0\,\mathbf{i}_{qds}^{s} + \mathbf{B}_1\frac{d\mathbf{v}_{qds}^{s}}{dt} + \mathbf{B}_0\mathbf{v}_{qds}^{s} \tag{5.43}$$

Note: The standard regression format would be $y = c_1\,x_1 + c_2\,x_2 + c_3\,x_3 + c_4\,x_4$. The coefficients for the induction motor model (at constant average velocity) are defined as

$$\mathbf{A}_0 = \frac{r_sL_r}{L_sL_r - L_m^2}\left(\frac{1}{\tau_r} - j\omega_r\right) \tag{5.44}$$

$$\mathbf{A}_1 = \frac{r_sL_r + r_rL_s}{L_sL_r - L_m^2} - j\omega_r \tag{5.45}$$

$$\mathbf{B}_0 = \frac{L_r}{L_sL_r - L_m^2}\left(\frac{1}{\tau_r} - j\omega_r\right) \tag{5.46}$$

$$\mathbf{B}_1 = \frac{L_r}{L_sL_r - L_m^2} \tag{5.47}$$

The coefficients for this continuous time, linear regression equation can be calculated from a data record using standard regression analysis:

$$y = c_1 x_1 + c_2 x_2 + c_3 x_3 + c_4 x_4 \ldots$$
$$\hat{y} = \hat{c}_1 x_1 + \hat{c}_2 x_2 + \hat{c}_3 x_3 + \hat{c}_4 x_4 \ldots$$

solve for $\hat{c}_1$, $\hat{c}_2$, $\hat{c}_4$ such that sum of $(y - \hat{y})^2 = $ SSE (sum of squared errors) is minimized; that is, solve

$$\frac{\partial \text{ SSE}}{\partial \hat{c}_1} = 0, \qquad \frac{\partial \text{ SSE}}{\partial \hat{c}_2} = 0, \qquad \frac{\partial \text{ SSE}}{\partial \hat{c}_3} = 0, \qquad \frac{\partial \text{ SSE}}{\partial \hat{c}_4} = 0$$

The standard matrix solution for the regression coefficients becomes

$$\{\hat{c}\} = [X^t X]^1 - X^t Y \tag{5.48}$$

To implement this simplified approach, the derivatives of the input voltage, stator current, and rotor position must be formed. This can best be done by application of simple observer filters (without feedforward) as shown in Figure 5-50.

Since feedforward is not implemented in this type of observer structure, it is important that nearly identical gains be used so that no net observer dynamics will appear in the estimated model. Significant phase differences will affect the values of the coefficients in the estimated transfer functions. While analog observer filters have been successfully constructed, digital filters are a more appropriate approach for obtaining matched characteristics.

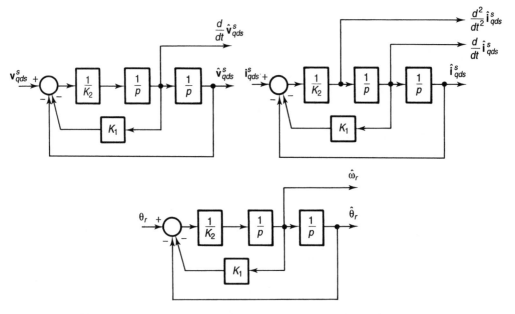

Figure 5-50. Observer filters for extracting the stator voltage, stator current, and rotor position derivatives needed for linear regression analysis.

5.7.3. Rotor Time Constant and Resistance and Inductance Parameter Extraction

After the complex coefficients $\mathbf{A}_0$, $\mathbf{A}_1$, $\mathbf{B}_0$, and $\mathbf{B}_1$ have been estimated, the physical parameters of interest for FO-IM control can be extracted as follows:

$$\hat{r}_s = \frac{\text{Re}(\mathbf{A}_0)}{\text{Re}(\mathbf{B}_0)}, \quad \text{where Re( ) is the real part of the complex coefficient} \quad (5.49)$$

$$\hat{\tau}_r = \frac{\text{Re}(\mathbf{B}_1)}{\text{Re}(\mathbf{B}_0)} \quad (5.50)$$

Given estimates for r_s and τ_r, it is possible to estimate L_s:

$$\hat{L}_s = \left[\frac{\text{Re}(\mathbf{A}_1)}{\text{Re}(\mathbf{B}_1)} - \hat{r}_s\right] \times \hat{\tau}_r \quad \text{or} \quad \hat{L}_s = \left[\frac{\text{Re}(\mathbf{A}_1)}{\text{Re}(\mathbf{B}_1)} - \frac{\text{Re}(\mathbf{A}_0)}{\text{Re}(\mathbf{B}_0)}\right] \times \frac{\text{Re}(\mathbf{B}_1)}{\text{Re}(\mathbf{B}_0)} \quad (5.51)$$

$$\hat{\sigma}L_s = \frac{1}{\text{Re}(\mathbf{B}_1)} \quad (5.52)$$

Once the parameters have been calculated, the field oriented part of the drive may be tuned for correct field orientation. A properly functioning field oriented torque control will then allow a simplified mechanical load parameter analysis as developed in the following section.

5.7.4. Statistical Regression Model Formulation— Mechanical Load Parameters

After the field orientation algorithms have the proper physical parameter gains installed, the mechanical load parameter estimation proceeds along similar steps except for the following caveats:

- Electromagnetic torque is assumed known (based on properly calibrated FO-IM).
- Constant speed is no longer appropriate; some acceleration is needed to estimate inertia.
- The load is often assumed to be linear.

The most common model used for the mechanical load is shown in Figure 5-51. The linear regression model for the mechanical system thus becomes

$$\frac{d^2\theta_r}{dt^2} = -A_{m1}\frac{d\theta_r}{dt} + B_{m0}T_m \quad (5.53)$$

To implement this simplified approach, the derivatives of rotor position must be formed. This can best be performed by application of simple observer filters as shown in Figure 5-52.

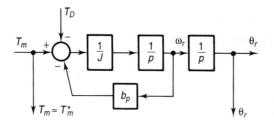

Figure 5-51. Mechanical load model used for linear regression analysis.

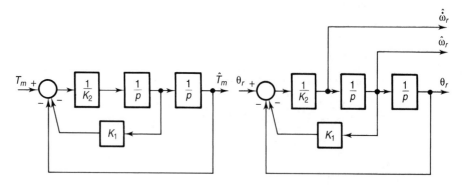

Figure 5-52. Observer filters for extracting the rotor position derivatives and matched filtering on torque input as needed for accurate linear regression analysis.

Both torque (command) and position must be identically filtered to avoid significant phase differences which would otherwise affect the values of the coefficients. Again, digital filters are a more appropriate approach for obtaining identically matched characteristics.

The mechanical load parameters can be simply estimated from the coefficients

$$\hat{J} = \frac{1}{B_{m0}} \tag{5.54}$$

$$\hat{b}_p = \frac{A_{m1}}{B_{m0}} \tag{5.55}$$

The remaining issues relate to how the operating conditions used for this method will determine the accuracy of the parameter estimation.

5.7.5. Operating Condition and Input Excitation Limitations for Statistical Estimation

The accuracy of parameter estimation using statistical techniques is dependent on several key factors:

- The spectral content and amplitude of the excitation.

- The level of response to the excitation of each characteristic part of the dynamic response, measured relative to the resolution of the A/D converters (for voltage and current) and of the angular encoder (for position).
- The sampling frequency and data record length relative to the dynamic response being estimated.

The spectral content of the excitation has a dramatic impact on the ability to estimate dynamic response properties. The excitation spectral content has a significant influence on induction motor parameters because the magnetic properties change as a function of frequency. This is especially true for high frequencies because the flux linkage paths and current distributions are substantively different for 20 kHz signals than for 5 to 300 Hz signals. This means that parameters estimated from very-high-frequency excitation signals would not be suitable for estimating the fundamental component properties. Techniques based on inverter switching frequency inputs such as those used in [41–43] are not likely to provide accurate fundamental component models unless the PWM switching frequency is low, such as 1–4 kHz.

The spectral amplitude of the excitation and the magnitude of response attributed to each characteristic dynamic part must be considered with respect to the resolution of the signal acquisition devices, that is, the A/D converters and the encoder counter least significant bit angular resolution. Clearly, if the response contribution of a certain characteristic is small, the signal-to-resolution (that is, signal-to-noise) ratio degrades. In general, the response of higher-frequency dynamics is of smaller amplitude than that of lower-frequency dynamics. Thus, it is often difficult to obtain accurate parameter estimates without preshaping the high-frequency content to provide appropriate signal-to-noise ratios for the high-frequency response. Furthermore, techniques which are based solely on operating conditions are unlikely to excite all the dynamics adequately for estimation.

The sampling frequency and data record length can limit the dynamics which can be estimated. If the data record contains less than one cycle of a given frequency input, it becomes difficult to extract model parameters for that frequency. Methods having data records which are shorter than one cycle of the fundamental component frequency [45] can be expected to have difficulties estimating the induction motor equivalent circuit parameters.

5.7.6. Summary of Statistical Methods for FO-IM

It should be noted that the statistical techniques described are based on batch processing of data records and thus are inherently compatible with self-commissioning. Furthermore, the requirement for appropriately formed excitation and measurement make these techniques ideally suited to off-line self-commissioning. One of the primary distinguishing factors between some of the existing approaches lies in the clever form of excitation used for the self-commissioning [53, 54]. By comparison, these same attributes make statistical approaches much less appropriate for continuous self-tuning.

5.7.7. Adaptive Control Approaches to Parameter Estimation for FO-IM

There are a relatively large number of parameter estimation techniques for FO-IM which are based on adaptive control techniques. Such methods include:

- Recursive, least squares parameter estimation [47–48].
- Model reference adaptive control (MRAC) [51–55].
- Deadbeat adaptive control [56–59].

Each of these techniques is appropriate for continuous self-tuning of FO-IM drives. The major issues distinguishing the techniques are the closed loop adaptation dynamics and the processing overhead which each method requires. In addition, not all of the techniques will work at low speeds, due to the nature of the models used.

5.7.8. Recursive, Least Squares

The recursive, least squares parameter estimation technique is shown conceptually in Figure 5-53. This method is a special form of adaptive control in which the parameter estimates are recursively revised based on an exponentially decaying response characteristic. While such methods are feasible, they are still sensitive to the same excitation limitations as the batch processing methods. Furthermore, the response of such systems still tends to be slow based on the total processing required.

5.7.9. MRAC Approaches

The MRAC approaches are based upon comparing actual measurements with a feasible model. For the induction machine, the most commonly used models are the voltage model and the current model discussed in Section 5.5 and shown in Figures 5-38 and 5-39, respectively. One form of this type of MRAC is shown in Figure 5-54 [35].

Different forms of this MRAC have been implemented by various researchers [55–59]. One general undesirable feature of MRAC is the nonlinear product nature

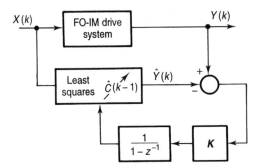

Figure 5-53. Recursive, least squares statistical method for parameter estimation.

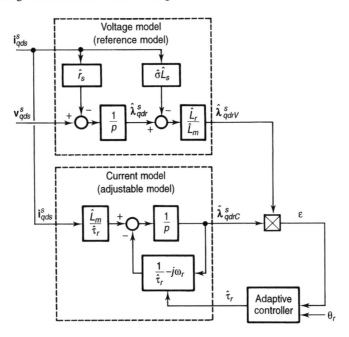

Figure 5-54. Model reference adaptive controller based on the voltage and current model.

of the error term, ϵ, used for the adaptation. This type of nonlinearity is used to reduce sensitivity to noise and load disturbances that are not correlated with the model. In effect, the error term is a cross-correlation product that inherently averages out the signals that are uncorrelated to the reference model. However, this same nonlinearity also forces the response of the closed loop adaptation dynamics to become operating point dependent. Such operating space variant dynamics are less desirable because the system must be tuned for the worst case circumstances and thus will be detuned for other operating conditions.

5.7.10. Deadbeat, Adaptive Control Approaches

The deadbeat adaptive control approach attempts to resolve the space variant dynamics problem by forming a controller which should inherently settle to its adapted value in one sample period of the controller. Such a control structure is well known in discrete time control theory because of its very desirable closed loop dynamics. Figure 5-55 shows the general model of a deadbeat adaptive controller as applied to an indirect FO-IM drive whereby the change in rotor resistance is modeled as a disturbance for the controller [56].

Deadbeat control usually refers to a closed loop discrete time system that achieves a one-sample period delay between command and response. This is equivalent to a closed loop command tracking transfer function of

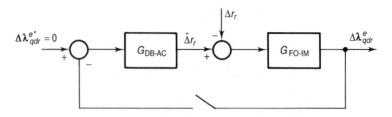

Figure 5-55. General form of deadbeat adaptive controller for indirect FO-IM.

$$\frac{\Delta\lambda_{qdr}^{e^*}}{\Delta\lambda_{qdr}^{e}} = z^{-1} \tag{5.56}$$

where $\Delta\lambda_{qdr}^{e^*} =$ is the desired change in rotor flux, normally = zero and

$\Delta\lambda_{qdr}^{e} =$ is the measured change in rotor flux, normally = observer estimate

or expressed as a difference equation model

$$\Delta\lambda_{qdr}^{e}(k) = \Delta\lambda_{qdr}^{e^*}(k-1) \tag{5.57}$$

The controller which would achieve this type of command response is formed as

$$G_{\text{DB-AC}}(z) = G_{\text{FO-IM}}(z)^{-1} \; \frac{z^{-1}}{1-z^{-1}} \tag{5.58}$$

where this deadbeat command response controller consists of two parts:

$G_{\text{FO-IM}}(Z)^{-1},$ \quad an inverse model of the FO-IM

$\dfrac{z^{-1}}{1-z^{-1}},$ \quad a one-step delayed integration process

It is important to note that this deadbeat controller uses

- The inverse model of the FO-IM to calculate the rotor resistance change, $\Delta\hat{r}_r,$ required to produce the desired output change, $\Delta\lambda_{qdr}^{e^*}$
- The integration process to hold that value of $\Delta\backslash o(r,\ s\backslash up\,4(\char`\^))_r,$ to maintain the desired output, $\Delta\lambda_{qdr}^{e^*}$

The closed loop disturbance rejection transfer function for this deadbeat control system would be

$$\frac{\Delta\hat{r}_r}{\Delta r_r} = z^{-1} \tag{5.59}$$

or expressed as a difference equation model

$$\Delta\hat{r}_r(k) = \Delta r_r(k-1) \tag{5.60}$$

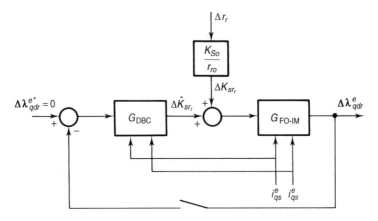

Figure 5-56. Operating point deadbeat adaptive controller for indirect FO-IM.

Thus, a deadbeat controller is capable of adapting correctly after just one time sample period.

The limitation of this approach is that the induction motor is a nonlinear model and thus cannot be inverted directly. However, an operating point (perturbation or small signal) model can be inverted. Such an inverted model will be a function of the operating point of the motor. Thus, the one sample period deadbeat adaptation dynamics are only strictly correct for small changes about an operating point. Figure 5-56 shows the operating point formulation of the deadbeat adaptive controller where the change in rotor resistance has been formulated as a change in slip gain to be consistent with indirect FO-IM. This controller can be implemented as shown in Figure 5-57 [59] with the following theoretical adaptation dynamics.

$$\frac{\Delta \hat{K}_{sr_r}}{\Delta K_{sr_r}} = z^{-1} \tag{5.61}$$

For this controller, two important issues must be addressed:

1. Selection of the sample rate for the deadbeat controller.
2. Formulation of the inverse model of the induction motor.

The sample rate can be selected to greatly facilitate the model inverse by choosing a very slow sample rate such that flux changes are largely settled out by the time a new sample is taken. This is consistent with the general need to form deadbeat controllers with sample rates such that complete response is physically realizable in one sample period. A 10-sample rate is sufficiently slow to make this adaptation a possibility.

The model inverse now looks like a simple gain relationship as [56]:

$$\Delta K_s = \Delta_m \, \Delta \lambda_{dr}^{e^*-1} \tag{5.62}$$

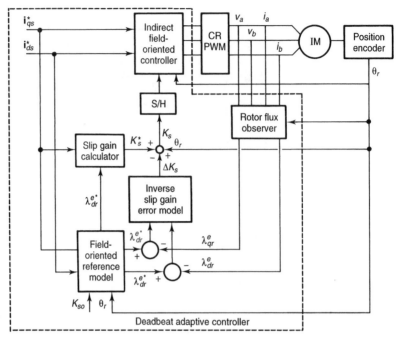

Figure 5-57. Operating point deadbeat adaptive controller implementation for indirect FO-IM.

where

$$\hat{m} \equiv \frac{L_m}{\tau_r}$$

and

$$\frac{\Delta_m}{\hat{m}} \equiv \frac{\Delta\lambda_{dr}^{e-1}}{\lambda_{dr}^{e-1}} + \frac{\Delta\lambda_{qr}^e}{\hat{L}_m i_{qs}^{e*}}$$

This approach yields a very simple algorithm and a very easily implemented, real-time control approach which yields nearly deadbeat response as long as the change in rotor resistance during any one sample period (0.1 sec) is small. Since the rotor thermal time constants are long relative to this sample rate, the approach has been shown to produce a very uniform dynamic adaptation.

The deadbeat adaptive control approach can be improved further by refinements to the flux estimator. One improvement which is applicable to FO-IM drives where the flux distribution shows some degree of saturation is that described in [59].

5.8. CONCLUSION

This chapter has developed modern motion control structures that place a clear emphasis on high fidelity torque control. Torque amplitude accuracy, wide bandwidth dynamic torque control, and low amplitudes of torque distortion were shown to be required to obtain dynamic stiffness of the motion controller and zero-tracking error to commands.

It was shown how field oriented induction motor control can enable the induction motor servo to produce the torque control properties needed for modern motion control. However, it was shown that errors in the field-oriented controller or in the current regulator would cause low-frequency torque dynamics, incorrect torque amplitudes, and torque distortion. In particular it was shown that indirect field orientation is particularly parameter sensitive. It was also shown that direct field orientation using simple open loop voltage model flux observers is not compatible with the low-speed torque control often required for motion control.

The chapter has also presented a series of current regulators which are generally compatible with field-oriented torque control, albeit at differing levels of performance. The performance of each of the current controllers ultimately determines the attainable level of motion control performance. Additionally, each had a different set of limitations and/or costs associated with the implementation. The predictive (nearly optimal) current regulators presented offer the highest performance option for motion control with field-oriented induction motors. However, such controllers are currently structured only for discrete pulse modulation approaches such as those used in resonant link inverters.

The flux and torque controllers presented in the chapter show how direct field orientation is feasible and appropriate for motion control. Zero-speed performance can easily be achieved thanks to the use of the position feedback sensor in flux observers. It was shown how zero-speed direct field orientation was achievable using either stator flux or rotor flux via open loop observers. It was furthermore shown how parameter sensitivity of direct field orientation can be improved by using closed loop observers. Finally, it was shown that rotor flux field orientation, with stator flux regulation is a very appropriate solution for induction motor motion control.

The chapter also has presented two basic methodologies for self-commissioning and self-tuning of field-oriented induction motor motion control. The statistical methods were shown to be very appropriate for initial self-commissioning of drives. The adaptive control methods, most especially the deadbeat adaptive controller were shown to be very realistic solutions for continuous self-tuning of field-oriented induction motor motion control. The methods presented in this chapter collectively demonstrate that induction servo motor technology is sufficiently well advanced for high-performance motion control.

ACKNOWLEDGMENT

The authors wish to thank the industrial sponsors of the Wisconsin Electric Machines and Power Electronics Consortium for their support of the basic research which has led to the development of much of the technology discussed in this chapter.

Nomenclature

Abbreviations

FO-IM field-oriented induction motor(s)

MRAC model reference adaptive control

PI proportional integral control algorithm

Symbols

b_p viscous process damping in mechanical system model

$\mathbf{i}_{qdr}$ rotor current in complex vector notation, that is, $\mathbf{i}_{qdr} \equiv i_{qr} - j i_{dr}$ or, real vector notation, that is, $\mathbf{i}_{qdr} \equiv [i_{qr}, i_{dr}]^T$

$\mathbf{i}_{qds}$ stator current

J inertia of motor plus load

L_{lr} rotor leakage inductance

L_{ls} stator leakage inductance

L_m magnetizing inductance

L_r rotor self inductance, $L_r = L_{lr} + L_m$

L_s stator self inductance, $L_s = L_{ls} + L_m$

p differential operator; $p = j\omega_e$ at steady state

P pole number

r_r rotor resistance

r_s stator resistance

r_s' stator transient resistance, $r_s' \equiv r_s + r_r L_m^2 / L_r^2$

S rotor slip in per unit

$S\omega_e$ rotor slip frequency

T_d disturbance (load) torque

T_m electromagnetic torque

T_m' normalized electromagnetic torque, $T_m \equiv 3/2 \; P/2 \; T_m'$

T sampling time

$\mathbf{v}_{qds}$ stator voltage in complex vector notation; that is, $\mathbf{v}_{qds} \equiv v_{qs} - j v_{ds}$, or real variable vector notation; that is, $\mathbf{v}_{qds} \equiv [v_{qs}, v_{ds}]^T$

λ_r rotor flux magnitude

λ_s stator flux magnitude

$\boldsymbol{\lambda}_{qds}$ stator flux in complex vector notation, that is, $\boldsymbol{\lambda}_{qds} \equiv \lambda_{qs} - j \lambda_{ds}$, or real vector notation, that is, $\boldsymbol{\lambda}_{qds} \equiv [\lambda_{qs}, \lambda_{ds}]^T$

$\boldsymbol{\lambda}_{qdr}$ rotor flux

σ leakage or coupling factor, $\sigma \equiv 1 - L_m^2 / L_r L_s$

σL_s stator transient inductance, $\sigma L_s \equiv L_s - L_m^2 / L_r$

τ_r rotor time constant, L_r / r_r

θ_r rotor position (elec. rad.)

θ_{rf} rotor flux angle (elec. rad.)

θ_{sf} stator flux angle (elec. rad.)

ω_{br} rotor break frequency, that is, in stationary frame $\omega_{br} \equiv r_r / L_r - j\omega_r$

ω_e fundamental excitation frequency (rad/sec)

ω_r rotor velocity (rad/sec)

ω_s slip frequency (rad/sec)

Superscripts

$\;\hat{}\;$ estimated quantity

$*$ commanded or reference quantity

e arbitrarily aligned synchronous frame quantity

i quantity synchronous to the injected signal

rr rotor current aligned synchronous frame quantity

rf rotor flux aligned synchronous frame quantity

s stationary frame quantity

sf stator flux aligned synchronous frame quantity

References

[1] Novotny, D. W., "A comparative study of variable frequency drives for energy conservation applications," University of Wisconsin Report #/ECE-81-4, 1981.

[2] Davis, R. M., "Inverter-fed induction machines," *Proc. of Drives/Motors/ Controls '82*, Leeds, England, pp. 66–75, 29 June–1 July 1982.

[3] Lipo, T. A., and F. G. Turnbull, "Analysis and comparison of two types of square wave inverter drives," *IEEE Trans. on Industry Applications*, Vol. IA-11, no. 2, pp. 137–147, March/April 1975.

[4] Houldsworth, J. A., and W. B. Rosink, "Introduction to PWM speed control system for 3-phase AC motors," *Electronics Components and Applications*, Vol. 2, no. 2, February 1980.

[5] Bowes, S. R., "New sinusoidal pulse width-modulated inverter," *Proc. IEE*, Vol. 122, pp. 1279–1285, November 1975.

[6] Lorenz, R. D., "Microprocessor control of motor drives and power converters," Chapter 4, *Microprocessor Motion Control of AC and DC Drives*, IEEE Tutorial Course Note Book from 1991, '92 & '93 IEEE, IAS Annual Meetings, IEEE Publishing Services #THO587-6.

[7] Lorenz, R. D., and K. VanPatten, "High resolution velocity estimation," *IEEE Trans. In. Appl.*, Vol. 27, no. 4, pp. 701–708, July/August 1991.

[8] Schmidt, P. B., and R. D. Lorenz, "Design principles and implementation of acceleration feedback to improve performance of DC drives," *IEEE Trans. Ind. Appl.*, pp. 594–599, May/June 1992.

[9] Moatemri, M. H., P. B. Schmidt, and R. D. Lorenz, "Implementation of a DSP-based, acceleration feedback robot controller: Practical issues and design limits," *IEEE-IAS Conf. Rec.*, pp. 1425–1430, 1991.

[10] Hasse, K., "Zur dynamik drehzahlgeregelter antriebe mit stromrichter-gespeisten asynchron-kurzschlublaufermaschinen," Ph.D. dissertation, Tech. Hochschule Darmstadt, July 17, 1969.

[11] Blaschke, F., "The principle of field orientation as applied to the new transvector closed loop control for rotating machines," *Siemens Review*, Vol. 39, no. 5, pp. 217–220, 1972.

[12] de Doncker, R., and D. W. Novotny, "The universal field oriented controller," *IEEE-IAS Annual Meeting Conf. Rec.*, pp. 450–456, October 1988.

[13] Xu, X., and D. W. Novotny, "Selection of the flux reference for induction machines in the field weakening region," *IEEE Trans. Ind. Appl.*, Vol. 28, no. 6, pp. 1353–1358, November/December 1992.

[14] Nordin, K. B., D. W. Novotny, and D. S. Zinger, "The influence of motor parameter deviations in feedforward field orientation drive systems," *IEEE-IAS Trans.*, Vol. IA-21, no. 4, pp. 1009–1015, July/August 1985.

[15] Lorenz, R. D., and D. W. Novotny, "Optimal utilization of induction machines in field oriented drives," *J. Electrical and Electronic Engin.*, Australia, Vol. 10, no. 2, pp. 95–100, June 1990.

[16] Wallace, I. T., D. W. Novotny, R. D. Lorenz, and D. M. Divan, "Increasing the dynamic torque per ampere capability of induction machines," *IEEE-IAS Conf. Rec.*, pp. 14–20, 1991.

[17] Brod, D. M., and D. W. Novotny, "Current control of VSI-PWM inverters," *IEEE Trans. Ind. Appl.*, Vol. IA-21, no. 4, pp. 562–570, May/June 1985.

[18] Brod, D. M., "Current control of VSI-PWM inverters," M.S.E.E. thesis, University of Wisconsin, 1984.

[19] Rowan, T., "Analysis of naturally sampled current regulated pulse width modulated inverters," Ph.D. thesis, University of Wisconsin, 1985.

[20] Salama, S., and S. Lennon, "Overshoot and limit cycle free current control method for PWM inverters," European Power Electronics Conference, Florence, pp. 3-247–3-251, 1991.

[21] Schauder, C. D., and R. Caddy, "Current control of voltage-source inverters for fast four-quandrant drive performance," *IEEE Trans. Ind. Appl.*, Vol. IA-18, no. 2, pp. 163–171, March/April 1982.

[22] Rowan, T., and R. Kerkman, "A new synchronous current regulator and an analysis of current-regulated PWM inverters," *IEEE Trans. Ind. Appl.*, Vol. IA-22, no. 4, pp. 678–690, July/August 1986.

[23] Holtz, J., and S. Stadtfeld, "A predictive controller for the stator current vector of AC machines fed from a switched voltage source," *Int. Power Electronics Conf.*, Tokyo, pp. 1665–1675, March 27–31, 1983.

[24] Nabae, A., S. Ogasawara, and H. Akagi, "A novel control scheme for current-controlled PWM inverters," *IEEE Trans. Ind. Appl.*, Vol. IA-22, no. 4, pp. 697–701, July/August 1986.

[25] Habetler, T. G., and D. M. Divan, "Performance characterization of a new, discrete pulse modulated current regulator," *IEEE-IAS Annual Meeting Conf. Rec.*, pp. 395–403, October 1988.

[26] Böcker, J., and J. Janning, "Discrete-time flux observer for PWM fed induction motors," *Proc. EPE*, Florence, pp. 171–176, 1991.

[27] Bottura, C. P., J. L. Silvino, and P. Resende, "A flux observer for induction machines based on a time-variant discrete model," *Proc. APEC*, 1991.

[28] Franceschini, G., C. Tassoni, and A. Vagati, "Flux estimation for induction servo motors," *Proc. IPEC-Tokyo*, pp. 1227–1234, 1990.

[29] Franceschini, G., M. Pastorelli, F. Profumo, C. Tassoni, and A. Vagati, "About the gain choice of flux observer in induction servo motors," *Proc. IEEE-IAS Annual Meeting*, pp. 601–606, 1990.

[30] Hori, Y., V. Cotter, and Y. Kaya, "A novel induction machine flux observer and its application to a high performance AC drive system," *IFAC 10th Triennial World Congress*, Munich, FRG, pp. 363–368, 1987.

[31] Hori, Y., and T. Umeno, "Implementation of robust flux observer based field orientation (FOFO) controller for induction machines," *Proc. IEEE-IAS Annual Meeting*, pp. 523–528, 1989.

[32] Hori, Y., and T. Umeno, "Flux observer based field orientation (FOFO) controller for high performance torque control," *Proc. IPEC-Tokyo*, pp. 1219–1226, 1990.

[33] Jansen, P. L., and R. D. Lorenz, "A physically insightful approach to the design and accuracy assessment of flux observers for field oriented induction machine drives," *Proc. IEEE-IAS Annual Meeting*, pp. 570–577, October 1992.

[34] Jansen, P. L., C. O. Thompson, and R. D. Lorenz, "Observer-based direct field orientation for both zero and very high speed operation," *Proc. PCC-Yokohama*, Japan, pp. 432–437, April 1993.

[35] Jansen, P. L., and R. D. Lorenz, "Accuracy limitations for velocity and flux estimation in direct field oriented induction machines," *Proc. EPE Conf.*, Brighton, UK, September 1993.

[36] Jansen, P. L., R. D. Lorenz, and D. W. Novotny, "Observer-based direct field orientation: Analysis and comparison of alternative methods," *Proc. IEEE-IAS Annual Meeting*, Toronto, pp. 536–543, October 1993.

[37] Kubota, H., and K. Matsuse, "Flux observer of induction motor with parameter adaption for wide speed range motor drives," *Proc. IPEC-Tokyo*, pp. 1213–1218, 1990.

[38] Kubota, H., K. Matsuse, and T. Nakano, "DSP-based speed adaptive flux observer of induction motor," *Proc. IEEE-IAS Annual Meeting*, pp. 380–384, 1991.

[39] Matsuse K., and H. Kubota, "Deadbeat flux level control of high power saturated induction servo motor using rotor flux observer," *Proc. IEEE-IAS Annual Meeting*, pp. 409–414, October 1991.

[40] Nilsen, R., and M. P. Kazmierkowski, "New reduced-order observer with parameter adaption for flux estimation in induction motors," *PESC*, pp. 245–252, 1992.

[41] Xu, X., R. de Doncker, and D. W. Novotny, "A stator flux oriented induction machine drive," *1988 Power Electronics Specialist's Conference*, Kyoto, Japan, pp. 870–876, April 1988.

[42] Xu, X., R. de Doncker, and D. W. Novotny, "Stator flux orientation control of induction machines in the field weakening region," *Proc. IEEE-IAS Annual Meeting*, pp. 437–443, October 1988.

[43] Xu, X., and D. W. Novotny, "Implementation of direct stator flux orientation control on a versatile DSP based system," *Proc. IEEE-IAS Annual Meeting*, October 1990.

[44] Bendat, J. S., and A. G. Piersol, *Random Data: Analysis and Measurement Procedures*, Wiley-Interscience, New York, 1971.

[45] Box, G. E. P., and G. M. Jenkins, *Time Series Analysis, Forecasting and Control*, Holden Day, San Francisco, 1970.

[46] Holtz, J., and T. Thimm, "Identification of machine parameters in a vector controlled induction motor drive," *Conf. Rec. IEEE-IAS Annual Meeting*, San Diego, pp. 601–606, October 1989.

[47] Vélez-Reyes, M., K. Minami, and G. C. Verghese, "Recursive speed and paramter estimation for induction machines," *Conf. Rec. IEEE-IAS Annual Meeting*, San Diego, pp. 607–611, October 1989.

[48] Borgard, D. E., G. Olsson, and R. D. Lorenz, "Accuracy issues for parameter estimation of field oriented induction machine drives," *Conf. Rec. of the IEEE-IAS Annual Meeting*, Denver, pp. 503–600, October 2–6, 1994.

[49] Matsuo, T., and T. A. Lipo, "A rotor parameter identification scheme for vector-controlled induction motor drives," *IEEE Trans. Ind. Appl.*, Vol. 21, no. 4, pp. 624–632, May/June 1985.

[50] Wang, C., D. W. Novotny, and T. A. Lipo, "An automated rotor time constant measurement system for indirect field-oriented drives," *IEEE Trans. Ind. Appl.*, Vol. 24, no. 1, pp. 151–159, January/February 1988.

[51] Lorenz, R. D., and D. B. Lawson, "A simplified approach to continuous, on-line tuning of field oriented induction machine drives," *IEEE Trans. Ind. Appl.*, Vol. 26, no. 3, pp. 420–425, May/June 1990.

[52] Garces, L. J., "Parameter adaption for the speed-controlled static AC drive with a squirrel-cage induction motor," *IEEE Trans. Ind. Appl.*, Vol. 16, no. 2, pp. 173–178, March/April 1980.

[53] Zai, L. C., and T. A. Lipo, "An extended kalman filter approach to rotor time constant measurement in PWM induction motor drives," *1987 IEEE Industry Applications Soc. Annual Meeting*, October 1987.

[54] Sugimoto, H., and S. Tamai, "Secondary resistance identification of an induction-motor applied model reference adaptive system and its characteristics," *IEEE Trans. Ind. Appl.*, Vol. 23, no. 2, pp. 296–303, March/April 1987.

[55] Rowan, T. M., R. J. Kerkman, and D. Leggate, "A simple on-line adaption for indirect field orientation of an induction machine," *1989 IEEE Industry Applications Soc. Annual Meeting*, pp. 579–587, 1989.

[56] Hung, K. T., and R. D. Lorenz, "A rotor flux error-based adaptive tuning approach for feedforward field-oriented induction machine drives," *1990 IEEE Industry Applications Soc. Annual Meeting*, pp. 589–594, October 1990.

[57] Hung, K. T., "A slip gain error model-based correction scheme of near-dead-beat response for indirect field orientation," Master's thesis, University of Wisconsin-Madison, 1990.

[58] Moreira, J. C., "A study of saturation harmonics with applications in induction motor drives," Ph.D. thesis, University of Wisconsin-Madison, 1990.

[59] Moreira, J. C., K. T. Hung, T. A. Lipo, and R. D. Lorenz, "A simple and robust adaptive controller for deturning correction in field oriented induction machines," *IEEE Trans. Ind. Appl.*, pp. 1359–1366, November/December 1992.

Chapter 6

Thomas M. Jahns

Variable Frequency Permanent Magnet AC Machine Drives

6.1. INTRODUCTION

Permanent magnet AC (PMAC) machines provide a unique set of advantages and opportunities to designers of modern motion control systems. PMAC machines can be designed in many different geometries as discussed in Chapter 2, and a few common examples are illustrated in Figure 6-1. The use of permanent magnets to generate substantial air gap magnetic flux without external excitation makes it pos-

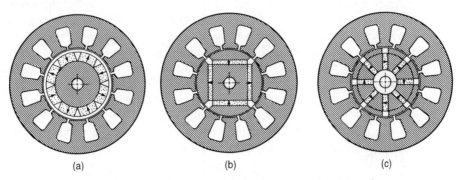

(a)   (b)   (c)

Figure 6-1. Alternative PMAC machine rotor geometries: (a) surface magnets; (b) interior radially-oriented magnets; (c) interior tangentially oriented magnets. Magnet material is identified by dotted areas, with arrows indicating polarity.

sible to design PMAC machines with unsurpassed efficiency characteristics. Such efficiency advantages are becoming increasingly valuable in many parts of the world where market forces and governmental mandates are focusing increased attention on reduced energy consumption in many types of electrical equipment. For example, minimum energy efficiency requirements are gradually being imposed by the U. S. government during the 1990s on major classes of residential appliances as well as heating, ventilating, and air conditioning (HVAC) equipment [1].

Alternatively, the low losses made possible by the appropriate application of permanent magnets can be used to achieve machine designs with impressively high values of power density and torque-to-inertia (T_e/J) ratios [2]. These characteristics make PMAC machines highly attractive for many actuator and servo applications that demand the fastest possible dynamic response.

On the other hand, PMAC machines present special challenges to the electrical equipment designer since they are synchronous machines that, in the absence of auxiliary rotor windings, absolutely require accompanying power electronics for operation. The drive electronics is necessary to perfectly synchronize the AC excitation frequency with the rotational speed as a prerequisite for generating useful steady-state torque, in contrast to an induction motor which typically operates from a fixed frequency AC source (50 or 60 Hz). As a result, PMAC machines are particularly well suited to motion control applications that can justify the cost of the power electronics to take advantage of the uniquely attractive efficiency and dynamic response characteristics provided by this type of machine.

The objective of this chapter is to review the principal motion control techniques that have been developed since the 1970s for PMAC machine drives. Two major classes of PMAC machines—trapezoidal and sinusoidal—were introduced in Chapter 2 that have some notable differences in their respective control requirements and performance characteristics. A major portion of this chapter will be devoted to a discussion of the commonalities and differences associated with the control of these two important families of PMAC machines. The special control requirements of PMAC machines to achieve extended speed operating ranges using flux-weakening techniques will also be addressed.

The particular requirements of PMAC machines for synchronization of the excitation waveforms with rotational speed have given rise to a considerable variety of control techniques for eliminating the shaft-mounted position sensor that will be reviewed later in this chapter. This discussion of advanced drive control techniques will be followed by a comparison of the strengths and weaknesses of PMAC machine drives compared to other major competing machine drive technologies. The chapter will conclude with summary descriptions of several PMAC machine drive systems that have been selected to illustrate the breadth of fielded applications, leading to a closing discussion of expected future trends.

6.1.1. Background

PMAC machine drives represent the convergence of at least two distinct threads of permanent magnet machine development. One of these threads is the early development of line-start PMAC motors with embedded rotor squirrel cage windings designed for operation directly from utility-supplied AC power. Work on this special class of hybrid PMAC induction machine dates back to the 1950s [3, 4] using Alnico magnets. These machines were widely applied in some important industrial applications such as textile manufacturing lines that require large numbers of machines operating at identical speeds [5].

Later during the 1970s, considerable research attention was focused on improved designs for line-start PMAC machines as a means of achieving significant energy savings in industrial applications. Integral horsepower versions of these line-start PMAC motors have been developed with impressive efficiency characteristics using both ferrite and rare-earth magnets [6], but their resulting manufacturing cost premiums over conventional induction machines prevented wide market acceptance. Nevertheless, significant technical progress has been demonstrated in the development of these high-power PMAC machines, and work in this area continues today [7].

Representing the second thread of development, permanent magnet DC (PMDC) servo motors began to displace conventional wound-field DC motors in high-performance machine tool servo applications in the 1960s when solid-state DC chopper circuits reached market maturity [8]. The availability of high-strength rare earth permanent magnets made it possible to develop compact fast-response PMDC servo motors [9] without the steady-state losses and additional circuit complications associated with traditional wound-field DC machines.

Finally, in the 1970s, these two development paths converged as PMAC machines (without rotor cages) were combined with adjustable frequency inverters to achieve high-performance motion control [10]. This approach had the desirable effect of eliminating the dual disadvantages of high rotor inertia and brushwear associated with PMDC motor commutators. The class of "brushless DC motors" using trapezoidal PMAC motors was developed first [11] in order to take advantage of the control simplifications that are achievable with this configuration, as discussed in more detail in Section 6.3. This has been followed by the evolution of high-performance sinusoidal PMAC machine drives during the late 1970s and 1980s which have been made possible by the rapid advances in digital real-time control hardware and vector control technology first applied to induction motors [12].

6.1.2. Motion Control Performance Requirements

Although several of the basic issues associated with machine drive specifications were introduced in Chapter 2, a few key points will be briefly reviewed here in order to set the stage for the following discussion of PMAC machine control techniques. Torque-speed envelope requirements for PMAC machine drives typically fall into one of two classes illustrated in Figure 6-2. "Constant torque" applications are

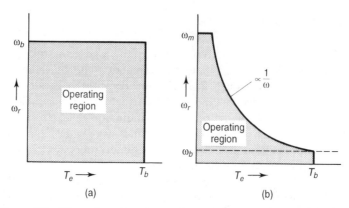

Figure 6-2. Alternative torque-speed operating envelopes: (a) constant torque; (b) constant horsepower.

characterized by the square torque-speed envelope shown in Figure 6-2a which requires that maximum torque T_b be available at all speeds up to the maximum rotor speed ω_b. Actuators and servo systems are typical examples of constant torque applications which rely on maximum torque availability at all speeds to ensure maximum dynamic response. PMAC machines are naturally well suited for constant torque applications because of the constant level of magnetic flux delivered by the rotor permanent magnets to the machine's air gap.

In contrast, "constant horsepower" applications are characterized by a torque-speed envelope which follows a hyperbolic constant power trajectory $(T_e \cdot \omega_r = P_o = \text{constant})$ over a wide speed range above base speed ω_b as illustrated in Figure 6-2b. One of the most familiar examples of a constant horsepower application is electric vehicle traction that requires high torque for low-speed acceleration and reduced torque for high-speed cruising. Constant horsepower operation poses special challenges for PMAC machines since there is no field winding which can be directly weakened as in a conventional separately excited DC motor. However, "flux-weakening" control techniques are available for PMAC machines that can achieve the same effect, as described later in this chapter.

Although the preceding discussion has been framed in terms of single-quadrant motoring operation with positive torque and positive speed, many applications require that the PMAC machine develop braking torque for controlled deceleration and that the machine be able to rotate in both directions. Such requirements give rise to two- and four-quadrant torque-speed operating envelopes as discussed previously in Chapter 2. The PMAC machine is well suited for such applications since it can operate equally well as a motor or generator, and direction of rotation has no impact on the machine's performance characteristics.

PMAC machine drive applications can be further characterized from a control standpoint by the nature of the primary control variable. The three principal types of drive control configurations are torque control, speed control, and position control.

Torque control represents the most basic control requirements, typified by the accelerator and brake pedals of an electric vehicle drive. Many other industrial and commercial applications such as pumps and process lines require speed control, representing an intermediate level of control performance requirements. Finally, position control systems give rise to some of the most demanding motion control requirements for high-performance applications such as machine tool servos.

Motion control performance requirements are also reflected in dynamic response specifications. For example, dynamic response requirements for a PMAC speed- or position-controlled servo system are typically expressed in terms of bandwidth specifications which define the ability of the drive's output shaft to faithfully track sinusoidal command signals as the frequency is increased. Development of high-bandwidth PMAC machine servo systems can be accomplished through a combination of design techniques that involve coordinated selection of the machine parameters as well as the controls configuration.

6.2. PMAC MACHINE CONTROL FUNDAMENTALS

6.2.1. Sinusoidal versus Trapezoidal PMAC Machines

As discussed in Chapter 2, PMAC motors fall into the two principal classes of sinusoidally excited and trapezoidally excited machines (referred to hereinafter as simply sinusoidal and trapezoidal machines). The differences in construction between these two classes of PMAC machines must also be reflected in their respective motion control requirements. While some of these differences are obvious and others are more subtle, they are all derived in one way or from their fundamentally different excitation waveforms.

Figure 6-3 provides a direct comparison of idealized current excitation waveforms for typical three-phase sinusoidal and trapezoidal PMAC machines. Despite their obvious differences in waveshape, some important similarities are also apparent. For example, these excitation waveforms form balanced three-phase sets for both PMAC classes, with 120 electrical degree separations between successive phases. Furthermore, the presence of a dominant fundamental-frequency component can be readily identified in the six-step current waveforms for the trapezoidal PMAC machine, although it should be noted that the higher-frequency time harmonics also contribute useful steady-state torque in this type of machine.

The two sets of excitation waveforms shown in Figure 6-3 reflect the fact that there are both notable similarities and differences between the motion control requirements of sinusoidal and trapezoidal PMAC machines. Control principles and components shared in common by the two PMAC machine classes will be described first in this section, followed by discussions of their major differences in Sections 6.3 and 6.4.

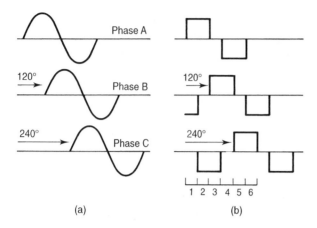

(a)   (b)

Figure 6-3. Basic current excitation waveforms for (a) sinusoidal and (b) trapezoidal PMAC motors.

6.2.2. Converter Configurations

For all PMAC motor drives supplied from utility power lines, some form of power electronic converter is required to transform the fixed frequency (50 or 60 Hz), fixed amplitude input power into the variable frequency, variable amplitude power needed to excite the PMAC motor. While some special power converter topologies are available which directly convert the incoming AC power into variable frequency AC output power as discussed in Chapter 3 (e.g., cycloconverters, matrix converters [13]), the most popular approach uses a two-stage configuration with an intermediate DC link between the input rectifier and output inverter stages, as sketched in Figure 6-4.

The three-phase PMAC machine is adopted as the baseline configuration for much of the discussion in this chapter, reflecting its predominance in commercially produced PMAC drive systems. Furthermore, it is worthwhile noting in the following discussion of inverter configurations that the basic power circuit topologies are often identical for sinusoidal and trapezoidal PMAC machines. In these cases, differences in the switch gating sequences for the two types of machines are entirely responsible for distinguishing the phase excitation waveforms shown in Figure 6-3.

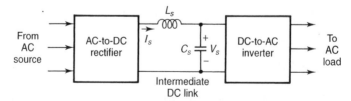

Figure 6-4. Basic AC-to-AC converter configuration using an intermediate DC link, showing typical inductive and capacitive link energy storage components.

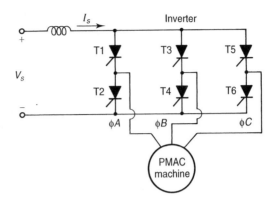

Figure 6-5. Three-phase current source inverter topology for a PMAC machine, using load-commutated thyristor switches.

Current versus Voltage Source Inverters. Both current source and voltage source topologies have been developed for DC-to-AC inverters used to excite PMAC machines. An example of a current source inverter for a three-phase PMAC machine drive is shown in Figure 6-5, which uses an inductor in the DC link as the current source and thyristors as the six inverter switches [14–16]. The thyristors in this inverter can be load commutated by the back-EMF voltage waveforms of the PMAC machine using identical control techniques developed originally for wound field synchronous machines [17]. The inverter and motor must be designed to sustain the classic inductive voltage pulses which are developed at the machine terminals each time the DC link current is switched from an off-going phase to the subsequent on-going phase. However, the amplitudes of these transient voltage pulses can be reduced significantly by adding a damper squirrel cage to the rotor [14], borrowing further from techniques applied originally in conventional wound-field synchronous motor drives.

A complementary three-phase voltage source inverter topology used to excite both sinusoidal and trapezoidal PMAC machines is shown in Figure 6-6. Although this familiar full-bridge inverter is sketched here using six insulated gate bipolar transistors (IGBTs) as the active power switches, any type of power semiconductor switch can be used that can be turned on and off from a control

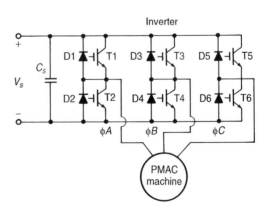

Figure 6-6. Three-phase voltage source inverter for a PMAC machine, using gate-controlled inverter switches with antiparallel diodes.

terminal (e.g., bipolar junction transistors, power MOSFETs, and MCTs) [18]. The link inductor in the current source inverter shown in Figure 6-5 is replaced here by a large DC link capacitor that provides low-impedance voltage source characteristics at the inverter's input. Each switch in the voltage source topology is shunted by an antiparallel diode that provides a freewheeling path for inductive motor currents and protects the switching devices from having to sustain large reverse blocking voltages that can appear in the current source inverter.

Voltage source inverters are used far more broadly than the current source topologies in PMAC machine drive applications for a combination of reasons. The cost, size, and weight of electrolytic capacitors tend to be significantly lower than DC link inductors in comparably rated voltage and current source configurations. Furthermore, new generations of gate-controlled power switches that lack reverse voltage blocking capability tend to be more naturally suited to voltage source inverter requirements. As a result, current source inverters are generally reserved for special applications such as high-power PMAC machine drives which can benefit from the thyristor's high current-handling capabilities.

The basic voltage and current source inverter topologies displayed in Figures 6-5 and 6-6 are by no means the only inverter configurations used with PMAC machines. For example, Figure 6-7 shows a member of one family of inverter topologies that has been developed with hybrid characteristics of current and voltage source configurations in an attempt to capture some of the best features of both approaches [19, 20]. This particular inverter is best suited for trapezoidal PMAC machines, behaving generally as a type of current source inverter except during phase-to-phase current commutation intervals when it reverts to operation as a voltage source inverter. As indicated in Figure 6-7, a seventh switch has been introduced in the buck converter stage preceding the inverter, which helps to simplify the drive's control requirements as discussed in more detail later in Section 6.3.

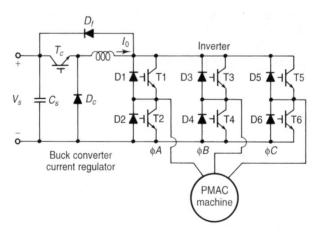

Figure 6-7. Alternative two-stage "quasi-current source" inverter for a
PMAC machine, consisting of current-regulated converter followed by three-phase inverter.

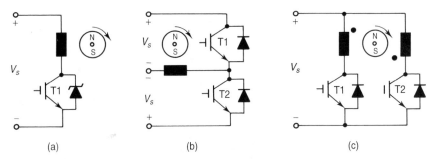

Figure 6-8. Examples of single-phase PMAC machine inverters: (a) single-switch unipolar drive; (b) half-bridge inverter; (c) bifilar-winding inverter.

Special PMAC Machine Converters. Although three-phase machines are used in the largest majority of PMAC drive applications, important product opportunities have also developed for PMAC machines with phase numbers both smaller and larger than three. For example, single-phase PMAC machines are finding wide acceptance in low power (subfractional and fractional hp) applications with modest control performance requirements such as heatsink cooling fans [102]. Cost is typically the dominant factor in such applications, and the range of specialty PMAC machine designs developed for these low-power applications is fascinating but beyond the scope of this chapter. The range of inverter configurations applied to these single-phase PMAC machines is similarly intriguing [21], and a few examples are illustrated in Figure 6-8. As illustrated by the simple unipolar current drive circuit in Figure 6-8a, acute cost pressures sometimes make it necessary to sacrifice available machine performance to strip the control electronics down to the absolute minimum required to make the rotor turn.

At the other end of the power spectrum extending up to the megawatt range for some applications, PMAC machine drive designers have found compelling reasons to increase the phase order number to six or more [22] to reduce the per phase power-handling requirements. Large PMAC machines designed to directly drive submarine propellers at low speeds (< 300 r/min) have been developed using the basic inverter architecture shown in Figure 6-9 for power ratings in the range of 1 to 5 mW [23, 24]. The use of a separate single-phase H-bridge to excite each machine phase enhances the modularity and fault tolerance of the overall drive system at the price of an increased total switch count compared to inverter architectures for alternative wye- or delta-winding connections.

6.2.3. Position Synchronization

Since all of the PMAC machines discussed here are specific examples of synchronous machines, average torque can be produced only when the excitation is precisely synchronized with the rotor frequency (i.e., speed) and instantaneous position [25]. The most direct and powerful means of ensuring that this

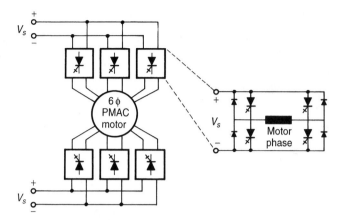

Figure 6-9. Basic six-phase PMAC machine drive configuration for a high-power marine propulsion system [23], using six modular H-bridge inverter units.

requirement is always met is to continuously measure the rotor's absolute angular position so that the excitation can be switched among the PMAC motor phases in exact synchronism with the rotor's motion (see Figure 6-10). This concept, known commonly as self-synchronization or self-controlled synchronization, uses direct feedback of the rotor angular position to ensure that the PMAC machine never experiences loss of synchronization (i.e., pull-out).

Another descriptive name for this control scheme is electronic commutation [26], which highlights the controller's responsibilities for directing excitation to the correct stator windings at each time instant. Since this action is functionally equivalent to the mechanical commutator's role in a DC motor, the electronic commutation terminology has been helpful in explaining the differences between PMDC and PMAC operating principles. It is particularly appropriate for trapezoidal PMAC machine drives for which direct analogs can be drawn between the rectangular current excitation pulses in Figure 6-3 and the periodic contacts between rotating commutator bars and stationary brushes in a conventional DC motor.

The most direct and popular method of providing the required rotor position information is to mount an absolute angular position sensor on the PMAC machine's rotor shaft. Alternatively, accurate information regarding rotor position can be derived indirectly from the motor's voltage or current waveforms using more sophisticated control algorithms discussed in Section 6.5. Limited opportunities for

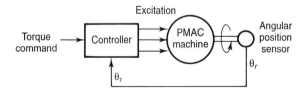

Figure 6-10. PMAC motor self-synchronization concept using rotor angular position feedback.

"open loop" excitation of sinusoidal PMAC motors without any type of position sensing will also be reviewed in Section 6.4.

6.2.4. Mechanical Drive Configurations

One of the most powerful means of reducing the volume and weight (and often the cost) of a PMAC machine for a given output power requirement is to increase its speed to inversely reduce the machine's torque requirement which determines its size. PMAC machines designed for speeds in the range of 10,000 to 20,000 revolutions per minute (rpm) are not unusual, and some special units have been designed for much higher speeds in the 50,000 to 200,000 rpm range [20]. The design of these high-speed PMAC machines requires special mechanical provisions to ensure that the magnets are safely contained at the highest speeds.

Since the required speed range of the load is often much lower than the speed of machine shaft, an intermediate gearbox or some other type of power transmission device is typically required to provide the mechanical interface. In many cases, the desired load output motion is linear rather than rotary, so ballscrews or localized hydraulic loops including rotary pumps and linear rams can be introduced to convert the high-speed rotary motion of the PMAC machine into high-force, low-speed linear motion.

Selection of the motor speed often involves multiple system issues that must be traded off in order to optimize the overall drive system [27]. For example, increasing the motor speed may reduce the size and weight of the motor, but the corresponding values for the mechanical gearbox are likely to grow as the gear ratio (β) is increased. The impact of the gear ratio value is particularly notable in high-bandwidth actuator applications since the peak motor power increases as the square of the gear ratio (β^2) for a fixed amplitude, fixed frequency sinusoidal load motion. As a result of this strong dependence combined with other factors, some actuator designers opt for direct drive mechanical configurations ($\beta = 1$) for improved system dynamic performance, focusing their efforts on the development of high-torque, low-speed PMAC machines [28].

6.2.5. PMAC Drive Control Structure

Regardless of whether the PMAC machine class is sinusoidal or trapezoidal, the control structures for a wide variety of PMAC motor drives share important characteristics in common. Figure 6-11 shows how control loops can be cascaded using classic techniques to achieve, successively, torque, speed, and position control in PMAC machine motion control systems.

Torque control, which constitutes the most motor basic control function, maps very directly into current control because of the intimately close association between current and torque production in any PMAC machine, described in more detail in Sections 6.3 and 6.4. In view of the fundamental importance of torque control as a foundation for achieving high-performance motion control, the majority of PMAC drive applications incorporate closed loop regulation of the motor phase currents as

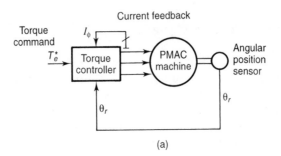

(a)

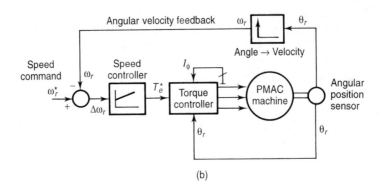

(b)

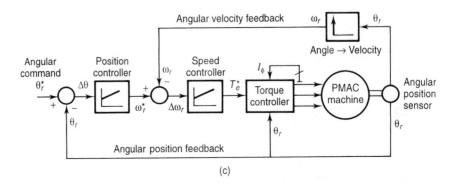

(c)

Figure 6-11. Typical cascaded controls structures for high-performance
PMAC machine drives using nested control loops: (a) current-regulated torque control; (b) closed loop speed control
with inner current loop; (c) closed loop position control with
inner speed and current loops.

shown in Figure 6-11a. Note that this basic torque control configuration also incorporates the self-synchronization concept using shaft position feedback discussed previously in Section 6.2.3.

In some cases the relationship between PMAC machine torque and current is almost perfectly linear so that the torque command maps directly into the current commands with only a simple proportionality constant, while other types of PMAC machine designs (e.g., with buried magnets) require a nonlinear mapping between the torque and current commands. Flux-weakening control techniques introduced to achieve extended constant horsepower operating ranges also tend to complicate the basic torque control scheme in Figure 6-11a as discussed in more detail in Sections 6.3 and 6.4.

It was noted earlier in this chapter that the majority of commercial PMAC drives use voltage source inverter topologies, so the addition of closed loop current control yields a configuration often referred to as a current-regulated voltage source inverter. Many of the key concepts associated with this type of inverter configuration are discussed in Chapter 4. This closed loop system behaves like a very fast current source inverter, depending on the internal motor phase inductances combined with closed loop control to provide the desired current source characteristics without the need for additional inductors.

Speed control can be conveniently achieved in the PMAC machine drive by closing a speed feedback loop around the inner torque/current loop as illustrated in Figure 6-11b. Often this speed feedback signal can be derived from the same shaft-mounted sensor used to detect the rotor position, eliminating the need for an additional sensor. In many cases, a simple proportional integral (PI) controller is sufficient to achieve the desired dynamic performance requirements.

Using a similar loop cascading procedure, position control can be accomplished by closing an outer position loop around the nested current and speed loops as shown in Figure 6-11c. If a gearbox or some other power transmission device couples the machine to the load, it is often necessary to introduce a second position sensor attached to the movable output load. Direct measurement of the load position makes it possible to compensate any nonlinear effects in the mechanical drivetrain (e.g., gear backlash) and frees the controls from the problems associated with tracking the absolute position of the load over multiple turns of the machine shaft. On the other hand, cost pressures or other considerations may make it necessary to sacrifice some of the positioning accuracy in order to eliminate the extra sensor.

As in the case of speed control, a basic PI controller may be sufficient in many cases for closure of the position loop as shown in Figure 6-11c. Using classic servo control techniques, the dynamic characteristics of the nested loops are typically designed so that the innermost current loop has the fastest dynamic response with bandwidths in the kilohertz range, while the outermost position loop is the slowest with bandwidths typically in the 1 to 10 Hz range for integral horsepower actuators. However, a wide variety of alternate controller configurations are also possible within the framework of classical control theory using combinations of feedback and feedforward techniques to achieve desired performance characteristics [29].

More sophisticated adaptive control algorithms have also been developed and will be reviewed briefly in Section 6.5

6.3. TRAPEZOIDAL PMAC MACHINE CONTROL

6.3.1. Machine Control Characteristics

The trapezoidal PMAC motor is specifically designed to develop nearly constant output torque when excited with six-step switched current waveforms of the type presented previously in Figure 6-3. As discussed in Chapter 2, trapezoidal PMAC motors are predominantly surface magnet machines of the type shown in Figure 6-1a designed with wide magnet pole arcs and concentrated stator windings. The resulting back-EMF voltage V_f induced in each stator phase winding during rotation can be modeled quite accurately as a trapezoidal waveform as shown in Figure 6-12. Ideally, the crest of each back-EMF half-cycle waveform should be as broad ($\geq 120°$ electrical) as possible to maximize the smoothness of the resulting output torque, and the back-EMF amplitude V_{f0} is proportional to the rotational frequency ω_r as follows,

$$V_{f0} = k_0 \omega_r \tag{6.1}$$

where k_0 is a machine constant proportional to the linked flux amplitude.

If a current with fixed amplitude I_0 is flowing into this machine phase during the time interval when the back-EMF is cresting as indicated in Figure 6-12, the instantaneous power converted by this phase from electrical form ($P_{e\phi}$) into mechanical form ($P_{m\phi}$) will be

$$P_{e\phi} = V_{f0} \cdot I_0 = P_{m\phi} = T_{e\phi} \cdot \omega_r \tag{6.2}$$

where ($T_{e\phi}$) is the instantaneous torque developed by the excited phase. Using equation (6.1), this leads to

$$T_{e\phi} = (V_{f0} \cdot I_0) / \omega_r = k_0 I_0 \tag{6.3}$$

This torque equation is particularly significant since it indicates that the torque developed by the machine can be controlled directly by varying the current amplitude.

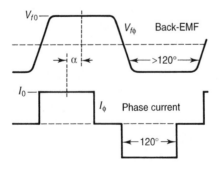

Figure 6-12. Typical back-EMF (V_f) and phase current (I_ϕ) waveforms for trapezoidal PMAC motor, including identification of excitation advance angle α.

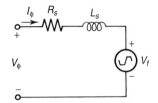

Figure 6-13. Basic per phase equivalent
circuit for a trapezoidal PMAC machine.

These dual direct proportionalities of back-EMF to speed and torque to current expressed in equations 6.1 and 6.3 are identical in form to expressions characterizing a conventional DC motor with constant field excitation. In fact, one of the most common generic trade names given to the trapezoidal PMAC motor is *brushless DC* motor, reflecting this close relationship. Despite the popularity of this name, *brushless DC* is actually quite misleading since the trapezoidal PMAC motor is fundamentally a synchronous AC machine, and *not* a DC machine as the name implies [30].

The standard per phase equivalent circuit model for a three-phase trapezoidal PMAC machine takes the simple form shown in Figure 6-13. The equivalent phase inductance in this model is constant since the rotor of a surface magnet machine presents virtually no net magnetic saliency which would result in inductance variations at the machine's stator terminals. The back-EMF voltage source included in this model delivers the signature trapezoidal voltage waveform sketched in Figure 6-12, corresponding to the open circuit back-EMF waveform measured across each of the machine's phase terminals.

6.3.2. Basic Control Approach

The inverter that excites a trapezoidal PMAC machine has two primary responsibilities. Electronic commutation as described in Section 6.2.3 is the first of the two, requiring the inverter to direct excitation to the proper phases at each time instant to maintain synchronization and to maximize torque output. Current regulation is the second major inverter responsibility, taking advantage of the direct proportionality between current amplitude and output torque during self-synchronized operation as indicated by equation 6.3.

Electronic Commutation. These inverter responsibilities can be explained most easily using the "quasi-current source" inverter topology presented previously in Figure 6-7, which decouples the execution of the two functions into separate converter stages. The electronic commutation function is accomplished by opening and closing the six inverter switches according to the six-step sequence shown in Figure 6-14 to produce the phase current excitation waveforms shown previously in Figure 6-3. (Note the coordinated sequence numbers in both figures.) There are only six discrete inverter switching events during each electrical cycle, and only two inverter switches—one in the upper inverter bank and one in the lower—are conducting at any instant. Under the assumption that the motor is connected in wye,

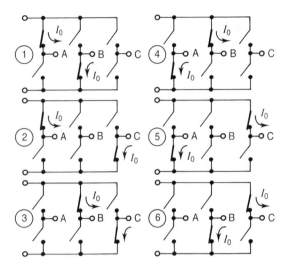

Figure 6-14. Switching sequence for a three-phase full-bridge inverter delivering current waveforms shown in Figure 6-3 to a trapezoidal PMAC motor (note coordinated numbering).

this means that the inverter input current I_0 is flowing through two of the three motor phases in series at all times, consistent with the Figure 6-3 waveforms.

In addition to the back-EMF waveform discussed previously, Figure 6-12 also includes the idealized six-step current excitation waveform for the same phase. The phase shift angle α between the back-EMF and the current waveform indicated in this figure is typically set by the physical alignment of rotor angular position sensor which performs the self-synchronization, thereby controlling the electronic commutation. Torque and output power are maximized for a given current amplitude I_0 at low rotor speeds by holding angle α as close to zero as possible so that the phase current and back-EMF waveforms are held in phase. This relationship corresponds to the standard practice of aligning the commutator brushes in a conventional DC motor to be orthogonal to the magnetic flux imposed by the stationary field. At higher speeds, it may be desirable to advance the angle α to increase the drive system's torque production for extended high-speed operation as discussed later in Section 6.3.4.

Current Regulation. Referring again to the quasi–current source inverter in Figure 6-7, the current regulation function is executed by the simple buck converter stage which precedes the six-switch inverter. By measuring the DC link current I_0 delivered to the inverter stage, the buck converter switch T_c can be readily controlled to provide closed-loop regulation of the machine's phase current using basic duty cycle control techniques. Note that even though the current is being regulated in each of the three stator phases, only a single current sensor in the DC link is sufficient to accomplish this function since the regulated bus current I_0 flows through the two active phases connected in series at each time instant.

While the two-stage quasi–current source inverter simplifies the explanation of the decoupled current regulation and electronic commutation functions, the generally preferred approach for most PMAC machine drive applications is to execute both functions in a single stage using the basic voltage source full-bridge inverter shown in Figure 6-6. This is accomplished by using pulse width modulation (PWM) control of one or both of two inverter switches which are "active" during each 60° (electrical) time interval shown in Figure 6-14 to regulate the current flowing through the two energized motor phases.

Although not shown in Figure 6-14, the freewheeling diodes shown previously in Figure 6-6 (D1–D6) provide crucial conduction paths for the inductive motor phase currents whenever the inverter switches are turned off to accomplish PWM regulation or to transfer current to the next "active" motor phase. In essence, the switch-diode combinations in each of the three inverter phase legs form elementary switching converter modules which can regulate the inductive motor phase current amplitude in either direction.

The achievement of current regulation using the inverter switches can be explained for motoring operation using the simplified circuit and waveforms shown in Figure 6-15. This figure corresponds to mode 1 in Figure 6-14 when switches T1 and T4 are the two active switches directing current through phases A and B. This series combination of two phases can be conveniently modeled for purposes of this description as an equivalent back-EMF voltage source V_{fab} and inductor L_{ab} as shown in the upper half of Figure 6-15. Source voltage V_s is larger than the back-EMF voltage V_{fab} so that current builds up in the motor phases (loop A) when both switches T1 and T4 are closed. Conversely, the current

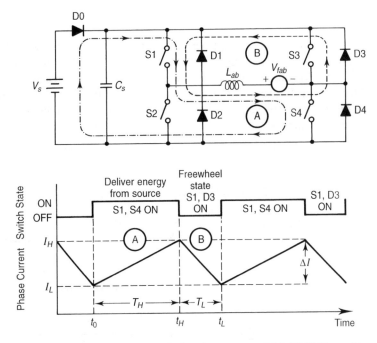

Figure 6-15. PWM current regulation in a trapezoidal PMAC machine during motoring operation, showing simplified drive equivalent circuit and associated steady-state waveforms.

decays due to the back-EMF voltage when only one of the two active switches (T4) is opened so that current freewheels through diode D3 and switch T1 (loop B).

The amplitude of the phase current is regulated by controlling the relative lengths of time intervals T_H and T_L in Figure 6-15 corresponding to the duty cycle of switch T4. Various PWM algorithms can be applied to achieve the desired current regulation [31], including straightforward "constant frequency" and "constant off-time" duty cycle control algorithms used widely in DC-to-DC switching converters [35, 72]. However, closer observation of the two current paths identified in Figure 6-15 reveals that the use of a single current sensor in the DC link constrains the choice of acceptable current regulation algorithms since current feedback is lost during the freewheeling intervals. Alternately, current regulation can also be achieved by turning off both switches T1 and T4 during the current decay intervals (T_L) which has the advantage of maintaining current flow through the DC link current sensor at all times (albeit with alternating polarities) at the price of higher PWM switching frequencies.

Figure 6-16 shows a typical phase current waveform measured in a 0.5 hp trapezoidal PMAC machine drive. The basic six-step current waveform associated with the electronic commutation function (refer to Figure 6-3) is readily apparent. In addition, the high-frequency ripple in the crest of the current waveform represents the action of the switching current regulator algorithm to hold the current close to the commanded value. Some additional features of this waveform will be discussed in Section 6.3.3.

Sensors. One of the attractive features of the trapezoidal PMAC motor drive is that its sensor requirements are generally less demanding than for counterpart sinusoidal PMAC motors discussed in Section 6.4. The electronic commutation function illustrated in Figure 6-14 requires that the rotor angular position be detected at only six discrete points per electrical cycle (60° electrical intervals). One of the most common means of achieving this sensing function is a set of three Hall sensor

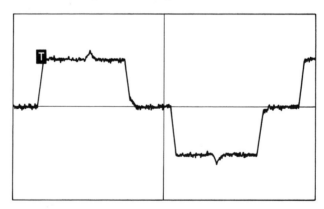

Figure 6-16. Typical phase current waveform for a 0.5 hp trapezoidal PMAC machine during low-speed current-regulated motoring operation.

switches mounted in or near the machine's air gap to detect the impinging magnetic field of the passing rotor magnets [32]. When mounted at 60° electrical intervals and aligned properly with the stator phase windings, these Hall switches deliver digital signals which can be decoded very simply into the desired three-phase switching sequence shown in Figure 6-14.

A wide variety of alternative rotor position sensor types and configurations can be used with trapezoidal PMAC motors to meet specific application requirements. For example, absolute encoders and resolvers [33] can deliver electronic commutation signals while also providing the high-resolution position and velocity feedback necessary to close high-performance speed and position servo loops as shown in Figure 6-11. In contrast, the basic Hall sensor configuration described earlier is very limited in the angular resolution it can provide for closing position servo loops. Relative encoders with marker pulses are appealing candidates for PMAC machine drives because of cost advantages over their absolute encoder counterparts, but special provisions must be made during start-up to ensure proper alignment of the electronic commutation transitions.

Current-sensing requirements for a current-regulated trapezoidal PMAC motor drive are typically reduced to a single sensor in the DC link, as discussed earlier in this section. Current shunt resistors are often used as the sensors in low-power (< 1 kW) drives where their low cost is very appealing and their power dissipation is within acceptable limits. In addition to power dissipation, one of the other significant limitations imposed by current shunts is that they are inherently nonisolated sensors which constrains the placement of the logic common point unless a costly isolation amplifier is provided. In contrast, nondissipative Hall effect current sensors are popular choices at higher power ratings where their cost premium may be more tolerable, and their output isolation characteristics are very appealing to the controls hardware designer [34].

Regenerative Braking Operation. Although much of the discussion up to this point has been framed in terms of motoring operation, the PMAC machine (trapezoidal or sinusoidal) operates equally well as a generator or as a motor. This inherent capability is critical to many high-performance motion control applications such as machine tool servos that require dynamic braking torque for rapid deceleration.

Referring back to the reference waveforms shown in Figure 6-12, the polarity of the delivered torque can be conveniently changed from motoring to generating by simply reversing the polarity of the phase current excitation waveform with respect to the back-EMF. Using the same voltage source inverter stage topology shown in Figure 6-6, only the PWM control algorithm must be changed [35] to reverse the direction of the current flow in the DC link. For example, Figure 6-17 shows one regeneration approach which alternately turns on the two active switches (T1 and T4 in loop A) and then switches them both off so that regenerative energy is returned to the DC link through freewheeling diodes D2 and D3 (loop B). Note that the polarity of the back-EMF in this figure is reversed compared to the motoring case depicted in Figure 6-15.

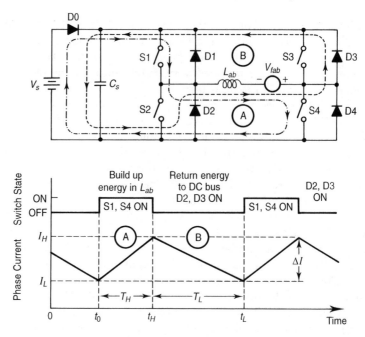

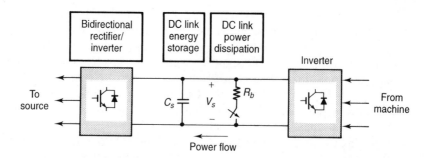

Figure 6-17. PWM current regulation in a trapezoidal PMAC machine during regenerative operation, showing simplified drive equivalent circuit and associated steady-state waveforms.

Depending on the converter topology and application, the regenerated energy flowing back into the DC link from the machine can be returned to the source, temporarily stored in DC link capacitors, or dissipated in auxiliary resistors as illustrated in Figure 6-18. The PMAC machine is generally too efficient to make it practical to dissipate the regenerated energy in the machine's internal phase winding resistances, analogous to "plugging" operation in induction motors.

Unfortunately, the typical AC-to-DC rectifier stage topologies used as the front-end converter in most voltage source PMAC machine drives cannot accommodate current flow back to the AC utility source, so auxiliary gated switches must be added in parallel with the rectifier stage diodes to perform this function. The situation is

Figure 6-18. Three alternative techniques for handling regenerative energy in a DC link inverter.

considerably simplified in applications such as electric vehicles with DC battery sources that can readily accept the regenerated power, making regenerative braking easily achievable in such cases.

In the absence of a convenient means of returning the regenerated energy to the source, resistive dissipation of the excess DC link energy is typically the most practical alternative. Although local energy storage on the DC link would seem to be an appealing energy-efficient alternative, closer examination indicates that electrolytic capacitors are seldom capable of storing sufficient energy unless the amount of regenerative energy is very modest. Instead, a simple shunt voltage regulator consisting of a switch and a power resistor is typically connected across the DC link, designed to turn on whenever the DC link voltage rises above its nominal value because of the regenerative energy being delivered back to the link capacitor.

6.3.3. Torque Ripple

One of the prices that must be paid for the relative simplicity of the torque control algorithm for a trapezoidal PMAC motor drive is the presence of periodic torque ripple components that may pose problems in some demanding motion control applications. Figure 6-19 shows a simulated instantaneous torque waveform for a three-phase trapezoidal PMAC motor drive with PWM-regulated six-step current waveforms, indicating that the basic periodicity of the nonsinusoidal torque ripple is six times the electrical excitation frequency.

There are three major contributing components to the torque ripple, one principally motor related while the other two are predominantly inverter related. The motor-related component causes the crest of the back-EMF waveform and the instantaneous torque to be somewhat rounded during the intervals between commutations in Figure 6-19 due to parasitic effects, including magnetic flux leakage paths between adjacent rotor magnet poles. This factor is a major contributor to the 6th harmonic component appearing in the instantaneous torque waveform. Various PM motor design improvements have been suggested to minimize this torque ripple contribution [36–38].

The first of the two inverter-related torque ripple components is directly proportional to the high-frequency (> 5 kHz typical) PWM ripple in the motor phase currents, which is visible in the measured current waveform in Figure 6-16 as well as the simulated waveform in Figure 6-19. This torque ripple component is generally not a problem because the mechanical load inertia is typically large enough to filter out its effect on speed.

The second inverter-contributed torque ripple component which tends to be more troublesome is caused by the phase current commutations when current is transferred from an off-going inverter phase to the next on-coming active phase at the end of each 60° electrical interval. It is this ripple component which is responsible for the pronounced torque spikes and dips which appear in the Figure 6-19 torque waveform. This commutation torque ripple develops because the sum of the currents in the off-going and on-coming motor phases is almost never constant during the

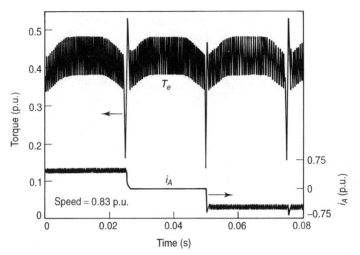

Figure 6-19. Simulated instantaneous torque (T_e) and phase current (i_A) waveforms for a 15 kW trapezoidal PMAC motor [41], illustrating torque ripple components.

transition intervals. In addition to the torque spikes and dips, this phenomenon can create transient overcurrents in the inverter switches which are undetectable using a single current sensor in the DC link [39]. These current spikes (visible in the Figure 6-16 current waveform) can become particularly large under regeneration conditions at elevated speed, with peak phase currents transiently exceeding their commanded values by more that 100% [40].

Various techniques have been proposed for modifying the current PWM regulation scheme in order to prevent these overcurrents and to reduce the amplitude of the resulting torque transients. One brute force approach moves the current sensing point from the DC link to the motor phase windings, requiring a minimum of two current sensors for a three-phase machine [40]. Other proposed approaches retain the single DC link current sensor, but modify the current regulation algorithm using either predictive or open loop chopping techniques to prevent any of the phase currents from exceeding safe limits during commutation intervals [39]. These techniques can be effective in minimizing any overcurrents and the associated torque spikes that would otherwise occur under these operating conditions.

The inverse conditions associated with transient dips in phase current and torque primarily occur during high-speed motoring operation. Under such conditions, the effects of these torque disturbances on motor speed tend to be less troublesome due to the effective low-pass filtering action of the rotor inertia.

6.3.4. High-Speed Operation

Baseline 120° Conduction Intervals. As discussed earlier in this section, the amplitude of the back-EMF voltage for any PMAC machine increases linearly with

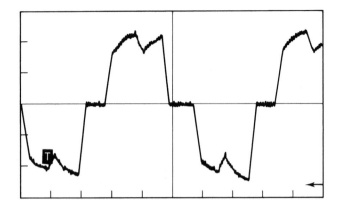

Figure 6-20. Typical phase current waveform for a 0.5 hp trapezoidal
PMAC machine during high-speed motoring operation with
saturated current regulators.

rotor speed. Current control (and thus torque control) for the trapezoidal PMAC
motor drive gradually degrades in quality when the speed increases into the regime
where the sum of the back-EMF voltages for the two conducting motor phases
$(2V_{f0})$ approaches the amplitude of the DC link source voltage V_s. Eventually the
current regulators "saturate," losing their ability to force the commanded current
into the motor phases, so that the inverter naturally reverts to its basic voltage source
operation. Figure 6-20 shows an example of a measured phase current waveform for
a 0.5 hp trapezoidal PMAC machine drive at an elevated speed at which the current
regulators have saturated.

Unless special steps are taken, the phase current amplitudes and the motor
torque both fall off quite abruptly as the rotor speed approaches the critical value
where $V_s = 2V_{f0}$. The solid curve marked $\alpha = 0°$ in Figure 6-21 corresponds to the
baseline case for a 15 kW trapezoidal PMAC machine drive in which each inverter
switch is active for 120° electrical intervals (see Figure 6-14 for reference), and the
phase currents are aligned to be in phase with the back-EMF waveforms for
maximum torque production at low-speeds [41]. This curve shows that the
available torque decays quickly at speeds above the threshold value where
$V_s = 2V_{f0}$ (corresponding to 1.0 per unit (p.u.) speed in Figure 6-21), with the
torque dropping to zero in the vicinity of 1.3 per unit speed for this particular
machine.

Some degree of torque production in this high-speed regime can be restored for
the trapezoidal PMAC motor drive by advancing the phase angle of the excitation
transition points relative to the back-EMF voltages (angle α in Figure 6-12). As this
angle is advanced, current in the on-coming phase is given a controlled time interval
to build up before the back-EMF voltage increases and chokes off further current
growth. The solid curves in Figure 6-21 for $\alpha = 15°$ and 30° show the benefits of this
angle advance on expanding the high-speed torque-speed envelope. Unfortunately,
the available envelope expansion with this approach is relatively modest and comes

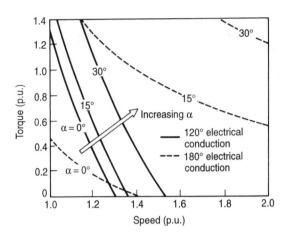

Figure 6-21. Simulated torque-speed curves for a 15 kW trapezoidal PMAC motor [41] comparing results for 120° (two-switch-on) and 180° (three-switch-on) excitation with increasing values of advance angle α.

at a price of increased peak phase current amplitude and significant increases in the torque ripple [41].

It should also be recognized that the ability to controllably advance the excitation angle α implies an increase in the sophistication of the rotor position sensing scheme. That is, the basic Hall sensor scheme discussed earlier in this section yields commutation signals at 60° electrical intervals with a fixed value of α. Suitable methods for electronically adjusting the excitation angle include the introduction of a phase-locked loop to interpolate angles between the discrete Hall sensor switching points. On the other hand, the value of α can be easily and accurately adjusted if the PMAC machine is designed to incorporate a high-resolution encoder or resolver.

Operation in the speed regime above 1.0 per unit means that the line-to-line back-EMF amplitude ($2 V_{f0}$) increasingly exceeds the source voltage as the speed is increased. This "overexcitation" condition poses no threat to the inverter switches as long as the drive system is operating normally so that the DC link voltage is maintained at its nominal value of V_s. However, a sudden removal of the inverter switch gating signals while the machine is operating at speeds above 1.0 per unit may transiently expose the inverter switches to elevated off-state voltages delivered by the machine's back-EMF sources as the machine decelerates in a generating mode. The presence of sufficient DC link energy storage or a shunt voltage regulator across the DC link which can absorb the regenerated energy will avert this potential overvoltage condition.

Impact of Extended Conduction Intervals. It was discussed earlier that the basic commutation control for a trapezoidal PMAC motor closes each inverter switch for 120° per electrical cycle (see Figure 6-14) so that each machine phase is open circuited for two 60° electrical intervals. In contrast, classic six-step operation of a voltage source inverter calls for each switch to be closed for 180° electrical per cycle corresponding to three (rather than two) closed switches at every time instant.

The difference between these two cases has a major impact on the torque production of a trapezoidal PMAC machine, as illustrated by the dashed-line curves in Figure 6-21 for 180° electrical conduction. These curves show that the high-speed torque production for a 15 kW trapezoidal PMAC motor can be considerably enhanced once the current regulators have saturated by converting the switch conduction intervals from 120° to 180° and advancing the excitation angle α [41]. Extensions of the high-speed operating range by 2:1 or more can be achieved by applying this technique.

An additional advantage of the conversion to 180° electrical conduction intervals at high-speeds is a reduction in the torque ripple amplitude compared to the baseline 120° electrical conduction case. On the other side of the ledger, the dynamic stability of the control loops suffers when the conversion to full 180° electrical conduction is made [42], as reflected in the reduced damping of the transient current and instantaneous torque waveforms which is a trademark of open loop voltage control.

Previous work has shown that the torque at a given speed in this high-speed regime increases monotonically as the switch conduction intervals are gradually extended from 120° to 180° [43]. Alternative algorithms have been proposed in the literature [42, 44] for actively controlling both the angular conduction intervals together with the advance angle in order to provide a smooth transition from current-regulated operation with 120° conduction intervals to six-step voltage excitation with 180° conduction intervals as the speed is increased. For example, one appealingly straightforward approach [44] holds the turn-off angle for each phase fixed while gradually advancing the turn-on angle. This approach has the desirable feature of simultaneously increasing the advance angle α and the conduction interval in a coordinated manner as the speed is raised, resulting in favourable high-speed preformance characteristics.

6.4. SINUSOIDAL PMAC MACHINE CONTROL

6.4.1. Machine Characteristics

While the stator windings of trapezoidal PMAC machines are concentrated into narrow phase belts, the windings of a sinusoidal machine are typically distributed over multiple slots in order to approximate a sinusoidal distribution, as described in Chapter 2. The resulting back-EMF waveforms generated by sinusoidal PMAC machines are, in fact, sinusoidally shaped in contrast to the characteristic trapezoidal waveforms discussed in Section 6.3.

Trapezoidal excitation strongly favors PMAC machines with nonsalient rotor designs (surface magnets) so that the phase inductances remain constant as the rotor rotates. In contrast, PMAC machines with salient rotor poles can offer appealing performance characteristics when excited sinusoidally, providing flexibility for adopting a variety of rotor geometries including inset or buried magnets (see Figure 6-1) as alternatives to the baseline surface magnet design [45–48].

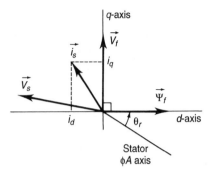

Figure 6-22. Basic phasor relationships for sinusoidal PMAC machine in synchronously rotating reference frame using d-axis alignment with rotor magnet flux $\vec{\Psi}_f$.

The most convenient manner of analyzing sinusoidal PMAC machine uses instantaneous current, voltage, and flux linkage phasors in a synchronously rotating reference frame locked to the rotor [45]. As indicated in Figure 6-22, the direct or d-axis has been aligned with the permanent magnet flux linkage phasor $\vec{\Psi}_f$, so that the orthogonal quadrature or q-axis is aligned with the resulting back-EMF phasor $\vec{V}_f$. The amplitude of the back-EMF phasor $\vec{V}_f$ can be expressed very simply as

$$V_f = p\,\omega_r\,\Psi_f \tag{6.4}$$

where p is the number of pole pairs and Ψ_f is the magnet flux linkage amplitude.

The sinusoidal three-phase current excitation can also be expressed in Figure 6-22 as an instantaneous current phasor $\vec{i}_s$ made up of d- and q-axis scalar components i_d and i_q, respectively, and the applied stator voltage phasor $\vec{v}_s$ can be similarly depicted. Coupled equivalent circuits can be developed for the direct and quadrature axes, as shown in Figure 6-23. Note that the magnet flux source represented by the equivalent current source I_f appears in the d-axis circuit while the resulting back-EMF appears as a dependent voltage source in the q-axis circuit, (i.e., a component of $\omega_e\,\Psi_{ds}$).

The values of magnetizing inductance L_{md} and L_{mq} are equal in a nonsalient PMAC machine. However, L_{md} will be smaller than L_{mq} in salient-pole PMAC machines using buried or inset magnets since the total magnet thickness appears

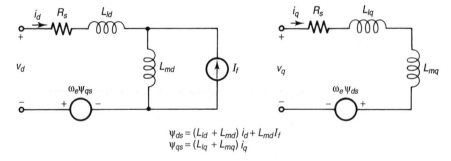

$$\Psi_{ds} = (L_{ld} + L_{md})\,i_d + L_{md}I_f$$
$$\Psi_{qs} = (L_{lq} + L_{mq})\,i_q$$

Figure 6-23. Coupled d–q equivalent circuits for a sinusoidal PMAC machine in the synchronously rotating reference frame as defined in Figure 6-22.

as an incremental air gap length in the d-axis magnetic circuit (i.e., $\mu_r \cong 1$ for ceramic and rare earth magnet materials) [45]. Interior PMAC machine designs of the type shown in Figure 6-1 with a single magnet barrier typically provide L_{mq}/L_{md} ratios in the vicinity of 3, while novel axially laminated designs have been reported with elevated inductance saliency ratios of 7 or higher [49].

This d–q phasor representation leads to the following general expression for the instantaneous torque developed in a sinusoidal PMAC machine,

$$T_e = 3p/2 \left[\Psi_f i_q + i_d i_q (L_d - L_q) \right] \tag{6.5}$$

where L_d and L_q are the d- and q-axis stator phase inductances, corresponding to $(L_{md} + L_{ld})$ and $(L_{mq} + L_{lq})$, respectively, in the Figure 6-23 equivalent circuits. Since L_q is typically larger than L_d in salient pole PMAC machines, it is worth noting that i_d and i_q must have opposite polarities for the second term to contribute a positive torque component.

Each of the two terms in the torque equation (6.5) has a useful physical interpretation. The first "magnet" torque term is independent of i_d but is directly proportional to stator current component i_q which is in phase with the back-EMF $\overrightarrow{V}_f$. In contrast, the second "reluctance" torque term in equation 6.5 is proportional to the $(i_d i_q)$ current component product and to the difference in the inductance values along the two axes $(L_d$–$L_q)$. This interpretation emphasizes the hybrid nature of the salient pole PMAC machine. Note that the torque is no longer linearly proportional to the stator current amplitude in the presence of magnetic circuit saliency.

The relative amplitudes of the magnet and reluctance torque terms in equation 6.5 are set during the machine design process by choosing the rotor topology and adjusting the magnet and rotor saliency dimensions in order to vary the relative amplitudes of Ψ_f and $(L_d$–$L_q)$. At one design extreme, the reluctance term naturally disappears in a nonsalient surface magnet PMAC machine since L_d equals L_q. At the other extreme, the machine design reverts to a pure synchronous reluctance machine if the magnet material is removed. The effects of these design choices are particularly apparent in the high-speed regime as will be discussed later in this section.

6.4.2. Basic Control Approach

Inverter Configuration. Torque control of sinusoidal PMAC machines is typically achieved using the same underlying concepts of self-synchronized current-regulated excitation that were introduced in Sections 6.2 and 6.3. However, the distinguishing characteristic of the sinusoidal excitation scheme is that the six inverter switches (Figure 6-5) have the added responsibility of sinusoidally shaping the applied excitation waveforms in addition to adjusting their amplitudes and frequency (that is, synchronization). A PWM switch control algorithm is typically used to eliminate all major low-order harmonics of the desired fundamental frequency excitation waveforms.

Alternative PWM algorithms can be used to perform the sinusoidal waveshaping on the applied phase voltage or phase current waveforms, depending on whether voltage source or current source characteristics are desired [50, 51]. The basic principles underlying several of these PWM techniques are reviewed in Chapter 4. In contrast to the trapezoidal PMAC drive, the sinusoidal version of closed loop current regulation conventionally requires at least two current sensors to directly measure the phase currents. (The third phase current can be inferred from the other two because of the assumed wye connection, so that $i_A + i_B + i_C = 0$.)

Open-Loop V/Hz Control. As discussed briefly in Section 6.1.1, buried-magnet PMAC machines can be designed with an induction motor squirrel cage winding embedded along the surface of the rotor as sketched in Figure 6-24. This hybridization adds a component of *asynchronous* torque production so that the PMAC machine can be operated stably from an inverter without position sensors. This simplification makes it practical to use a simple constant volts-per-hertz (V/Hz) control algorithm as shown in Figure 6-24 to achieve open loop speed control for applications such as pumps and fans that do not require fast dynamic response. According to this approach, a sinusoidal voltage PWM algorithm is implemented which linearly increases the amplitude of the applied fundamental voltage amplitude in proportion to the speed command to hold the stator magnetic flux approximately constant.

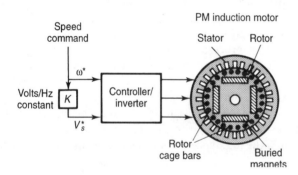

Figure 6-24. Cross section of a hybrid PM induction machine, showing simplified block diagram of open loop volts-per-hertz control scheme.

The open loop nature of this control scheme makes it necessary to avoid sudden large changes in the speed command or the applied load to avoid undesired loss of synchronization (pull-out) of the PMAC machine. However, an appealing aspect of this drive configuration is that the same constant volts-per-hertz control approach is used in many packaged induction motor drives for general-purpose industrial speed control applications. Thus, hybrid PM induction motors can be selected to replace conventional induction motors in some adjustable speed drive applications to improve system operating efficiency without changing the drive control electronics. An additional advantage of this open loop scheme is that the speed of the PMAC synchronous machine can be precisely controlled by the excitation frequency without suffering from the slip frequency differential which must be tolerated (or compensated) with asynchronous induction motors.

High-Performance Closed Loop Control. In contrast to the open loop operation described above, a rotor position sensor is typically required to achieve high-performance motion control with the sinusoidal PMAC machine. The rotor position feedback needed to continuously perform the self-synchronization function for a sinusoidal PMAC machine is significantly more demanding than for a trapezoidal machine which needs only six discrete switching points per electrical cycle. As result, an absolute encoder or resolver providing an equivalent digital resolution of 6 bits per electrical cycle (5.6° electrical) or higher is typically required with a sinusoidal PMAC machine, depending on the specific drive performance specifications. Resolution of the position sensor is important in some high-performance applications since it is one of the contributing factors to torque ripple production in a sinusoidal PMAC machine.

High-quality phase current regulation is the second key ingredient for achieving high-performance motion control with a sinusoidal PMAC machine. Referring back to the instantaneous phasor diagram in Figure 6-22, responsive current regulation plus self-synchronization make it possible to place the instantaneous current phasor $\vec{i}_s$ anywhere within the rotor-referenced d–q plane on command (limited only by maximum current and regulator saturation constraints). This ability to independently command the values of orthogonal current components i_d and i_q is tantamount to commanding the output torque in light of torque equation 6.5. The speed of the torque response is equivalent to that of the fast current regulators which is limited only by the source voltage and stator inductance values, assuming that there is no damper squirrel cage winding embedded in the rotor.

One baseline approach for implementing this type of high-performance torque control for sinusoidal PMAC machines is shown in Figure 6-25 [45]. According to this approach, the incoming torque command T_e^* (asterisk designates command) is mapped into commands for d–q axis current components i_d^* and i_q^* using functions f_d and f_q, respectively, which are extracted from torque equation 6.5. These current commands in the rotor d–q reference frame (DC quantities for a constant torque command) are then transformed into the instantaneous sinusoidal current commands for the individual stator phases (i_A^*, i_B^*, and i_C^*) using the rotor angle feedback θ_r and the inverse vector rotation equations [45] as follows;

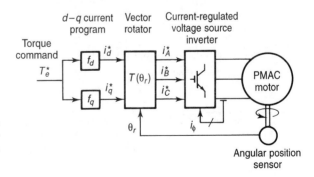

Figure 6-25. Block diagram of high-performance torque control scheme for sinusoidal PMAC motor using vector control principles [45].

$$\begin{bmatrix} i_A^* \\ i_B^* \\ i_C^* \end{bmatrix} = \begin{bmatrix} \cos\theta_r & -\sin\theta_r \\ \cos(\theta_r - 2\pi/3) & -\sin(\theta_r - 2\pi/3) \\ \cos(\theta_r - 4\pi/3) & -\sin(\theta_r - 4\pi/3) \end{bmatrix} \begin{bmatrix} i_d^* \\ i_q^* \end{bmatrix} \qquad (6.6)$$

where θ_r is the angle (in elec. degrees) between the stator phase A axis and the rotating d-axis (see Figure 6-22) and asterisks designate command values. Current regulators for each of the three stator current phases then operate to excite the phase windings with the desired current amplitudes.

The most common means of defining functions f_d and f_q to map the torque command T_e^* into the d–q current component commands i_d^* and i_q^* is to set a constraint of maximum torque-per-amp operation which is nearly equivalent to maximizing operating efficiency. Trajectories of the stator current phasors $\vec{i}_s$ in the rotor d–q reference frame which obey this maximum torque-per-amp constraint are plotted in Figure 6-26a and 6-26b for nonsalient and salient PMAC machines, respectively. These phasor trajectories are plotted over a range of torque amplitudes ranging from negative (generating/braking) to positive (motoring) values [52].

The differences between the maximum torque-per-amp trajectories for the two types of sinusoidal PMAC machine shown in these two plots highlight the differences between the torque production mechanisms in the nonsalient and salient machines. Torque-per-amp in the nonsalient PMAC machine is maximized by setting i_d to zero for all values of torque so that the sinusoidal phase currents are always exactly in phase (or 180° out of phase) with the sinusoidal back-EMF

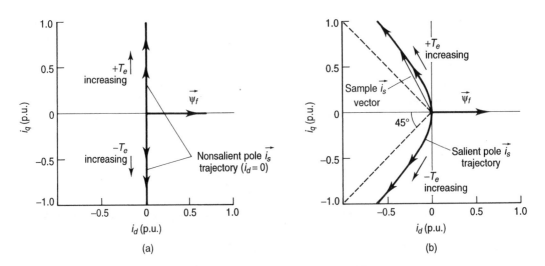

Figure 6-26. Trajectories of stator current phasors $\vec{i}_s$ in synchronously rotating i_d–i_q reference frame for maximum torque-per-amp control of (a) nonsalient pole and (b) salient pole sinusoidal PMAC machines.

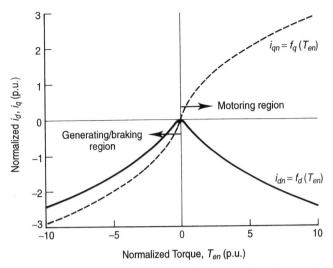

Figure 6-27. Plots of f_d and f_q functions for achieving maximum torque-per-amp operation in salient pole sinusoidal PMAC machines, using normalized torque and current variables with base quantities defined as follows: $i_b = \Psi_f/(L_q - L_d)$, and $T_{eb} = 1.5 p \Psi_f i_b$.

voltages. The close similarity of this control scheme to that described earlier for the nonsalient trapezoidal PMAC machines in Section 6.3 is readily apparent.

In contrast, the maximum torque-per-amp trajectory for a salient pole PM reluctance machine shown in Figure 6-26b initially moves along the q-axis for low values of torque before swinging symmetrically into the second and third quadrant along 45° asymptotes. For motoring operation, this means that maximum torque for a given amount of stator current is developed by advancing the phase angle of the stator phase currents so that they lead their respective back-EMF waveforms by angles between zero and 45° electrical. The maximum torque-per-amp trajectory in Figure 6-26b reflects the hybrid nature of the PM reluctance machine [45] since this trajectory asymptotes to the indicated 45° lines in the second and third quadrant for a pure synchronous reluctance machine (i.e., no magnets). The corresponding nonlinear f_d and f_q functions (see Figure 6-25) which define the maximum torque-per-amp trajectory for the salient pole PMAC machine are plotted in Figure 6-27.

Alternative field-oriented formulations of the sinusoidal PMAC machine control algorithm have been reported that are also capable of achieving high performance by aligning the rotating reference frame with the stator flux phasor rather than with the rotor magnet flux [53, 54]. Although the performance characteristics of these two control formulations are quite similar at low speeds, their differences may become more apparent at higher speeds during the transition from constant torque to constant horsepower operation [53, 55].

Regenerative Braking Operation. The regenerative braking operation in a sinusoidal PMAC machine generally follows the same principles as in the trapezoidal machines discussed in Section 6.3. For example, Figure 6-26a shows that negative braking torque is developed in the nonsalient sinusoidal machine when i_q is negative ($i_d = 0$), so that the phase currents are 180° out of phase with the back-EMF waveforms, just as in the trapezoidal machines. For the salient pole PMAC machine, Figure 6-26b shows that the maximum torque-per-amp trajectories are mirror images in the second and third quadrants. Stated in a different way, the f_q function plotted in Figure 6-27 is sensitive to the polarity of the torque command, while the f_d function depends only on its absolute value.

Just as in the case of the trapezoidal PMAC machine, the current-regulated voltage source inverter used with the sinusoidal PMAC machines naturally returns the regenerated braking energy to the DC link. The same techniques for handling this regenerated energy described earlier in Section 6.3 for the trapezoidal PMAC machine—local storage, dissipation, or returning it to the source—apply equally well to its sinusoidal counterpart.

6.4.3. Torque Ripple

The torque ripple associated with sinusoidal PMAC machines is generally less than that developed in trapezoidal machines, providing one of the major reasons that sinusoidal machines are preferred in high-performance motion control applications such as machine tools. The principal reason for this difference is that the sinusoidal PMAC machines do not experience the abrupt phase-to-phase current commutations that characterize the trapezoidal machine's excitation waveforms.

Although secondary sources of torque ripple exist in sinusoidal PMAC machines (e.g., slot effects, winding harmonics), standard techniques such as skewing and chorded windings that are used to attenuate torque ripple in other types of AC machines [25] can generally be applied to PMAC machines as well. While trapezoidal PMAC machines need highly concentrated phase windings for smooth torque production, the objective in sinusoidal PMAC machines is to distribute the stator phase windings in as many slots per pole as practical.

These well-established AC machine design techniques can be supplemented by special control algorithms to further suppress the residual sources of pulsating torque. One particularly popular approach is to supplement the basic fundamental frequency current commands with harmonic current components which are specifically adjusted in amplitude to actively compensate the remaining pulsating torque components [56–59]. Unfortunately, such compensation control techniques tend to be rather sensitive to the machine parameter values, making it necessary to introduce adaptive control techniques [60, 61] to accommodate parameter changes during normal operation.

Notwithstanding such practical difficulties, careful application of a combination of machine design and control techniques can yield sinusoidal PMAC machine drives which deliver very low levels of ripple torque ($< 2\%$ rated) for the most demanding servo applications.

6.4.4. High-Speed Operation

High-speed operation of sinusoidal PMAC machines is constrained by the same linear proportionalities of the back-EMF amplitude and inductive voltage drops to speed that characterize the trapezoidal PMAC machines discussed in Section 6.3.4. However, in contrast to the situation with trapezoidal PMAC machines, the current regulators for sinusoidal machines naturally saturate to six-step voltage excitation with 180° switch conduction intervals (i.e., three inverter switches on at all times). This difference makes it somewhat easier to achieve extended high-speed operation with sinusoidal PMAC machines, although special control provisions are still required.

Current and Voltage Limit Circles. To understand the limits of high-speed operation in sinusoidal PMAC machines, it is useful to begin by considering the impact of the voltage and current constraints imposed by the inverter. The limited current ratings of the inverter switches can be represented in the rotor-referenced i_d–i_q plane as a maximum current limit circle with radius $I_0 = 1$ centered at zero as shown by the solid line in Figure 6-28. That is, the current regulators can safely generate any current phasor $\vec{i_s}$ terminating inside this limit circle with a normalized amplitude of 1.0 or less.

The availability of a finite DC voltage source V_s at the inverter's input requires the stator phase current components to obey the following inequality [55],

$$\left[\frac{(2/\pi)V_s}{p\omega_r L_q}\right]^2 \geq i_q^2 + \left(\frac{L_d}{L_q}\right)^2 \left(i_d + \frac{\Psi_f}{L_d}\right)^2 \tag{6.7}$$

where ω_r is the rotor speed.

For a nonsalient machine $(L_d = L_q)$, this expression reduces to a set of voltage limit circles in the i_d–i_q plane as shown in Figure 6-28 which are centered at $(-\Psi_f/L_d)$ with radii proportional to V_s but inversely proportional to the rotor

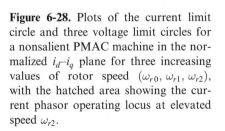

Figure 6-28. Plots of the current limit circle and three voltage limit circles for a nonsalient PMAC machine in the normalized i_d–i_q plane for three increasing values of rotor speed $(\omega_{r0}, \omega_{r1}, \omega_{r2})$, with the hatched area showing the current phasor operating locus at elevated speed ω_{r2}.

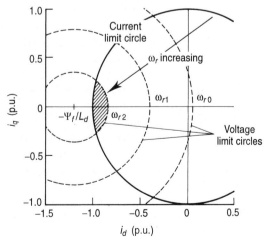

speed ω_r. Ignoring current limits for a moment, the current vector $\vec{i_s}$ can reach anywhere inside or on the voltage limit circle at a given rotor speed during steady-state operation—but not outside it. Since the voltage and current limits simultaneously constrain the machine drive's operation, the achievable set of current phasors at any speed is formed by the intersection of the voltage limit and current limit circles defined earlier. An example of such a locus at one particular rotor speed ω_{r2} is indicated by the hatched area in Figure 6-28.

It is important to note in Figure 6-28 that the size of the voltage limit circle shrinks as the rotor speed increases, progressively reducing the locus of achievable steady-state current phasors. At low speeds, the radius of the voltage limit circle is so large that the machine drive can operate anywhere within the fixed radius current limit circle. However, as the speed increases so that the peak line-to-line back-EMF amplitude approaches the DC link source voltage V_s, the voltage limit plays an increasingly dominant role in shrinking the size of the current phasor operating locus. This has the effect of simultaneously reducing the maximum torque which can be developed by the machine as the speed increases.

When the speed increases to the key threshold value ω_b where the line-to-line back-EMF equals the source voltage (i.e., equal fundamental frequency component amplitudes), the voltage limit circle has shrunk to the point that its outer periphery passes directly through the origin of the i_d-i_q plane $(i_d - i_q = 0)$. At speeds above this ω_b threshold, current must flow in the machine at all times to satisfy the terminal voltage requirements, even if the torque being produced is zero.

The same basic concepts apply equally well to a salient pole PMAC machine as to the nonsalient machine described earlier. The major difference is that the voltage limit circles for the nonsalient machine turn into voltage limit ellipses for the salient pole machines as a result of the different values of d-axis and q-axis inductance [55].

Flux Weakening Principles. Having introduced the concepts of current limit and voltage limit circles, the ability of the sinusoidal PMAC machine to develop useful torque in the elevated speed regime can be shown to depend entirely on the current phasor trajectory commanded by the controller. Consider the alternative current phasor trajectories plotted in the normalized i_d-i_q plane in Figure 6-29a for a non-salient PMAC machine during high-speed operation. Both of these candidate current phasor trajectories (A and B) meet the inverter-imposed constraints associated with maximum current ($i_s \leq I_0 = 1$ p.u.) and source voltage limits as were just discussed.

As indicated by trajectory A in Figure 6-29a, the controller can be designed to hold i_d at zero just as in the low-speed regime to maintain maximum torque-per-amp operation at all speeds. However, the voltage limit circle forces i_q to decrease rapidly as the speed is increased toward the ω_b threshold value, resulting in a rapid drop in the high-speed torque as shown in Figure 6-29b. Voltage limit circles are shown as dashed lines for three speed values to show how the current phasor is forced down the q-axis as speed is increased.

In comparison, an alternative approach (trajectory B in Figure 6-29a) is to control the current phasor to move into the second quadrant along the current limit circle ($| \vec{i_s} | = 1.0$ p.u.), corresponding to an advance of the current phase

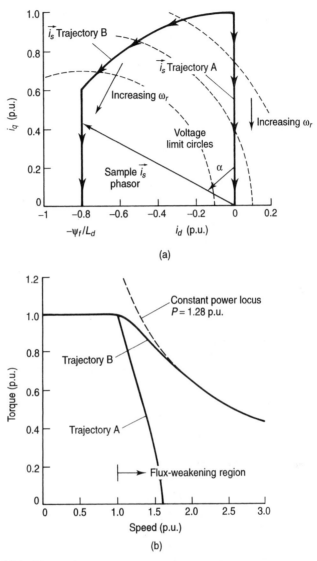

Figure 6-29. Alternative stator current phasor trajectories (A and B) are defined in the normalized i_d-i_q plane shown in (a) for a non-salient sinusoidal PMAC motor, yielding the contrasting torque-speed trajectories shown in (b) which illustrate the benefits of flux weakening (trajectory B) in the high-speed regime.

angle α. Compared to the preceding case with trajectory A, trajectory B allows the stator current to continue to flow and more torque to be produced at higher speeds even though the torque-per-amp values are lower than for trajectory A along the q-axis. This operating regime expansion can be confirmed by observing that the intersection of each high-speed voltage limit circle with the two trajectories occurs at a higher q-axis current level for trajectory B than for trajectory A, thereby generating higher-output torque. The new torque-speed envelope corresponding to trajectory B (see Figure 6-29b) is considerably expanded compared to that of trajectory A, following a constant-power hyperbola ($T_e \omega_r = P_0$) in the high-speed regime.

The technique of advancing the current angle at higher speeds is generally known as flux weakening since negative d-axis stator flux ($L_d i_d$) is added to counteract the positive d-axis magnet flux Ψ_f. It is worth emphasizing that flux weakening requires that the controller abandon the maximum torque-per-amp trajectory to add the necessary d-axis current to neutralize the magnet flux. Although this concept has been introduced in Figure 6-29 using a nonsalient PMAC machine, flux weakening can also be used very effectively in a salient PM reluctance machine by advancing the current phasor angle (counterclockwise) from the maximum torque-per-amp trajectory (see Figure 6-26b) in the second quadrant [55].

Several alternative control algorithms for implementing flux weakening have been proposed in the literature since the transition does not occur naturally in current-regulated PMAC machine drives [53, 55, 62–64]. One distinguishing feature among these various algorithms is whether the current regulators are allowed to fully saturate following the activation of flux-weakening control. An advantage of algorithms which prevent the current regulators from fully saturating (referred to as quasi-saturation) is that the drive system retains some of the faster dynamic response characteristics associated with current-regulated operation [55]. In contrast, algorithms which force a full transition into six-step voltage excitation [53] sacrifice some of this response speed in favor of increased inverter efficiency since the switching frequency of the inverter switches is minimized.

Impact of PMAC Machine Parameters. It is important to recognize that the maximum torque-speed operating envelope which can be achieved by a PMAC machine drive system is ultimately determined by the machine parameters. The control algorithm only determines whether the full potential of the PMAC machine for high-speed operation is realized or not. As a result, it is critically important that the high-speed operating requirements be taken into consideration when the machine is initially designed, since the capabilities of different PMAC designs vary enormously, and no clever control algorithm can overcome the fundamental limitations which are embodied in the machine's design.

The most important single machine parameter determining the size and the shape of the PMAC machine's torque-speed operating envelope is the ratio of the magnet flux Ψ_f to the d-axis inductance L_d [65], which has the units of current and is defined here as I_m where $I_m = \Psi_f/L_d$. In fact, analysis reveals that the constant-power speed range is fundamentally limited whenever I_m is greater than the maximum inverter current I_0.

Referring back to Figure 6-28, this design relationship means that center of the family of voltage limit circles (which occurs at $i_d = -I_m$) must fall inside the current limit circle as a prerequisite for achieving wide constant power speed ranges. This observation appeals to logic since the intersection of the contracting voltage limit circles and the current limit circle will shrink to the null set (meaning zero torque) at high speeds if the point $i_d = -I_m$ falls outside the current limit circle. Interestingly, the highest level of constant output power for a given source voltage can be achieved by designing the machine so that $I_m = -I_0$, causing the voltage limit circles in Figure 6-28 to be centered right on the periphery of the current limit circle.

One important machine design trade-off that must be considered when designing a PMAC machine for wide ranges of flux-weakening operation is that reducing the value of I_m compared to the maximum inverter current I_0 exposes the magnets to greater demagnetizing coercive force (MMF) during the course of flux-weakening operation [52]. In fact, reducing I_m to values less than I_0 can lead to a reversal of the flux polarity in the magnets during heavy flux-weakening operation (i.e., large negative values of i_d) corresponding to magnet operation in its third quadrant. This demanding type of operation makes it advisable to choose a magnet material with a straight characteristic throughout its second quadrant such as one of the rare earth magnet materials (neodymium-iron-boron or samarium-cobalt) if the widest possible range of flux-weakening operation is desired.

In practice, designing either surface magnet machines or interior magnet with a single barrier (as shown in Figure 6-1) to meet the prerequisite conditions for unlimited constant horsepower operation is difficult. As a result, high-speed flux-weakening operation for these types of machines is typically limited to ranges of 3:1 above base speed or less [55]. On the other hand, novel axially laminated PM reluctance machines have recently been reported that make it possible to meet the prerequisite conditions to achieve extended speed ranges of at least 7:1 [66]. Such results confirm the observation that the magnetic saliency associated with hybrid PM reluctance machines provides a crucial extra degree of machine design freedom which can be usefully exercised to achieve wide ranges of constant power operation [67].

6.5. ADVANCED CONTROL TECHNIQUES

6.5.1. Position Sensor Elimination

One of the most active areas of controls development during recent years involving PMAC machine drives has been the rapid evolution of new techniques for eliminating the rotor angular position sensor conventionally used for self-synchronization. Elimination of the shaft-mounted position sensor is a very desirable objective in many applications since this transducer is often one of the most expensive and fragile components in the entire drive system.

Trapezoidal PMAC Machines. The trapezoidal PMAC machine has been a particularly appealing target for this effort because one of its three stator phases is unexcited during each $60°$ electrical interval, making it possible to conveniently use the open circuit back-EMF generated in the unexcited phase as a position sensing

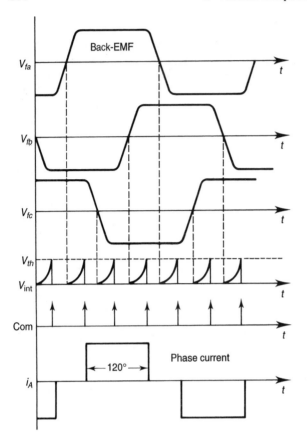

Figure 6-30. Waveforms illustrating operating principles of position sensor elimination algorithm for a trapezoidal PMAC drive [72] including (from top): back-EMF waveforms $V_{f\phi}$, integrated magnet flux signal V_{int}, inverter commutation signal *Com*, and phase A current i_A.

signal. A variety of specific algorithms have been developed [68–72] that use back-EMF voltage waveform measurements to determine the electronic commutation instants for trapezoidal machines without shaft-mounted position sensors.

An example of one of these algorithms which is based on magnet flux thresholds [72] can be explained using the simplified waveforms shown in Figure 6-30. According to this algorithm, the control begins to integrate the back-EMF in the unexcited winding as soon as it crosses zero during each 60° electrical time interval, thereby providing a measure of the incremental magnet flux which is shown as signal V_{int} in the figure,

$$V_{int} = \int_0^t V_{f\phi}(\tau)\,d\tau \qquad (6.8)$$

where $V_{f\phi}(\tau)$ represents the back-EMF of each phase and $\tau = 0$ corresponds to the instant when the back-EMF crosses zero.

The quadratic shapes of the V_{int} pulses result from the fact that the instantaneous back-EMF voltages vary approximately linearly in the vicinity of each zero crossing. The instant of the next commutation event (signal *Com* in

Figure 6-30) occurs when V_{int} reaches the preset fixed threshold voltage V_{th}, at which time the inverter switches change conduction states and the integrator is reset until the next back-EMF crossing.

One of the advantages of this back-EMF integration scheme is that the delay interval between the zero crossing and the comparator transition stays constant in terms of electrical degrees as the speed varies despite the fact that the back-EMF amplitude is increasing in proportion to speed. Stated in different terms, the development of the flux signal via integration eliminates the need to vary the comparator threshold as a function of speed to maintain the desired in-phase alignment of the current and back-EMF waveforms. In addition, adjustment of the comparator threshold V_{th} provides a convenient means of advancing the excitation phase angle α at elevated rotor speeds.

One of the special control issues that must be addressed in any such scheme which uses back-EMF sensing is starting from standstill at which point the back-EMF is zero. A typical approach for handling this condition is to initially override the position signal by sequentially stepping the commutation state machine at a fixed rate in the desired direction of rotation. This open loop excitation sequence provides initial acceleration to the rotor and generates a back-EMF signal with sufficient amplitude for the back-EMF sensing scheme to smoothly take over control of the inverter switch commutation sequencing.

Such back-EMF sensing schemes for trapezoidal PMAC machines are sufficiently mature and straightforward that they have been successfully implemented in integrated circuits [71–73], some of which are now in commercial production. Applications for this type of sensorless control include computer disk drives, compact disk (CD) players, blowers, and pumps.

Sinusoidal PMAC Machines. Position sensor elimination in sinusoidal PMAC machines is more challenging because all three machine phases are continuously excited, preventing direct access to the open circuit back-EMF voltages. As a result, more sophisticated estimation and observer techniques are generally required to extract position information from measurements of the phase current and terminal voltages [68].

A variety of alternative algorithms have been proposed and investigated for rotor position estimation in sinusoidal PMAC machines, and several of these involve the manipulation of magnetic flux amplitudes and phasor orientations [74, 75]. The computational requirements of some of these algorithms can be significant, calling for the application of high-speed digital signal processors [76]. Sensitivity to machine parameter variations can be important in some cases, requiring the addition of on-line parameter adaptation schemes for high-performance operation [77]. As in the case of the back-EMF sensing schemes, special open loop starting schemes are required at standstill because of the absence of back-EMF voltages at zero speed.

Hybrid PM reluctance machines provide the basis for a different approach to indirect position sensing in PMAC machines using the angular variation of the phase inductances as the shaft rotates. One of the advantages of this scheme over alternative approaches used with nonsalient PMAC machines is that the sensing algorithm can be designed to detect inductance variations at low speeds down to standstill, avoiding the problems associated with the disappearance of the back-EMF voltages at zero speed. Alternative algorithms have been proposed for extracting the position estimates from interior PM machines based on either active probing of the machine phases with test pulses [78] or passive measurements of the voltage and current waveforms during normal operation [79].

6.5.2. Current Sensor Elimination and Advanced Regulators

Attention has also been addressed to eliminating or minimizing the need for discrete current sensors in current-regulated PMAC drives. For example, it has been shown that the three-phase current values for a sinusoidal PMAC machine can be reconstructed from DC link current measurements using a single sensor with the appropriate control [80]. The success of such schemes depends on modifications of the basic PWM algorithm to ensure that the single current sensor is granted access to each of the three-phase currents on a sufficiently frequent basis to avoid undesired current deviations. The net result is a modest degradation of the current regulator performance which appears in the form of increased waveform distortion, even though the system dynamic response is still very fast.

For trapezoidal PMAC machine drives, discrete current sensors can be eliminated entirely using current sensors embedded in three of the six inverter switches. MOS-gated power semiconductor devices including power MOSFETs and IGBTs are available from several manufacturers under a variety of trade names that incorporate current sensors integrated into the monolithic power device structures. By using such devices for the three lower inverter switches (T2, T4, and T6 in Figure 6-6), the phase currents in the trapezoidal PMAC machine can be controlled very tightly under all operating conditions [81]. In fact, the presence of a separate current sensor integrated into each inverter phase leg makes it possible to reduce the opportunities for transient overcurrents during phase-to-phase commutations that were described earlier in Section 6.3.

In addition, work has been reported on the application of more sophisticated current regulators to PMAC drives [82, 83] using space vector modulation techniques. Such advanced regulators use predictive algorithms combined with an internal model of the PMAC machine to choose the best inverter switches to close at each switching instant to minimize the instantaneous current errors. Attractive features of this approach include the fast dynamic response characteristics achievable using a sampled-data configuration that is directly compatible with an all-digital current regulator implementation. On the other hand, the current regulator performance may become more dependent on the actual machine parameters and the DC link voltage amplitude when using these predictive techniques, so parameter adaptation may be necessary to maintain high performance under all operating conditions.

Neural networks are also under investigation as a means of implementing self-learning current regulators [84] and "optimal" converter control [85]. This work has succeeded in demonstrating that neural network structures based on back-propagation techniques with a single layer can achieve high-performance current regulation characteristics. Such neural networks lend themselves to digital implementations which have the potential for high-speed execution, permitting the adoption of high switching frequencies. However, neural networks also pose some special requirements such as network training which must be appropriately accommodated in practical applications. More discussion of the future role of neural networks in power electronics can be found in Chapter 11.

6.5.3. Robust Control

Considerable effort has also been devoted to the development of various classes of robust control algorithms for PMAC machine drives. Since the topic of robust control applied to AC drives is thoroughly discussed in Chapter 9, attention here is limited to highlighting a few of the specific applications of these techniques to PMAC machine drives. For example, there have been several papers describing applications of variable structure (i.e., sliding mode) and adaptive control algorithms for position and/or velocity control of PMAC drives [86–89]. Generally, the objective of these alternative control configurations is to improve the ability of the PMAC servo drive to tolerate wide swings in load inertia or motor parameter values without suffering dynamic performance degradation. Fuzzy logic is also receiving attention [90–92] as a promising alternative controls approach for achieving fast adaptation to load changes in PMAC drives.

6.6. PMAC DRIVE APPLICATION ISSUES

6.6.1. Motor Drive Comparisons

Table 6-1 provides a qualitative comparison of key drive application characteristics for trapezoidal and sinusoidal PMAC motor drives drawn from the discussions in Sections 6.3 and 6.4. Induction motor drives are included in this table as a baseline for comparison because of their wide popularity for many adjustable speed applications. Switched reluctance (SR) motor drives are also included, recognizing the considerable attention they have drawn during recent years as candidates for brushless drive applications [93]. The table has been arranged to highlight the attractive features of each of the machine drive technologies without attempting to assign quantitative scores which depend too much on application requirements.

Despite the risks of oversimplification, some general observations can be made from the Table 6-1 entries. PMAC machines are particularly appealing in drive applications which benefit from the efficiency advantages held by PMAC machines (both trapezoidal and sinusoidal) over induction and SR machines. Trapezoidal PMAC machines are particularly attractive in drive applications that can take advantage of the simple controls and minimum sensors required by this type

TABLE 6-1 APPLICATION CHARACTERISTIC COMPARISONS (A "+" IN A COLUMN DENOTES NET ADVANTAGE)

	Trapezoidal PMAC	Sinusoidal PMAC	Induction AC	Switched Reluctance
Motor efficiency	+	+		
Torque smoothness		+	+	
Open loop control		+*	+	
Closed loop simplicity	+			+
Minimum control sensors	+			
Extended speed range		+	+	+
Motor robustness		+†	+	+

*With rotor cage.

†For buried-magnet rotor.

of machine as discussed in Section 6.3. SR machines also lend themselves to simple closed loop control in relatively undemanding speed-control applications that can tolerate fixed commutation angles and simple current regulation of the excitation pulse amplitudes [94].

However, one of the prices for the drive controls simplicity of the trapezoidal PMAC and SR machine is their higher torque ripple compared to the other two sinusoidally excited machines. An additional weakness of the trapezoidal PMAC machine compared to the other three candidates is its more limited extended speed range in the constant horsepower regime. Each of the other three types of machine drive can be designed to achieve wide speed operating ranges above base speed. Interestingly, a sinusoidal PM reluctance machine drive designed with just the right combination of machine parameters may be able to achieve the widest constant horsepower speed range of all [67].

While SR machines excel in the category of mechanical ruggedness due to the absence of spinning magnets or conductors, sinusoidal PMAC machines can also be designed with enhanced robustness characteristics by burying the magnets inside the rotor. Squirrel cage induction machines likewise benefit from their undeniably rugged rotor construction while also sharing the advantage of smooth torque production with sinusoidal PMAC machines. As was pointed out in Section 6.3, PMAC and induction machines can be combined by adding an auxiliary squirrel cage to the rotor of an interior magnet machine, making it possible to achieve open loop volts-per-hertz control, which is one of the attractive features of the basic induction motor.

6.6.2. PMAC Drive Application Trends

The specific positive characteristics of PMAC machines summarized earlier make them highly attractive candidates for several classes of drive applications, a few of which will be discussed in this section.

Figure 6-31. Prototype dual-tandem electrohydrostatic actuator (EHA) incorporating two 12 hp vector-controlled PMAC machines [98] for positioning an aircraft flight control surface. (Photo courtesy of Lockheed Martin Control Systems.)

Servo Actuators. The superior power density of PMAC machines is invaluable for applications which require maximum dynamic response including machine tool servos [95] and robotic actuator drives [96]. The absence of rotor losses is another feature that makes PMAC machines attractive candidates for servos, particularly in applications that require significant intervals of low-speed positioning operation. The desire for the highest possible performance tends to drive these PMAC machine designs to incorporation of rare-earth magnet materials (neodymium-iron-boron or samarium-cobalt) in order to minimize both the machine size and rotor inertia.

One of the promising areas for PMAC development includes electric-powered actuators for aircraft flight control surfaces for which high efficiency and minimal weight are critical specifications. Although trapezoidal PMAC machines were applied in some of the earlier development efforts [19, 97] in order to take advantage of the associated controls simplicity, more recent designs are using sinusoidal PMAC machines as well. Figure 6-31 shows an example of a large dual-tandem electrohydrostatic actuator (EHA) consisting of two 12 hp sinusoidal PMAC machines driving separate hydraulic loops with mechanical summing of the output forces in the hydraulic piston ram visible in the foreground [98].

The PMAC machines in the Figure 6-31 system use an interior magnet configuration with neodymium-iron-boron magnets and have been tested at speeds up to 18,000 rpm. The efficiencies of these machines have been measured in the vicinity of 95%. The machines are equipped with resolvers to provide high-resolution angular position feedback, and the controllers are designed with high-speed digital signal processors (DSPs) to execute high-performance vector control based on the algorithm described earlier in Figure 6-25. Small-signal velocity loop bandwidths in excess of 100 Hz have been demonstrated using this equipment, sufficient to meet the dynamic response requirements for the intended aircraft control surfaces.

Commercial-Residential Speed Control Applications. Opportunities for PMAC motor drives are also growing in a wide range of commercial and residential applications. For example, use of such drives is expanding in domestic appliances and heating, ventilating, and air conditioning equipment as a result of consumer and regulatory pressures for higher operating efficiencies in several parts of the world [99]. The basic ability to control compressor and blower speeds can provide a significant boost to appliance and HVAC equipment efficiency over a range of operating conditions, and the attractive efficiency characteristics of PMAC machines further enhance this advantage. Although PMAC machine drives have been under development for such applications for several years [26, 100], cost reduction is the single largest challenge for achieving wider market acceptance of these drives in such high-volume commercial-residential applications.

Trapezoidal PMAC motors have generally been the most popular choice for this class of cost-sensitive application to take advantage of the simpler controls and minimum sensor requirements associated with this type of machine. Controls for trapezoidal PMAC machines have been further simplified over the years by the development of special integrated circuits which can perform the majority of the necessary logic-level controls functions on a single chip [71–73, 101]. The availability of such controller circuitry improvements has accelerated a trend toward higher levels of physical integration of the motor and drive controls. Figure 6-32 shows an example of a self-contained 0.5 hp PMAC motor drive product for HVAC blower applications that packages the controller electronics in a concentric shell fastened to the back of the machine. Such products are being designed with increasing levels of programmability and controls sophistication, using microprocessors to implement such advanced features as closed loop air flow control in HVAC equipment.

Such physical integration concepts are pushed even further in low-power (< 50W) tube-axial fans for cooling of electronics equipment. In one documented case [102], cost has been reduced by adopting a single-phase PMAC machine configuration using bifilar windings and an external rotor which can be driven by a single integrated circuit combined with a minimum of additional resistors and capacitors. Although some efficiency performance is sacrificed in the name of cost, such simplification makes it possible to package the controller on a small printed circuit card mounted immediately adjacent to the stator windings inside the fan hub. External rotor PMAC machines are also widely used in computer disk drives which demand the maximum possible acceleration in ever-smaller volumes [103].

Automotive Applications. Another application area for efficient variable speed drives with high market volume potential is electric vehicles. Sinusoidal PMAC machines are the clear favorite over their trapezoidal counterparts for this traction application because of the need for wide ranges of constant horsepower

Figure 6-32. Example of PMAC motor drive (0.5 hp, case: 19 × 14 cm OD) designed for residential-commercial blower applications showing integration of drive electronics into motor housing. (Photo courtesy of GE Motors.)

operation. Although the low-loss characteristics of PMAC machines make them appealing candidates for traction motors in battery-supplied vehicles [104–106], this advantage can be compromised by the need to maintain high efficiency levels over the complete speed range. This can be particularly difficult near the top end of the constant horsepower speed range where the d-axis stator current needed to neutralize the magnet flux via flux weakening generates losses without contributing useful output torque.

Proper choice of rotor geometry and parameters can minimize such high-speed efficiency disadvantages, but the drive designer must struggle with often-conflicting requirements to simultaneously maximize motor efficiency and mechanical robustness while minimizing drive system cost. For example, a promising path to improved high-speed performance is to design the PMAC machine as an interior magnet geometry with multiple rotor flux barriers to increase the magnetic saliency [66], but this tends to degrade the machine's mechanical robustness while increasing its manufacturing cost. As a result of such trade-offs, the PMAC machine faces intense competition from induction motors and other machine types to demonstrate overall superiority for this high-visibility application.

Despite these challenges, there is no question that high-performance electric vehicle drives can be successfully built using PMAC machines. Figure 6-33 shows an example of an electric vehicle transaxle assembly which incorporates a 70 hp interior PM machine and a two-speed step-down gearbox [106]. The machine uses neodymium-iron-boron magnets in a single-barrier rotor geometry (closely related to that in Figure 6-1b), designed for a maximum operating speed of 11,000 rpm. The constant horsepower regime covers a speed range of approximately 2:1, and the

Figure 6-33. Transaxle assembly for a battery-powered electric van incorporating a 70 hp interior-magnet PMAC machine [106] and a
two-speed transmission. (Photo courtesy of Ford Motor Co.
This assembly was developed jointly by the Ford Motor Co.
and GE under a contract sponsored by the U.S. Department
of Energy.)

microprocessor-based vector control scheme uses a reference frame synchronized
with the stator flux [53], as discussed briefly in Section 6.4. One of the special
features of this controls design is tight regulation of the instantaneous torque
amplitude that makes it possible to achieve extremely smooth gear shifts which
are barely noticeable to the driver or passengers.

In addition to the traction drives, PMAC machine drives are also being actively
developed for important auxiliary functions in electric vehicles such as electric air
conditioning. As with traction drives, high efficiency is critically important for the
air-conditioning system which can consume 3 hp or more of precious battery power
[107]. PMAC machines are also being seriously considered for nearer-term accessory
applications such as electric power steering and active suspension in conventional
internal combustion vehicles. One of the enticing advantages of electric accessory
systems is that they are freed from the belts and pulleys which constrain the physical
placement of conventional hydraulic pumps and accessories around the engine.

High-Power Industrial and Propulsion Drives. Although the large majority
of reported applications for PMAC machine drives are in the range below 50 kW,
development activity has also been reported at much higher power levels. One example is the 1.1 mW six-phase PMAC machine drive mentioned earlier in Section 6.2
(refer to Figure 6-9) which has been built and tested for a low-speed (230 rpm) direct
drive marine propulsion application [23]. The stator of this trapezoidal PMAC
machine is made in eight 45° angular segments (stator ID = 1.25 m), and the phase

windings are broken into halves to permit series or parallel excitation depending on the speed range. Savings of approximately 40% in machine weight, length, and volume are claimed compared to a conventional DC propulsion motor it replaces.

Another interesting approach to achieving output power levels as high as 2 MW with PMAC machines has been reported using axial-airgap disk machines [108]. These machines are designed with neodymium-iron-boron magnets and aggressive stator cooling in order to achieve high specific power densities in the range of 2 to 3 kW/kg. Measured machine efficiency values in the range of 95 to 96% have been reported for a 430 hp unit tested at 3800 rpm. Intended applications for this PMAC drive technology include land and marine propulsion systems, as well as industrial pumping.

On the other hand, it is worth pointing out that PMAC machine drives can be manufactured at power levels above 100 kW without resorting to exotic materials or unusual machine geometries. For example, trapezoidal PMAC machine drives have been commercially offered with ratings up to at least 448 kW (600 hp) using ferrite magnets in conventional radial air gap configurations [109]. PWM transistor inverters are supplied with these machines to operate them at speeds up to 1750 rpm. Both air-cooled and liquid-cooled versions of these PMAC machines have been announced, each equipped with an integral encoder for commutation control.

6.6.3. Future Application Trends

Although PMAC machines are not likely to challenge induction machines for dominance of the general-purpose adjustable speed drive market, growing demands for high efficiency and high power density in a wide range of specialty applications bode well for the future of PMAC machine drives. Thus, significant international development activity on all aspects of PMAC machines and drives should be expected to continue. This work will undoubtedly involve continued development of improved permanent magnet materials, including new grades of rare-earth magnets with still higher energy products. New PMAC machine configurations should also be expected which expand the drive system's operating envelope to either higher low-speed torque [38] or higher maximum speeds [20]. As a result of such improvements, PMAC machine drives will continue to be formidable candidates for future high-performance brushless positioning servo applications.

As the demand for high-efficiency motor control increases in commercial-residential products, international competition is growing for development of low-cost fractional horsepower (< 1 kW) PMAC drives for high-volume OEM products, including appliances and HVAC equipment. As a result, the trend toward increasing levels of physical integration of the motor and drive controls as reflected in Figure 6-32 should be expected to accelerate further. New logic-level and power integrated circuits [110] are likely to appear that further simplify the PMAC motor controls and contribute to additional system cost reductions.

Opportunities for PMAC machines in automotive accessory applications for conventional vehicles will also grow, benefiting from the same integration and cost-cutting improvements that are propelling the technology into commercial-residential products. However, the prospects for PMAC machines as traction drives in future electric vehicles are more difficult to predict in light of intense competition from induction machines and other drive technologies. Success in this demanding application will hinge on the development of PMAC machines which combine features of low manufacturing cost, wide constant horsepower speed ranges, and high operating efficiencies under all operating conditions. Axially laminated PMAC machines could be serious contenders for meeting these multiple requirements if manufacturing costs can be reduced to the levels of induction motors that provide the standard baseline for comparison.

6.7. CONCLUSION

This chapter has reviewed many of the principal motion control techniques developed during recent years for PMAC machines. Key conclusions include the following:

- Control techniques are available to extract the high-efficiency operation and fast dynamic response that are inherent characteristics of well-designed PMAC machines.
- Trapezoidal PMAC machine (brushless DC) drives are particularly well suited for high-efficiency speed control applications that require simple controls with minimum sensors for cost-sensitive applications.
- Sinusoidal PMAC machine drives are capable of achieving high-performance servo control plus extended high-speed operation using vector control techniques.
- Interior magnet PMAC machines provide an extra degree of design freedom that can be used to increase rotor robustness while achieving wide ranges of constant horsepower operation.
- Significant progress has been made on the development of advanced control techniques that eliminate the need for rotor position sensors in many applications.
- Availability of higher-strength magnet materials, improved power switches, and new integrated circuits are providing the ingredients for intriguing new PMAC machines and drive configurations with important commercial implications.

In view of these developments, PMAC machine drives can be expected to enjoy a bright future in a wide variety of adjustable speed drive applications where high efficiency and high dynamic performance are valued—applications including machine tool servos, HVAC blowers and compressors, major appliances, automotive accessories, and vehicle propulsion drives.

References

[1] "Manufacturers, environmentalists agree on national appliance standards," *Air Conditioning, Heating and Refrigeration News*, pp. 1, 21, August 18, 1986.

[2] Zijlstra, H., "Application of permanent magnets in electromechanical power converters: The impact of Nd-Fe-B magnets," *J. Physique*, pp. C6-3–C6-8, September 1985.

[3] Strauss, F., "Synchronous machines with rotating PM fields," *AIEE Trans.*, Vol. 70, Part II, pp. 1578–1581, 1952.

[4] Herschberger, D. D., "Design considerations of fractional horsepower size permanent magnet motors and generators," *AIEE Trans.*, Vol. 72, Part III, pp. 581–585, 1953.

[5] Volkrodt, W., "Machines of medium-high rating with a ferrite-magnet field," *Siemens Review*, Vol. 43, no. 6, pp. 248–254, 1976.

[6] Richter, E., T. Miller, T. Neumann, and T. Hudson, "The ferrite permanent magnet AC motor—a technical and economical assessment," *IEEE Trans. Ind. Appl.*, Vol. IA-21, no. 4, pp. 644–650, May/June 1985.

[7] Rahman, M. A., and A. M. Osheiba, "Performance of large line-start PM synchronous motors," *IEEE Trans. Energy Conv.*, Vol. 5, no. 1, pp. 211–217, 1990.

[8] *DC Motors, Speed Controls, Servo Systems*, 1st Ed., Electro-Craft Corp., Hopkins, MN, 1972.

[9] Noodleman, S., "Application of rare earth magnets to DC machines," *Proc. 2nd Intl. Workshop on Rare-Earth Permanent Magnets*, Dayton, paper no. IV-1, June 1976.

[10] Gaede, H., "A new brushless DC drive for industrial power ranges and applications," *Proc. IEE Conf. Electr. Var. Speed Drives*, Vol. 93, pp. 132–136, 1972.

[11] Sawyer, B., and J. Edge, "Design of a samarium-cobalt brushless DC motor for electromechanical actuator applications," *IEEE Nat. Aerospace & Electronics Conf. (NAECON)*, Dayton, pp. 1108–1112, May 1977.

[12] Pfaff, G., A. Weschta, and A. Wick, "Design and experimental results of a brushless AC servo drive," *IEEE-IAS Annual Meeting Rec.*, pp. 692–697, 1982.

[13] Neft, C. L., and C. D. Schauder, "Theory and design of a 30 hp matrix converter," *IEEE Trans. Indus. Appl.*, Vol. 28, no. 3, pp. 546–551, May/June 1992.

[14] Slemon, G. R., and A. V. Gumaste, "Steady-state analysis of a PM synchronous motor drive with current source inverters," *IEEE Trans. Indus. Appl.*, Vol. 19, no. 2, pp. 190–197, March/April 1983.

[15] Ooi, B., P. Brissoneau, and L. Brugel, "Optimal winding design of a permanent magnet motor for self-controlled inverter operation," *Electric Machines & Electromechanics*, Vol. 6, no. 5, pp. 381–389, September/October 1981.

[16] Lajoie-Mazenc, M., R. Carlson, F. Giraud, and Y. Surchamp, "An electric machine with electronic commutation using high-energy ferrite," *Proc. IEE Conf. Small Electr. Machines*, pp. 31–34, 1976.

[17] Bose, B. K., *Power Electronics and AC Drives*, Prentice Hall, Englewood Cliffs, NJ, pp. 173–177, 1980.

[18] Bose, B. K., "Power electronics and motion control—technology status and recent trends," *IEEE Trans. Indus. Appl.*, Vol. 29, no. 5, pp. 902–909, September/October 1993.

[19] Demerdash, N., and T. Nehl, "Dynamic modeling of brushless DC motor-power condition unit for electromechanical actuator application," *Proc. IEEE Power Elec. Spec. Conf. (PESC)*, pp. 333–343, 1979.

[20] Takahashi, I., T. Koganezawa, G. Su, and K. Oyama, "A super high speed PM motor drive system by a quasi-current source inverter," *Rec. IEEE Ind. Appl. Soc. Annual Meeting*, Toronto, pp. 657–662, 1993.

[21] Acarnley, P. P., and A. G. Jack, "Power circuits for small PM brushless DC drives," *Proc. IEE Pow. Elec. & Var. Spd. Drives (PEVSD) Conf.*, Pub. no. 291, pp. 237–240, 1988.

[22] Bolton, H., Y. Liu, and N. Mallinson, "Investigation into a class of brushless DC motor with quasisquare voltages and currents," *IEE Proc.*, Vol. 133, Part B, no. 2, pp. 103–111, March 1986.

[23] Bausch, H., "Large power variable speed AC machines with PM excitation," *J. Elec. & Electronics Eng.*, Australia, Vol. 10, no. 2, pp. 102–109, June 1990.

[24] Letellier, M., "Les propulsions electriques a bord des sous-marins," *La technique moderne*, July/August 1989.

[25] Slemon, G. R., and A. Straughen, *Electric Machines*, Addison-Wesley, Reading, MA, 1980.

[26] Erdman, D., H. Harms, and J. Oldenkamp, "Electronically commutated DC motors for the appliance industry," *Rec. IEEE Ind. Appl. Soc. Annual Meeting*, Chicago, pp. 1339–1345, 1984.

[27] Johnson, T., "Primary flight control actuation with electric motors," *Proc. IEEE NAECON*, Dayton, pp. 1102–1107, 1977.

[28] Williams, S., "Direct drive system for an industrial robot using a brushless DC motor," *Proc. IEE*, Vol. 132, Part B, no. 1, pp. 53–56, 1985.

[29] D'Azzo, J., and C. Houpis, *Feedback Control System Analysis and Synthesis*, McGraw-Hill, New York, 1966.

[30] Kusko, A., and S. M. Peeran, "Definition of the brushless DC motor," *IEEE-IAS Annual Meeting Rec.*, pp. 20–22, 1988.

[31] Patni, C. K., and B. W. Williams, "The effect of modulation techniques & electromagnetic design on torque ripple in brushless DC motors," *Proc. IEE PEVSD Conf.*, Pub. no. 264, pp. 76–79, 1986.

[32] Miller, T. J. E., *Brushless PM and Reluctance Motor Drives*, Clarendon Press, Oxford, 1989.

[33] Dote, Y., *Servo Motor and Motion Control Using Digital Signal Processors*, Prentice-Hall, Englewood Cliffs, NJ, pp. 41–56, 1990.

[34] Lorenz, R. D., "Sensors and signal converters in PMAC motor drives," *Tutorial Notes: Performance & Design of PMAC Motor Drives*, THO408-5, IEEE Pub. Serv., 1991.

[35] Becerra, R. C., M. Ehsani, and T. M. Jahns, "Four-quadrant brushless ECM drive with integrated current regulation," *IEEE Trans. Indus. Appl.*, Vol. 28, no. 4, pp. 833–841, July/August, 1992.

[36] Li, T., and G. R. Slemon, "Reduction of cogging torque in PM motors," *IEEE Trans. Magnetics*, Vol. 24, no. 6, pp. 2901–2903, 1988.

[37] Bolton, H., and R. Ashen, "Influence of motor design and feed-current waveform on torque ripple in brushless DC drives, *IEE Proc.*, Vol. 131, Part B, no. 3, pp. 82–90, May 1984.

[38] Leonardi, F., M. Venturini, and A. Vismara, "Design and optimization of very high torque, low ripple, low cogging PM motors for direct driving optical telescopes," *Rec. IEEE Ind. Appl. Soc. Annual Meeting*, Denver, 1994.

[39] Schulting, L., and H.-Ch. Skudelny, "A control method for permanent magnet synchronous motors with trapezoidal electromotive force," *Proc. European Pow. Elec. Conference (EPE)*, Vol. 4, pp. 117–122, 1991.

[40] Carlson, R., M. Lajoie-Mazenc, and J. C. Fagundes, "Analysis of torque ripple due to phase commutations in brushless DC machines," *IEEE-IAS Annual Meeting Rec.*, pp. 287–292, 1990.

[41] Jahns, T. M., "Torque production in permanent magnet synchronous motor drives with rectangular current excitation," *IEEE Trans. Indus. Appl.*, Vol. 20, no. 4, pp. 803–813, July/August 1984.

[42] Schaefer, G., "Field weakening of brushless PM servomotors with rectangular current," *Proc. EPE Conf.*, Vol. 3, pp. 429–434, 1991.

[43] Becerra, R. C., and M. Ehsani, "High-speed torque control of brushless PM motors," *IEEE Trans. Indus. Elec.*, Vol. 35, no. 3, pp. 402–405, August 1988.

[44] Fratta, A., A. Vagati, and F. Villata, "Extending the voltage saturated performance of a DC brushless drive," *Proc. EPE Conf.*, Vol. 4, pp. 134–138, 1991.

[45] Jahns, T. M., G. B. Kliman, and T. W. Neumann, "Interior PM synchronous motors for adjustable-speed drives," *IEEE Trans. Indus. Appl.*, Vol. 22, no. 4, pp. 738–747, July/August 1986.

[46] Sebastian, T., and G. R. Slemon, "Operating limits of inverter-driven PM motor drives," *IEEE Trans. Indus. Appl.*, Vol. 23, no. 2, pp. 327–333, March/April, 1987.

[47] Alasuvanto, T., and T. Jokenen, "Comparison of four different permanent magnet rotor constructions," *Proc. Int. Conf. Electr. Machines (ICEM)*, Boston, pp. 1034–1039, 1990.

[48] Russenschuck, S., and E. Andresen, "Comparison of different magnet configuration in the rotor of synchronous machines using numerical field calculation and vector-optimization methods," *Proc. Int. Conf. Electr. Machines (ICEM)*, Boston, pp. 1027–1033, 1990.

[49] Fratta, F., A. Vagati, and F. Villata, "Design criteria of an IPM machine suitable for field-weakened operation," *Proc. Int. Conf. Electr. Machines (ICEM)*, Boston, pp. 1059–1065, 1990.

[50] Enjeti, P. N., P. D. Ziogas, and J. L. Lindsay, "Programmed PWM techniques to eliminate harmonics: A critical evaluation," *IEEE Trans. Indus. Appl.*, Vol. 26, no. 2, pp. 302–316, March/April 1990.

[51] Brod, D. M., and D. W. Novotny, "Current control of VSI-PWM inverters," *IEEE Trans. Indus. Appl.*, Vol. 21, no. 3, pp. 562–570, May/June 1985.

[52] Morimoto, S., Y. Takeda, T. Hirasa, and K. Taniguchi, "Expansion of operating limits for PM motor by optimum flux weakening," *IEEE Trans. Indus. Appl.*, Vol. 26, no. 5, pp. 866–871, September/October 1990.

[53] Bose, B. K., "A high-performance inverter-fed drive system of an interior PM synchronous machine," *IEEE Trans. Indus. Appl.*, Vol. 24, no. 6, pp. 987–998, November/December 1988.

[54] Bilewski, M., A. Fratta, L. Giordano, A. Vagati, and F. Villata, "Control of high performance interior PM synchronous drives," *IEEE-IAS Annual Meeting Rec.*, 1990, pp. 531–538.

[55] Jahns, T. M., "Flux weakening regime operation of an interior PM synchronous motor drive," *IEEE Trans. Indus. Appl.*, Vol. 23, no. 4, pp. 681–689, July/August 1987.

[56] Favre, E., L. Cardoletti, and M. Jufer, "Permanent-magnet synchronous motors: a comprehensive approach to cogging torque suppression," *IEEE Trans. Ind. Appl.*, Vol. 29, no. 6, pp. 1141–1149, November/December 1993.

[57] Hanselman, D., "Minimum torque ripple, maximum efficiency excitation of brushless permanent magnet motors," *IEEE Trans. Ind. Elec.*, Vol. 41, no. 3, pp. 292–300, June 1994.

[58] Le-Huy, H., R. Perret, and R. Feuillet, "Minimization of torque ripple in brushless DC motor drives," *IEEE Trans. Ind. Elec.*, Vol. 22, no. 4, pp. 748–755, July/August 1986.

[59] Jahns, T., and W. Soong, "Pulsating torque minimization techniques for permanent magnet AC motor drives—A Review," *IEEE Trans. Ind. Elec.*, Vol. 43, no. 2, pp. 321–330, April 1996.

[60] Matsui, N., T. Makino, and H. Satoh, "Auto-compensation of torque ripple of DD motor by torque observer," *Rec. IEEE Ind. Appl. Soc. Annual Meeting*, Dearborn, MI, pp. 305–311, September 1991.

[61] Holtz, J., "Identification and compensation of torque ripple in high-precision permanent magnet motor drives," *IEEE Trans. Ind. Elec.*, Vol. 43, no. 2, pp. 309–320, April 1996.

[62] Morimoto, S., Y. Takeda, and T. Hirasa, "Flux-weakening control method for surface permanent magnet sychronous motors," *Proc. Int. Power Elec. Conf. (IPEC)*, Tokyo, pp. 942–949, 1990.

[63] Kumamoto, A., and Y. Hirane, "A semi-closed loop torque control of a buried permanent magnet motor based on a new flux weakening approach," *IEEE-IAS Annual Meeting Rec.*, pp. 656–661, 1989.

[64] Adnanes, A., and T. Undeland, "Optimum torque performance in PMSM drives above rated speed," *IEEE-IAS Annual Meeting Rec.*, pp. 169–175, 1991.

[65] Schiferl, R., and T. A. Lipo, "Power capability of salient pole PM synchronous motors in variable speed drive applications," *IEEE Trans. Indus. Appl.*, Vol. 26, no. 1, pp. 115–123, January/February 1990.

[66] Soong, W. L., D. A. Staton, and T. J. E. Miller, "Design of a new axially-laminated interior PM motor," *IEEE-IAS Annual Meeting, Rec.*, pp. 27–36, 1993.

[67] Soong, W., and T. Miller, "Theoretical limitations to the field-weakening performance of the five classes of brushless synchronous AC motor drive," *Proc. IEE Conf. Elec. Machines and Drives*, 1993.

[68] Lee, P. W., and C. Pollock, "Rotor position detection techniques for brushless PM and reluctance motor drives," *IEEE-IAS Annual Meeting Rec.*, pp. 448–455, 1992.

[69] Jufer, M., and R. Osseni, "Back EMF indirect detection for self-commutation of synchronous motors," *Proc. EPE Conf.*, pp. 1125–1129, 1987.

[70] Kuo, B., W. Lin, and U. Goerke, "Waveform detection of PM step motors, Parts I and II," *Proc. Ann. Symp. Incremental Motion Control*, pp. 227–256, 1979.

[71] Bahlmann, J., "A full-wave motor drive IC based on the back EMF sensing principle," *IEEE Trans. Consumer Elec.*, Vol. 35, no. 3, pp. 415–420, August 1989.

[72] Becerra, R. C., T. M. Jahns, and M. Ehsani, "Four-quadrant sensorless brushless ECM drive," *Proc. Applied Pow. Elec. Conf. (APEC)*, pp. 202–209, 1991.

[73] Berardinis, L., "Good motors get even better," *Machine Design*, pp. 71–75, November 21, 1991.

[74] Ertugrul, N., and P. P. Acarnley, "A new algorithm for sensorless operation of PM motors," *IEEE-IAS Annual Meeting Rec.*, pp. 414–421, 1992.

[75] Wu, R., and G. Slemon, "A permanent magnet motor drive without a shaft sensor," *IEEE-IAS Annual Meeting Rec.*, pp. 553–558, 1990.

[76] Matsui, N., and M. Shigyo, "Brushless DC motor control without position and speed sensors," *IEEE-IAS Annual Meeting Rec.*, pp. 448–453, 1990.

[77] Sepe, R., and J. Lang, "Real-time observer-based (adaptive) control of a permanent magnet synchronous motor without mechanical sensor," *IEEE-IAS Annual Meeting Rec.*, pp. 475–481, 1991.

[78] Schroedl, M., "An improved position estimator for sensorless controlled permanent magnet synchronous motors," *Proc. European Power Elec. Conf. (EPE)*, Vol. 3, Florence, pp. 418–423, 1991.

[79] Kulkarni, A. B., and M. Ehsani, "A novel position sensor elimination technique for the interior PM synchronous motor drive," *IEEE-IAS Annual Meeting Rec.*, pp. 773–779, 1989.

[80] Moynihan, J. F., R. C. Kavanagh, M. G. Egan, and J. M. D. Murphy, "Indirect phase current detection for field oriented control of a permanent magnet synchronous motor drive," *Proc. EPE Conf.*, Vol. 3, pp. 641–646, 1991.

[81] Jahns, T. M., R. C. Becerra, and M. Ehsani, "Integrated current regulation for a brushless ECM drive," *IEEE Trans. Pow. Elec.*, Vol. 6, no. 1, pp. 118–126, January 1991.

[82] Le-Huy, H., K. Slimani, and P. Viarouge, "A predictive current controller for synchronous motor drives," *Proc. EPE Conf.*, Vol. 2, pp. 114–119, 1991.

[83] Matsui, N., and H. Ohashi, "DSP-based adaptive control of a brushless motor," *IEEE-IAS Annual Meeting Rec.*, pp. 375–380, 1988.

[84] Lorenz, R.D., M. R. Buhl, and D. M. Divan, "Design and implementation of neural networks for digital current regulation of inverter drives," *IEEE-IAS Annual Meeting Rec.*, pp. 415–421, 1991.

[85] Tryzynadlowski, A. M., and S. Legowski, "Application of neural networks to the optimal control of 3-Ph. voltage-controlled power inverters," *Proc. IECON*, pp. 524–529, 1992.

[86] Consoli, A., A. Raciti, and A. Testa, "Experimental low-chattering sliding-mode control of a PM motor drive," *Proc. EPE Conf.*, Vol. 1, pp. 13–18, 1991.

[87] Sepe, R. B., and J. H. Lang, "Real-time adaptive control of the PM synchronous motor," *IEEE Trans. Indus. Appl*, Vol. 27, no. 4, pp. 706–714, July/August 1991.

[88] Lee, C. K., and N. M. Kwok, "Reduced parameter variation sensitivity with a variable structure controller in brushless KC motor velocity control system," *IEEE-IAS Annual Meeting Rec.*, pp. 746–756, 1993.

[89] Chern, T. L., and Y. C. Wu, "Design of brushless DC position servo system using integral variable structure approach," *IEE Proc.*, Vol. 140, Part B, no. 1, January 1993.

[90] Cerruo, E., A. Consoli, A. Raciti, and A. Testa, "Adaptive fuzzy control of high performance motion systems," *Proc. IEEE Ind. Electr. Conf. (IECON)*, pp. 88–94, 1992.

[91] Kovacic, Z., S. Bogdan, and P. Crnosija, "Fuzzy rule-based model reference adaptive control of a permanent magnet synchronous motor drive," *Proc. IEEE Ind. Electr. Conf. (IECON)*, pp. 207–212, 1993.

[92] Ko, J. W., J. G. Hwuang, and M. J. Youn, "Robust position control of BLDD motors using integral-proportional plus fuzzy logic controller," *Proc. IEEE Ind. Electr. Conf. (IECON)*, pp. 213–218, 1993.

[93] Miller, T., *Switched Reluctance Motors and Their Control*, Clarendon Press, Oxford, 1993.

[94] Miller, T., C. Cossar, and D. Anderson, "A new control IC for switched reluctance motor drives," *Proc. IEE Conf. Power Elec. & Var. Speed Drives*, London, pp. 331–335, July 1990.

[95] Zimmerman, P., "Electronically commutated DC feed drives for machine tools," *Proc. Motorcon Conf.*, pp. 69–86, 1982.

[96] Katayama, M., S. Nara, and K. Yamaguchi, "Newly developed AC servomotor RA Series," *Power Conversion International*, pp. 12–20, May 1984.

[97] Venturini, M., "Integrated high power density brushless servo drives for avionic, fly-by-wire and satellite applications," *Proc. MotorCon Conf.*, pp. 162–171, 1982.

[98] Jahns, T. M., and R. C. Van Nocker, "High-performance EHA control using an interior PM motor," *IEEE Trans. Aero. & Elec. Sys.*, Vol. 26, no. 3, pp. 534–542, May 1990.

[99] Sulfstede, L., "Applying power electronics to residential HVAC—the issues," *Proc. APEC Conf.*, pp. 615–621, 1991.

[100] Serizawa, Y., K. Iizuka, and M. Senou, "Inverter controlled rotary compressors," *Hitachi Review*, Vol. 36, no. 3, pp. 177-184, 1987.

[101] Kinniment, D., P. Acarnley, and A. Jack, "An integrated circuit controller for brushless DC drives," *Proc. European Power Elec. Conf. (EPE)*, Florence, Vol. 4, pp. 111–116, 1991.

[102] Newborough, L., "Electronically commutated DC motor for driving tube-axial fans: A cost-effective design," *Applied Energy*, Vol. 36, pp. 167–190, 1990.

[103] Kenjo, T., and S. Nagamori, *PM and Brushless DC Motors*, Clarendon Press, Oxford, 1985.

[104] Sneyers, B., G. Maggetto, and J. Van Eck, "Inverter fed permanent magnet motor for road electric traction," *Proc. Int. Conf. on Elec. Machines*, Budapest, pp. 550–553, September 1982.

[105] Miller, R., T. Nehl, N. Demerdash, B. Overton, and C. Ford, "An electronically controlled permanent magnet synchronous machine conditioner system for electric passenger vehicle propulsion," *IEEE-IAS Annual Meeting Rec.*, pp. 506–510, 1982.

[106] King, R. D., "ETX-II 70 Hp electric drive system performance—component tests," *Proc. 10th Int. Elec. Vehicle Symposium*, pp. 878–887, 1990.

[107] Oldenkamp, J. L., and D. M. Erdman, "Automotive electrically driven air conditioning system," *Proc. IEEE Automotive Power Electronics Workshop*, IEEE Cat. No. 89TH0299-8, pp. 71–72, August 1989.

[108] Millner, A. R., "Multi-hundred horsepower permanent magnet brushless disk motors," *Proc. APEC Conf.*, pp. 351–355, 1994.

[109] "Brushless DC motor line extended to 600 hp," *Control Eng.*, p. 79, December 1992.

[110] Jahns, T., "Designing intelligent muscle into industrial motion control," *IEEE Trans. Ind. Electronics*, Vol. 37, no. 5, pp. 329–340, October 1990.

Chapter 7

Herbert Stemmler

High - Power Industrial Drives

7.1. INTRODUCTION

When high-power thyristors for large converters and control methods for AC drives were ready for use at the end of the 1960s, large frequency-controlled converters could be built to meet the requirements for large drives with adjustable speed. This resulted in a first breakthrough for large drives in industrial applications during the 1970s.

The incentive to use variable speed drives has various reasons:

1. The cost for maintenance can be reduced and the lifetime increased when mechanical parts in the drive equipment are replaced by fully static converters. In the cement industry, gearless mill tubes with converter-fed wrap-around AC motor instead of the gear are a typical example for this substitution.

2. Large inrush currents of high-power motors and generators during start-up can be avoided by soft starting via frequency converters. Start-up converters for gas turbine generators were among the first applications of this type. Here the converters replaced complicated equipment with a starting motor.

3. Cutting energy use in power plants, chemical, and industrial plants by reducing losses in the process equipment is becoming more and more important. In rotating equipment such as pumps, fans, blowers, and

    compressors, much energy is thrown away by controlling the flow of fluid or gas with the help of throttling valves, dampers, and adjustable guide vans. With adjustable speed drives, this unnecessary dissipation can be eliminated [1]. In many U.S. power plants, existing fixed speed motors therefore have been retrofitted with frequency converters.

4. The speed of frequency-controlled motors is not linked to 3000 or 3600 rpm corresponding to the AC system frequencies 50 or 60 Hz. They can drive high-speed compressors up to 6000, 10,000, or even 18,000 rpm.

5. High-torque, low-speed drive applications such as reversing rolling mills in the steel industry or large hoists in the mining industry is another field for large drives.

6. High-power drives are also necessary in sea and rail transport systems such as large passenger liners or locomotives with on-board diesel generator or with overhead contact wire for AC or DC supply. These applications are not directly in the scope of Chapter 7. They, however, will be mentioned.

7. Fixed frequency, variable speed machines fed from or feeding into the AC system also need large-frequency converters. Converter-controlled large motor generators, up to about 400 MW, in pumped storage hydro power plants are the most recent development in this category of applications. In the generating mode, energy can be saved by operating the machine at that speed, where the hydraulic efficiency is at its optimum. In the pumping mode, load control can be achieved only by speed control.

    In this chapter, after the classification (Section 7.2) and a short review of the evolution (Section 7.3), Sections 7.4 and 7.5 present a survey of the motors and converters used for large drives. Then the different drive systems which became important in industrial applications are described in Sections 7.6 to 7.10. It is shown how motor, converter, and controls form a drive system. The specific characteristics of each system are emphasized in order to show which system is best suited for which kind of applications. Based on the development potential the paper also gives a projection for the future.

7.2. CLASSIFICATION WITH SPEED AND POWER RATINGS

The power ratings of large drives reach from just below 1 MW up to 10 MW and even 100 MW and more in special cases. The upper limit is rather given by the requirements of the applications than by the technology of the converters and machines.

    The maximum speed reaches from about 10 rpm for low-speed drives up to 1500 or 3000 rpm for normal-speed drives and 6000 and 12,000 rpm or even 18,000 rpm for high-speed and very-high-speed drives.

    The attainable maximum speed decreases with increasing power ratings. The 10 MW / 6000 rpm are typical ratings for high-power, high-speed drives. The attainable limit ratings can roughly be marked by the figures 80 MW / 3000 rpm, 15 MW / 6000 rpm, and 3 MW / 18,000 rpm.

7.3. SHORT REVIEW OF THE EVOLUTION OF LARGE DRIVES

During the 1960s *line-commutated, controlled rectifiers* using new high-power turn-on thyristors substituted for rotating converters and mercury arc valves which have been used before to control torque and speed of the *DC collector motors*. This was a real progress at the converter side: a fully static solution with fewer auxiliaries, which offered a higher flexibility with respect to the applications. But the disadvantages at the motor side were not removed: due to their mechanical collector, DC motors are not only expensive and subject to wear, but also limited in their ratings to about 1000 V, 1000 rpm, 1000 kW (increasing of one of these ratings means reduction of the others). DC motors, therefore, cover only the lowest range of large drives. These restrictions called for AC motors.

Starting in the middle of the 1960s, every effort was made to develop AC drive systems with variable frequency. New modulation methods such as pulse width modulation (PWM) emerged [2], and new control methods such as vector control [3] and field-oriented control [4, 5] were introduced. These investigations and developments resulted in a real *breakthrough for large drives in the 1970s. A variety of different large-drive systems with specific advantages opened the way for new applications.*

Cycloconverter-fed synchronous motors, well suited for low-speed drives with high torque even at standstill, had been the first large AC drives of this generation, which were put into commercial operation at the end of the 1960s. They were introduced in gearless cement mill drives and had already a new vector control [6]. Cycloconverter-fed synchronous motors are now state of the art for all kinds of low-speed high-torque drives (see Section 7.8).

In normal- and high-speed drives for all kinds of pumps, fans, blowers, and compressors and as start-up equipment for large sychronous generators and motors, the *synchronous motor fed by a line and motor-commutated current source converter became the most successful large-drive system.* It was introduced at the beginning of the 1970s mainly as start-up equipment for gas turbine generators [7]. Maximum speed of 6000 rpm and power ratings close to 100 MW are now state of the art. (See Section 7.6.)

Slip ring motors fed by subsynchronous converter cascades with slip power recovery were introduced at the same time. They are well suited for a limited speed range between 70% and 100% of the synchronous speed. The smaller the speed range, the lower the rated power for which the converter cascade has to be designed. This system is successful as a low-cost design for applications with reduced requirements, where power transfer via sliprings is accepted and where auxiliaries for the start-up to the normal speed range can be tolerated. Its attractivity goes down in the same degree as the cost of other drive systems with better characteristics decreases. (See Section 7.10.)

Slip-ring motors with sub- and hypersynchronous cascades, using three-phase cycloconverters at the slipring side, found only special applications with, however, high-rated power up to 80 MVA per single unit. Especially in 50–16 2/3 Hz interties to supply the 16 2/3 Hz railway system in Europe, they were used since the middle of the 1970s at the 50 Hz side to drive the single-phase synchronous generators feeding the 16 2/3 Hz side. Such interties are frequency elastic. Even with deviations of some percentage in the frequency of the 16 2/3 Hz system, they can be operated without restrictions [8]. But there is a new application coming up at a still higher power level of some 100 MW per single unit: in pumped storage power plants with extreme variations of the turbine heads, the optimal hydraulic efficiency at a given head depends on the speed of the turbine. The system is well suited to operate the generator, feeding the 50 Hz or 60 Hz system at variable speed of about ± 10% or 20% around the synchronous speed. In addition load control even during pumping operation can be achieved by varying the speed [9]. (See Section 7.10.)

Many efforts have been made during the 1960s and the 1970s to use the simple, robust, and cheap induction motor for large variable speed drives. But whereas all the drive systems mentioned above can be operated with line or load commutated conventional turn-on thyristors, induction motors need inverters, the thyristors of which can be turned off. Since turn-off thyristors had not been available at that time, conventional turn-on thyristors with auxiliary circuits for forced commutation had to be used—a complicated and expensive technology. Forced-commutated turn-on thyristors were used in voltage source and in current source inverters.

Because of their complexity such forced commutated voltage source inverters feeding induction motors were only successfully used in railway applications. There it was attractive to substitute the smaller and simpler induction motor for the DC motor in the narrow space of the bogies of the locomotives [10, 11]. Even when large-gate turn-off thyristors (GTOs) became available in the first half of the 1980s, it took until the end of the 1980s before large induction motors fed by GTO inverters were delivered for industrial applications. There is, however, no doubt that inverter-fed induction motors will more and more become strong competitors for the synchronous motor drives. They benefit more from the progress made in the field of microprocessors and the development of modern high-power gate turn-off elements. In addition, compared with thyristor converter drives, they offer more economical drive solutions, since they make use of the simplest electrical motor and have a higher line power factor in the whole speed range. They will cover all known high-speed requirements and most of the required power ratings [12]. (See Section 7.9.)

Induction motor drives with forced-commutated current source inverters had more success in industrial applications. We have to distinguish between two different subsystems:

1. Drive systems with forced-commutated turn-on thyristors in the motor-side current source inverter were widely used in the power range around 100 kW and partly also at the low end of the large drives [13]. Efforts made to

substitute GTOs for the forced commutation did not lead to large drives solutions, which entered commercial operation—with some exceptions especially in the United States [14].

2. Drive systems with forced commutation in the current DC link as an aid for start-up, which are equipped with compensating condensers between converter and induction motor, were successfully used since the middle of the 1980s especially in power plants in the United States. Their purpose was to provide existing fixed speed induction motors with variable frequency to save energy [15]. Whereas the voltage source inverters generate PWM voltages, the high dv/dt of which stresses the insulation of the windings, this system produces sinusoidal voltages. It is therefore best suited to supply existing motors, designed for sinusoidal voltages. (See Section 7.7.)

7.4. MOTORS FOR LARGE DRIVES

7.4.1. Types of Motors Used

The business activities with high-power drives in general are characterized by rather smaller numbers of however large orders for specific applications. The motors for large drives are three-phase AC machines with conventional design at the electrical and magnetical side:

- Synchronous motors
- Slip ring asynchronous motors
- Induction motors

In many cases, however, they have to be adapted at the mechanical side to meet the specific requirements of the load machines.

DC motors are not well suited for large drives. Due to their mechanical collector, they are expensive and subject to wear, and their voltage, speed, and power ratings are too much limited.

Synchronous motors and induction motors (Figure 7-1) are converter controlled via their three-phase stator windings. Synchronous motors for variable speed drives have, in most cases, solid turbo rotors, especially when used in high-speed applications. The rotor carries a DC excitation winding that has to be fed via slip rings or—in high-speed applications—brushless via a rotating transformer and a rotating diode rectifier. Due to the DC excitation the rotor of the synchronous motors can be built with a wider air gap. Therefore, they are mechanically robust. Synchronous motors are well proven as high-power and high-speed drives. The wound rotor with its inherent dissymmetry, however, limits the maximum speed to about 7000 rpm [34]. Induction motors are cheap and robust because of their simple squirrel cage rotor. They have the potential for high-power and high-speed drives. Solid rotor bodies with buried cages and magnetic bearings are means to extend the maximum speed up to about 10,000 rpm at 6 MW and 18,000 rpm at 3 MW.

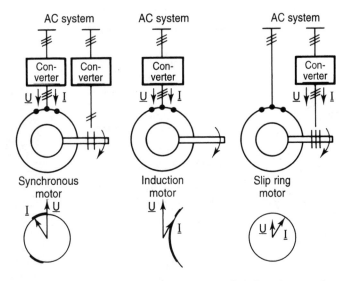

Figure 7-1. Motors for large-drive systems and their voltage and current characteristics.

Slip ring asynchronous motors (Figure 7-1) are connected to the AC system with their three-phase stator, whereas the three-phase wound rotor is converter controlled via slip rings. They can be built for very high power ratings. The wound rotor and the slip rings, however, limit the rated speed.

7.4.2. Mathematical Representation of AC Motors

When designing converter controls for AC drives one has to focus on the motors: the selection of the converter circuits and the design of the control structure strongly depend on the characteristics of the motors. How three-phase motors work is not easy to understand, when looking only at the three-phase quantities of the windings. The usual methods to describe AC machines therefore are not only based on

- Three-phase quantities, but also on their transformation to
- Rotating phasors and to
- Stationary phasors.

This is required for two reasons:

1. AC machines are easier to describe mathematically and easier to understand physically when using phasors instead of three-phase quantities.
2. Modern controls are directly based on phasors and their real and imaginary components. The measured three-phase quantities have to be transformed to

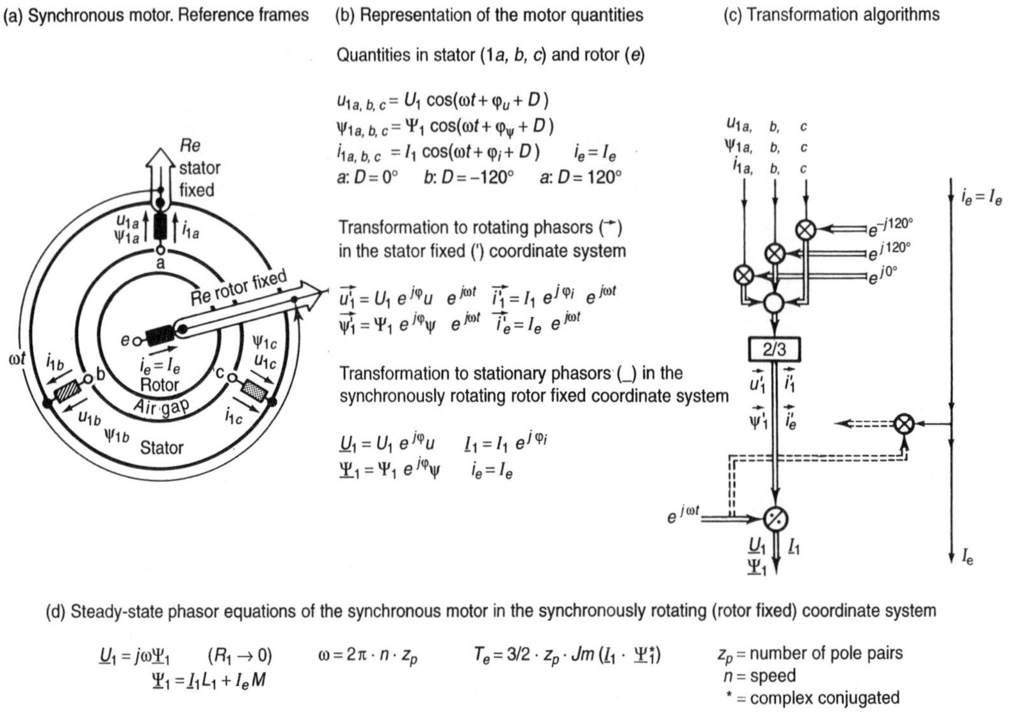

(a) Synchronous motor. Reference frames

(b) Representation of the motor quantities

(c) Transformation algorithms

Quantities in stator (1a, b, c) and rotor (e)

$u_{1a,b,c} = U_1 \cos(\omega t + \varphi_u + D)$
$\psi_{1a,b,c} = \Psi_1 \cos(\omega t + \varphi_\psi + D)$
$i_{1a,b,c} = I_1 \cos(\omega t + \varphi_i + D)$ $i_e = I_e$
$a: D = 0°$ $b: D = -120°$ $a: D = 120°$

Transformation to rotating phasors ($\rightarrow$)
in the stator fixed (') coordinate system

$\vec{u_1} = U_1\, e^{j\varphi_u}\, e^{j\omega t}$ $\vec{i_1} = I_1\, e^{j\varphi_i}\, e^{j\omega t}$
$\vec{\psi_1} = \Psi_1\, e^{j\varphi_\psi}\, e^{j\omega t}$ $\vec{i_e} = I_e\, e^{j\omega t}$

Transformation to stationary phasors (_) in the
synchronously rotating rotor fixed coordinate system

$\underline{U_1} = U_1\, e^{j\varphi_u}$ $\underline{I_1} = I_1\, e^{j\varphi_i}$
$\underline{\Psi_1} = \Psi_1\, e^{j\varphi_\psi}$ $i_e = I_e$

(d) Steady-state phasor equations of the synchronous motor in the synchronously rotating (rotor fixed) coordinate system

$\underline{U_1} = j\omega\underline{\Psi_1}$ $(R_1 \rightarrow 0)$ $\omega = 2\pi \cdot n \cdot z_p$ $T_e = 3/2 \cdot z_p \cdot Jm\,(\underline{I_1} \cdot \underline{\Psi_1^*})$ z_p = number of pole pairs
$\underline{\Psi_1} = \underline{I_1}L_1 + I_e M$ n = speed
                                                                                                    * = complex conjugated

Figure 7-2. Synchronous motor. Reference frames. Representation of the motor quantities. Transformation algorithms. Simplified steady-state motor equations.

phasors at the inputs of the controls, and an inverse transformation at the outputs has to generate the three-phase quantities to act upon the real "three-phase-world" of the high-power converters and motors.

This is well known from the literature [16–21]. In our context we therefore don't need to derive this theory. It should, however, be interpreted and illustrated in a physically plausible manner. This will be done with Figures 7-2 and 7-3 for the synchronous motor and Figures 7-4 and 7-5 for the asynchronous motor with slip rings or with squirrel cage rotor.

The Synchronous Motor. The synchronous motor with three stator windings a, b, c and the excitation winding e in the rotor is shown in Figure 7-2a. This picture also contains the real axis (Re) of two complex coordinate systems:

- A stator fixed stationary coordinate system
- A rotor fixed coordinate system turning with the rotor.

These reference frames are required to transform the following:

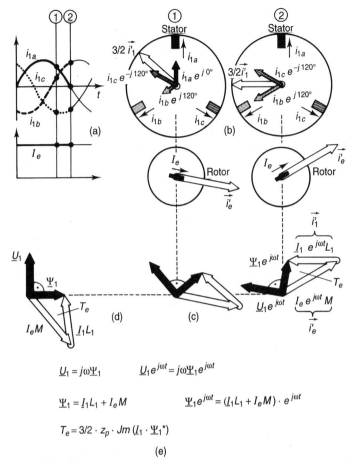

Figure 7-3. Synchronous motor. Illustration of the mode of operation.
Three-phase quantities (a); rotating phasors (b) and rotating
phasor diagrams (c) for two instants (1) and (2); stationary
phasor diagram (d); simplified steady-state motor equations
(e).

- Three-phase stator quantities of the voltages $u_{1a,b,c}$, the flux linkages $\psi_{1a,b,c}$, the currents $i_{1a,b,c}$ and the DC rotor current i_e to
- Stator-related ($'$) rotating phasors ($\rightarrow$) $\vec{u}'_1$; $\vec{\psi}'_1$; $\vec{i}'_1$; $\vec{i}'_e$; or
- Rotor-related stationary phasors (_) $\underline{U}_1$; $\underline{\Psi}_1$; $\underline{I}_1$; $\underline{I}_e = I_e$.

This is represented in Figure 7-2b. The line diagram in Figure 7-2c shows the
transformation algorithms. Figure 7-2d shows the *steady-state* equations for the
stator voltage $\underline{U}_1$, the stator flux linkage $\underline{\Psi}_1$, and the electromagnetic torque T_e
based on stationary phasors.

The simplified steady-state motor equations in Figure 7-2d are easy to interpret
physically: the torque T_e is generated by the force of the magnetic field (represented

by $\underline{\Psi}_1$) on the current (represented by $\underline{I}_1$). $Jm(\underline{I}_1 \cdot \underline{\Psi}_1^*)$ means the product of $\underline{\Psi}_1$ and the perpendicular component of $\underline{I}_1$. The stator flux linkage $\underline{\Psi}_1$ is determined by the voltage ($\underline{U}_1$) and frequency (ω) for the stator. To obtain a good magnetic utilization of the motor the absolute value Ψ_1 of the flux linkage $\underline{\Psi}_1$ has to be kept at its rated value by increasing the amplitude U_1 of the stator voltage $\underline{U}_1$ proportional to the angular stator frequency ω. The rotor follows synchronously to the angular stator frequency $\omega = 2\pi n \cdot z_p$. The stator flux linkage $\underline{\Psi}_1$ consists of two components which are generated by the stator current ($\rightarrow \underline{I}_1 L_1$) and the excitation current ($\rightarrow I_e M$). L_1 is the stator inductance and M is the stator-rotor coupling inductance.

Equations 7.1–7.7 and the diagrams in Figures 7-3a, b serve the purpose to illustrate the transformation of the three stator currents $i_{1a,b,c}$ and of the DC rotor current i_e to rotating phasors $\overrightarrow{i}\,'_1$ and $\overrightarrow{i}\,'_e$ for two instants (1) and (2). The transformations are based on the physical idea of three-phase motors where three-phase AC currents in three-phase windings generate a rotating field. The DC excitation current i_e generates a field which is fixed on the rotor.

Mathematically this can be described as follows: the three phase currents $i_{1a,b,c}$ in equation 7.1 are added geometrically in equation 7.3 (each in the direction of its winding axis) and generate a rotating phasor $\overrightarrow{i}\,'_1 = (I_1 e^{j\varphi_i}) \cdot e^{j\omega t}$ in equation 7.4. The DC excitation current i_e in equation 7.2 generates a rotor-fixed phasor $\overrightarrow{i}\,'_e = I_e \cdot e^{j\omega t}$ in equation 7.6, which turns together with the synchronously rotating rotor. Both phasors rotate with the same angular frequency $\omega = 2\pi n \cdot z_p$ (n: speed; z_p: number of pole pairs) relative to the stator (equations 7.4 and 7.6).

$$i_{1a,b,c} = I_1 \cos(\omega t + \varphi_i + D); \qquad D = 0°; \ -120°; \ +120°; \tag{7.1}$$

$$i_e = I_e; \tag{7.2}$$

$$\overrightarrow{i}\,'_1 = 2/3(i_{1a}e^{j0°} + i_{1b}e^{+j120°} + i_{1c}e^{-j120°}) \tag{7.3}$$

$$= 2/3\sum_D I_1 \cos(\omega t + \varphi_i + D)e^{-jD}$$

$$= I_1 e^{j\varphi_i}e^{j\omega t} \qquad \underline{I}_1 = I_1 e^{j\varphi_i} \tag{7.4}, (7.5)$$

$$\overrightarrow{i}\,'_e = I_e \cdot e^{j\omega t} \qquad \underline{I}_e = I_e \tag{7.6}, (7.7)$$

$$i_{1a,b,c} = Re(\overrightarrow{i}\,'_1 e^{+jD}) = Re(I_1 \cdot e^{j\varphi_i} \cdot e^{j\omega t} \cdot e^{jD})$$
$$= I_1 \cos(\omega t + \varphi_i + D) \tag{7.4 \rightarrow 7.1}$$

Equations 7.1 and 7.4 show clearly in which way amplitude I_1, frequency ω, and phase angle φ_i of the three-phase currents in equation 7.1 determine the absolute value I_1, the stationary unit phasor $e^{j\varphi_i}$, and the rotating unit phasor $e^{j\omega t}$ of the stator-related rotating phasor $\overrightarrow{i}\,'_1$ in equation 7.4. Therefore it is easy to recover the three-phase currents $i_{1a,b,c}$ from the rotating current phasor $\overrightarrow{i}\,'_1$ by the inverse transformation in equations 7.4 → 7.1.

As shown in equations 7.4, 7.5 and 7.6, 7.7 the rotating phasors $\overrightarrow{i}\,'_1 = I_1 e^{j\varphi_i} \cdot e^{j\omega t}$ and $\overrightarrow{i}\,'_e = I_e \cdot e^{j\omega t}$ can be regarded as stationary phasors $\underline{I}_1 = I_1 e^{j\varphi_i}$ and I_e multiplied with the rotating unit phasor $e^{j\omega t}$.

In the phasor diagrams in Figures 7-3c, d the stator respectively rotor current phasors are multiplied with the stator inductance L_1 and the stator-rotor coupling inductance M and summed up geometrically to form the resultant stator flux linkage $\underline{\Psi}_1 e^{j\omega t} = (\underline{I}_1 L_1 + \underline{I}_e M)e^{j\omega t}$. This illustrates the phasor equations in Figure 7-2d for the voltage $\underline{U}_1$ and the flux linkage $\underline{\Psi}_1$ of the stator under the assumption that the stator resistance can be neglected ($R_1 \rightarrow 0$).

The phasor diagrams can also serve to illustrate the torque equation. The torque T_e is, according to the torque equation, proportional to the area enclosed by the phasor triangle formed by the flux linkages $\underline{I}_1 L_1, \underline{I}_e M$, and $\underline{\Psi}_1$.

The Asynchronous Motor. The asynchronous motor with three-phase stator windings is shown in Figure 7-4a. As slip ring motor, it has a three-phase wound rotor accessible via slip rings. As squirrel cage motor it can be regarded as a special case with short-circuited three-phase rotor windings.

Figure 7-4a contains the real axis (*Re*) of three complex coordinate systems:

1. A stator fixed stationary coordinate system.
2. A rotor fixed coordinate system turning with the rotor, the electrical angular speed of which is ω (where $\omega = 2\pi n z_p$, $n =$ speed and $z_p =$ number of pole pairs).
3. A rotating coordinate system, turning synchronously with the angular frequency ω_1 of the stator.

It is assumed that the variations of the angular rotor speed ω are slow compared to those of the electrical angular frequencies of the stator ω_1 and the rotor ω_2.

$$\int \omega_1 dt - \omega t = \int \omega_2 dt \tag{7.8}$$

These reference frames are required to transform the three-phase quantities $x_{1a,b,c}$ and $x_{2a,b,c}$ (where $x = u$ or ψ or i) of the stator (index 1) and rotor (index 2) to rotating phasors ($\rightarrow$) in the stator fixed frame ($'$): $\vec{x}'_1$ and $\vec{x}'_2$ and in the rotor fixed frame ($''$): $\vec{x}''_2$ and $\vec{x}''_1$ or to stationary phasors ($\_$) $\underline{X}_1$ and $\underline{X}_2$ in the rotating frame.

The representation of these quantities is shown in Figure 7-4b. Nonsinusoidal transient three-phase quantities of any wave shape can be represented as "sinusoidal" quantities with varying amplitude $X_{1(2)} = X_{1(2)}(t)$, frequency $\omega_{1(2)} = \omega_{1(2)}(t)$, and phase angle $\varphi_{i_{1(2)}} = \varphi_{i_{1(2)}}(t)$. The transformation algorithms are shown in the line diagram in Figure 7-4c. Figure 7-4d shows the *dynamic* phasor equations for the voltages $\underline{U}$ and flux linkages $\underline{\Psi}$ in stator (index 1) and rotor (index 2) and for the instantaneous value of the electromagnetic torque t_e as a function of the stator current and flux linkage phasors $\underline{I}_1$ and $\underline{\Psi}_1$.

Here again we start our illustration with a physical interpretation of the *dynamic* motor equations in Figure 7-4d: the generation of the torque t_e is identical to the synchronous motor. $Jm(\underline{I}_1 \cdot \underline{\Psi}_1^*)$ means the product of $\underline{\Psi}_1$ representing the magnetic field and the perpendicular component of the current $\underline{I}_1$. The stator flux

(a) Asynchronous motor, reference frames

(b) Representation of the motor quantities

(c) Transformation algorithms

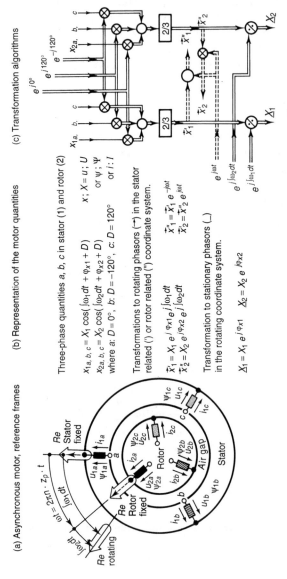

Three-phase quantities a, b, c in stator (1) and rotor (2)

$$x_{1a,b,c} = X_1 \cos\left(\int \omega_1 dt + \varphi_{x1} + D\right) \qquad x;\ X = u;\ U$$
$$x_{2a,b,c} = X_2 \cos\left(\int \omega_2 dt + \varphi_{x2} + D\right) \qquad \text{or } \psi;\ \Psi$$

where a: $D = 0°$, b: $D = -120°$, c: $D = 120°$ $\qquad$ or i; I

Transformations to rotating phasors ($\rightarrow$) in the stator related (') or rotor related (") coordinate system.

$$\vec{x}_1 = X_1\, e^{j\varphi_{x1}} e^{j\int\omega_1 dt} \qquad \vec{x}_1'' = \vec{x}_1'\, e^{-j\omega t}$$
$$\vec{x}_2 = X_2\, e^{j\varphi_{x2}} e^{j\int\omega_2 dt} \qquad \vec{x}_2' = \vec{x}_2''\, e^{j\omega t}$$

Transformation to stationary phasors ($\underline{\ }$) in the rotating coordinate system.

$$\underline{X}_1 = X_1\, e^{j\varphi_{x1}} \qquad \underline{X}_2 = X_2\, e^{j\varphi_{x2}}$$

(d) Dynamic phasor equations of the asynchronous motor in the synchronously rotating coordinate system

$$\underline{U}_1 = \underline{I}_1 R_1 + j\omega_1 \underline{\Psi}_1 + d\underline{\Psi}_1/dt \qquad \underline{\Psi}_1 = \underline{I}_1 L_1 + \underline{I}_2 M \qquad \int\omega_1 dt = \int\omega_2 dt + \omega t \qquad t_e = 3/2 \cdot z_p \cdot Jm(\underline{I}_1 \cdot \underline{\Psi}_1^*)$$
$$\underline{U}_2 = \underline{I}_2 R_2 + j\omega_2 \underline{\Psi}_2 + d\underline{\Psi}_2/dt \qquad \underline{\Psi}_2 = \underline{I}_2 L_2 + \underline{I}_1 M \qquad \omega = 2\pi n \cdot z_p$$

z_p = number of pole pairs
n = speed
* = complex conjugated

Figure 7-4. Asynchronous motor. Reference frames. Representation of the motor quantities. Transformation algorithms. Dynamic motor equations.

linkage $\underline{\Psi}_1$— determined by the voltage $\underline{U}_1$ and angular frequency ω_1 applied to the stator (assumption $R_1 \to 0$)—is made up of two components generated by the stator current ($\to \underline{I}_1 L_1$) and by the rotor current ($\to \underline{I}_2 M$). L_1 is the stator inductance, M is the stator-rotor coupling inductance. The rotor flux linkage $\underline{\Psi}_2$ can be interpreted in a quite analog manner. Here $\underline{U}_2$ and ω_2 are the voltage and angular (slip) frequency applied to the rotor via slip rings. Induction motors have a (short-circuited) cage in their rotor which is not accessible from outside. Therefore, $\underline{U}_2 = 0$. When controlling the induction motor with a linear voltage frequency characteristic $| \underline{U}_1 | \sim \omega_1$ the stator flux linkage $\underline{\Psi}_1$ can be kept constant at its rated value. This ensures a good magnetic utilization of the induction motor. ω_2 is the angular slip frequency by which the angular rotor speed $\omega = 2\pi n \cdot z_p$ is lagging behind the angular stator frequency ω_1 when the motor is in the driving mode.

The following *steady-state* equations 7.9 to 7.12 and the diagrams in Figures 7-5a, b serve the purpose to illustrate the transformation of the three-phase stator and rotor currents $i_{1a,b,c}$ and $i_{2a,b,c}$ in equation 7.9 to rotating phasors $\vec{i}_1$ and $\vec{i}_2$ in equation 7.11 for two instants (1) and (2) by geometrical addition in equation 7.10.

$$i_{1(2)a,b,c} = I_{1(2)} \cos(\omega_{1(2)}t + \varphi_{i_{1(2)}} + D) \tag{7.9}$$

$$D = 0°; \ -120°; \ +120°;$$

$$\vec{i}\,'^{(")}_{1(2)} = \frac{2}{3}\left(i_{1(2)a}\,e^{j0°} + i_{1(2)b}e^{+j120°} + i_{1(2)c}e^{-j120°}\right) \tag{7.10}$$

$$= 2/3 \sum_D I_{1(2)} \cos(\omega_{1(2)}t + \varphi_{i_{1(2)}} + D)e^{-jD}$$

$$= I_{1(2)}e^{j\varphi_{i_{1(2)}}} \cdot e^{j\omega_{1(2)}t} \qquad \underline{I}_{1(2)} = I_{1(2)} \cdot e^{j\varphi_{i_{1(2)}}} \tag{7.11), (7.12}$$

$$i_{1(2)a,b,c} = Re(\vec{i}\,'^{(")}_{1(2)}e^{+jD}) = Re(I_{1(2)} \cdot e^{j\varphi_{i1(2)}} \cdot e^{j\omega_{1(2)}t} \cdot e^{+jD})$$

$$= I_{1(2)} \cos(\omega_{1(2)}t + \varphi_{i1(2)} + D) \tag{7.11 \to 7.9}$$

The rotating phasor $\vec{i}\,'^{(")}_{1(2)} = I_{1(2)}e^{j\varphi_{i_{1(2)}}} \cdot e^{j\omega_{1(2)}t}$ can be regarded as stationary phasor $\underline{I}_{1(2)} = I_{1(2)}e^{j\varphi_{i_{1(2)}}}$ multiplied with the rotating unit phasor $e^{j\omega_{1(2)}t}$ (equations 7.11, 7.12).

The equations 7.9 and 7.11 show clearly in which way amplitude $I_{1(2)}$, frequency $\omega_{1(2)}$ and phase angle $\varphi_{i_{1(2)}}$ of the three-phase quantities determine the absolute value $I_{1(2)}$, the stationary unit phasor $e^{j\varphi_{i1(2)}}$ and the rotating unit phasor $e^{j\omega_{1(2)}t}$ of the stator (rotor)–related rotating phasor $\vec{i}_{1(2)} 1(2)$. Equation 7.11 $\to$ 7.9 shows the inverse transformation $\vec{i}_{1(2)} \to i_{1(2)a,b,c}$.

In the phasor diagrams in Figures 7-5c, d the stator respectively rotor current phasors are multiplied with the stator inductance L_1 and the stator-rotor coupling inductance M and summed up geometrically to form the resultant stator flux linkage

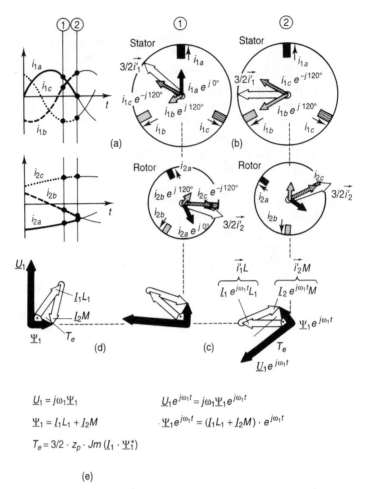

Figure 7-5. Asynchronous motor. Illustration of the mode of operation: three-phase quantities (a); rotating phasors (b) and rotating phasor diagrams (c) for two instants (1) and (2); stationary phasor diagram (d); simplified steady-state motor equations (e).

$\underline{\Psi}_1 e^{j\omega_1 t} = (\underline{I}_1 L_1 + \underline{I}_e M) e^{j\omega_1 t}$. This illustrates the phasor equation of the stator in Figure 7-4e in the steady state $(d/dt\,\underline{\Psi}_1 = 0;\ \int \omega_1 dt = \omega_1 t)$ under the assumption that the stator resistance can be neglected $(R_1 \to 0)$.

Here again, the area within the phasor triangle of the flux linkages $\underline{I}_1 L_1, \underline{I}_2 M$ and $\underline{\Psi}_1$ is proportional to the generated torque T_e —according to the torque equation in Figure 7-5e.

7.5. CONVERTERS FOR LARGE DRIVES

7.5.1. Basic Circuits

The converters connect the three-phase AC system with the motors and fulfill two main tasks:

1. They convert the voltages and currents of the AC system according to the requirements of the motors.
2. They control the power flow from the AC system to the motor in the driving mode, and vice versa, in the regenerative breaking mode.

A variety of converter configurations for large AC motors has been introduced to meet the requirements in different applications. But all these converters are based on only two basic circuits:

1. AC voltage source rectifiers (AC VSRs)
2. DC voltage source inverters (DC VSIs)

Both circuits are shown in Figure 7-6. Each of them can be used at both sides of the converters: at the AC system side and at the motor side. These circuits are well known from the literature [22–25] and therefore don't need to be explained in detail in this chapter. The intention is rather to recall their main characteristics with Figure 7-6 and to demonstrate with Figure 7-7 how these two circuits can be combined to form a large variety of converters which fit the requirements of the different kinds of motors.

Unfortunately different and improperly defined namings which are in use for these circuits might cause confusions. Therefore the namings in this chapter are strictly oriented either on the voltage conversion or on the current conversion of the circuits.

AC Voltage Source Rectifiers. In this way the basic circuit in Figure 7-6, right is

- An AC voltage source rectifier, AC VSR, (Figure 7-6 right top) that rectifies an impressed AC voltage and generates a switched DC voltage ($u_{a,b,c} \rightarrow u_d$). But at the same time it is also
- A DC current source inverter DC CSI, (Figure 7-6 right bottom) that "inverts" an impressed DC current and generates a switched AC current ($I_d \rightarrow i_{a,b,c}$).

THE DC SIDE. At the DC side the current I_d has only one polarity which is given by the thyristors whereas the generated DC voltage u_d and its polarity can be changed by the firing angle α of the modulator. U_d the mean value of u_d is

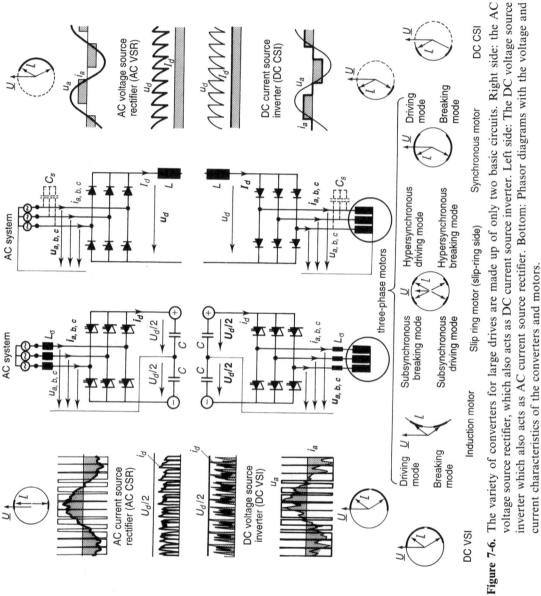

Figure 7-6. The variety of converters for large drives are made up of only two basic circuits. Right side: the AC voltage source rectifier, which also acts as DC current source inverter. Left side: The DC voltage source inverter which also acts as AC current source rectifier. Bottom: Phasor diagrams with the voltage and current characteristics of the converters and motors.

$$U_d = U_{d\,\max} \cos \alpha \quad \alpha \ \to \ 0° \qquad U_d = +U_{d\,\max}$$
$$\alpha \ \to \ 180° \qquad U_d = -U_{d\,\max} \tag{7.13}$$

This allows energy transfer in both directions AC → DC and DC → AC. Since the DC voltage u_d is a switched voltage, the inductance L is required to smooth the DC current I_d.

When equipped with diodes, voltage u_d and current I_d at the DC side have only one polarity, allowing energy transfer only in the direction AC → DC.

THE AC SIDE. When looking at the AC side one has to distinguish between external commutation and forced commutation.

AC VSRs (= DC CSIs) with conventional turn-on thyristors (as shown in Figure 7-6 right) need external commutation (line or machine commutation). External commutated circuits are still less expensive and easier to build than forced commutated circuits. Their characteristics, however, are restricted:

- Regarded as energy *consumer*, the AC VSR (= DC CSI) can only commutate when the 120° *square wave* AC current $i_{a,b,c}$, which it is *drawing* from the AC side is restricted to both *inductive* quadrants (Figure 7-6 right top). The lagging angle φ_i between this current $i_{a,b,c}$ and the AC system voltage $u_{a,b,c}$ is determined by the firing angle α which controls the DC voltage u_d.

$$\varphi_i \approx \alpha \qquad 0° < \alpha < 180° \tag{7.14}$$

- Regarded as energy *source*, the DC CSI (= AC VSR) can *deliver* only 120° *square wave* AC current $i_{a,b,c}$ in both *capacitive* quadrants (Figure 7-6 right bottom). Therefore, it can only feed synchronous motors and slip ring motors (usually in the subsynchronous driving mode). In both cases the motors are able to consume capacitive current. Induction motors, however, need an inductive current component (Figure 7-6 bottom.)

For forced commutation GTOs have to be substituted for turn-on thyristors. This is however not shown in Figure 7-6 right. In this case the AC current $i_{a,b,c}$ can be controlled with PWM. Its phase angle with respect to the AC voltage $u_{a,b,c}$ can be adjusted freely in all *four quadrants* according to the requirements of the AC side (dotted lines in the phasor diagrams in Figure 7-6 right). They can feed any kind of motors Figure 7-6 bottom. The forced-commutated current switching requires switching capacitors C_s at the AC side (shown with dotted lines in Figure 7-6 right). These capacitors also work as filters for the PWM AC currents $i_{a,b,c}$. The result is a relatively good *sinusoidal* wave shape of the current at the AC side (not shown in Figure 7-6 right).

DC Voltage Source Inverters. The basic circuit in Figure 7-6 left is

- A DC *voltage* source inverter, (Figure 7-6 left bottom), which "inverts" an impressed DC voltage and generates a switched AC voltage ($U_d \to u_{a,b,c}$) and at the same time it is also

- An AC *current* source rectifier, (Figure 7-6 left top), which rectifies an impressed AC current and generates a switched DC current $(i_{a,b,c} \rightarrow i_d)$. This, however, is a rather unusual naming.

THE DC SIDE. At the DC side, the voltage U_d has a fixed polarity which is given by the GTOs, whereas the DC current i_d can be positive or negative. Energy transfer in both directions, DC $\rightarrow$ AC and AC $\rightarrow$ DC, is therefore possible. Since the DC current i_d is a pulsed current, the capacitors C are required to smooth the DC voltage U_d.

THE AC SIDE. DC VSIs (= AC CSRs) are always operated with forced commutation. This requires GTOs and antiparallel diodes. At the AC side the voltages $u_{a,b,c}$ normally are PWM controlled. Four-quadrant operation with any phase angle between current $i_{a,b,c}$ and voltage $u_{a,b,c}$ is possible. They can feed any kind of motors (Figure 7-6 bottom). The inductances L_σ work as filters for the PWM AC voltages $u_{a,b,c}$. This results in a relatively good sinusoidal wave shape of the current $i_{a,b,c}$. The AC inductance L_σ normally is either the leakage inductance of the transformer (Figure 7-6 left top) or of the motor (Figure 7-6 left bottom).

7.5.2. Converter Configurations

The table presented in Figure 7-7 shows the most important large-drive systems (numbered from 1 to 13). They are made up of only two basic circuits—those shown in Figure 7-6.

The configurations drawn in bold lines are used in large numbers in practical applications (1, 2, 3, 6, 12, 13).

Configurations that are occasionally used (7, 11) or which are going to be introduced for high-power drives (4, 8) are indicated with normal lines.

Fine lines indicate future systems with promising characteristics, which however are still too expensive or need further development efforts (5, 9, 10).

Combinations of converters and motors which are theoretically possible, the introduction of which is however rather improbable, are not listed in Figure 7-7.

Figure 7-7 bottom indicates the sections in which the drive systems are described.

AC Voltage Source Rectifiers. Line and motor (load) commutated AC VSRs (= DC CSIs) are the most commonly used basic circuits for large drives. They are simple and easy to build for high-power applications. With regard to their characteristics they, however, are inferior: When implemented at the *mains side* they draw a 120° square wave current, which lags the voltage by 0° to 180° (1, 2, 3, 4, 5, 6, 7, 8, 11, 12, 13). Implemented at the *motor side* as single bridges, they can feed only synchronous motors (1) and—as diode bridges—they can be used to feed energy back from subsynchronous slip-recovery drives (2). Induction motors would require an inductive current component, which cannot be delivered by load commutated AC VSRs (= DC CSIs). Only when arranged as three-phase cycloconverters—with two

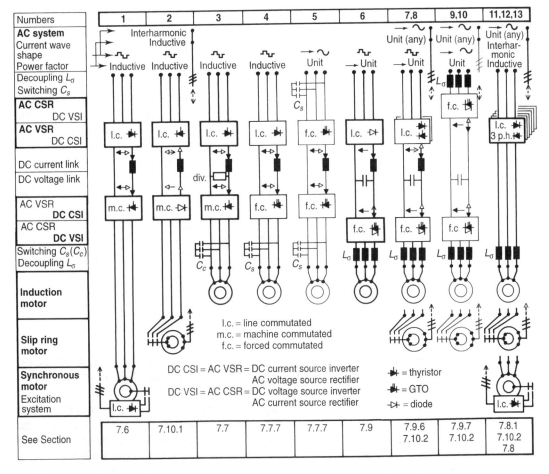

Figure 7-7. Systematic representation of the most important large drive
        systems.

antiparallel bridges per phase, the "DC" output of which is controlled sinusoidally—
AC VSRs can feed all kinds of motors (11, 12, 13).

  If single DC CSI bridges at the motor side have to feed induction motors, they
need a *diverter* in the DC link (3) or they have to be equipped with GTOs, for forced
commutation (4, 5). At the mains side, forced-commutated GTOs, controlled with
PWM, allow the take of a (nearly) sinusoidal current at unit power factor out of the
AC system (5).

DC Voltage Source Inverters. DC VSIs (=AC CSRs) can only be operated
with forced commutation. They are equipped with GTOs and antiparallel diodes.
Controlled with PWM they can take a nearly sinusoidal current at unit power factor
out of the AC system (9, 10) and they can feed any of the high-power motors (6, 7, 8,
9, 10; synchronous motor, not shown in Figure 7-7).

Direction of the Energy Transfer. The current and voltage arrows in the DC links in Figure 7-7 show in which way the direction of the energy transfer can be controlled. Filled arrows indicate the polarities of the DC voltage and of the DC current for energy transfer from the AC system to the motor. Unfilled arrows stand for energy feedback into the AC system.

AC VSR (= DC CSI) bridges with thyristors or GTOs have only one DC current direction. For power flow reversal the DC voltage polarity has to be changed by the controls (1, 2, 3, 4, 5, 7, 8 line side, 1, 3, 4, 5 motor side).

Complementary to that the DC VSI (=AC CSR) bridges have a fixed DC voltage polarity whereas the DC current can be positive or negative (9, 10 line side, 6, 7, 8, 9, 10 motor side).

Diode bridges (2 motor side, 6 line side) can only transfer energy from the AC to the DC side.

The combination of a line commutated AC VSR with a DC VSI at the motor side (7, 8) needs two antiparallel AC VSR bridges, which allow the change in polarity of the DC current for power flow reversal. The DC VSI does not allow the change in polarity of the DC voltage.

Cycloconverters (11, 12, 13) can be operated with both voltage and current polarities at their sinusoidally controlled "DC" side. Therefore, they are four-quadrant converters.

Forced Communication at Both Converter Sides. Large drive configurations with either forced commutated AC VSRs (= DC CSIs) or DC VSIs (= AC CSRs) at the line *and* motor side (5, 9, 10) have ideal characteristics: they take a sinusoidal current of unit power factor out of the AC system, they can transfer energy in both directions and they can feed any of the high-power motors (slip ring motor and synchronous motor not shown in Figure 7-7 for 5. Synchronous motor not shown in Figure 7-7 for 9, 10). Their introduction in industrial high-power applications is however a question of cost. In AC fed locomotives single-phase AC VSIs at the line side (9) are state of the art since the end of the 1980s.

7.6. SYNCHRONOUS MOTORS, FED BY EXTERNALLY COMMUTATED CURRENT SOURCE CONVERTERS

This system is the most important and successful large drive system in industrial applications. Therefore, many general aspects, also of importance for other large-drive systems, are presented in this section especially in Section 7.6.3.

7.6.1. Basic Principle

Figure 7-8(1) shows the basic principle: the stator of the synchronous machine is connected to the AC system via a three-phase frequency converter with an intermediate DC current link. The rectifier of the excitation system supplies a DC current to the rotor windings.

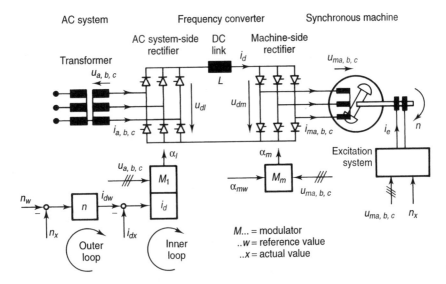

Figure 7-8. The synchronous machine fed by a frequency converter with intermediate DC current link. (a)–(f) Currents and voltages (idealized) during linear accelertion of the motor from low speed to maximum speed (in favor of the illustration of the operation mode, the assumed acceleration rate is higher than usual in practice).

7.6.2. Operation Modes

Motor. When describing the drive system, it is sufficient to consider the synchronous motor as a three-phase AC voltage source. The stator voltages $u_{ma,b,c}$ are controlled by the excitation system. To obtain a good magnetic utilization while avoiding saturation, the flux linkage Ψ_m must be held at the nominal value by varying the amplitude U_m of the voltages $u_{ma,b,c}$ proportional to the frequency f_m; that means proportional to the speed n (Figure 7-8f and Figure 7-2d).

$$U_m = \omega_m \, \Psi_m \quad \text{with} \ \Psi_m = \text{const.} \quad \text{follows} \ U_m \sim \omega_m \sim n \qquad (7.15)$$

$$\omega_m = 2\pi f_m = 2\pi n \cdot z_p \qquad z_p = \text{number of pole pairs}$$

The machine can be excited via slip rings by an external, controllable rectifier, or it can be equipped with a brushless excitation system. In this case a rotating diode bridge on the rotor, supplied by a rotating transformer, feeds the excitation winding. It is externally controlled by an AC-to-AC converter.

Converter. The frequency converter consists of two controlled rectifiers (AC VSR) of equal structure, one at the AC system side and a second at the machine side. Both are connected back to back with their DC sides and linked together via the inductance L of an intermediate DC current link. The mains-side rectifier rectifies the AC system voltage ($u_{a,b,c} \rightarrow u_{dl}$) and controls the DC current in the intermediate DC

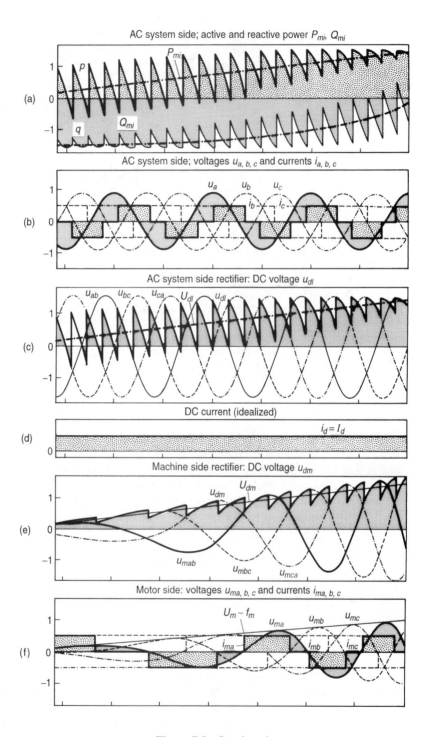

AC system side; active and reactive power P_{mi}, Q_{mi}

(a)

AC system side; voltages $u_{a,b,c}$ and currents $i_{a,b,c}$

(b)

AC system side rectifier: DC voltage u_{dl}

(c)

DC current (idealized)

(d)

Machine side rectifier: DC voltage u_{dm}

(e)

Motor side: voltages $u_{ma,b,c}$ and currents $i_{ma,b,c}$

(f)

Figure 7-8. Continued.

link against the opposing rectified AC voltage of the machine ($u_{ma,b,c} \rightarrow u_{dm}$) (index l: line side; index m: motor side).

Looking at the operation mode of the frequency converter, the following characteristics of its rectifiers must be called to mind:

- Their DC output voltage U_d (mean value) can be controlled continuously between a positive and a negative maximum value with the firing angle α (equation 7.13).

$$U_d = U_{d\max} \cdot \cos \alpha; \quad \alpha \rightarrow 0° \qquad \rightarrow +U_{d\max}$$
$$\alpha \rightarrow 180° \qquad \rightarrow -U_{d\max} \tag{7.16}$$

- The DC current i_d, smoothed by the reactance L ($i_d = I_d$), can only flow in the direction given by the thyristors. This means that for power flow reversal, the DC voltage polarity has to be changed.
- Since both rectifiers are externally commutated (line respectively machine commutated), they need external AC voltages for commutation and their commutation is guaranteed only for firing angles within the range $0° < \alpha < 180°$ with a safety margin to $180°$.
- The phase angle φ_i of the AC current with respect to the AC voltage is determined by the control angle α.

$$\varphi_i \approx \alpha (0° < \alpha < 180°) \tag{7.17}$$

That means that the rectifier on the mains side *takes current with an inductive component out* of the AC system (Figure 7-8a, b) and the rectifier on the machine side *delivers current with a capacitive component into* the machine (Figure 7-8f). The AC currents on both sides of the frequency changer are rectangular in shape because they are composed of DC current sections (Figure 7-8b and f): the converters, which—looking at the voltage—are AC voltage rectifiers, act as DC current inverters with respect to the current.

System: Motor plus Converter [7, 26]. When operating in the driving mode, the control angle of the machine rectifier is kept constant at $\alpha_m \rightarrow 180°$ to transfer energy from the DC link to the motor with minimal reactive current load at the motor. In this case the DC voltage U_{dm} (mean value) is proportional to the speed, since the AC voltage amplitude U_m of the motor is proportional to the speed. (See equations 7.15 and 7.16 and Figures 7-8e and f).

$$U_{dm} \sim -U_m \cos \alpha_m \quad \text{with } \alpha_m \rightarrow 180° = \text{const. follows}$$
$$U_{dm} \sim U_m \sim n \tag{7.18}$$

The AC system rectifier operates with a control angle between $0° < \alpha_l < 90°$ in order to be able to transfer energy from the AC system to the DC link. The firing angle α_l is adjusted in such a way that the DC current I_d, controlled with the generated DC voltage U_{dl} against the speed proprotional DC voltage U_{dm} of the motor rectifier, follows the set value (Figure 7-8(1)).

$$U_{dl} - U_{dm} = L\frac{d}{dt}I_d \rightarrow U_{dl} = U_{dm} \quad \text{for } I_d = \text{const.} \tag{7.19}$$

Therefore, the firing angle α_1 has to be varied from $\alpha_l \rightarrow 90°$ at low speed to $\alpha_l \rightarrow 0°$ at full speed. This leads to a line power factor, which is proportional to the speed, varying from zero to close to 1 (Figure 7-8c, b, a).

$$U_{dl} \sim \cos\alpha_1 \quad \text{with equations 7.19 and 7.18 follows} \quad \cos\alpha_1 \sim n \tag{7.20}$$

The motor can also be operated in the regenerative breaking mode. For power flow reversal, the control angle of the machine rectifier has to be changed from $\alpha_l \rightarrow 180°$ to $\alpha_l \rightarrow 0°$ and kept constant. The angle of the AC system rectifier has to be operated in the range between $90° < \alpha_l < 180°$.

The intermediate DC link decouples the frequencies of the AC supply system and the motor. Therefore, the motor speed is not limited by the AC system frequency. The converter is able to drive high-speed motors.

A special operation mode is required to run up the motor from standstill. As already mentioned, the converter is externally commutated: the motor-side rectifier needs the externally motor voltages $u_{ma,b,c}$ to commutate. At standstill, however, these voltages are zero. The commutation, therefore, has to be done via the mains side rectifier, using a special method, called "current pulsing in the DC link." To commutate the current from a previously conducting thyristor to the next thyristor, connected to the next motor phase, the controls of the AC system rectifier take the DC current to zero, thus extinguishing all thyristors in the motor converter. Then, after firing the new thyristor, the DC current again is raised to its normal value. In this way, the commutation takes place up to 3% to 5% of the rated speed where the motor voltage is high enough for commutation. Figure 7-9 shows the transition from "current pulsing in the DC link" (I) to the normal commutation mode (II) [7].

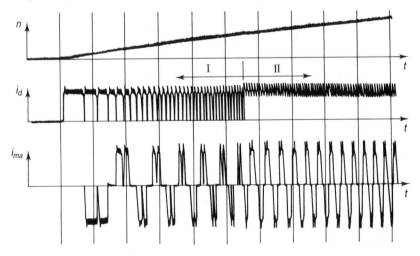

Figure 7-9. Transition from "current pulsing in the DC link" (I) to the normal commutation mode (II). Top to bottom: speed, DC current, motor current.

To keep the duration of the commutation short, the motor needs damper windings at the rotor, which reduce the commutation inductance.

Four-quadrant torque-speed operation is principally possible with this system. Because of the special operation mode required to start up the motor from standstill, it is, however, not an optimal four-quadrant drive.

Controls. The basic principle of the closed loop controls shown in Figure 7-8(1) is relatively simple. The main task is speed control. Parts of the controls have already been mentioned before.

- The excitation system controls the motor voltages $u_{ma,b,c}$.
- The outer control loop of the AC side rectifier is the speed control. It generates the reference value i_{dw} for the DC current.
- The inner control loop of the AC system rectifier controls the DC current I_d acting upon the firing angle α_l,
- This DC current I_d, distributed to the motor windings as a three-phase current, generates the electromagnetic torque T_e and acts upon the speed n:

$$U_{dm} \cdot I_d = T_e \cdot 2\pi \cdot n; \quad \text{with } U_{dm} \sim n \quad \text{follows } I_d \sim T_e \qquad (7.21)$$

- The current distribution to the motor windings is carried out by the controls related to the motor converter. Based on the motor voltages $u_{ma,b,c}$ as a reference, these controls keep the firing angle α_m constant at $\alpha_m \to 180°$ in the drive mode (Figure 7-8f, e) and at $\alpha_m \to 0°$ in the regenerative breaking mode. The reasons were explained earlier. When the motor voltage is zero at standstill, the shaft position serves as reference for the firing angle.

Keeping the firing angle α_m constant, leads to an additional advantage: synchronous motors directly connected to the AC system tend to long-lasting mechanical oscillations of their rotor relative to the rotating magnetic field in the stator. The motor can fall out of step. This oscillation also appears in the phase angle of the stator current relative to the voltage. The converter control with constant firing angle α_m, however, fixes this angle between AC current and AC voltage in the stator of the motor to $\varphi_i = \alpha_m = \text{const.}$, thus preventing oscillations completely!

7.6.3. Practical Implementations of the System

The externally commutated converters used in the system are well proven and reliable converters. There is no practical limit at all for their power and voltage ratings. In HVDC transmission systems, externally commutated converters of up to hundreds of kilovolt and thousands of megawatt per system have been in commercial operation for many years.

Series Connection. For higher power ratings, series connection of semiconductors is the preferred technique:

Figure 7-10. Detail of a water-cooled converter. Top: Series connected disk
cell thyristors alternating with water-cooled heat sinks.
Bottom: Oil-insulated pulse transformers.

- Series connection of thyristors (Figure 7-10)
- Series connection of converters (Figure 7-11)

Series connection has advantages compared with parallel connection. High-voltage,
low-current systems cause lower (current) losses in the converters, motors,
transformers, and cables and facilitate cost-effective design of these components.

Well-proven solutions for gate control in high-voltage applications are
available:

- Oil-insulated pulse transformers for medium voltage applications transfer the
  triggering signal and the triggering energy to fire the thyristors (Figure 7-10).

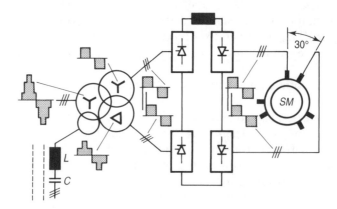

Figure 7-11. Series connection of 6-pulse bridges to form a 12-pulse configuration.

- Indirect light-triggered systems are state of the art in high-voltage converters. The triggering signal is transferred via fiber optics to a firing circuit directly allocated to each thyristor. The triggering energy to fire the thyristor is taken out of the snubber circuit. Operation status signals and fault status signals, indicating defective thyristors or firing circuits, can be transferred via fiber optics to control, monitoring, alarm, or protection devices.

Cooling System. In high-power converters, water cooling is preferred to air cooling, because it has many advantages:

- It allows a compact, space-saving design.
- It has a very low noise level.
- It can operate with high ambient temperature.

Disk cell thyristors alternating with water-cooled heat sinks form a stack pressed together with high pressure (e.g., 8 tons) by disk springs to ensure good heat transfer from thyristor to cooling water (Figure 7-10). The cooling water is conducted in plastic pipes to the heat sinks of the thyristors and to the snubber resistors, which are the main sources of losses. The cooling water circulates in a closed loop. It contains a deionizing equipment, operating in bypass, to keep corrosion and water conductivity at such a low level that the electric components to be cooled do not need an insulation from the cooling water. Recooling can be done by water-to-water (or water-to-air) heat exchangers.

Air cooling normally is only used at the low end of the high-power range.

Fault-Tolerant Design. The requirements on the reliability and availability increase with increasing power ratings. Power electronic systems have the advantage of modular design. Therefore, they have short repair times and are best suited for fault-tolerant solutions. Fault tolerance means that in case of a single fault, the operation of the system goes on without interruption and without restrictions. Major subsystems influencing the reliability and availability are the converters, the controls, and the auxiliaries.

Fault tolerance of the converter can be provided primarily by connecting one thyristor more than really needed in series $(N+1)$. Failures of single thryistors, snubber, or firing circuits cause a short-circuiting of the thyristor and therefore do not affect the functionality of the system as a whole.

Fault tolerance of the controls can be provided by a two-channel system. One channel acts upon the system, while the other is waiting in hot standby mode. After detecting a fault in the controls which operate in the closed loop mode, a changeover logic immediately disables the faulty channel and switches the hot standby channel into the closed loop operation mode. The fault detection normally is based on the self-test facilities of the controls and on a save area control which indicates variables operating out of their normal operation area (Figure 7-13) [27].

Redundant pumps and heat exchangers in the cooling system and redundant converters for excitation are additional means to increase the reliability and availability of the system.

Harmonics Reduction. Harmonics reduction in the motor and in the AC supply system is a usual requirement in high-power applications. The usual means to bring these currents closer to a sinusoidal shape is the installation of 12-pulse configurations (Figure 7-11).

At the AC system side two converter bridges are connected in series, each fed by separate transformer secondary windings, one of them in star-star (0°) and one in star-delta (30°) configuration with common primary windings [13]. Remaining harmonics can be eliminated by *LC* filters.

The motor can be equipped with two three-phase windings with a 30° phase-shifted position on the stator (Figure 7-11). The current fed in each of the three-phase windings remains, of course, a six-pulse current; the resulting magnetic field, however, which penetrates the rotor, has only 12-pulse harmonics. This reduces the torque riple and—more important in most cases—the rotor temperature rise caused by the losses of induced current harmonics in the rotor and its damper windings.

In cases, where an existing motor with only one three-phase winding is fed, or where the machine is a generator, which has to be started, 12-pulse currents in these machines can be reached by connecting them via transformers in the same way as mentioned before for the line side.

Line Power Factor. Since the power factor of converter-fed synchronous machines is proportional to the speed—as mentioned above (Figure 7-8a)—it sometimes needs to be compensated, at least for the normal operation speed range. This task can additionally be taken by the capacitors *C* of the harmonic *LC* filters (Figure 7-11).

Programmable High-Speed Controls. Converters are not only high power but also high-speed power stages and therefore put high requirements on their controls. It took until the first half of the 1980s until fast digital function plan controllers were ready to fulfill such requirements, combining high processing speed with graphic programming, even for the fastest closed loop controls. Based on identical hardware modules, the application specific orientation can be established only by software. The required control structures can be composed with a limited, but adequate set of predefined, pretested, ready-made software function blocks [28]. Graphic programming automatically delivers the documentation in form of a function plan (in case of closed loop controls) or as a state transition representation (in case of sequence controls). This leads to substantial cost reductions for engineering and documentation without reducing the flexibility!

Figure 7-12 shows the hardware configuration of a drives control system. It consists of the following:

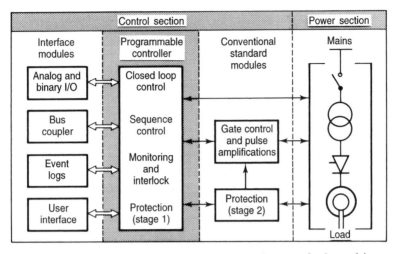

Figure 7-12. Hardware configuration of a control system for large drives.

- The programmable part for closed loop and sequential controls and for protection and monitoring.
- The interface modules to analog and binary inputs and outputs, to superimposed control systems, and to man machine communication systems.
- The system specific part for the gate controls and physically separated backup protection devices.

7.6.4. Applications

The basic principle of Figure 7-8 (top) has been well known for more than 50 years. But not until the beginning of the 1970s did thyristor converters and their controls reach a state making it possible to put this system on the market. In the meantime synchronous machines fed by frequency converters with intermediate DC current link found a wide range of different applications in the power range between 1 and 100 MW.

Start-Up Equipment. One of the first applications, which is still important today, was to run up large synchronous machines to their synchronous operation speed:

- In gas turbine generator sets the synchronous machine is used as a motor and run up to a speed high enough for the turbine to take the set over and synchronize it to the AC system.
- Synchronous condensers are not driven by a turbine. Used as a motor they are run up to synchronism with the converter.

- Large synchronous motors for compressors and blowers in chemical plants, in wind tunnels, and for blast furnaces in the steel industry are soft started from the converter with the starting current limited to rated current or less.
- In pumped storage power plants the converter runs the synchronous machine up to the generating or to the pumping operation mode.

Drive Systems. Converter-fed synchronous motors are best suited for large single drives at medium and high speed:

- Fans, blowers, and pumps in industrial processes
- High-speed compressors in power plants and chemical processes (Figure 7-13) and reciprocating compressors.

Continuous load control improves the efficiency especially in part load operation. It is also suitable for

- Continuous rolling mills.
- The French "Train à Grande Vitesse" (TGV) [29] with an overhead contact wire for AC supply.
- Large passenger liners like *Queen Elizabeth II* with on-board diesel generators for AC supply.

These are examples for rail and sea transport systems, which use the converter-fed synchronous motor as main drives.

Interties Between Machine and AC System. In pumped storage power plants, the turbine speed with the best hydraulic efficiency depends on the load and the head and is different in generating and pumping mode. Coupling the constant frequency AC system via a static frequency changer to the synchronous generator/motor with variable speed can optimize the efficiency in all operation modes and also allows control of the pumping water flow by speed control.

In wind power plants the optimal efficiency of the wind turbine depends on the speed when the wind conditions change. It is, therefore, advantageous to vary the speed of the generator and link it via a frequency converter to the AC system.

7.6.5. Future Trends

Synchronous motors fed by externally commutated current source converters are mature systems, which have only a limited development potential. See, however, Section 7.7.

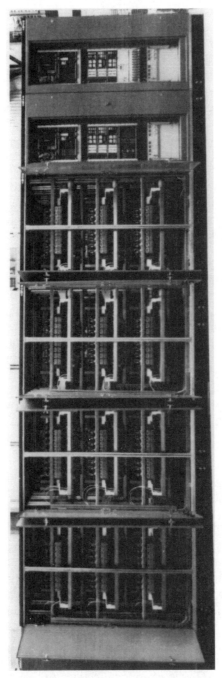

Figure 7-13. Twelve-pulse 13 MW converter with fault-tolerant controls (right side) for a high-speed (107 Hz) compressor drive.

7.7. INDUCTION MOTORS FED BY CURRENT SOURCE CONVERTERS

7.7.1. Basic Principle

As shown in Section 7.6.2 (Figure 7-8f), externally commutated converters with intermediate DC current link can only feed an AC current with a capacitive component into the motor. Induction motors, however, need an inductive current component. Therefore, it seems to be impossible to combine the simplest electrical motor (the induction motor) with the cheap and best proven converter (the externally commutated current source converter). But there is a solution, which, despite restrictions, could be introduced successfully into the practical applications: AC capacitors between converter and motor provide the necessary capacitive current which compensates for the inductive current of the motor, shifting the converter output current to the capacitive side, where load commutation is possible (Fig. 7-14 top).

7.7.2. Operation Modes

The amplitude I_c of the current $\underline{I}_c$ (phasor) delivered by the compensating capacitors C is proportional to the amplitude U and frequency f of the motor voltage $\underline{U}$ (phasor) (equation 7.22). To obtain a good magnetic utilization of the motor, the stator flux linkage $\underline{\Psi}$ must be held at its nominal value $\underline{\Psi} = \underline{\Psi}_n$. The amplitude of the motor voltage U, therefore, has to be controlled proportional to the stator frequency f (equation 7.23). The unpleasant result of this necessary measure is a drastic reduction of the capacitive compensating current I_c at low speed n (equation 7.24). The resistances are neglected in these equations.

$$\underline{I}_c = \underline{U} \cdot j2\pi f \cdot C \qquad (7.22)$$
$$\underline{U} = j2\pi f \cdot \underline{\Psi}_n \qquad (7.23)$$
$$I_c \sim f^2 \sim n^2 \qquad (7.24)$$

We therefore have to distinguish two operation modes:

1. *Load commutation* at high speed where the compensation current I_c is high enough to shift the phasor of the converter output current $\underline{I}_{\text{conv}}$ to the capacitive side where load commutation is possible
2. *Forced commutation* at lower speed where the current I_c is too low for compensation

The following terms will be used to describe the motor side converter: looking for the voltage it is a (AC voltage source) rectifier AC VSR, which rectifies the AC motor voltage; looking for the current it is a (DC current source) inverter DC CSI, which "inverts" the DC link current.

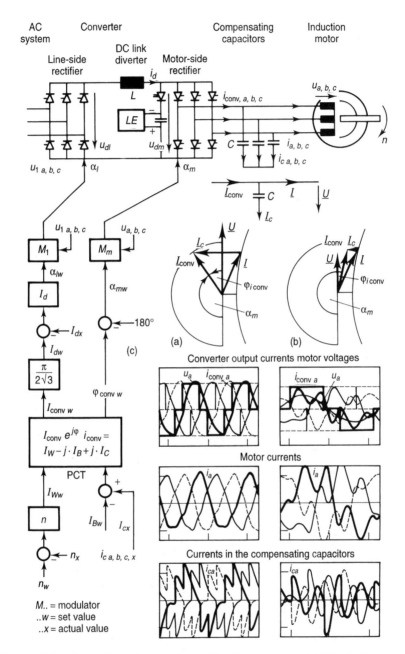

Figure 7-14. Top: Induction motor with C-compensated CSI. Left: Controls—c. Right: Operation at full speed $f = 100\% f_n$—a, and medium-speed $f = 50\% f_n$—b.

Load Commutation at High Speed. Figure 7-14a illustrates the operation mode with load commutation at nominal speed (100%) and nominal torque. The phasor diagram shows how the inductive component of the motor current $\underline{I}$ is compensated by the capacitors $(\underline{I}_c)$. The motor side rectifier can be operated with a firing angle $\alpha_m < 180°$. This is within the range $0° < \alpha_m < 180°$ which allows load commutation. The line diagrams in Figure 7-14a show that the waveshapes of the three-phase motor voltages $u_{a,b,c}$ and currents $i_{a,b,c}$ are more or less sinusoidal despite the rectangularly shaped converter output currents $i_{\text{conv } a,b,c}$ due to the smoothing effect of the compensating capacitors C.

Forced Commutation at Medium and Low Speed. Figure 7-14b illustrates the operation mode at medium speed (50%) and nominal torque. It can be seen in the phasor diagram and in the line diagrams that now the reduced compensation current $\underline{I}_c$ is too low to shift the converter output current $\underline{I}_{\text{conv}}$ at the capacitive side. The motor side rectifier has to be operated with a firing angle $\alpha_m > 180°$. This is no longer possible with load commutation. The converter, therefore, has to be equipped with an installation for forced commutation. Basically there are three alternatives [13, 30, 31, 14].

1. A forced commutated motor side rectifier with turn-on thyristors. The thyristor current is forced to zero by discharging a commutation capacitor, in opposite direction to the thyristor. Known as the forced-commutated DC current source inverter (DC CSI), this system has been built in large numbers, especially in Germany, for medium-power drives, but also for the low end of high-power drives.

2. A diverter in the DC link. There are different ways to realize diverters using LC circuits, thyristors and diodes. They all have to perform the forced commutation in basically the following way: the diverter—according to its name—diverts the current from the DC CSI to the diverter, thus extinguishing the previously conducting pair of thyristors in the DC CSI. Then, after blocking the diverter, and firing the new pair of DC CSI thyristors, the current again has to take its newly opened way through the DC CSI. At the present time, this is the most successful approach for large drives—especially in the United States. The diverter in Figure 7-14 (top), which we use to illustrate the principle, consists of a GTO, which diverts the DC current through a negatively charged capacitor. *LE* is the loading equipment.

3. A motor-side DC CSI in which turn-off thyristors (GTOs) substitute for the turn-on thyristors and the diverter (see Section 7.7.7). In this case the capacitor C between converter and motor can be reduced in size and its purpose can be changed from a compensating capacitor to a switching capacitor, which is needed to allow the forced commutated current switching at the AC side.

In the following we refer to the diverter in the DC link for forced commutation. It is the most widely used solution for large drives at the present time.

7.7.3. Resonance Problems

The insertion of compensating capacitors C, however, has also a rather unpleasant consquence: the capacitors form a resonant circuit together with the inductances of the motor. During start-up of the motor, the harmonics in the rectangular converter output currents $i_{\text{conv } a,b,c}$ excite this resonant circuit and cause a resonance step-up in the voltages $u_{a,b,c}$ and currents $i_{a,b,c}$ of the motor. This leads to an emergency switch off of the drive as soon as the motor voltages (or currents) exceed the protection levels.

The following considerations may give an idea about the numerical value of the frequency of resonance in practical applications and about the speed of the motor at which the resonant circuit will be excited: with regard to the harmonics, the motor behaves like a transformer with short-circuited secondary side. That means, when neglecting the resistances in stator and rotor the equivalent circuit of the motor can be represented by its total leakage inductance σL. The frequency of resonance f_{res}, therefore, is given by equation 7.25 and determined by C, the compensating capacitance, L, the no-load inductance of the motor, and σ, its total leakage factor, which is $\sigma \approx 5\%$ for large drives (equation 7.26). For reasons explained shortly, the nominal (n) compensating current I_{cn} at nominal speed ($\hat{=}$ nominal frequency f_n) should be $I_{cn} > \approx I_n$ (equation 7.27). In large drives I_n, the nominal motor current, is about 4 times the no-load (magnetizing) current I_m (equation 7.28). Both currents I_{cn} and I_m depend on the nominal values of the motor voltage U_n and the motor frequency f_n according to equations 7.29 and 7.30. From equations 7.25 to 7.30 follows equation 7.31 which shows that the frequency of resonance f_{res} is about twice the nominal frequency f_n of the motor:

$$f_{\text{res}} = \frac{1}{2\pi\sqrt{\sigma L \cdot C}} \tag{7.25}$$

$$\sigma \approx 5\% \tag{7.26}$$

$$I_{cn} \approx 1.25 \, I_n \tag{7.27}$$

$$I_n \approx 4 \, I_m \tag{7.28}$$

$$I_m = U_n/(2\pi f_n L) \tag{7.29}$$

$$I_{cn} = U_n \cdot (2\pi f_n C) \tag{7.30}$$

$$\rightarrow f_{\text{res}} \approx 2f_n \tag{7.31}$$

Therefore, during the start-up of the motor, the harmonics of the rectangular converter output current $i_{\text{conv } a,b,c}$ with the frequencies $5f$, $7f$, $11f$, $13f$ meet the frequency of resonance f_{res} at the following fundamental frequencies f:

$$
\begin{aligned}
f &= 40 \ \% f_n & \rightarrow & \quad 5f = 2f_n = f_{res} \\
f &= 28.6\% f_n & \rightarrow & \quad 7f = 2f_n = f_{res} \\
f &= 18.2\% f_n & \rightarrow & \quad 11f = 2f_n = f_{res} \\
f &= 15.4\% f_n & \rightarrow & \quad 13f = 2f_n = f_{res}
\end{aligned}
\tag{7.32}
$$

These figures can be considered as representative for practical applications. There is not much freedom left for the dimensioning of the compensating capacitors C: increasing the compensating current I_{cn} compared to equation 7.27 means a higher reactive current at the converter output at full speed. Decreasing it means a smaller upper speed range with load commutation.

7.7.4. How to Avoid Resonance Problems

Since the resonant circuit can't be avoided, one has to concentrate on how to reduce or to avoid its excitation. This can be achieved, because most of the practical applications put reduced requirements on the system: a limited operation speed range of $60\% \div 100\% \ f_n$ and low torque at low speeds.

The commonly used method, therefore, is to simply ride through the resonances. This has to be done at one side with low current but at the other side with high acceleration rate in order to keep the intensity of the excitation low and the duration short. The higher the rated power of the drive, the lower are the damping resistances in the stator and rotor windings! Fortunately the resistance for the harmonics exciting the resonance is increased due to the skin effect. Figure 7-15a shows the "ride-through" method, when passing $f = 40\% \ f_n$ during start-up. At

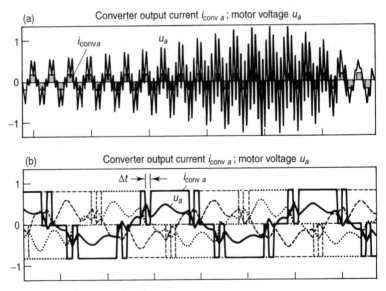

Figure 7-15. How to avoid resonance problems during start-up. (a) "Ride through" method. (b) SHE method.

$f = 40\% \ f_n$ the 5th harmonic meets the resonance frequency f_{res}. In Figure 7-14b with $f = 50\% \ f_n$ the excitation of the resonance is still visible.

Selected harmonics elimination (SHE) [32, 33] is another method to avoid the excitation of resonances. This means cutting gaps into the rectangular AC current (Fig. 7-15b) where only the width Δt is such that only those harmonics which would excite resonances are completely eliminated. The required width of the gaps can easily be calculated by means of Fourier analysis. Equation 7.33 shows the remarkable and simple result: to avoid harmonic frequencies, which are equal to the frequency of resonance $f_{res} = 2f_n \ (\to f_{res} = 120$ Hz for $f_n = 60$ Hz), the gap width Δt has to be kept constant at

$$\Delta t = 1.33 \text{ ms } \frac{60 \text{Hz}}{f_n}; \ \Delta t = 1.33 \text{ ms for } f_n = 60 \text{ Hz} \tag{7.33}$$

This is illustrated for $f = 40\% \ f_n$ in Figure 7-15b, where the 5th harmonic is eliminated, which otherwise would excite the resonance: $5f = 2f_n = f_{res}$. Compared to the "ride-through" method, the SHE method causes higher losses and puts higher stresses on the diverter because of its three times higher switching frequency. It is, however, only used in the lower-speed range during start-up (equations 7.32).

7.7.5. Basic Control Structure

Figure 7-14c shows a suitable basic structure of the speed control. The output of the speed controller n is regarded as set value I_{Ww} of the active component I_W of the stator current $\underline{I}$ which acts upon the electromagnetic torque T_e of the motor (equation 7.34) (z_p = number of pole pairs). The set value I_{Bw} for the reactive component I_B has to keep the flux linkage Ψ of the motor at its nominal value (equation 7.35):

$$T_e \approx 3/2 \cdot z_p \ \Psi \cdot I_W \tag{7.34}$$

$$\Psi \approx L_1 I_B \tag{7.35}$$

I_W, the torque-generating component of the stator current, is perpendicular to the stator flux linkage phasor $\underline{\Psi}$, whereas I_B, the magnetizing component, is in line. L_1 is the stator inductance. Both set values I_{Ww} and I_{Bw} are regarded as set values of the components of the stator current phasor $\underline{I} = I_W - jI_B$. Adding the measured (or estimated) compensation current $+jI_c$ leads to the set value $\underline{I}_{conv \ w}$ for the phasor of the current $\underline{I}_{conv}$ of the motor-side DC CSI, which we use in polar coordinate representation (PCT, polar coordinate transformer):

$$\underline{I}_{conv} = I_W - jI_B + jI_c = \underline{I}_{conv} = I_{conv} \cdot e^{j\varphi_{i conv}} \tag{7.36}$$

The amplitude $\underline{I}_{conv \ w}$ —multiplied with $\pi/(2\sqrt{3})$—is regarded as set value I_{dw} for the current I_d in the DC link. It has to act upon the firing angle α_1 of the line-side rectifier via the DC current controller I_d to control the DC current against the opposing DC voltage u_{dm} of the motor-side rectifier.

The phase angle $\varphi_{i conv \ w}$ acts on the firing angle $\alpha_{mw} = \varphi_{i conv \ w} - 180°$ of the motor-side CSI, which distributes the DC current as a three-phase current to the motor windings. $\alpha_m \approx \varphi_{i\,conv} - 180°$ determines the phase angle of this current with respect to the motor voltage (Figure 7-14a).

7.7.6. Applications—Practical Implementations

Voltage source PWM inverters put high dv/dt stresses on the insulation of the motor windings. Compensated current source inverters, however, generate more or less sinusoidal motor voltages—due to their compensation capacitors. Therefore, it was not by chance that retrofitting of existing motors is the main application: in many U.S. power plants existing fixed speed motors, designed for sinusoidal AC system voltages, have been equipped with current source converters of this type to save energy by varying the speed according to the load requirements [15].

High-speed compressor drives in gas pipelines (e.g., 10,000 rpm, 6MW) are new applications of this system. The advanced design, with motor and compressor contained in one case, is using dry seals and magnetic bearings. High horse power capability of the motor at a given volume is achieved due to the high speed and due to the use of pipeline gas to cool the motor very efficiently [34]. As in other large drive systems, water-cooled, series-connected turn-on thyristors (in $N+1$ configuration) with indirect light triggering are used (see Section 7.6). Line power factor and AC system harmonics are similar to those of the converter-fed synchronous motor in Section 7.6. They can be improved by basically the same measures.

The system can also be used to feed the recovered energy of gas expansion compressors, the speed of which is high and variable, into the AC system.

7.7.7. Future Trends

In traction applications the induction motor fed by voltage source GTO inverters is winning through against other drive systems (see Figure 7-7, 9). In stationary applications however, it seems that two systems are going into competition to succeed the converter fed synchronous motor (Section 7.6 and Figure 7-7, 1): voltage source GTO inverters (Section 7.9 and Figure 7-7, 6, 7, 9) and current source GTO inverters (Section 7.7 and Figure 7-7, 3, 4, 5) both feeding induction motors.

Today's large current source inverters are still using conventional turn-on thyristors and a diverter in the DC link for the operation modes, which require forced commutation. In the next generation GTOs will substitute for the turn-on thyristors and the diverters. The mode of operation, however, is basically the same as described. This development has already started for medium-sized drives and will be continued for large drives with series connection of GTOs [14].

GTOs should also be used at the line side of the converter to improve the power factor.

7.8. THE CYCLOCONVERTER-FED SYNCHRONOUS MOTOR

7.8.1. Basic Principle

Figure 7-16 shows the basic principle: the stator of the three-phase motor is connected to the AC system via a three-phase cycloconverter, the input transformer of which has three different secondary sides. The motor can be an induction motor. In most cases, however, synchronous motors are preferred. They are more robust because of their wider air gap. The rotor is excited via sliprings by a rectifier.

7.8.2. Mode of Operation

Converter. Cycloconverters for large drives are made up of line-commutated controlled six-pulse rectifier bridges (AC VSRs), the "DC" output of which is sinusoidally controlled. They consist of two antiparallel bridges per phase, one for each direction of the generated AC current. Three single-phase cycloconverters form a

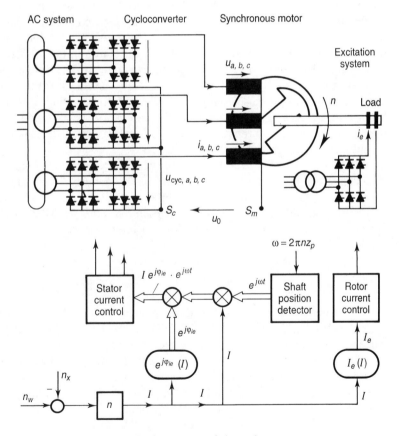

Figure 7-16. Cycloconverter-fed synchronous motor.

three-phase cycloconverter. The cycloconverter of each motor phase cuts individual sections out of the 50 Hz (respectively, 60 Hz) AC supply voltage and recombines them into a new voltage, the short-term mean value of which has a sinusoidal form (Figure 7-18a). This voltage generates a sinusoidal motor current with low harmonic components. The bridge, which conducts the current, also generates the output voltage. The other one is blocked until the polarity of the current changes. Since the voltage is composed of sections of the AC supply voltage its frequency is limited to about 40% of this frequency.

The AC side transformer needs three separated secondary sides with the leakage inductances mainly concentrated at the secondary sides (Figure 7-16). This is to decouple the commutation loops of the three single-phase cycloconverters.

Cycloconverters are able to deliver output currents $i_{a,b,c}$ with any phase angle relative to the output voltage $u_{cyc\ a,b,c}$. Therefore, energy transfer in both directions is possible, enabling the drive to be operated in the four-quadrant torque speed mode.

As we will see, the line power factor is proportional to the speed.

Motor. In synchronous motors fed by externally commutated current source converters with intermediate DC link (Section 7.6), the output voltage is generated by the motor and impressed to the motor-commutated rectifier. The motor, therefore, must be equipped with rotor damper windings. They reduce the commutation inductance and keep the duration of the motor commutation short.

In complete contrast to the foregoing, cycloconverters are line commutated and generate an output voltage, which is impressed to the stator of the motor. The motor must be built without rotor damper windings. Damper windings would reduce the leakage inductance. The harmonics in the voltage of the cycloconverter—for which the equivalent circuit of the motor consists (nearly) only of its leakage inductance— would generate increased current harmonics.

Further harmonics reduction can be achieved with a star connection in cycloconverter and motor (see S_c and S_m in Figure 7-16): the zero-sequence component u_0 in the cycloconverter output voltage $u_{cyc\ a,b,c}$ can be kept away from the motor and shifted between both star points S_c and S_m. There it cannot generate current harmonics, if the star points are not interconnected.

$$u_{a,b,c} = u_{cyc\ a,b,c} - u_0 \qquad (7.37)$$

$$u_0 = 1/3\ (u_{cyc\ a} + u_{cyc\ b} + u_{cyc\ c}) \qquad (7.38)$$

To understand the way the drive system works, it is useful to regard the cycloconverter and the excitation rectifier as current sources which impress their currents onto the motor by means of their controls.

The excitation DC current $i_e = I_e$ generates a magnetic field in the stator which is turning with the rotor. The angular frequency of the rotor is $\omega = 2\pi \cdot n \cdot z_p$, where n is the speed and z_p the number of pole pairs. The flux linkage of this field is represented in Figure 7-17 as a rotating phasor $I_e M e^{j\omega t}$ given by the current I_e and the stator-rotor coupling inductance M. The phasor is in line with the excitation field axis.

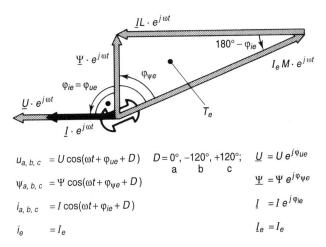

$$u_{a,b,c} = U\cos(\omega t + \varphi_{ue} + D) \qquad D = 0°, -120°, +120°; \qquad \underline{U} = U e^{j\varphi_{ue}}$$
$$\phantom{u_{a,b,c} = U\cos(\omega t + \varphi_{ue} + D) \qquad} a \qquad b \qquad c$$

$$\psi_{a,b,c} = \Psi\cos(\omega t + \varphi_{\psi e} + D) \qquad\qquad\qquad \underline{\Psi} = \Psi e^{j\varphi_{\psi e}}$$

$$i_{a,b,c} = I\cos(\omega t + \varphi_{ie} + D) \qquad\qquad\qquad \underline{I} = I e^{j\varphi_{ie}}$$

$$i_e = I_e \qquad\qquad\qquad\qquad\qquad\qquad\qquad I_e = I_e$$

Figure 7-17. Phasor diagram.

The three-phase currents $i_{a,b,c}$ also form a magnetic field in the stator, which turns synchronously at the same angular frequency ω. The flux linkage of this field is represented in Figure 7-17 as a rotating phasor $\underline{I}L e^{j\omega t}$ with $\underline{I} = I e^{j\varphi_{ie}}$. It is determined by the amplitude I of the stator current multiplied with the stator inductance L and has a phase shift φ_{ie} to the excitation field axis.

The triangle formed by the flux linkage phasors $\underline{I_e}M e^{j\omega t}$ and $\underline{I}L e^{j\omega t}$ and of their resultant phasor $\underline{\Psi}e^{j\omega t}$ (Figure 7-17 and equation 7.39) encloses an area, which is proportional to the electromagnetic torque. That means, the torque T_e can be influenced by the currents I, I_e, and the phase angle φ_{ie}.

The relation between the resultant flux linkage and the stator voltage is given by the steady-state phasor equation 7.40 in which the stator resistors are neglected.

$$\underline{\Psi} = \underline{I}L + \underline{I_e}M \qquad\qquad\qquad (7.39)$$

$$\underline{U} = j\omega\underline{\Psi} \qquad\qquad\qquad (7.40)$$

To obtain good magnetic utilization of the motor and to work under all load conditions with the lowest possible current in converter and motor, the currents I and I_e and the phase angle φ_{ie} should be combined in such a way, that

- The resultant magnetic field (Ψ) is always kept constant at its nominal value.
- The stator current $\underline{I} = I e^{j\varphi_{ie}}$ is a purely active current, the phasor of which is in line with the voltage phasor $\underline{U} = U e^{j\varphi_{ue}}$ (U: amplitude). See Figure 7-17.

In this case the electromagnetic torque T_e is given by equation 7.43. Equation 7.43 follows from the equation for the electric power P supplied by the converter (equation 7.41) which is transferred to the rotor (equation 7.42). (The losses are neglected.)

$$P = \frac{3}{2} UI \tag{7.41}$$

$$P = \frac{\omega}{z_p} T_e \tag{7.42}$$

$$\xrightarrow{(7.40)} T_e = z_p \frac{3}{2} I\Psi \tag{7.43}$$

Controls. The main objective of the controls is speed control by means of torque control. This has to be achieved with rated stator flux and a purely active stator current in order to keep converter and motor as small as possible at the given power and speed ratings.

The simplified basic control structure shown in Figure 7-16 should be regarded rather as an illustration of the control tasks than as a real control structure.

The output of the superimposed speed controller n is used to derive the set values for the rotor and stator current controls, which have to act upon the torque.

The rotor current controls have to impress the DC excitation current I_e on the rotor via a controlled rectifier.

The stator current controls, which are not shown in detail in Figure 7-16, have to impress the stator ampere turns to the motor with a given intensity (I) and position relative to the shaft (φ_{ie}) via the cycloconverter.

Both currents I and I_e and the phase angle φ_{ie} determine the torque T_e. But they have to be coordinated (e.g., by means of function generators $I_e(i)$ and $\varphi_{ie}(I)$) in such a way that they keep the stator flux constant at its rated value and that they generate a purely active stator current. In this way, with the converters operated as current sources, the torque can be controlled independently from the speed. High torque at low speed and even at standstill is easily attainable.

Mechanical oscillations of the rotor relative to the rotating magnetic field in the stator, known from synchronous motors directly connected to the AC system, cannot occur, when impressing the phase angle φ_{ie} of stator current phasor $\underline{I} = Ie^{j\varphi_{ie}}$ relative to the excitation field axis of the rotor!

The controls which are in practical use are very different and sophisticated [35, 36]. But basically all of them have to follow the foregoing rules. The vector control structure in Figure 7-16 is excitation field oriented and has parameter sensitive function generators $I_e(I)$ and $\varphi_{ie}(I)$. This structure proved to be good enough for high-torque, low-speed drives with low dynamic requirements [6]. For high dynamic requirements, which can also include heavy load torque peaks and operation in the field-weakening range, this structure needs additional loops to ensure

- A more direct field orientation of the stator current
- A more direct stator flux control with a less parameter sensitive flux estimation

Line Power Factor. Keeping the magnetic field at the nominal value means that the amplitude of the motor voltage is proportional to the frequency (equation

7.40). Adjusting the motor voltage proportional to the frequency means that the control angles α of the three-phase cycloconverter, starting from $\alpha \to 90°$ at standstill, decrease with increasing frequency. Therefore, the line power factor is proportional to the frequency (Figure 7-18a, b). At maximum frequency the power factor comes close to "1" if the cycloconverter then operates in the trapezoidal mode with control angles $\alpha \to 0°$ (Figure 7-18b).

7.8.3. Practical Implementations of the System

The general statements made in Section 7.6.3 can basically also be applied to the cycloconverter synchronous motor as well. But one has to take into consideration the following:

- Frequency variable cycloconverters produce frequency variable interharmonics at the AC system side (Figure 7-18a and b show clearly that the wave shape of the currents drawn from the AC system is not periodic). Passive *LC* filters are not well suited to suppress interharmonics with variable frequencies. If necessary, active filters have to be used for this purpose [37].
- When feeding two different three-phase winding systems at the same stator with two three-phase cycloconverters, one has to take into account that the cycloconverters impress their *voltages* onto the windings. (This is unlike the DC CSIs in Section 7.6.3, which impress *currents* onto the windings.) Therefore, both three-phase winding systems have to be decoupled magnetically to such an extent that their voltages do not produce high interacting current harmonics.

7.8.4. Applications

Four-quadrant torque-speed operation at a high power level with high torque at low speed and at standstill but with a rather low maximum speed are drive requirements for which cycloconverter-fed synchronous motors are best suited:

- Gearless cement mill drives were the first applications of this kind. The mill tube is driven from a low-speed wrap around motor with a high number of poles (Figure 7-19).
- Drives for mine hoists with similar high-power ratings became a typical application of this system.
- Reversing rolling mills are applications with extremely high dynamic requirements for torque and speed reversal.
- Many ice breakers and some other ships are equipped with diesel generator fed cycloconverter synchronous motors with power ratings up to about 20 MW per unit.

(a)   Output voltages; output currents (idealized)

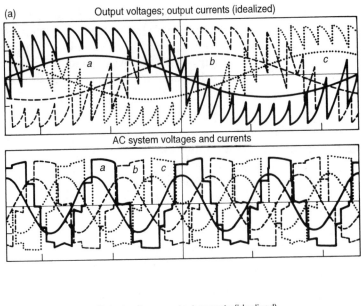

AC system voltages and currents

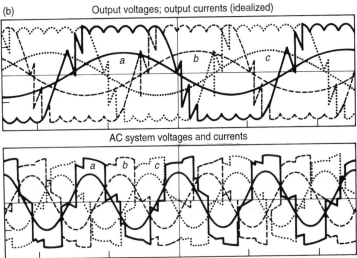

Figure 7-18. Three-phase voltages and currents at the output and at the input side of the cycloconverter. AC system frequency 50 Hz. Output frequency (a) 12.5 Hz, (b) 20 Hz (trapezoidal mode).

7.8.5. Future Trends

Despite the fact that cycloconverter-fed synchronous machines are mature drive systems, which do not draw advantages from new power semiconductors, they will remain a standard solution for high-torque, low-speed drives.

Figure 7-19. Gearless cement mill drive 6 MW 15 rpm.

7.9. LARGE VOLTAGE SOURCE INVERTER DRIVES

7.9.1. Characteristics of Today's Voltage Source Inverters

Despite their successful use in traction vehicles, voltage source inverters (DC VSIs) still suffer from many restrictions which limit their application areas. Because of the high switching losses of the GTOs and especially of the snubber circuits, the basic three-phase two-level DC VSI (Figure 7-21A) has a relatively poor efficiency, a limited switching frequency, produces high current and torque harmonics and allows only a limited maximum speed. Series connection of GTOs to increase the rated power is still difficult to master. These restrictions may be illustrated by the following figures.

Typical characteristics of commercially available high-power GTOs are 4.5 kV blocking voltage, 3 kA turn-off current, $di/dt_{max} \leq 300$ A/μs, $dv/dt_{max} \leq 500$ V/μs. With such GTOs, three-phase, two-level DC VSIs can be built with

DC supply voltage	$U_d = 2.6 \ldots 3\,\text{kV}$
phase current amplitude	$I \geq 1\,\text{kA}$
rated power	$S \approx 3/2 \cdot U_d/2 \cdot I \geq 2\,\text{MVA}$

Snubber circuits (Figure 7-20), which are necessary to keep the dv/dt and di/dt stresses within the permissible limits, need capacitors $C > I/(dv/dt) \rightarrow C \approx 4\ \mu F$ per GTO and inductances $L \approx U_d/(di/dt) \approx 10\ \mu H \rightarrow L \approx 5\ \mu H$ per GTO. At each switching cycle of the six GTOs in a three-phase two-level VSI, the energy $6 \cdot 1/2 \cdot Li^2$ and $6 \cdot 1/2 \cdot CU_d^2$ stored in the snubber elements L and C has to be dissipated in the snubber resistors R_L and R_C. When we calculate the snubber

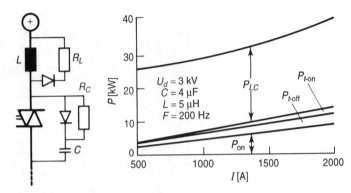

Figure 7-20. Snubber circuits per GTO. Losses in a three-phase two-level inverter.

losses P_{LC}, we have to take into account that the current $i(t)$ to be switched off, follows a sinusoidal course and therefore is not at its maximum I each time. The mean value of $i(t)^2$ is $I^2/2$.

$$P_{LC} = F\left\{6 \cdot \frac{1}{2}\left(\frac{LI^2}{2}\right) + 6 \cdot \frac{1}{2}CU_d^2\right\} \qquad (7.44)$$

$$F - \text{switching frequency per GTO}$$

In Figure 7-20 the main components of the total losses P in a three-phase, two-level VSI with the earlier-mentioned data and a switching frequency of only $F = 200$ Hz are plotted as a function of the amplitude I of the output current: the on-state losses P_{on}, turn-on and turn-off losses $P_{t\text{-}on}$, $P_{t\text{-}off}$ within the semiconductors are small compared to the snubber losses P_{LC}. Since P_{LC}, the largest loss component, as well as $P_{t\text{-}on}$ and $P_{t\text{-}off}$ increase proportionally to the switching frequency F, it becomes plainly recognizable why high-power GTO inverters can be operated only with a maximum switching frequency of $F \approx 300$ Hz per GTO or rather less (< 200 Hz) in practical applications [38].

With conventional *RLCD* snubber circuits (Figure 7-20), these disadvantages become clearly obvious. Even if improved snubber circuits (Figure 7-24) [39, 40] with reduced losses (about 60%) are used, the problems still cannot be solved properly. The use of energy recovering snubber circuits leads to further improvements, but also to voluminous and expensive solutions and—if common to all three phases—to restrictions in the PWM pulse patterns.

Therefore, one main objective for large drives configurations must be, to reach a given maximum speed with the lowest possible switching frequency and the lowest possible current and torque harmonics (see Section 7.9.3). Facing these objectives we have to focus mainly on two configurations: two-level and three-level inverter drives.

7.9.2. Two-Level Inverter Drives

Innumerable investigations have been carried out in the recent years, dealing with modulation methods to generate the pulse patterns for the inverter output voltages [41, 42]. Even to find a classification for all these methods is not easy:

- Carrier modulation [2]
- Space vector modulation [43]
- Off-line optimized modulations [44–46]
- Modulations directly based on feedback signals like current and flux [47].

Many of them are not well suited for large drives with low switching frequency. Our further considerations will be based on the carrier modulation method. With this method, the switching frequency is clearly determined. It also shows in the most clear way the advantages of using two (or more) pulse patterns with staggered switching instants. They are required to reduce the harmonics, to extend the maximum speed and to switch the GTOs with lower frequency. (See Section 7.9.4).

Figure 7-21 parts A and B show a three-phase, two-level configuration using the PWM method to generate three-phase voltages, with a linear voltage frequency characteristic for the fundamental component. At each intersection of the three-phase sinusoidal control signal ($u_{St\ a,b,c}$) and the triangular carrier signal (u_H), the inverter is changing the polarity of its output voltages $u_{a,b,c}$

$$u_{St\ a,b,c} > u_H \quad \rightarrow \quad u_{a,b,c} = + U_d/2$$
$$u_{St\ a,b,c} < u_H \quad \rightarrow \quad u_{a,b,c} = - U_d/2 \tag{7.45}$$

The sinusoidal control signals $u_{St\ a,b,c}$ determine amplitude U_{GS} and frequency f_1 of the fundamental components $u_{GS\ a,b,c}$ of the output voltages $u_{a,b,c}$.

$$U_{GS} = a_0 \cdot U_d/2 \qquad \text{with } a_0 = U_{St}/\hat{U}_H$$

where
$$U_{St} = \text{amplitude of } u_{St\ a,b,c}$$
$$\hat{U}_H = \text{peak value of } u_H$$
$$a_0 = \text{modulation index } (0 \div 1) \tag{7.46}$$

The triangular carrier signal u_H determines the switching frequency F per GTO.

The voltages across the motor windings $u_{1\ a,b,c}$ differ from the inverter output voltages $u_{a,b,c}$ by the zero-sequence voltage u_0 and have a remarkably lower harmonic content.

$$u_{1\ a,b,c} = u_{a,b,c} - u_0 \tag{7.47}$$
$$u_0 = 1/3(u_a + u_b + u_c) \tag{7.48}$$

The fundamental no-load (magnetizing) currents $i_{1mGS\ a,b,c}$ are given by equation 7.49. L_1 is the stator inductance. R_1 the stator resistance is neglected.

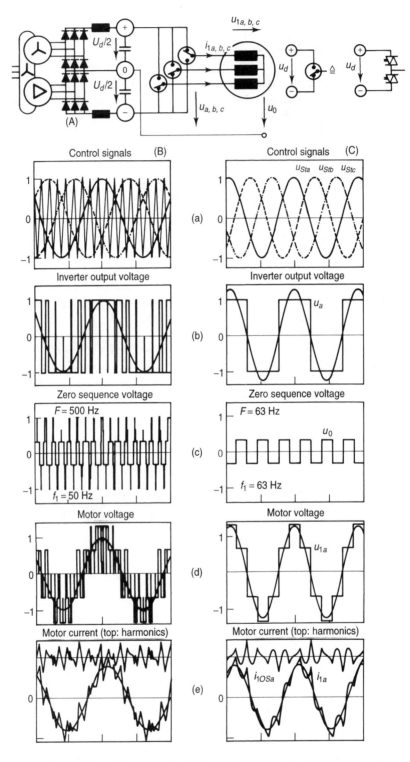

Figure 7-21. Three-phase, two-level inverter: (A) circuit; (B) PWM mode 50 Hz; (C) square wave mode 63 Hz.

$$i_{1mGS\,a,b,c} = \frac{1}{L_1} \int u_{GS\,a,b,c}\,dt \tag{7.49}$$

For large drives the nominal values of the fundamental currents $i_{1GS\,a,b,c}$ can be assumed to be about 4 times higher than the no-load magnetizing currents $i_{1mGS\,a,b,c}$ with a lagging angle of about 25° with respect to the fundamental voltages $u_{GS\,a,b,c}$.

The harmonics of the motor voltages $u_{1\,OSa,b,c} = u_{1\,a,b,c} - u_{GSa,b,c}$ generate the curent harmonics $i_{1\,OSa,b,c}$.

$$i_{1OS\,a,b,c} = \frac{1}{\sigma L_1} \int (u_{1\,a,b,c} - u_{GS\,a,b,c})\,dt \tag{7.50}$$

For harmonics the equivalent circuit of the motor is reduced to its leakage inductance σL_1 when the resistances are neglected. The current harmonics $i_{1OS\,a,b,c}$ are independent from the load and superimposed to the fundamental current. The total leakage factor is about $\sigma = 5\%$ for large motors.

The resultant motor currents are

$$i_{1\,a,b,c} = i_{1GS\,a,b,c} + i_{1OS\,a,b,c} \tag{7.51}$$

This is illustrated for the PWM mode in Figure 7-21 part B for $f_1 = 50$ Hz, $F = 500$ Hz, $a_0 = 100\%$, and for the square wave mode at full speed in Figure 7-21 part C for $f_1 = f_{1\,max} = 63$ Hz, $F = f_1 = 63$ Hz, $a_0 = 127\%$. Figure 7-21 represents the control signals $u_{St\,a,b,c}$ and u_H (a), the inverter output voltage u_a (b) the zero-sequence voltage u_0 (c), the motor voltage u_{1a} (d), and the rated motor current i_{1a} with its harmonics $i_{1OS\,a}$ (e).

7.9.3. Optimization Goal

Coming back to our optimization goals—high speed ($\hat{=}f_1$), low harmonics ($i_{1OS\,a,b,c}$), and low switching frequency (F) (see end of Section 7.9.1)—we define the total harmonic distortion (THD) of the no-load current as a measure for the quality of the generated pulse patterns of the output voltages $u_{a,b,c}$ (see equations 7.49 and 7.50).

$$\mathrm{THD} = \frac{\sqrt{\frac{1}{T}\int_0^T i^2_{1OS\,a,(b,c)}\,dt}}{\sqrt{\frac{1}{T}\int_0^T i^2_{1mGS\,a,(b,c)}\,dt}} \tag{7.52}$$

At maximum frequency and maximum modulation index $f_{1\,max} \sim a_{0\,max} = 4/\pi = 127\%$ when the inverter generates the maximum output voltage and switches only with fundamental frequency $F = f_{1\,max}$ (Figure 7-21 part B), we get $\mathrm{THD}(a_{0\,max}) = 0.928$ (for $\sigma = 5\%$). THD$=1$ would mean that the rms value of the current harmonics is equal to the rms value of the magnetizing current at no load.

In search of optimal pulse patterns at given fundamental frequency we regard the switching frequency $F \approx 300$ Hz and the THD $\approx \mathrm{THD}(a_{0\,max}) = 0.928 \approx 1$ as tolerable upper limits for the whole speed range. Than we have to compromise about the lowest possible THD and the lowest possible switching frequency F.

Doing this, we mainly focus on two cases:

1. normal-speed drives with $f_{1\max} = 50/60$ Hz and
2. high-speed drives, where we detect the highest possible maximum frequency $f_{1\max}$ within the limits given.

We will not take into account the extension of the speed range with field weakening.

Looking at the three-phase, two-level inverter drives (Figure 7-21) from these aspects reveals some rather unpleasant facts: as shown in Figure 7-23(1) the current harmonics can only be kept in the region of THD ≈ 1 with switching frequencies exceeding the set limit $F = 300$ Hz—even for *normal-speed drives* with a maximum frequency of only $f_{1\max} \approx 60$ Hz. Up to now high-power, *high-speed* drives cannot be built with two-level inverters.

7.9.4. Three-Level Inverter Drives

A three-phase, three-level inverter consists of 12 GTOs—4 GTOs per phase. Compared to two-level inverters, they can be operated with twice the DC voltage and therefore with twice the rated power. In Figure 7-22 part A, each phase is represented by a three-point change-over switch the output of which can be connected either to the positive pole, the zero point or the negative pole of the DC supply voltage.

One three-level inverter can be regarded as an inverter which can be operated with *two* independent pulse patterns!

In the *lower speed range* the inverter is operated in the *PWM mode* (Figure 7-22 part B): both pulse patterns I and II generate the same fundamental voltage, using however staggered switching instants to reduce the harmonic content in the resultant voltage $u_{a,b,c} = u_{\text{I }a,b,c} + u_{\text{II }a,b,c}$ [2]. The control algorithms are

$$
\begin{aligned}
u_{St\,a,b,c} \quad &>/< \quad u_{H\text{I}} \quad \rightarrow \quad u_{\text{I }a,b,c} = \pm U_d/2 \\
u_{St\,a,b,c} \quad &>/< \quad u_{H\text{II}} \quad \rightarrow \quad u_{\text{II }a,b,c} = \pm U_d/2 \\
&\qquad\qquad\qquad\qquad u_{a,b,c} = u_{\text{I }a,b,c} + u_{\text{II }a,b,c}
\end{aligned}
\tag{7.53}
$$

The three-phase sinusoidal control signal $u_{St a,b,c}$—common to both pulse patterns—determines the fundamental (GS) components $u_{\text{I}GS\,a,b,c} = u_{\text{II}GS\,a,b,c}$ of the resultant inverter output voltage $u_{a,b,c} = u_{\text{I }a,b,c} + u_{\text{II }a,b,c}$. The 180° phase shift between both triangular carrier signals $u_{H\text{I}}$ and $u_{H\text{II}}$ enforces the staggered switching.

In the *upper speed range* it is favourable to operate the inverter in the *square wave mode* (Figure 7-22 part C). That means to use pulse patterns, each switching with the fundamental frequency $F = f_1$. They are phase shifted against each other in order to control the resulting (frequency proportional) fundamental (GS) $u_{GS\,a,b,c} = u_{\text{I}GS\,a,b,c} + u_{\text{II}GS\,a,b,c}$ ($u_{\text{I}GS\,a,b,c} \neq u_{\text{II}GS\,a,b,c}$) of the voltage $u_{a,b,c}$ applied to the motor. The control algorithms are

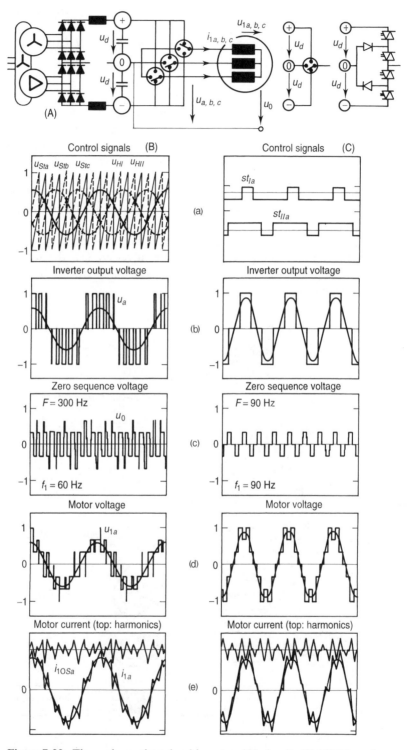

Figure 7-22. Three-phase, three-level inverter: (A) circuit; (B) PWM mode 60 Hz; (C) square wave mode 90 Hz.

$$st_{I\,a,b,c} \quad > / < 0 \quad \rightarrow \quad u_{I\,a,b,c} = \pm U_d/2$$

$$st_{II\,a,b,c} \quad > / < 0 \quad \rightarrow \quad u_{II\,a,b,c} = \pm U_d/2$$

$$u_{a,b,c} = u_{I\,a,b,c} + u_{II\,a,b,c} \qquad (7.54)$$

In Figures 7-22 part B and C, the control signals $u_{St\,a,b,c}$ and u_{HI}, u_{HII} (a) respectively $st_{I\,a,(b,c)}$, $st_{II\,a,(b,c)}$, the inverter output voltage u_a (b), the zero-sequence voltage u_0 (c), the resultant motor voltage u_{1a} (d), and the rated motor current i_{1a} with its harmonics $i_{1\,OS\,a}$ (e) are shown for the frequencies $f_1 = 60$ Hz (PWM mode) and $f_1 = 90$ Hz (square wave mode).

For three-level inverters the switching frequency F is defined in the same way as for the two-level inverters: as the frequency by which each individual GTO is turned on and off.

Figure 7-23 presents the results, which show the improved performance of three-level inverters (2) compared to two-level inverters (1): the upper part shows the THD, and the lower part shows the switching frequency F as functions of the fundamental frequency f_1.

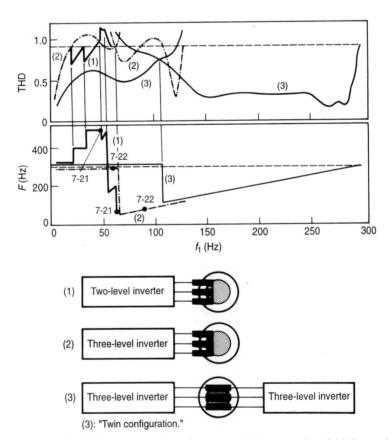

Figure 7-23. THD and switching frequency F for normal and high-speed drives as function of the fundamental frequency f_1.

For *normal-speed drives* with $f_{1\,\mathrm{max}} \approx 60$ Hz the switching frequency $F = F_{\mathrm{max}} = 300$ Hz can be used for further THD reduction in current and torque. In this case (which is not shown in Figure 7-23), the PWM mode should be extended to fundamental frequencies f_1 close to $f_{1\,\mathrm{max}} \approx 60$ Hz.

For *high-speed drives* (Figure 7-23 (2)) the maximum frequency can be increased up to about $f_{1\,\mathrm{max}} \approx 130$ Hz (which is 7800 rpm for a two-pole machine) without exceeding the set limit $F_{\mathrm{max}} = 300$ Hz of the switching frequency F and without using field weakening. The control method should be changed at $f_1 \approx 65$ Hz from PWM switching with $F = 300$ Hz to switching with fundamental frequency $F = f_1 = 65$–130 Hz.

7.9.5. Low-Inductance Design

Voltage source inverters need a careful mechanical design. Due to the physical size of the circuit components, to the space needed for electrical insulation and to the increased distances to the hot snubber resistors, all current loops must be regarded as parasitic inductances. Currents, switched off within these loops by semiconductors cause high dv/dt and voltage stresses which reduce the attainable rated power. Low-inductance design therefore is a must!

Figure 7-24 left shows the four loops a, b_1, b_2, and c in a two-level GTO inverter circuit with $RLCD$ snubbers [39] in which currents are switched with high di/dt. These loops have to be kept as small as possible. Figure 7-24 right shows how this has to be done basically:

- Low-inductance capacitors C_{DC} with sandwiched connections to the semi-conductors (loop a)
- Snubber capacitors C and diodes located as close as possible to the GTOs (loops b_1, b_2)
- Snubber resistors R with low-inductance design and sandwiched connections to the other snubber elements (loop c).

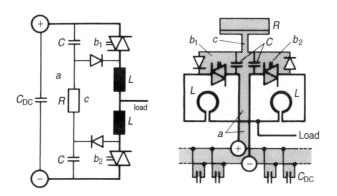

Figure 7-24. Low-inductance design of large VSIs.

Snubber inductors L in toroidal form keep the magnetic field inside the choke, thus avoiding inductive heating of neighboring metallic components.

7.9.6. Controls

In many low- and medium-power drive systems, where good dynamic behavior of the torque control or the elimination of the speed sensor is advantageous (e.g., servo drives), vector control and field orientation became standard solutions.

Large drives, however, are normally not operated without speed sensors and most of them, for example, those for pumps, fans, blowers, and compressors, don't need dynamic controls at all.

High dynamic requirements are normally combined with the need for fast energy reversal (e.g., steel mill control). But in these cases DC VSIs are still not competitive enough because they need antiparallel thyristor bridges or DC VSIs at the AC system side (see Section 7.5.2) instead of the simple low-cost diode bridges shown in Figure 7-21 and 7-22. That doesn't mean that field-oriented vector control should not be used in large VSI drives. But its advantages up to now did not count too much in today's industrial applications.

7.9.7. Future Trends

Induction motors, fed by VSIs, have the potential for real high-power and high-speed drives. When trying to look into the future from this point of view, advances in semiconductor technology as well as new motor-converter configurations have to be considered.

Advances in Semiconductor Technology. By means of improved GTO gate control, which drastically reduces individual differences in the turn-on and turn-off delay of single GTOs, *series or parallel connection of GTOs* will become easy. This leads to significant advantages:

- Higher power per VSI unit permits building large drives at reduced cost.
- Series or parallel connections with one GTO more than needed in each branch allow the introduction of a new quality: fault tolerance.

Together with the "twin configuration" (discussed shortly), this will open up VSIs for all power levels and speed ranges required in high-power and/or high-speed drives.

Looking farther into the future, MOS-controlled high-power GTOs (MCTs) with a much simpler and less expensive gate control can be expected.

However, there is also another more probable trend: Insulated Gate Bipolar Transistors (IGBTs) will substitute for the GTOs and allow the manufacturers to build cheaper and better inverters with reduced harmonics. They have three attractive advantages: higher switching frequency, easy and simple gate control, and no need for snubber circuits. GTOs—like flip-flops—snap shut and snap off when they are turned on and turned off because of their internal positive feedback. In contrast to that, IGBTs are continuously controllable during turn-on and turn-off. This makes overcurrent limitation much easier and allows dv/dt control to reduce the dv/dt stresses on motor and transformer windings.

The transition from GTOs to IGBTs takes place in low- and medium-power drives. How long it will take for large drives, depends on how soon currents and voltages of IGBTs can be increased to an adequate level comparable to that of today's GTOs. It also depends on how easy it will be to design series and parallel configurations of IGBTs. The reliability of IGBTs and their fault behavior in series connections still need further investigations before IGBTs are ready for use in large drives. Whereas GTOs are short-circuiting when they fail, IGBTs open the circuit. Fault-tolerant design for high-power *high-voltage* converters therefore is much easier with GTOs (see Section 7.6.3) than with IGBTs. It is to be expected that IGBTs will be introduced during the next few years at the low end of large drives.

Design and Configuration of VSIs. Rated power and frequency of today's GTO-VSIs are still limited to about 2 MW / 60 Hz for two-level VSIs and 4 MW / 130 Hz for three-level VSIs.

The obvious mean to increase the rated power—but not the rated frequency—is *parallel connection of two or more VSIs.*

It could also be considered *to feed a motor, carrying two different three-phase windings on its stator, with two different VSIs.* But, unlike CSIs which impress currents (see Section 7.6.3), VSIs impress voltages onto the windings. It is difficult to decouple both winding systems magnetically to such an extent that the voltages of both VSIs do not produce high interacting current harmonics.

The "*twin configuration*" shown in Figure 7-23(3) and described in more details in [12] is a new approach to extend both, the power and the frequency limitations. It is an unconventional motor-inverter configuration, based, however, on existing VSIs and on a motor with conventional three-phase stator windings, accessible at both ends: two three-phase VSIs feed the stator windings from both sides. The "twin configuration" based on existing 4 MW three-level inverters allows to increase the rated power to about 8 MW without series connections of GTOs. If necessary the maximum frequency f_1 can also be extended up to about $f_{1max} \approx 300$ Hz, which means 18,000 rpm for a two-pole machine. This is combined with a low total harmonic distortion of current and torque and a low switching frequency $F \leq f_{1max}$, which doesn't exceed the maximum fundamental frequency f_{1max} (see Figure 7-23(3)).

In a further development step *VSIs at the AC supply side*—which are state of the art for AC-fed locomotives—will also be introduced for high-power industrial applications. They allow drawing sinusoidal active currents with low harmonics from the AC system.

7.10. SLIP POWER–CONTROLLED DRIVES

7.10.1. Introduction

Slip power-controlled drives use AC motors with a three-phase wound rotor. The stator is connected to the AC system. The rotor side is controlled by a converter cascade, which is connected to the rotor via slip rings. Two different kinds of converter cascades, the *subsynchronous* and the *sub- and hypersynchronous* cascade, gained significance in practical applications. Both systems have attractive features, but also inherent disadvantages, which hinder a more widespread use.

Subsynchronous Cascades. Drives with subsynchronous converter cascades are low-cost systems with limited characteristics for reduced requirements. Figure 7-25 shows the basic principle. The externally commutated current source converter needs only a diode bridge rectifier at the rotor side. Speed control is achieved by energy feedback from the rotor via the AC side current source inverter to the AC supply system. Drives with subsynchronous cascades are suited for applications with limited speed range Δn of about 30% below the synchronous speed n_{syn}. The smaller the slip $\Delta n/n_{syn}$, the lower the slip power P_{slip} to be recovered at the rotor side and the smaller the rated power of the converter $P_{conv} = P_{slip}$ relative to the rated motor power P_{mot}.

$$\Delta n/n_{syn} \approx P_{conv}/P_{mot} \tag{7.55}$$

Especially at the beginnings of the converter-fed AC motors, this was an attractive feature, because the converter, at that time, was by far the most expensive part of the whole drive system. But in the long run this advantage doesn't offset the inherent disadvantages of the system:

- Power transfer via slip rings
- A wound rotor, which is larger and more expensive than a squirrel cage rotor
- Auxiliaries like switched resistors for start-up to the normal-speed range
- A rather poor power factor

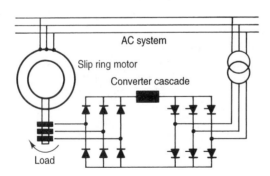

Figure 7-25. Slip power–controlled drive with subsynchronous converter cascade.

- Interharmonics at the AC supply side, the frequencies of which vary with the speed
- A high-capacity overvoltage and overcurrent protection for the converter cascade

The combination of so many drawbacks reduced the attractiveness of this system more and more. In the meantime, the cost reduction of power electronic equipments favors other drive systems with better characteristics.

Therefore, since drives with subsynchronous cascades cannot be regarded as a system of the future, we refer to the literature [13, 48] and concentrate in the following on sub- and hypersynchronous cascades. Despite the fact that they also are subject to some of the earlier-mentioned drawbacks, they seem to find new applications as ultrahigh-power motor generators.

7.10.2. Sub- and Hypersynchronous Cascades

Basic Principle. The system (Fig. 7-26) consists of an AC slip ring machine with a three-phase excitation converter with variable frequency. Three-phase excitation is required, when the speed n of large AC machines, the stator of which is connected to the AC system, should be varied in a limited speed range $\pm\Delta n = \pm n_{\text{slip}}$ around the synchronous speed n_{syn}. These machines can be operated as motors, driving a load machine, or as a generator, driven by a motor or a turbine.

Mode of Operation. To illustrate how the system works, it is useful first to start with a physically plausible idea, then to derive the equations and to show the modes of operation by means of phasor diagrams.

When running with sub- or hypersynchronous speed $n = n_{\text{syn}} \mp n_{\text{slip}}$, the fixed-frequency variable speed machine cannot be a normal synchronous machine. It needs an excitation field, which rotates relative to the rotor to compensate for the difference (n_{slip}) between actual speed n and synchronous speed n_{syn}.

$$n \pm n_{\text{slip}} = n_{\text{syn}} \tag{7.56}$$

The excitation system, therefore, has to feed a three-phase current with slip frequency $f_{\text{slip}} = z_p \cdot n_{\text{slip}}$ ($z_p =$ number of pole pairs) into the three-phase rotor windings. This can be achieved by a three-phase cycloconverter. The cycloconverter can be regarded as a sinusoidal three-phase current source, adjustable in frequency, amplitude, and phase angle, which forms a rotating excitation field in the three-phase rotor windings. The angular speed of this excitation field—moving relative to the rotor and rotating together with the turning rotor— appears as synchronous angular speed relative to the stator—only if the slip frequency is derived from the measured speed by the following equation.

$$\underbrace{2\pi f_{\text{slip}}}_{\omega_2} + \underbrace{2\pi n \cdot z_p}_{\omega} = \underbrace{2\pi \cdot 50 \text{ Hz}}_{\omega_1} \text{ or } 2\pi \cdot 60 \text{ Hz} \tag{7.57}$$

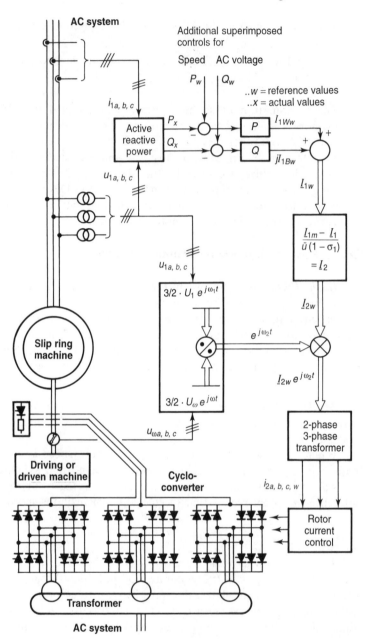

Figure 7-26. Fixed frequency, variable speed motor generator. Control structure.

The excitation field acts via the air gap on the stator and gives rise to back ampereturns in the stator. In this way, the stator current drawn from or fed into the 50 or 60 Hz system and its active and reactive components can be controlled freely via the rotor current.

To control the stator current via the rotor current makes sense only if the rated power of the rotor-side converter can be kept low compared to the rated power of the machine. The following considerations show that this is the case: the AC system voltage generates a rotating magnetic field in the stator, the angular frequency ω_1 of which is determined by the AC system frequency $f_1 : \omega_1 = 2\pi f_1 = 2\pi \cdot 50$ or 60 Hz. Relative to the rotor—turning itself with the speed n and the angular frequency $\omega = 2\pi \cdot n \cdot z_p$—this magnetic field turns only with the angular slip frequency $\omega_2 = \omega_1 - \omega$. The voltage which it induces in the rotor is therefore correspondingly low. The same is the converter output voltage, needed to control the rotor current against the induced voltage. That means, if the speed deviation n_{slip} from the synchronous speed n_{syn} remains within narrow limits, the capacity of the cycloconverter is small compared with the motor ratings.

The cycloconverter consists of two antiparallel line-commutated thyristor bridges (AC VSRs) per phase, one for the positive and one for the negative half wave of the rotor current. Its mode of operation is shown in Section 7.8.2: it cuts individual sections out of the 50 or 60 Hz supply voltage and recombines them into a new voltage, the short-term mean value of which has a sinusoidal wave shape.

Taking a closer look on the system, we use the steady-state phasor equations of the AC machine (see Section 7.4.2):

voltages

$$\underline{U}_1 = j\omega_1\underline{\Psi}_1 \tag{7.58}$$
$$\underline{U}_2 = j\omega_2\underline{\Psi}_2 \tag{7.59}$$

flux linkages

$$\underline{\Psi}_1 = \underline{I}_1 L_1 + \underline{I}_2 M \tag{7.60}$$
$$\underline{\Psi}_2 = \underline{I}_2 L_2 + \underline{I}_1 M \tag{7.61}$$

speed

$$\omega = 2\pi \cdot n \cdot z_p \tag{7.62}$$

angular frequencies

$$\omega_2 = \omega_1 - \omega \tag{7.63}$$

Index "1"-stator; L-inductance
Index "2"-rotor M-coupling inductance
                    z_p-number of pole pairs

The resistances in stator and rotor can be neglected: their voltage drop in the stator is very small compared to the stator voltage $\underline{U}_1$. They play only an insignificant role in the rotor, if we regard the cycloconverter as a current source.

Using equations 7.58 to 7.63, it is easy to show how the system works:

- According to equation 7.58 the stator flux linkage $\underline{\Psi}_1$ is determined by the voltage $\underline{U}_1$ and the angular frequency ω_1 of the AC system. Since both are constant, the flux linkage is constant too.

- With $\underline{\Psi}_1$ = const., equation 7.60 shows that with the amplitude I_2 and the phase angle φ_{i2} of the rotor current phasor $\underline{I}_2 = I_2 e^{j\varphi_{i2}}$ the stator current phasor $\underline{I}_1$ can be adjusted freely in all four quadrants.
- In accordance with equations 7.62 and 7.63, the rotor current has to be impressed with slip frequency $\omega_2 = \omega_1 - \omega$.
- The converter output voltage $\underline{U}_2$ can be determined by eliminating $\underline{\Psi}_1$, $\underline{\Psi}_2$ and $\underline{I}_2$ in equations 7.58–7.61. (See equation 7.64.)

Equations 7.58, 7.60, and 7.64, which we use to describe the operation of the three-phase excited AC machine, can be transferred in a more general form (see equations 7.69–7.71) by introducing the definitions for the magnetizing current $\underline{I}_{1m}$, that stator/rotor windings ration $\ddot{u}$, the leakage factor for stator and rotor σ_1 and σ_2, and the total leakeage factor σ.

$$\underline{U}_1 = j\omega_1 \underline{\Psi}_1 \tag{7.58}$$

$$\underline{\Psi}_1 = \underline{I}_1 L_1 + \underline{I}_2 M \tag{7.60}$$

$$\underline{U}_2 = \frac{\omega_2}{\omega_1} \cdot \underline{U}_1 \cdot \frac{L_2}{M} - j\omega_2 L_1 \left(\frac{L_2}{M} - \frac{M}{L_1} \right) \cdot \underline{I}_1 \tag{7.64}$$

$$\sigma_1 \equiv 1 - \frac{M}{(L_1 \ddot{u})} \tag{7.65}$$

$$\sigma_2 \equiv 1 - \frac{M \ddot{u}}{L_2} \tag{7.66}$$

$$\sigma \equiv (1 - \sigma_1)(1 - \sigma_2) \tag{7.67}$$

$$\underline{I}_{1m} \equiv \frac{\underline{\Psi}_1}{L_1} \tag{7.68}$$

$$\underline{U}_1 = j\omega_1 L_1 \cdot \underline{I}_{1m} \tag{7.69}$$

$$\underline{I}_{1m} = \underline{I}_1 + \underline{I}_2 \ddot{u}(1 - \sigma_1) \tag{7.70}$$

$$\underline{U}_2 = (\underline{U}_1 - j\omega_1 \sigma L_1 \cdot \underline{I}_1) \cdot \frac{\omega_2}{\omega_1} \cdot \frac{\ddot{u}}{(1 - \sigma_2)} \tag{7.71}$$

    The phasor diagram and the quivalent circuits for the voltages and currents are shown in Figure 7-27. They are based on equations 7.69–7.71. The figures for the nominal current $I_{1n} \approx 4I_{1m}$ and the leakage factors $\sigma_1 \approx \sigma_2 \approx 5\%$ and $\sigma \approx 10\%$ are representative for large slip ring AC machines. The phasor diagram is represented for purely active nominal current $I_1 = I_{1n}$.

    The diagram of the voltages is represented by unfilled phasors. It follows from the equivalent circuit which is based on equation 7.71. The current diagram is made up with filled phasors and based on the equivalent circuit which follows from equation 7.70 and equation 7.69. Equation 7.69 determines the relation between stator voltage $\underline{U}_1$ and no-load magnetizing current $\underline{I}_{1m}$.

    The current phasor diagram and the corresponding equivalent circuit show that the cycloconverter output current $(\underline{I}_2)$ determines the stator current $(\underline{I}_1)$. The magnetizing current $(\underline{I}_{1m})$ is constant for a given voltage $(\underline{U}_1)$ and frequency (ω_1) of the AC system.

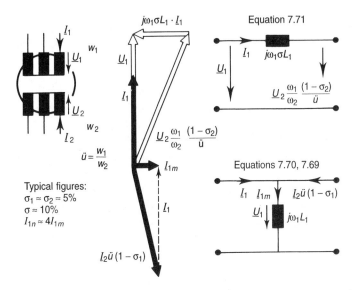

Figure 7-27. Phasor diagram and equivalent circuits according to equations 7.69, 7.70, and 7.71.

The voltage phasor diagram and the corresponding equivalent circuit show that if voltage ($\underline{U}_1$) and frequency (ω_1) of the AC system are constant, the amplitude U_2 of the cycloconverter output voltage ($\underline{U}_2$) necessary to generate a particular stator current ($\underline{I}_1$) is proportional to the slip frequency (ω_2).

$$\underline{U}_1, \omega_1, \underline{I}_1 = \text{const.} \rightarrow \underline{U}_2 \frac{\omega_1}{\omega_2} \frac{(1-\sigma_2)}{\ddot{u}} = \text{const.} \rightarrow U_2 \sim \omega_2 \qquad (7.72)$$

The current and voltage phasor diagrams and their equivalent circuits show that when the stator is drawing active stator current out of the AC system ($\underline{I}_1$ in line with $\underline{U}_1$), the cycloconverter also takes active power out of the rotor if the slip is positive ($\omega_2 > 0$). It feeds active power into the rotor if the slip is negative ($\omega_2 < 0$). This can be seen from the directions of the voltage and current phasors $\underline{U}_2$ and $\underline{I}_2$. They point mainly in the opposite direction in the first case and mainly in the same direction in the second case. (Note that $U_2 \omega_1 / \omega_2 \cdot (1 - \sigma_2)/u = \text{const.}$ means that U_2 is positive when ω_2 is positive and U_2 is negative when ω_2 is negative.)

Protection. During normal operation the voltage $\underline{U}_2$ at the slip rings is low. This can be shown in equation 7.71, and it becomes especially clear for no-load condition $\underline{I}_1 = 0$;

$$\underline{U}_2 = \frac{\omega_2}{\omega_1} \cdot \underline{U}_1 \cdot \ddot{u}, \qquad \text{for } \underline{I}_1 = 0 \text{ (and } 1 - \sigma_1 \approx 1) \qquad (7.73)$$

The slip ring voltage is given not only by the stator rotor windings ratio $\ddot{u}$ but also by the slip ω_2/ω_1, which is low in the normal operation range: $|\omega_{2\,\text{max}}|/\omega_1 < 10\%$.

Voltage steps $\Delta \underline{U}_1$, however, generated, for example, by single-phase or three-phase short circuits in the AC system lead to problems (Δ-deviation from steady state):

- They cause high transient overvoltages $\Delta \underline{U}_2$ at the slip rings, if the current is kept constant $\Delta \underline{I}_2 = 0$ by the controls. Without voltage protection the cyclo-converter had to be overdimensioned in order to be able to counteract against the dynamic overvoltage $\Delta \underline{U}_2$ of the rotor and to limit the current.
- They give raise to a high transient overcurrent $\Delta \underline{I}_2$ at the rotor side, if the overvoltage is limited to $\Delta \underline{U}_2 = 0$ by the cycloconverter output voltage applied to the slip rings. Without protection device the cycloconverter had to be overdimensioned to be able to carry the dynamic overcurrent $\Delta \underline{I}_2$ coming out of the rotor. The rotor current raise $d\Delta \underline{I}_2/dt$, in this case, is limited only by the total leakage inductance σL_2 at the rotor side:

$$\Delta \underline{U}_2 = \Delta \underline{U}_1 \cdot \ddot{u}, \qquad \text{for } \Delta \underline{I}_2 = 0 \qquad (7.74)$$
$$d\Delta \underline{I}_2/dt = -\Delta \underline{U}_1 \cdot \ddot{u}/(\sigma L_2) \qquad \text{for } \Delta \underline{U}_2 = 0 \qquad (7.75)$$

These equations—valid for the first moments after a voltage step ΔU_1 in the AC system—can be derived by using the *dynamic* phasor equations of the machine, represented in equation 7.76 and 7.77 for deviations "Δ" from the stationary values (see Section 7.4.2). The stator and rotor resistors R_1 and R_2 can be neglected:

$$\Delta \underline{U}_1 = j\omega_1 \Delta \underline{\Psi}_1 + d\Delta \underline{\Psi}_1/dt; \qquad \Delta \underline{\Psi}_1 = \Delta \underline{I}_1 L_1 + \Delta \underline{I}_2 M \qquad (7.76)$$
$$\Delta \underline{U}_2 = j\omega_2 \Delta \underline{\Psi}_2 + d\Delta \underline{\Psi}_2/dt; \qquad \Delta \underline{\Psi}_2 = \Delta \underline{I}_2 L_2 + \Delta \underline{I}_1 M \qquad (7.77)$$

In the *first moments* after a stator voltage step $\Delta \underline{U}_1$, we can assume that the flux linkages still did not change: $\Delta \underline{\Psi}_1 = 0$ and $\Delta \underline{\Psi}_2 = 0$ (but $d\Delta \underline{\Psi}_1/dt \neq 0$ and $d\Delta \underline{\Psi}_2/dt \neq 0$). If we eliminate $\Delta \underline{I}_1$ in both equations 7.76 and 7.77 under these assumptions and use the definitions in equations 7.65–7.67, we obtain equations 7.74 and 7.75, in which $1 - \sigma_1$, is simplified to $1 - \sigma_1 = 1$.

The converter cascade at the slip ring side, therefore, needs a carefully designed high-capacity protection. Normally, it consists of thyristor-switched resistors (see Figure 7-26 left), which are turned on in case of overvoltage. Limiting the overvoltages in this way, they also take the subsequent overcurrent and keep it away from the converter.

The Controls. The control structure, shown in Figure 7-26, was one of the first vector control schemes introduced in practical applications. The superimposed control loops with a P and a Q controller take the main task. They control the active and reactive power P_x and Q_x—calculated with equations 7.78 and 7.79 from the measured stator voltages and currents—according to the set reference values P_w and Q_w.

$$P_x = u_{1a} \cdot i_{1a} \qquad + u_{1b} \cdot i_{1b} \qquad + u_{1c} \cdot i_{1c} \qquad (7.78)$$

$$Q_x = \frac{u_{1b} - u_{1c}}{\sqrt{3}} i_{1a} + \frac{u_{1c} - u_{1a}}{\sqrt{3}} i_{1b} + \frac{u_{1a} - u_{1b}}{\sqrt{3}} i_{1c} \qquad (7.79)$$

Since the P and Q controllers should act on the stator current $\underline{I}_1$, their output signals are regarded as set values (index w) for the active (index W) and reative (index B) stator current components $\underline{I}_{1w} = I_{1Ww} + jI_{1Bw}$. The stator current, however, can only be controlled via the rotor current. Therefore, we need an equivalent set value for the rotor current phasor, which we gain by means of a model for the machine based on equation 7.70: $\underline{I}_{2w} = (\underline{I}_{1m} - \underline{I}_{1w})/\ddot{u}(1 - \sigma_1)$. To act on the three phases of the cycloconverter, the phasor $\underline{I}_{2w} = I_{2w} \cdot e^{j\varphi_{i2}}$ has to be transformed. This can be achieved by multiplication with the slip frequency unit vector $e^{j\omega_2 t}$ and by subsequent three-phase transformation:

$$i_{2a,b,cw} = Re(I_{2w}e^{j\varphi_{i2}}e^{j\omega_2 t}e^{+jD}) \qquad (7.80)$$

$$= I_{2w} \cos(\omega_2 t + \varphi_{i2} + D) \qquad (7.81)$$

where $\qquad\qquad D = 0° \ (a); -120° \ (b); +120°(c)$

The rotor current control—not shown in detail in Figure 7-26—impresses the rotor currents according to these set values.

The slip frequency unit vector $e^{j\omega_2 t}$ can be derived from the stator voltages $u_{1\,a,b,c}$ and the output signals $u_{\omega\,a,b,c}$ of, for example, a little three-phase tacho-machine with permanent magnets mounted at the shaft:

$$u_{1\,a,b,c} = U_1 \cos(\omega_1 t + D); \qquad D = 0°; -120°; +120° \qquad (7.82)$$

$$u_{\omega\,a,b,c} = U_\omega \cos(\omega t + D); \qquad\qquad a \qquad b \qquad c \qquad (7.83)$$

$$u_{1a}e^{j0°} + u_{1b}e^{-j120°} + u_{1c}e^{+j120°} = \frac{3}{2} \cdot U_1 e^{j\omega_1 t} \qquad (7.84)$$

$$u_{\omega a}e^{j0°} + u_{\omega b}e^{-j120°} + u_{\omega c}e^{+j120°} = \frac{3}{2} \cdot U_\omega e^{j\omega t} \qquad (7.85)$$

$$\frac{e^{j\omega_1 t}}{e^{j\omega t}} = e^{j(\omega_1 - \omega)t} = e^{j\omega_2 t} \qquad (7.86)$$

The set values P_w and Q_w for active and reactive power can be derived from additional superimposed control loops for the speed and the AC system voltage level.

Conventional and New Applications. Interties between the three-phase 50 Hz AC system and the single-phase 16 2/3 Hz railway supply system in Central Europe are the main applications in which slip power-controlled motors with sub- and hypersynchronous speed are used [8]. Figure 7-28 shows the principle: The slip power–controlled motor, the stator of which is connected to the 50 Hz system, is driving a single-phase synchronous generator feeding into the 16 2/3 Hz system. The rated power is between 5 and 80 MVA per unit. Such interties are frequency elastic:

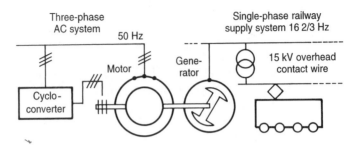

Figure 7-28. The 50–16 2/3 Hz intertie.

the motor can follow speed deviations caused by frequency deviations in the 16 2/3 Hz system with its variable slip. The frequency of the railway system can be controlled with the active power transfer. The 50 Hz system voltage can be controlled—if desired—by reactive power control. The cycloconverter can also be used as start-up equipment in a rearranged configuration.

A similar system with, however, a three-phase synchronous generator is used to feed the large magnets in particle accelerators. There it decouples the large-power pulsations to magnetize and demagnetize the magnets from the AC system. This can be achieved by delivering kinetic energy from the rotating masses and from an additional flywheel when the speed decreases and by absorbing energy when the speed increases.

A new application with still higher power ratings of some ten or even some hundred megawatt is on the horizon: fixed frequency, variable speed motor generators in pumped-storage hydropower plants using reversible pump turbines [9, 49–51]. With extreme variations of the turbine or pumping head, for example, between 60 m and 200 m, the full head range cannot be utilized with a fixed speed. With variable slip, however, the speed can be adapted around the synchronous speed and the machine can always be operated at the speed at which the electric energy is generated with the optimal hydraulic efficiency. In the pumping mode, load control can only be achieved with speed variation.

Pumped storage power plants are used in combination with constant power generators (e.g., nuclear power plants) to adapt the generated power to varying loads and to keep the AC system frequency constant (see Figure 7-29). For this purpose the fixed-frequency, variable speed machines have to be operated in the generating mode as well as the pumping mode.

Fixed frequency, variable speed generators are well suited to work as *wind energy generators*. Speed adaptation around the synchronous speed allows gain of maximum of energy under all wind conditions [52].

Future Trends at Converter and Control Side. The line-commutated cyclo-converters are going to be replaced by forced-commutated GTO converters with an intermediate DC voltage link (No. 8 with GTOs only at the slip-ring side and 10 in Figure 7-7). This leads to a significant advantage: fixed frequency, variable speed motor generators should be able to ride through most of the line disturbances with-

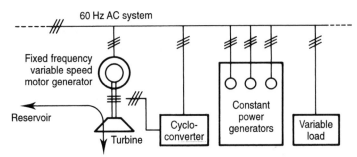

Figure 7-29. Fixed frequency, variable speed motor generators in pumped-
storage power plants

out switch off. This is much easier with a DC voltage source inverter DC VSI at the
mains side (No. 10 in Figure 7-7): whereas cycloconverters tend to fail during line
disturbances because of their line commutation, the forced-commutated DC VSIs
can easily follow even fast line voltage drops by adapting their PWM AC voltage.

Vector control structures (as shown, for example, in Figure 7-26) containing
complex functions with phasors, have to be decomposed in real and imaginary
components for processing in today's control units. This calls for user interfaces
with graphic function plan programming, which also handle complex functions
directly without decomposition in real and imaginary components.

7.11. CONCLUSION

Chapter 7 gives a brief but comprehensive survey of the large drive systems in
industrial applications.

After an introduction in which the main incentives to use high-power drives
with adjustable speed are mentioned it starts with a classification of the large drives
and with a short review of their evolution. The maximum power and speed range of
the motors are discussed and the mathematical description of their behavior is
illustrated in a physically plausible manner. It is shown that the large variety of
high-power converters for drives is made up of only two basic circuits. Their main
characteristics are briefly recalled and it is demonstrated how they can be combined
to form converters which meet the requirements of the motors and the supply
system.

The main drive systems, which are in practical use, are described. Emphasis is
given to explain in an illustrative manner how the converters, motors and controls
work together to form the drive system. It is shown how the system characteristics
meet the requirements of the applications. Emphasis is also given to the practical
implementation. Development potential and future trends of each system are also
discussed.

It has been shown that for all the various applications with their different
requirements, suitable drive systems are available.

Nomenclature

List of Key Symbols

Three-phase quantities	Rotating phasors stator rotor related	Stationary phasors	
$u_{a,b,c}$;	U; φ_u;	$\vec{u}'\ \vec{u}''$	$\underline{U} = Ue^{j\varphi_u}$ = voltages
$i_{a,b,c}$;	I; φ_i;	$\vec{i}'\ \vec{i}''$	$\underline{I} = Ie^{j\varphi_i}$ = currents
$\psi_{a,b,c}$;	Ψ; φ_ψ;	$\vec{\psi}'\ \vec{\psi}''$	$\underline{\Psi} = \Psi e^{j\varphi_\psi}$ = flux linkages

Phase angles

Amplitudes

Instantaneous values

u_0 zero-sequence voltage

u_d, i_d switched DC voltage respectively DC current

U_d, I_d mean value of the DC voltage respectively DC current or constant DC voltage, respectively, DC current

f frequency

F inverter switching frequency

ω angular frequency

t_e instantaneous value of the electromagnetic torque

T_e mean value of the electromagnetic torque

P, Q active respectively reactive power

S apparent power

α control angle

a_0 modulation index (PWM)

u_{st} control signal (PWM)

st control signal (square wave mode)

u_H carrier signal

D Three-phase angle $0°, -120°, +120°$

t time

L inductance

M stator-rotor coupling inductance

σ leakage factor

C capacitance

R resistance

z_p number of pole pairs

$\ddot{u}$ stator-rotor winding ratio

Indices

$1, 2, e$ stator, rotor, excitation

e electromagnetic (torque)

n nominal, rated

m magnetizing (currents)

m, l machine respectively line side (voltages, frequencies, angles)

w, x reference, respectively actual value

GS, OS fundamental, respectively, harmonic components

W, B active, respectively, reactive components

References

[1] Hickok, Herbert N., "Adjustable speed—a tool for saving energy, losses in pumps, fans, blowers, and compressors," *IEEE-IAS Trans.* Vol. 21, January/February 1985.

[2] Schönung, A., and H. Stemmler, "Static frequency changers with subharmonic control in conjunction with reversible variable speed AC drives," *Brown Boveri Rev.*, August/September 1964.

[3] Stemmler, H., "Verfahren und Anordnung zur frequenz- und amplitudenabhängigen Steuerung oder Regelung eines über Umrichter gespeisten Wechselstrommotors," Deutsches Patentamt, Patentschrift 15 63 980, Anmeldetag 7. 1. 67.

[4] Hasse, K., "Zur Dynamik drehzahlgeregelter Antriebe mit Stromrichtergespeisten Asynchron-Kurzschlussläufermaschinen," Ph.D. Dissertation, Darmstadt, Techn. Hochsch., Darmstadt, Germany, 1969.

[5] Blaschke, F., "The principle of field orientation as applied to the New TRANSVECTOR closed loop control system for rotating field machines," *Siemens Rev.*, Vol. 34, pp. 217–220, May 1972.

[6] Stemmler, H., "Drive systems and electric control equipment of the gearless tube mill," *Brown Boveri Rev.*, pp. 120–128, March 1970.

[7] Peneder, F., R. Lubasch, and A. Voumard, "Static equipment for starting pumped-storage plant, synchronous condensers and gas turbine sets," *Brown Boveri Rev.*, no. 9/10, pp. 440–447, 1974.

[8] Stemmler, H., "Active and reactive load control for converters interconnecting 50 and 16 2/3 Hz systems, using a static frequency changer cascade," *Brown Boveri Rev.*, Vol. 65, no. 9, pp. 614–618, 1978.

[9] Sugimoto, O., E. Haraguchi, et al., "Developments targeting 400 MW class adjustable speed pumped hydro plant and commissioning of 18.5 MW adjustable speed hydro plant," *Fluid Machinery* (The American Society of Mechanical Engineers), *FED*, Vol. 68, (book no. G00468), 1988.

[10] Bitterberg, F., and W. Teich, "Henschel-BBC-DE 2500 Ein Wendepunkt in der Lokomotivtechnik," *ETR*, no. 11, 1971.

[11] Brechbühler, M., and H. Stemmler, "Probleme bei der Entwicklung und Auslegung eines Oberleitungsversuchsfahrzeuges mit Asynchronfahrmotoren," *Elek. Bahnen*, no. 5, 1972 (ICEB 1971 München).

[12] Stemmler, H., and P. Guggenbach, "Configurations of high-power voltage source inverter drives," *EPE 93*, 1993.

[13] Bose, B. K., *Power Electronics and AC Drives*, Prentice Hall, Englewood Cliffs, NJ, 1986.

[14] Oliver, J. A., "Does GTO make a Difference? Current source Inverter of 1984 VS Current Source GTO—PWM inverter of 1988. Results of ASD tests on 2000 hp Boiler pump motors," in *Proc. EPE 89, 3rd European Conf. on Power Electronics and Applications*, Aachen, Germany, pp. 657–661, 1989.

[15] Oliver, J. A., R. K. McCluskey, H. W. Weiss, and M. J. Samotyj, "Adjustable-speed drive retrofit for Osmond Beach FD fans," *IEEE Trans. Energy Conversion*, Vol. 7, no. 3, pp. 580–586, September 1992.

[16] Kovácz, K. P., and I. Rács, "Transiente Vorgänge in Wechselstrommaschinen," Akadémia Kiadó, Budapest, 1954.

[17] Bühler, H., *Einführung in die Theorie geregelter Drehstromantriebe*, Birkhäuser Verlag, Basel und Stuttgard, 1977.

[18] Leonhard, W., *Control of Electrical Drives*, Springer-Verlag, Berlin, 1985.

[19] Vas, P., *Vector Control of AC Machines*, Oxford University Press, Oxford, 1990.

[20] Vas, P., *Electrical Machines and Drives, A Space-Vector Theory Approach*, Oxford University Press, Oxford, 1992.

[21] Späth, H., *Steuerverfahren für Drehstrommaschinen*, Springer-Verlag, Berlin/ Heidelberg, 1983.

[22] Heumann, K., *Grundlagen der Leistungselektronik*, B. G. Teubner, Stuttgart (4., überarbeitete und erweiterte Auflage), 1989.

[23] Meyer, M., *Leistungselektronik*, Springer-Verlag, Berlin/Heidelberg, 1990.

[24] Bühler, H., *Convertisseurs statiques*, Presses polytechniques et universitaires romands, Lausanne, 1991.

[25] Michel, M., *Leistungselektronik*, Springer-Verlag, Berlin/Heidelberg, 1992.

[26] Schmidt, I., "Analysis of converter-fed synchronous motors," *Control in Power Electronics and Electrical Drives*, IFAC Symposium, Düsseldorf, pp. 571–85, 1974.

[27] Steimer, P. K., "Redundant, fault-tolerant control system for a 13 MW high speed drive," *3rd Int. Conf. Power Electronics and Variable Speed Drives* (IEE in association with IEEE), London, July 1988.

[28] Stemmler, H., H. Kobi, and P. Steimer, "A new high-speed controller with simple program language for the control of variable-speed converter-fed AC-drives," paper presented at the IEEE Power Electronics Specialists Conf., Gaithersburg, MD, June 1984.

[29] Cossié, A., "Evolution de la locomotive à thyristors à la S. N. C. F.," *Elektrische Bahnen*, Vol. 79, nos. 1 and 2, 1981.

[30] Vorser, Ph. Latair, N. F. W. O., "Digital simulation of the six-pulse current-fed converter with asynchronous motor," *IEEE Ind. Appl. Soc. Conf.*, pp. 824–30, 1979.

[31] Graham, A. D., R. L. Lattimer, and J. G. Steel, "Capacitor compensated variable speed drives," *IEEE Conf. EMD*, pp. 339–343, 1989.

[32] Turnbull, F. G., "Selected harmonic reduction in static DC and AC inverters," *IEEE Trans. Communications and Electronics*, Vol. 83, pp. 374–378, 1964.

[33] Patel, H. S., and R. G. Hoft, "Generalized techniques of harmonic elimination and voltage control in thyristor inverters, Pt. I. Harmonic elimination," *IEEE Trans. Ind. Appl.*, Vol. IA-9, pp. 310–317, 1973.

[34] Oliver, James, et al., "Application of high speed, high horsepower, ASD controlled induction motors to gas pipelines," *Proc. EPE.*, 1993.

[35] Bayer, K. H., H. Waldmann, and M. Weibelzahl, "Field-oriented close-loop control of a synchronous machine with the new transvector control system," *Siemens Rev.*, Vol. 39, pp. 220–223, 1972.

[36] Nakano, T., H. Ohsawa, and K. Endoh, "A high performance cycloconverter-fed synchronous machine drive system," *Conf. Rec. IEEE-IAS Int. Sem. Power. Conv. Conf.*, pp. 334–341, 1982.

[37] Nakajima, T., M. Tamura, and E. Masada, "Compensation of non-stationary harmonics using active power filter with Prony's spectral estimation," *PESC 88 Record*, pp. 1160–1167, April 1988.

[38] Stemmler, H., "Power Electronics in Electric Traction Applications," *IECON '93 Rec.*, pp. 707–713, 1993.

[39] McMurray, W., "Efficient snubber for voltage-source GTO inverter," *IEEE Trans. Power Electron.*, Vol. PE-2, no. 3, pp. 264–272, July 1987.

[40] Undeland, T. M., "Snubber for pulse width modulation bridge converters with power transistors or GTO," *Int. Power Electr. Conf. (IPECD)*, Tokyo, pp. 313–323, March 1983.

[41] Holtz, J., "Pulsewidth modulation—a survey," *IEEE Trans. Ind. Electron.*, Vol. 39, no. 5, December 1992.

[42] Malesani, L., and P. Tomasin, "PWM current control techniques of voltage source converters—a survey," *IEEE Conf. Ind. Elec.*, Control and instrumentation, Record, pp. 670–673, 1993.

[43] Holtz, J., and E. Bube, "Field-oriented asynchronous pulsewidth modulation for high performance AC machine drives operating at low switching frequencies," *IEEE Ind. Appl. Soc. Annual Meeting Rec.*, pp. 412–417, 1988.

[44] Holtz, J., and B. Beyer, "Off-line optimized synchronous pulsewidth modulation with on-line control during transients," *EPE Journal*, Vol. 1, no. 3, pp. 193–200, 1991.

[45] Takahashi, I., and H. Mochikawa, "A new control of PWM inverter waveform for minimum loss operation on an induction motor drive," *IEEE Trans. Ind. Appl.*, Vol. 21, no. 4, pp. 580–587, 1985.

[46] Stemmler, H., and T. Eilinger, "Spectral analysis of the sinusoidal PWM with variable switching frequency for noise reduction in inverter-fed induction motors," *IEEE Power Electronics Specialist Conf. Rec.*, pp. 269–277, 1994.

[47] Depenbrock, M., "Direct self-control (DSC) of inverter-fed induction machine," *IEEE Trans. Power Electronics*, Vol. 3, no. 4, pp. 420–429, 1988.

[48] Franz, P., and A. Meyer, "Digital simulation of a complete subsynchronous converter cascade with 6/12-pulse feedback system," *Trans. IEEE*, PAS-100, pp. 4948–4957, 1981.

[49] Stemmler, H., and A. Omlin, "Converter controlled fixed-frequency variable-speed motor/generator," *Proc. 1995 Int. Power Electronics Conference*, IPEC-Yokohama, pp. 170–176, 1995.

[50] Tang, Y., and L. Xu, "A flexible active and reactive power control strategy for a variable speed/constant frequency generating system," *PESC*, 1993.

[51] Kudo, K., "Japanese experience with a converter-fed variable speed pumped-storage system," *Hydropowers and Dams*, March 1994.

[52] Arsudis, D., "Doppeltgespeister Drehstromgenerator mit Spannungszwischen-kreis-Umrichter im Rotorkreis für Windkraftanlagen," Dissertation, Braunschweig, 1989.

Chapter 8

Ned Mohan
William P. Robbins
Lars A. Aga
Mukul Rastogi
Rajendra Naik

Simulation of Power Electronic and Motion Control Systems

8.1. INTRODUCTION

8.1.1. Power Electronics Environment

Power electronics and motion control systems process electrical power in a broad range of applications utilizing a large class of electrical circuits [1]. This field has been experiencing a vigorous evolution for almost 30 years [2]. As of late, these systems seem to have approached maturity in their topological form and control, while the hardware continues to evolve [3]. This maturity is partially responsible for the large degree of penetration in the application areas of the 1970s and the 1980s. In the 1990s, the marketplace has expanded: computer and communication equipment requiring very low DC supply voltages and electromagnetic interference (EMI) [4], electrical transportation in the form of electrical vehicles requiring adjustable speed drives and battery chargers, and utility-related applications such as the flexible AC transmission systems (FACTS) requiring high-power electronics [5, 6]. This chapter presents a user's perspective to simulation in power electronics and motion control systems. The rationale for simulation in research, in education, and in the design process is discussed. The hierarchical approach to simulation is stressed. A brief overview of some of the widely used simulation programs is provided by means of simple examples.

8.1.2. Need for Simulation

New designs of power electronics systems are the norm due to new applications and the lack of standardization in specifications is because of varying customer demands. Accurate simulation is necessary to minimize costly repetitions of designs and breadboarding and, hence, reduce the overall cost and the concept-to-production time lag [7].

There are many benefits of simulation in the design process, some of which are listed here [8]:

1. Simulation is well suited for educational purposes. It is an efficient way for a designer to learn how a circuit and its control work.

2. Simulation may give a comprehensive and an impressive documentation of system performance that gives a competitive edge to a company using the simulation.

3. It is normally much cheaper to do a thorough analysis than to build the actual circuit in which component stresses are measured. A simulation can discover possible problems and determine optimal parameters, increasing the possibility of getting the prototype "right the first time" [9]. Simulation can be used to optimize the performance objective by letting the simulation search over a large number of variables.

4. New circuit concepts and parameter variations (including tolerances on components) are easily tested. Changes in the circuit topology are implemented with no cost. There is no need for components to be available on short notice.

5. In the initial phase of a study, parasitic effects such as stray capacitances and leakage inductances are best omitted. They are important and must be considered, but they often cause confusion until the fundamental principles of a concept are understood. In a physical circuit, it is not possible to remove the stray capacitance and leakage inductances in order to get down to the fundamental behavior of the system.

6. Simulated waveforms at different places in the circuit are easily monitored without the hindrance of measurement noise (and other noise sources). As switching frequencies increase, the problem of laboratory measurements becomes increasingly difficult. Thus, simulations may become more accurate than measurements [9].

7. Destructive tests that cannot be done in the laboratory, either because of safety or because of the costs involved, can easily be simulated. Responses to faults and abnormal conditions can also be thoroughly analyzed.

8. Specification of component voltage and current ratings is difficult before the working principles of the circuit and the current and voltage waveforms are fully determined. In a simulation, component ratings are not needed.

9. It is possible to simplify parts of circuits in order to focus on a specific portion of the circuit. This may not be possible in a laboratory setup.

10. There may be a need to predict the interaction between several converters connected to the same distribution network. This will be almost impossible and certainly very expensive to do without simulations [10].

It is important to construct the simulation so that it is possible to answer appropriate questions. Each question requires specific emphasis in the simulation. For example, in calculating component stresses due to parasitics, the components and parasitics need to be modeled in detail for only a few switching cycles, and the dynamic behavior of the system in response to control inputs can be neglected. It is usually not advisable to answer all questions at once in one simulation.

8.2. SIMULATION IN THE DESIGN PROCESS

Figure 8-1 shows a typical design process where simulation is utilized to complement the hardware prototype [9]. The process starts with a power converter concept. This is based on system performance specifications, technology limitations, and past experience. The concept leads to a topology selection whose appropriateness is verified by means of simulation.

The next step is the circuit design phase where the designer performs more detailed analysis to select appropriate components. This includes various subcircuit

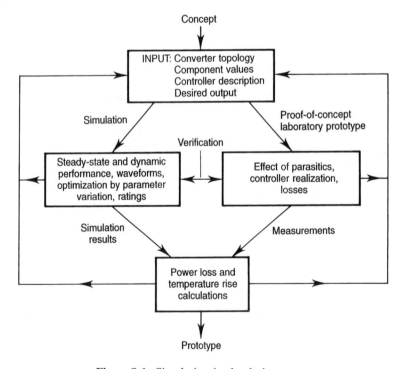

Figure 8-1. Simulation in the design process.

analyses by simulation, stability studies, transient and dynamic responses, start-up and shut-down predictions, and preliminary thermal calculations. These manual calculations are often aided by computer-based tools.

The layout and realization of a laboratory prototype are carried out simultaneously with the simulations. After building the prototype, it is tested and verified against the original design specifications, sometimes also making comparison with analysis results. This testing is often part of a debugging process that is used to either modify the original design or in many cases to complete the design details that were originally neglected.

Before a functional prototype is realized, another step in the design process must be carried out. This final step involves calculation and measurement of the power loss and the temperature rise in the circuit. This is necessary to ensure that proper cooling is incorporated into the design.

8.3. FREQUENCY DOMAIN VERSUS TIME DOMAIN ANALYSIS

Both the time domain and frequency domain analysis are usually required to understand the system behavior. For instance, consider the motor drive example shown in a block diagram form in Figure 8-2. If the effects of the switching operation of the power electronic converter on the motor are to be investigated, then a time-domain analysis is needed. Due to the switchings, the network has a time-varying structure that results in simulation challenges discussed in Section 8.4.

On the other hand, if the objective of the simulation is to analyze and design the controller, then it may be possible to linearize the converter and the motor around their steady-state operating point and represent them by their frequency-domain transfer functions using Laplace transforms. Such a frequency-domain analysis applies to small-signal perturbations for which the assumption of linearity is valid. It is often used for designing the controllers and analyzing the stability and the dynamic performance of the overall system. It also allows well-established control theory techniques to be applied. Simulation capability in the frequency domain is now a standard feature of almost all general-purpose simulation packages. Many include design capabilities where, for example, by specifying the gain and phase margin, the simulation program is able to design the controller.

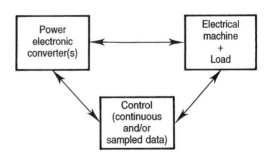

Figure 8-2. Block diagram of a motor drive.

However, to study the effects of nonlinearities and saturations during large disturbances, time-domain simulations are necessary to verify the control performance. It is the simulation in time domain that requires the bulk of the effort and makes simulation challenging. Therefore, the rest of the discussion in this chapter focuses on the time-domain simulation.

8.4. CHALLENGES IN SIMULATION

Simulation of power electronics and motion control systems poses many challenges to the simulation program and to its user. These are discussed in the following subsections.

8.4.1. Requirements of a Simulation Program

User-Friendly Interface. A simulation program must have an easy-to-use interface for data entry and for output data processing.

Multilevel Modeling Capability. In simulating the motor drive example of Figure 8-2, the power electronics converters are described by the interconnection of circuit element models. On the other hand, the electrical machine and the load are best described by differential equations formulated in terms of state variables. Finally, the continuous and sampled data control systems are often represented in their functional form by transfer functions and/or logic statements that describe the behavioral properties between their inputs and outputs. A good simulation program should allow these various blocks to be easily implemented.

Accurate Models. For a detailed analysis, a simulation program needs accurate models of all circuit elements, including transformers and transmission lines for utility-related applications. Even if accurate models are available, it is difficult to know their parameter values. Parasitic inductances and capacitances are often difficult to estimate.

Robust Switching Operations. Switching actions due to solid-state switches (diodes, thyristors, and transistors) must be appropriately handled. Depending on how the switches are modeled, their on-off transitions either represent an extreme nonlinearity or lead to a time-varying structure of the network.

Execution Time. The simulation may need to cover a sufficiently large time span to observe the effect of slowly changing variables with large time constants. At the same time, the simulation often needs to proceed with a small time step in order to accurately represent rapidly changing variables with small time constants. A small time step is also required to represent switching times with good resolution in the presence of high switching frequencies. In such simulations, this leads to a large number of time steps. To keep the execution time within reasonable limits, an efficient equation solver with variable time steps is normally required.

Initial Conditions. Unlike small-signal electronics, where the steady-state operating conditions are established by a DC bias analysis, there is no easy way to establish the steady-state operating conditions in a power electronics system. Often, an initial estimate of various state variables is provided by the user, and the time required to bring the system to its steady state may be substantial, depending on the accuracy of the initial estimates. Therefore, simulation programs for power electronics must allow users to set initial conditions.

8.4.2. Challenges to the User of Simulation Tools

In view of the requirements on the simulation tools discussed in the previous section, it is obvious that the objective of each simulation be carefully considered prior to undertaking the task of simulation. An ambitious attempt to answer too many questions at once is a common mistake made by novice and experienced users alike. This results in large computation time, loss of accuracy, and an overwhelming amount of information that is difficult to digest.

A good discussion of a hierarchical approach is presented in [11], where the analysis of a system is carried out by means of several simulations. Each simulation addresses a specific set of issues, where detailed models are used only for the essential part of the system under focus, and the rest (major part) of the system is represented by approximate models. This results in a reasonable simulation time, and the amount of output information is limited to a size that the user can efficiently utilize. This is illustrated by means of a simulation of a unity-power-factor single-phase rectifier shown in Figure 8-3. It is a commonly used topology where the output current of the diode rectifier is shaped by a boost DC-to-DC converter to produce an essentially sinusoidal line current at unity power factor with a regulated output DC voltage.

In the hierarchical approach, the system is simulated from the device (component) level to the system level, resulting in five levels of simulation [11]:

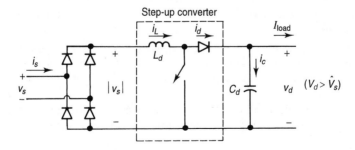

Figure 8-3. Block diagram of a unity power factor rectifier.

1. Detailed circuit modeling where elaborate device models are used to obtain device stresses and to evaluate snubber topologies. Only a few cycles are included in the open loop (without the current and voltage control loops) simulation, where the input voltage to the boost converter is assumed to be constant and the load and the input voltage are chosen to represent the peak stresses.

2. At the next level up in the hierarchy of the simulation, the detailed device models are replaced by ideal diodes and switches. One of the objectives here is to evaluate the effect of slope compensation on the current mode control scheme. The input voltage is assumed to be a rectified sine wave, and the simulation is carried out over several line cycles. The current control loop is closed, but the voltage loop is open.

3. At a higher level yet, an averaged PWM switch model [12] is used, and the controller is represented by its functional model, with the objective of looking at the steady-state waveforms of the input current and the voltage ripple in the output voltage. Since the switching frequency behavior is averaged out, the simulation is carried out for tens of line cycles. The current and the voltage control loops are both closed.

4. For evaluating the transient response and the action of the voltage control loop, which is much slower than the inner current control loop, it is assumed that the inner current control loop works ideally. Based on this assumption, a power balance equation is derived [13]. The current loop is idealized, and the voltage control loop is modeled. Now the simulation can be easily carried up to hundreds of line cycles to evaluate the transient response of the system.

5. At the highest level, for designing the feedback compensation in the outer voltage loop, a model averaged over the rectified line frequency period is used [13]. This linearization results in a time invariant circuit that can be simulated in frequency domain to design the feedback controller. The current control loop is ignored and the voltage control loop is modeled. It can then be tested in time domain for large-signal transient response, for example, to step changes in load.

The foregoing discussion shows clearly that if the entire system were simulated at the detailed circuit level, the computation time and the amount of information generated would be prohibitively large. Therefore, the simulation must be carried out in segments. However, this is easier said than done. The reason is that a significant amount of experience and insight are needed to develop approximate models or to utilize existing ones. This points to the need of the application-specific simulation packages [14] or modules within general-purpose packages, as discussed later on in this chapter.

A simulation will be a success only if the simulation tools meet the requirements of Section 8.4.1 and if the user adopts the hierarchical approach discussed in this section.

8.5. CLASSIFICATION OF SIMULATION TOOLS AND HISTORICAL OVERVIEW

At the core of each simulation are differential and algebraic equations that describe the system. Simulation programs primarily differ in how these equations are solved. The other differences such as the program-user interface (pre- and postprocessor, graphical capabilities) are also unique to a simulation program. In the following subsections, a brief historical perspective and a classification of various simulation programs are provided.

8.5.1. Transient Network Analyzers and DC (HVDC) Simulators

As discussed in Section 8.1.2, breadboarding has its shortcomings and is impractical at high power levels. Historically, transient networks analyzers (TNAs) were developed to study the effects of switching and lighting surges on power systems [15]. TNAs are small-scale models of actual components operating at much lower voltages and currents. A great effort is made to ensure that the second-order parasitics in the actual system do not become significant in the scaled-down model. Transmission lines, for example, are represented by lumped elements, and the thyristors may be represented by transistors. If a thyristor is used, a compensation voltage is inserted in series to partially overcome the voltage drop across it in order to keep the per unit voltage drop realistic. The advantage of TNAs is that they operate in real time, and therefore a lot of "runs" can be made for statistical "Monte Carlo" type studies, where for example, the switching time of a circuit breaker is varied. The circuit breakers in modern TNAs are represented by a solid-state switch controlled by a computer that dictates its closing and opening times (the actual opening occurs at the next current zero). Computers are also used for data acquisition and postprocessing.

In power systems modeling, TNAs continue to be used. Their real-time simulation capability also makes them ideal for testing control hardware before it is applied to the real system in the field. For example, consider the thyristor-controlled series capacitors (TCSC) for rapid adjustment of network impedance by series compensation of transmission lines [16]. The controller for this application is first tested for its performance on a TNA.

A high-voltage DC (HVDC) simulator is an example of a TNA built for a specific application. Before applying the control to a multimillion-dollar HVDC converter terminal, it is important to evaluate the performance of the control hardware in real time.

For such applications, TNAs will continue to be maintained at central locations where they will be used by their owners, as well as rented out for a fee to others. In the future, TNAs will face increasing competition from real-time digital simulators being developed [17].

8.5.2. Analog and Hybrid Computers

Prior to the development of digital computers for rapid calculations of equations describing a system, a common practice was to use analog computers [18]. In an analog computer, all system variables are treated as voltages and the system equations are solved by means of integrators, summers, and multipliers. As digital computers became more powerful, hybrid computers were developed where the continuous system which lends itself to description by differential equations was solved on the analog computer, and the algebraic and logical equations describing the control portion of the system were solved on the digital computer. The interface in a hybrid computer passed the variable values back and forth between the analog and the digital computer.

The demise of analog and hybrid computers can be attributed to their inflexibility, poor reliability, and the high cost of maintenance. Also, a large amount of setup time was necessary, and the user experience played a major role due to a lack of well-established procedures.

In the late 1960s, a significant effort was made to overcome the disadvantages of analog and hybrid computer simulations and to achieve similar benefits of breadboarding or using TNAs. The basic idea was to represent each element, such as an inductor, by a two-terminal electronic model in hardware, where its inductance value could be varied under a computer control. Then, these hardware models could be connected together almost as components on a breadboard. This approach, called the parity [19] simulation, is no longer in use due to increased competition from purely digital simulations.

8.5.3. Digital Simulators

As rapid strides continue to be made in computational speeds and memory storage, coupled with a downward pressure on prices, digital simulation is where the "action" is and will be.

Given that each simulator must ultimately solve a set of differential/algebraic equations, the digital simulation programs can be classified into two broad categories at the user interface level:

1. Equations solver programs
2. Circuit-oriented programs.

Equations Solver Programs. Equations solver programs are very useful in many power electronic simulation problems. For example, consider the modeling of the motor drive example shown in Figure 8-2, with an emphasis on the motor-load control dynamics. To study low-frequency dynamics over a long period of time under a large disturbance condition, the inverter need not be modeled in detail. Therefore, the inverter switches can be assumed to be ideal, which allows the inverter output voltages v_{aN}, v_{bN}, and v_{cN} (with respect to the negative DC bus N) to be defined explicitly in terms of the DC bus voltage and the inverter switch status, as illustrated in Figure 8-4. This simplified system is then described by equations in a state variable form. The controller may be either digital and/or analog. Once the system is described by means of differential/algebraic equations, there is a choice of several equations solver programs discussed in the paragraphs that follow.

It is possible to solve these equations using any one of the several higher-level computer languages such as FORTRAN, C, and so on. There are libraries available for solving the differential equations, matrix manipulations, and graphics.

However, it is often easier to use one of the many available simulation programs, such as MATLAB [20], SIMNON [21], ACSL [22], and MATRIXx [23]. These programs are all designed to solve a general class of problems rather than just the simulation of power electronics and motor drives. To utilize these programs, the user must provide the system description in terms of differential equations in a state variable form. These programs are able to accept controller transfer functions, for example, in a Laplace transform form. They have built-in integration routines; in fact, the user may have a choice of numerical methods to match the type of problem being solved.

Programs such as SIMNON allow modularity, where each portion of the system can be simulated in modules and the interconnection between various modules is specified by an interconnection module [24].

Packages such as MATLAB contain sophisticated toolboxes for control analysis and design based on neural networks and fuzzy logic [25]. In programs such as MATLAB and MATRIXx, the controller can be designed based on the performance specifications and then it is translated into C and finally into assembly language for programming the digital signal processors (DSPs). Most of these programs now have a graphical interface to make them more user-friendly, for example, the MATLAB-related modular program SIMULINK.

One of the serious drawbacks of using equations solvers in the simulation of power electronic and motion control systems in general is the large initial setup time. Also, slight changes in the circuit can be very costly in terms of the setup time. As an example, consider the three-phase, phase-controlled thyristor converter shown in Figure 8-5a.

It will be extremely difficult to define all possible combinations of thyristor conduction including current commutation intervals as well as operation under abnormal conditions. One possible technique to simplify this simulation, called

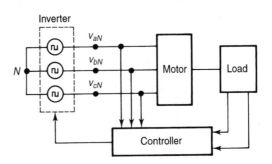

Figure 8-4. Simplified motor-drive system of Figure 8-2.

the *central process method*, repeatedly uses an equivalent subcircuit shown in Figure 8-5b where voltage and current variables depend on the time interval during the simulation [26]. This example illustrates the difficulty in the setup of such a simple circuit using equations solver programs.

Equations solver programs are very useful when the simulations need to be "tailored" for a specific application, where the speed and accuracy are of primary concern and the time for the initial setup is of relatively little importance. It also needs to be emphasized that these programs should be used when the user is generally more interested in the system behavior rather than the detailed device-level operation.

Circuit-Oriented Programs. In circuit-oriented programs, the users need to specify the interconnection of the circuit element models. From this information, these programs themselves develop the system equations. This results in a very short setup time, and it is easy to make changes in the circuit topology. Good circuit-oriented programs are multilevel, which in addition to the circuit-oriented description of the converters, also allow convenient incorporation of user-defined models in terms of differential equations and the description of controllers with the same ease as in a higher-level computer language.

There are many circuit-oriented programs available. Some are designed specifically with power electronics and motion control in mind. However, two of the most commonly used programs at present, SPICE and EMTP, were developed for integrated circuits and power systems, respectively.

The user interface of each of these programs is unique. They also differ in more fundamental respects in terms of the following:

1. Numerical integration used to solve differential equations
2. Treatment of nonlinearities
3. Variation in the time step
4. Search for breakpoints

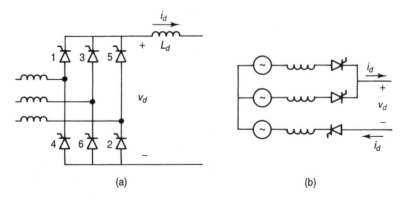

(a)   (b)

Figure 8-5. (a) Three-phase thyristor converter; (b) central process equivalent circuit.

5. Ease with which the controllers and external models can be incorporated for multilevel simulations

6. Treatment of switch models.

In the following sections, a brief review of the issues in numerical solution is presented. A brief synopsis of circuited-oriented programs then follows.

8.6. ISSUES IN NUMERICAL SOLUTION

In any simulation, a system of differential/algebraic equations needs to be solved as a function of time. Therefore, the problem can be stated in general terms as follows: given the solution (values of circuit variables) at the present time step $t_n \, (= hn$ sec) and before, obtain the solution at the next time step $t_{n+1} \, (= t_n + h$ sec), where the time step h may be fixed (specified by the user) or may be determined by the program itself.

Using the numerical integration routines discussed in the following section and including the nonlinear and time-dependent elements, the algebraic equations for the network are obtained by a nodal or a modified nodal analysis [27]. These nonlinear equations, combined with other algebraic equations (for the controller, for example) are solved by an iterative technique, such as the modified Newton combined with Gaussian elimination or by LU (where L stands for lower triangular and U for upper triangular) factorization for the linear sparse part of the system matrix.

8.6.1. Numerical Integration Methods

A system of differential equations can in general be written as

$$\dot{x} = f(x, t) \tag{8.1}$$

Typically, differential equations for power electronics and motion control systems are extremely nonlinear and stiff. A dynamic system is "stiff" if there is a large spread in the time constants of the circuit (large spread in eigenvalues). There are many elegant numerical methods developed over the years for "stiff" systems, but not all of them have been found to be useful in simulation of power electronics circuits. Only a few methods are efficient in terms of accuracy, stability, and the speed of execution. Some of these are Backward Differential methods (BDf methods) such as the backward Euler, the trapezoidal method, and the implicit two-step second-order BDf method (named Gear-2 in [27]). Alternatively, one can use implicit Runge-Kutta methods, like the SDIRK method (singly diagonal implicit Runge-Kutta). Detailed information on such methods is found in [28] and [29].

Table 8-1 summarizes the solutions as well as the derivatives at the time step t_{n+1}. For comparison purposes, the forward Euler method is also included.

The backward Euler, trapezoidal, and Gear-2 methods are implicit because they require X_{n+1} for the solution, where X_{n+1} is precisely what is being calculated. This will not be a problem in a linear circuit, but in a nonlinear circuit, it is necessary to

TABLE 8-1   SUMMARY OF FORWARD EULER (FE), BACKWARD EULER (BE), TRAPEZOIDAL (TR), AND GEAR-SHICHMAN (GEAR-2, GS), ROUTINES [30]

FE	$X_{n+1} = X_n + hX_n'$	$X_n' = \dfrac{X_{n+1} - X_n}{h}$	Explicit, first order, single step
BE	$X_{n+1} = X_n + hX_{n+1}'$	$X_{n+1}' = \dfrac{(X_{n+1}) - X_n}{h}$	Implicit, first order, single step
TR	$X_{n+1} = X_n$ $\quad + \dfrac{h}{2}(X_{n+1}' + X_n')$	$X_{n+1}' = 2\left(\dfrac{X_{n+1} - X_n}{h}\right)$ $\quad -X_n'$	Implicit, second order, single step
GS	$X_{n+1} = \dfrac{4}{3}X_n$ $\quad -\dfrac{1}{3}X_{n-1} + \dfrac{2h}{3}X_n' + 1$	$X_{n+1}' = \dfrac{3}{2h}X_{n+1}$ $\quad -\dfrac{2}{h}X_n + \dfrac{1}{2h}X_{n-1}$	Implicit, second order, two step

use iterative methods such as Newton-Raphson to find the solution of the nonlinear algebraic equations.

To make an estimate of the truncation error, let X_{n+1} be expressed in the form of a Taylor series as

$$X_{n+1} = X_n + h X_n' + \frac{h^2}{2} X_n'' + \frac{h^3}{6} X_n''' + \cdots \tag{8.2}$$

where h is the time step of integration. Let $\hat{X}_{n+1}$ be the approximate value calculated by the integration routine at t_{n+1}. Therefore, the error can be stated as

$$\varepsilon_{n+1} = X_{n+1} - \hat{X}_{n+1} \tag{8.3}$$

Reference [30] shows that the trapezoidal and Gear-2 methods, being second-order methods, have errors proportional to the third derivative, and the trapezoidal method has a slightly less error than the Gear-2 method. The first-order methods (forward and the backward Euler) have larger truncation errors, proportional to the second derivative.

Stability is a global property that indicates whether the error introduced at any time step will grow or decay later on in the simulation. Integration methods that are not stable will amplify an error (if a sufficient large time step is used). Some methods, like the trapezoidal method, are on the limit of stability. In such methods, the error will neither grow nor decay if time steps larger than a certain limit are used. However, if the solution oscillates, for example, around a correct average value, then such a simulation in power electronics is often meaningless.

To illustrate this phenomenon of oscillation [31], consider the simple hypothetical circuit of Figure 8-6a. The circuit parameters, initial values, and the time step (kept constant) are chosen so that the current zero coincides with the time step labeled t_n. The switch is opened exactly at the current zero instant t_n, beyond which the current in the circuit is zero and the voltage v_L across the inductor should also be zero. Applying the trapezoidal rule of Table 8-1 to the inductor equation

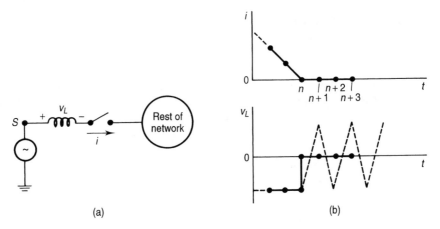

Figure 8-6. (a) Trapezoidal rule of integration; (b) oscillations.

$$i_L(t_{n+1}) = i_L(t_n) + \frac{1}{L} \int_{t_n}^{t_{n+1}} v_L(\zeta) \cdot d\zeta \qquad (8.4)$$

results in

$$i_L(t_{n+1}) = i(t_n) + \frac{h}{2L} [v_L(t_n) + v_L(t_n + 1)] \qquad (8.5)$$

Substituting $i_L(t_n)$, $i_L(t_{n+1})$, ... to be zero subsequent to the switch opening at t_n,

$$v_L(t + h) = -v_L(t), \qquad (8.6)$$

which results in the oscillatory response shown in Figure 8-6b.

Various solutions to this usually unacceptable oscillatory response due to the trapezoidal rule of integration have been proposed. A circuit-level "fix," which does not involve any change in the integration routine, is to add a damping resistor across the inductor as a "numerical snubber" to damp these oscillations [31, 32]. The damping resistor of course needs to be small to be effective, yet it must be large in order not to alter the circuit characteristic. Reference [31] includes suggestions for the values of damping resistors in parallel with inductors and in series with capacitors to effectively damp out these oscillations, in terms of the component (L and C) values and the time step of integration.

Modifications in the numerical integration procedure at the point of discontinuity have also been proposed and implemented. One of these involves using the backward Euler method at the switch opening over one-half time step $h/2$ and then reverting back to the trapezoidal rule of integration which is more accurate than the backward Euler method [33]. Solutions for the oscillation problem due to the trapezoidal rule of integration may be found in [34, 35].

8.6.2. Nonlinear Differential Equations

In power electronics systems, nonlinearities are introduced by component saturation (due to component values which depend on the associated currents and voltages), and limits imposed by the controller. An example is the output capacitance of a MOSFET which is a function of the voltage across it. In such systems, the differential equations can be written as

$$\dot{x} = f(x(t),\, t) \tag{8.7}$$

The solution of equation 8.7 can be written as

$$x(t_{n+1}) = x(t_n) + \int_{t_n}^{t_{n+1}} f(x(\zeta),\, \zeta)d\zeta \tag{8.8}$$

For example, applying the trapezoidal rule to the integral in equation 8.8 results in

$$x(t_{n+1}) = x(t_n) + \frac{h}{2}\,\{f[x(t_n), t_n] + f[x(t_{n+1}), t_{n+1}]\} \tag{8.9}$$

Equation 8.9 is nonlinear and cannot be solved directly. This is because in the right side of equation 8.9, $f[x(t_{n+1}), t_{n+1}]$, depends on $x(t_{n+1})$. Such equations are solved by iterative procedures that converge to the solution within a reasonably small number of iterations. One commonly used solution technique is the Newton-Raphson iterative procedure [30, 31].

8.6.3. Automatic Time-Step Control

The use of automatic time step control is found in many simulation programs. The objective is to reduce computation time without significant loss of accuracy. This is achieved by intelligent choice of the time step during a simulation. The time step is selected such that

1. The local truncation error (LTE) is kept less than a user-specified tolerance.
2. Convergence problems in the iterative solution techniques are avoided.

The local truncation error is the error introduced in one time step by the numerical integration method. The simulation program needs a measure or an estimate of the LTE and of the convergence problems to be able to select the proper time steps. Convergence problems are easily "measured." For example, the number of iterations is a direct measure of possible convergence problems. The LTE is calculated based on error estimation formulas. These formulas can be derived for any numerical integration method (see, for instance, [28] or [29]). These give an upper bound for the error in the last computed time step. The simulation program compares this estimate with a user-specified (or default) tolerance.

The basic procedure in an automatic time step control algorithm can now be outlined as follows:

1. If the iterative procedure fails to converge, try again with a decreased time step.
2. If the estimated LTE in the last time step is greater than a specified tolerance, compute the last solution again with a decreased time step.
3. If the number of iterations in the last time step is high, accept the new solution and reduce the time step before continuing.
4. If the LTE is (much) smaller than the specified tolerance, accept the solution and increase the time step before continuing.
5. If the number of iterations are "normal" and the LTE is slightly less than the specified tolerance, accept the solution and continue with the same time step.

Different programs use different strategies for how much the time steps are to be increased or decreased in the foregoing cases. There may be other possible improvements in the time step selection method. Nevertheless, the description just given illustrates the fundamental principles of automatic time-step control.

It should be noted that the global error, that is, the total error of the simulation from the beginning to the end, can be controlled only indirectly by controlling the LTE in each time step. However, this gives no guarantee that the global error is less than the LTE multiplied by the number of time steps. This is a general property of all numerical integration methods.

8.6.4. Treatment of Switches

Switches are the most widely used elements in power electronic simulations because all semiconductor devices are modeled as switches. Switch models are the origin of most of the difficulty for the numerical routines used in simulation programs. They may cause convergence errors, floating node voltages (singular systems), and extremely small time steps in automatic time-step selection routines and can also initiate numerical oscillations like those illustrated in Section 8.6.1. It is therefore of great importance that the models of switches be carefully handled by the simulation program.

There are four basic classes of switch models:

1. Ideal
2. Two-valued resistors
3. Subcircuits
4. Equations.

In the first category of ideal models, a switch is simply a short circuit in its on-state and an open circuit in the off-state. From a user's point of view, these are the simplest possible models, but from a numerical point of view, they are not. If no special care is taken, the system of equations will easily end up as a singular system, that is, a system with no solution. The typical problems are electrically isolated subcircuits which cause floating node voltages. Another problem that often arises is ideal short circuit or parallel connection of voltage sources and capacitors.

However, these problems can be overcome by topological checks implemented in the simulation program or by very careful circuit modeling.

The second class models the switch as a low resistance in the on-state and a high resistance in the off-state. From a user's point of view, a two-valued resistor will be identical to an ideal switch if proper on-state and off-state resistances are selected. These models usually do not introduce problems like the ideal switches. However, care must still be taken because very low on-state and very high off-state resistances combined with numerical round-off errors may lead to the same problems as with ideal switches.

A serious problem for both of the foregoing switch models is the discontinuity introduced when the state of a switch changes. Discontinuities cause difficulties in all simulation programs if they are not given special attention. An example is the oscillation problem associated with the trapezoidal rule of integration described in Section 8.6.1. If these models are to be used, special care in detection of breakpoints is needed to avoid such problems.

The third class of switch models are subcircuits made up of available models in the simulation program. These can typically be subcircuits consisting of a number of controlled sources, resistors, and nonlinear capacitors. These models usually describe the switch behavior more accurately, especially because the transition from one state to another is modeled.

The fourth class of switch models are those which are described by some nonlinear equation either implemented in the program or written by the user in some modeling language understood by the simulation program. The accuracy of these models is limited only by the physical understanding of the switch behavior and the limits of the simulation program itself.

It is difficult to give any general comments about possible numerical problems related to the two last modeling approaches. They depend very much on the model itself. However, much effort is usually made in modeling of the transition interval. The transition interval consists of phenomena with very small time constants. This forces the programs to use very small time steps when passing through the switch instants, thus reducing the speed of simulations. For example, such phenomena as reverse recovery and high-frequency ringing certainly require small time steps and extra computation compared to simplified models that neglect these phenomena.

The last two modeling approaches can also cause convergence problems. State changes still represent quasi-discontinuities, even if the transition from one state to another is modeled. Unlike the ideal and the two-valued resistor switch models, the problems for the more advanced models are not easily reduced by introducing break-point detection.

None of the foregoing approaches for switch modeling is "the best" for all kinds of simulations. The model to use must be selected based on the phenomena which are to be studied. From a numerical point of view, they are all difficult.

Nevertheless, either ideal or two-valued resistor models of switches should be available in any simulation program intended for power electronics. They are as fundamental as the ideal linear resistor. The more advanced models are useful for more detailed simulations. Good implementations of ideal or two-valued resistor models will be the most efficient and result in the fastest simulations.

8.7. OVERVIEW OF SOME WIDELY USED
SIMULATION PROGRAMS

The objective of this section is to give an overview of some of the widely used simulation programs in power electronics.

8.7.1. SPICE

SPICE [27] was developed at the University of California-Berkeley in the early 1970s for the simulation of integrated circuits. The acronym SPICE stands for Simulation Program with Integrated Circuit Emphasis. It is the most widely used program in power electronics at low power levels, for example, in the simulation of switch-mode DC power supplies. There are several commercial versions of SPICE. All of them make use of the original, and continuing development of SPICE because of the availability of the code of the program from the UC-Berkeley. The main differences between the various versions are in the available models and subcircuit libraries, graphical preprocessors and postprocessors, ease of specifying the behavioral models of the controller, and so on. Some commercial versions have attempted to improve the convergence of the program in dealing with "stiff" systems such as power electronics and motor drives.

SPICE offers both time domain and frequency domain simulations. In the time domain analysis, the circuit equations are built on a nodal basis. The dynamic elements (capacitors and inductors) are modeled by the companion models [36]. This gives a system of algebraic equations to be solved at each simulation point. LU factorization is used to solve the sparse matrix of the algebraic equation system. The solution technique depends on the type of the system to be solved:

1. A time-invariant linear system is solved by Gaussian elimination.
2. A time-invariant nonlinear system is solved by repeated linearization and use of the previous method iteratively.
3. A time-variant nonlinear dynamic system is solved by repeatedly transforming to a time-invariant nonlinear system and using method 2 iteratively.

The time steps are automatically adjusted by the program based on the error estimation formulas. Time steps are always adjusted such that a simulation point is placed at all "corners" of piecewise linear sources. In SPICE, the voltage-dependent switches are modeled by a nonlinear resistor. This program can handle nonlinearities in the circuit elements and within the controller. The program does not search for breakpoints. There is a choice of using the trapezoidal or the Gear method of integration.

The following comments can be made that apply to most of the commercial versions of SPICE. There are several reasons for the popularity of SPICE. It is easy to use and graphical preprocessors are available. Many university courses and undergraduate textbooks in circuits and electronics incorporate examples and exercises based on SPICE. This familiarity with SPICE makes it ideal for use in power electronics courses. After graduation, the former students continue using SPICE in their workplace.

There are several disadvantages associated with the use of SPICE in power electronics because it was developed with the integrated circuit emphasis. Often, the solution in power electronics simulation fails to converge if there are sharp discontinuities in the circuit or in the controller. In this respect, the choice of the Gear method is found to be superior than the trapezoidal method of integration. There are very few rules to remedy this problem [37]. Sharp discontinuities, for example, due to switching action, cause the program to reduce the time step to very small values, resulting in intolerably long execution times. If the switching operation is not to be analyzed in detail and the system-level behavior is of interest, then one solution is to "soften" the discontinuity, for example by putting a "numerical R-C snubber" in parallel with the switch [1]. As discussed in the following section, the device models within SPICE are not sufficiently accurate for power semiconductor devices. It is difficult to specify behavioral models of the controllers and to link the user-defined models in terms of differential equations.

SPICE will continue to be used extensively in the near future, based on familiarity, ease of usage, large library of available device models, and affordability.

8.7.2. EMTP

EMTP [38] was developed for the analysis of electromagnetic transients in power systems at the Bonneville Power Administration (BPA). The acronym EMTP stands for Electro-Magnetic Transients Program. To simulate high-voltage DC transmission systems, the capability to model diodes and thyristors as switches was added. The EMTP development carried out at BPA is in the public domain. Like SPICE, now there are several versions of EMTP, some on personal computers, all of which incorporate most of the original features of the BPA's EMTP.

EMTP uses companion models for dynamic elements based on the trapezoidal rule of integration. Equations are built by use of the nodal approach. The algebraic equations are solved by use of sparse matrix and LU factorization methods [39]. The time step of integration is specified by the user and is kept constant throughout the simulation. Unlike SPICE, where the switches are modeled by nonlinear resistors, switches in EMTP truly represent an open circuit when "off" and a short circuit when "on" by coalescing the two associated nodes as one. EMTP has a powerful, yet relatively easy, way to specify controllers within a module called TACS, and now within another module called MODELS.

As expected, EMTP is the most widely used program for simulating power electronics in power systems at high power levels where the system-level operation, rather than the details of each individual switching, is of interest. There are several reasons for EMTP's popularity. EMTP has been an industry standard for power systems modeling for decades with many active user groups. Therefore, it has models for transformers and transmission lines which have been validated by field tests. It also has models for various electrical machines. The availability of diodes, thyristors, and switch models in combination with an easy-to-use controller has made EMTP a very powerful tool for this application.

There are some drawbacks that have limited the use of EMTP in the modeling of power electronics, especially at the lower power levels where the EMTP capability of power systems modeling is usually not important. One of these limitations is the use of a fixed time step for integration that does not make use of the circuit "inactivity" to increase the time step. As discussed earlier, the trapezoidal rule of integration can result in numerical oscillations following switching actions. Some of the remedies discussed earlier for suppressing these oscillations must be used, otherwise, the simulation may be erroneous. Also, the way EMTP incorporates the controller module in the solution introduces unit time-step delay from the input to the output.

EMTP is a continuously evolving program with a large amount of resources pushing its development. Therefore, it is expected to remain the industry standard for simulation of power electronics in power systems.

8.7.3. MATLAB/SIMULINK [20, 40]

If we choose an equations solver, we must ourselves write the differential and algebraic equations to describe various circuit states, the logical expressions, and the controller that determine the circuit state. Then, these differential/algebraic equations are simultaneously solved as a function of time.

In the most basic form, we can solve these equations by programming in any one of the higher-level languages such as FORTRAN, C or PASCAL. It is also possible to access libraries in any of these languages which consist of subroutines for specific applications such as to carry out integration or matrix inversion. However, it is far more convenient to use a package such as MATLAB or a host of other packages where many of these convenience features are built in. Each of these packages uses its own syntax and also excels in certain applications.

The program MATLAB can easily perform array and matrix manipulations, where, for example, $y = a \cdot * b$ results in y, which equals cell-by-cell multiplication of two arrays a and b. Similarly, to invert a matrix, all one needs to specify is $Y = \text{inv}(X)$. Powerful plotting routines are built in. MATLAB also features various libraries, called toolboxes, which can be used to solve particular classes of problems. For example, the neural network toolbox enables the simulation of an unlimited number of layers and interconnections. Neurons can be modeled with sigmoid, linear, limit, or competetive transfer functions. The toolbox contains functions for implementing a number of networks, including back-propagation, Hopfield, and Widrow-Hoff networks.

Simulink is another toolbox for graphical entry and simulation of nonlinear dynamic systems. It consists of a large number of building blocks that enables the simulation of control-based systems. Some of the other features include seven integration routines and determination of equilibrium points.

MATLAB is widely used in industry. Also, such programs are used in the teaching of undergraduate courses in control systems and signal processing.

Therefore, the students are usually familiar with MATLAB prior to taking power electronics courses. Even if this is not the case, it is possible to learn its use quickly, especially by means of examples.

8.8. OVERVIEW OF SIMULATOR CAPABILITIES BY EXAMPLES

In this section, a few simple examples are presented to illustrate the capabilities of several widely used programs.

8.8.1. Representation of Switching Action Using PSpice

The first example illustrates the simulation of a switch mode power electronics converter using SPICE. There are several commercial versions of SPICE that operate on personal computers under several popular operating systems. One commercial version of SPICE is called PSpice [41]. In PSpice, many additional features are included to make it a multilevel simulator where the controllers can be represented by their behavior models, that is, by their input-output behavior, without resorting to a device-level simulation. There is an option for entering the input data by drawing the circuit schematic. PSpice, in addition to its use in industry, has also become very popular in teaching of undergraduate core courses in circuits and electronics. Therefore, many students are familiar with PSpice. One of the reasons for the popularity of PSpice is the availability and the capability to share its free evaluation (classroom) version. This evaluation version itself is very powerful for power electronics simulations.

To illustrate how the information about a circuit is put into a circuit-oriented program in general and PSpice in particular, a very simple example is presented. We will consider the circuit of Figure 8-7, where the pulse width modulated control signal for the switch under an open loop operation is generated by comparing a control voltage with a high-frequency sawtooth waveform. Note that we explicitly include the representation of the diode and the switch. If the inductor current in this

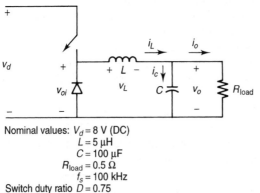

Nominal values: $V_d = 8$ V (DC)
$L = 5$ µH
$C = 100$ µF
$R_{load} = 0.5$ Ω
$f_s = 100$ kHz
Switch duty ratio $D = 0.75$

Figure 8-7. Step-down DC-to-DC converter.

circuit becomes discontinuous, this circuit-oriented simulator will automatically take that into account. In the present simulation, the diode is represented by a simple built-in model within PSpice, and the switching device is represented by a simple voltage-controlled switch. In a circuit-oriented simulator like PSpice, detailed device models can be substituted, if we wish to investigate switching details.

As a first step, we must assign node numbers as shown in the diagram of Figure 8-8a, where one of the nodes has to be selected as the ground ("0") node.

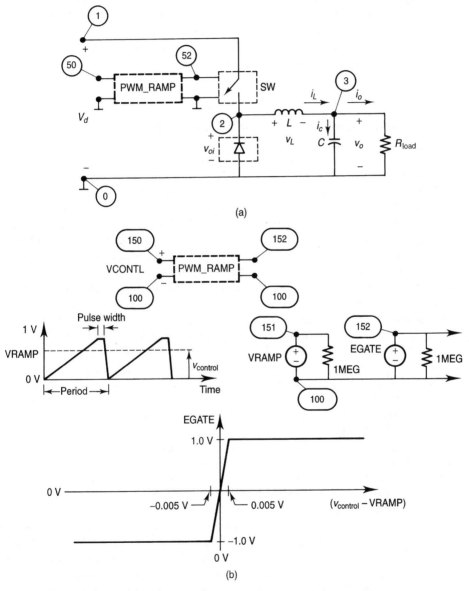

Figure 8-8. (a) PSpice circuit for Figure 8-7; (b) sawtooth waveform generator in PSpice.

The transistor is modeled by a voltage-controlled switch SW whose state is determined by the voltage at its control terminals. In the on-state with a control voltage greater than VON (= 1 V default value), the switch has a small on-state resistance RON (= 1 Ω default value). In the off-state with a control voltage less than VOFF (= 0 V default value), the switch is in its off-state and is represented by a large resistance ROFF (= 10^6 Ω default value). Of course, the default values are optional and the user can use any other values that are more appropriate.

The subcircuit to generate the switching-frequency saw-tooth waveform is shown in Figure 8-8b. The listing of the input circuit file to PSpice is shown in the appendix as A1. The output waveforms from the simulation can be plotted by a graphical postprocessor (called Probe) within PSpice. The resulting waveforms are well known and therefore are not included.

8.8.2. Thyristor Converter Representation Using EMTP

The ATP (Alternative Transients Program) is a version of EMTP, which is also available for personal computers under MS-DOS operating system [42]. Similar to SPICE, EMTP uses a trapezoidal rule of integration, but the time step of integration is kept constant. Because of the built-in models for various power system components such as three-phase transmission lines and so on, EMTP is a very powerful program for modeling power electronics in power systems. Compared to SPICE, the switches in EMTP are treated quite differently. When a switch is closed, the row and column (in the network matrix) corresponding to the terminal nodes of the switch are added together. There is a very powerful capability to represent analog and digital controllers that can be specified with almost the same ease as in a high-level language. The electrical network and the controller can pass values of various variables back and forth at each time step. The control over the time step Δt generally results in reasonable execution (run) times. EMTP is very well suited for analyzing complex power electronic systems at a system level, where it is adequate to represent switching devices by means of ideal switches and the controller by its transfer function or by logical expressions.

A simple thyristor converter shown in Figure 8-9a is modeled using the ATP version of EMTP which has a built-in model of an idealized thyristor. The firing angles of the thyristors are obtained by comparing the control voltage with the sawtooth waveform, synchronized with the input AC voltage, as shown in Figure 8-9b. The modeling of the electrical network in EMTP, similar to SPICE, requires numbering of nodes and specifying the circuit components as in the program listing included in the appendix as A2. The controller to determine the firing angle is represented as a distinct block. This block accepts inputs from the electrical network, which are the voltage and current values and the switch states. The outputs to the electrical network are the thyristor gate pulses. As shown in Figure 8-10, the controller can be divided into various submodules: a ramp (sawtooth) generator with a file name of RAMPGEN.MOD, a module where the sawtooth waveform is compared with the control voltage to generate thyristor gate pulses (file name of

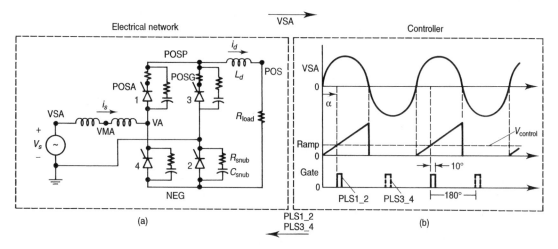

Figure 8-9. Thyristor converter simulation using ATP version of EMTP.

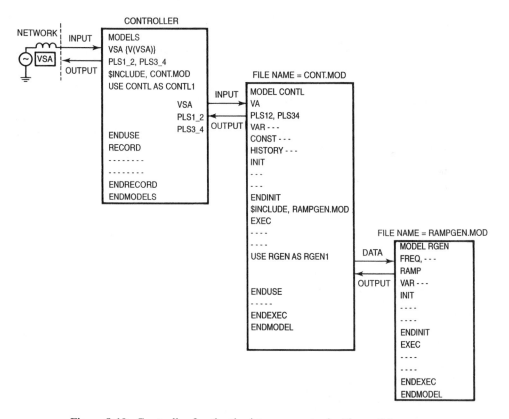

Figure 8-10. Controller for the thryistor converter in Figure 8-9.

CONT.MOD), and the outer shell that interfaces the controller with the electrical circuit. Once again, the well-known results of such a simulation are not shown.

8.8.3. Field-Oriented Control of Induction Motor Drives

To show the capabilities of PSpice, MATLAB/SIMULINK and the ATP version of EMTP for simulating motor drive systems, we have chosen a field-oriented induction motor drive as an example. Before presenting the programs and results, the system equations are described.

Introduction. The following simplifying assumptions are made while modeling an induction machine:

1. The air gap flux is sinusoidally distributed.
2. The magnetic circuit of the motor is operating in its linear region, without any magnetic saturation.
3. The stator windings are wye connected with an isolated neutral.
4. A two-pole motor is assumed, but the results can be easily extended to any number of poles.
5. The number of windings on the stator are equal to the number of rotor windings.

Figure 8-11 shows a cross-sectional view of the motor where the squirrel cage rotor is represented by means of equivalent phase windings. The stator and the assumed equivalent rotor windings are shown in Figure 8-12, along with the voltage polarities and the assumed current directions.

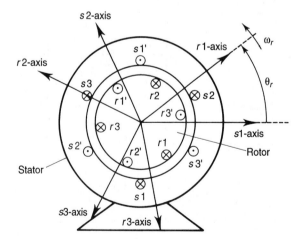

Figure 8-11. Cross-section of a two-pole induction motor.

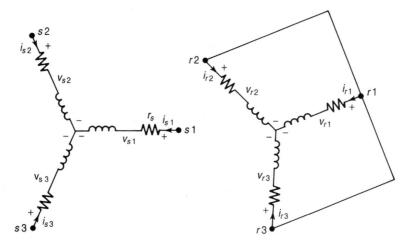

Figure 8-12. Stator and rotor equivalent circuits.

Space Vector Representation. The stator currents can be expressed as follows by the space vector

$$\overrightarrow{i}_s^{s1}(t) = \overrightarrow{i}_{s1}^{s1}(t) + \overrightarrow{i}_{s2}^{s1}(t) + \overrightarrow{i}_{s3}^{s1}(t)$$

$$= i_{s1}(t) + i_{s2}(t)e^{j120^\circ} + i_{s3}(t)e^{-j120^\circ} \tag{8.10}$$

where the superscript s_1 denotes the axis used as the reference axis with a phase angle of zero degrees. Similarly, we can express the stator voltage space vector and the stator flux-linkage space vector.

In the rotor circuit, we can express rotor currents by a space vector with respect to the r_1-axis as the reference axis:

$$\overrightarrow{i}_r^{r1}(t) = \overrightarrow{i}_{r1}^{r1}(t) + \overrightarrow{i}_{r2}^{r1}(t) + \overrightarrow{i}_{r3}^{r1}(t)$$

$$= i_{r1}(t) + i_{r2}(t)e^{j120^\circ} + i_{r3}(t)e^{-j120^\circ} \tag{8.11}$$

Similarly, we can express the space vectors $\overrightarrow{v}_r^{r1}(t)$ and $\overrightarrow{\lambda}_r^{r1}(t)$.

Some of the stator and the rotor circuit space vectors are shown in Figure 8-13 at an arbitrary time $t = t_1$. Note that $\overrightarrow{i}_s(t)$ and $\overrightarrow{\lambda}_r(t)$ in Figure 8-13 are shown with no superscripts because they represent the physical position of the vector in space at a given time instant. Then, they can be expressed mathematically with respect to any axis chosen as a reference, for example, $\overrightarrow{i}_s^{s1}(t)$ with the s_1-axis as the reference as in equation 8.10.

Transformation to a Common Reference—Rotor Field Axis as the Reference Axis. To simplify the equations and to provide a control over the torque production in a straightforward (as explained later) manner, both the stator and the rotor equations are expressed with respect to a common reference axis. This axis is

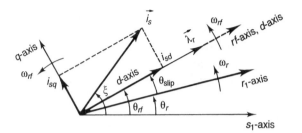

Figure 8-13. *d-* and *q-*axes.

along the rotor field λ_r, at a rotor field angle θ_{rf} with respect to the s_1-axis, as shown in Figure 8-13. The representation of $\vec{i}_s(t)$, transferred from the stationary s_1-axis to the rotor field axis (*rf*-axis) reference is as follows:

$$\vec{i}_s^{rf}(t) = \vec{i}_s^{s1}(t)\, e^{-j\theta_{rf}} = \vec{i}_s(t) \tag{8.12}$$

The superscript *rf* is common to all quantities hereon and, hence, is dropped in the following text.

To express the torque in terms of the rotor flux $\vec{\lambda}_r$ and the stator current $\vec{i}_s$, an orthogonal *d–q* axes reference frame, as shown in Figure 8-13, is defined. The *d*-axis is aligned with the rotor flux $\vec{\lambda}_r$ at all times and the *q*-axis is always 90° ahead of the *d*-axis. Therefore, this reference frame must rotate with the speed of the rotor field λ_r, that is, at a speed ω_{rf}. The current space vector can be decomposed along the *d–q* axes as

$$\vec{i}_s = i_{sd} + j i_{sq} \tag{8.13}$$

The electromagnetic torque developed by the motor can then be shown to be given by

$$T_{em} = k_t \lambda_r i_{sq} \tag{8.14}$$

where k_t is the motor torque constant.

Control of the Rotor Flux and the Electromagnetic Torque.

The rotor flux λ_r obtained from the *d*-axis component of the current space vector is

$$T_r \frac{d\lambda_r}{dt} + \lambda_r = L_m i_{sd} \tag{8.15}$$

where T_r is the rotor time constant.

The slip frequency at which the rotor current space vector lags behind the rotor field vector is given as

$$\omega_{slip} = \frac{L_m i_{sq}}{T_r \lambda_r} \tag{8.16}$$

Summing the slip speed calculated above with the measured rotor speed ω_r, we obtain the speed ω_{rf}. The rotor field position θ_{rf} is obtained by integrating ω_{rf}.

$$\omega_{rf} = \omega_r + \omega_{\text{slip}} \tag{8.17}$$

$$\theta_{rf}(t) = \int_0^t \omega_{rf}(t)\, dt + \theta_{rf}(0) \tag{8.18}$$

where $\theta_{rf}(0)$ is the position at the time origin.

The rotor speed is determined from the load equation as

$$\omega_r(s) = \frac{1}{Js + B}(T_{\text{em}} - T_{\text{load}}) \tag{8.19}$$

where J is the moment of inertia of the motor and the load, B is the coefficient of friction, and T_{load} is the load torque.

Transformation Between Reference Frames. The foregoing discussion shows that we need to control independently the stator current components i_{sd} and i_{sq}. These components are in terms of the d–q reference frame rotating at a speed ω_{rf}. In reality, we are able to control only the three physical phase currents i_{s1}, i_{s2}, and i_{s3} as functions of time. Therefore, there is a need to transform quantities between the rotating d–q reference frame and the stationary s_1–s_2–s_3 reference frame, and vice versa. Often, a stationary α–β reference frame with orthogonal axes, as shown in Figure 8-14, is used as an intermediary in this transformation.

The following equations are used when quantities in one reference frame are transformed to another reference frame:

$$\begin{bmatrix} i_{s\alpha} \\ i_{s\beta} \end{bmatrix} = \begin{bmatrix} \frac{3}{2} & 0 & 0 \\ 0 & \frac{\sqrt{3}}{2} & -\frac{\sqrt{3}}{2} \end{bmatrix} \begin{bmatrix} i_{s1} \\ i_{s2} \\ i_{s3} \end{bmatrix} \tag{8.20}$$

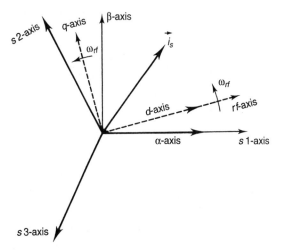

Figure 8.14. Intermediary alpha-beta axes.

$$(\alpha{-}\beta) \leftarrow 3\phi$$

$$\begin{bmatrix} i_{sd} \\ i_{sq} \end{bmatrix} = \begin{bmatrix} \cos\theta_{rf} & \sin\theta_{rf} \\ -\sin\theta_{rf} & \cos\theta_{rf} \end{bmatrix} \begin{bmatrix} i_{s\alpha} \\ i_{s\beta} \end{bmatrix} \tag{8.21}$$

$$(d{-}q) \leftarrow (\alpha{-}\beta)$$

The inverse tranformations are as

$$\begin{bmatrix} i_{s1} \\ i_{s2} \\ i_{s3} \end{bmatrix} = \begin{bmatrix} \frac{2}{3} & 0 \\ -\frac{1}{3} & \frac{1}{\sqrt{3}} \\ -\frac{1}{3} & -\frac{1}{\sqrt{3}} \end{bmatrix} \begin{bmatrix} i_{s\alpha} \\ i_{s\beta} \end{bmatrix} \tag{8.22}$$

$$3\phi \leftarrow (\alpha{-}\beta)$$

and

$$\begin{bmatrix} i_{s\alpha} \\ i_{s\beta} \end{bmatrix} = \begin{bmatrix} \cos\theta_{rf} & -\sin\theta_{rf} \\ \sin\theta_{rf} & \cos\theta_{rf} \end{bmatrix} \begin{bmatrix} i_{sd} \\ i_{sq} \end{bmatrix} \tag{8.23}$$

$$(\alpha{-}\beta) \leftarrow (d{-}q)$$

Therefore, if the reference currents i_{sd}^* (to control the flux λ_r in the machine) and i_{sq}^* (to control the torque produced) are given, the calculation of rotor flux and slip speed using equations 8.15 and 8.16 enables the reference currents i_{s1}^*, i_{s2}^*, and i_{s3}^* to be generated. The power electronics converter should supply these currents to the motor as illustrated in the block diagram in Figure 8-15.

In the block diagram of Figure 8-15, the measured stator currents i_{s1}, i_{s2}, and i_{s3} are used to calculate i_{sd} and i_{sq} and eventually the rotor field position θ_{rf}. It should be noted, however, that θ_{rf} is needed in the process. Also, we should note that if the current-regulated power electronics converter was ideal such that the actual currents i_{s1}, i_{s2}, and i_{s3} were instantaneously equal to their reference values i_{s1}^*, i_{s2}^*, and i_{s3}^*, respectively, then the inverse transformation (from i_{s1}, i_{s2}, and i_{s3} to i_{sd} and i_{sq}) could have been avoided in the block diagram of Figure 8-15.

Speed and Position Control Loops. The block diagram of Figure 8-15 shows the system, given the current references i_{sd}^* and i_{sq}^*. These current references are generated by the speed and the position control loops, as shown in the block diagram of Figure 8-16.

Initial Startup. Initially, the speed and the torque requirements are both zero. Therefore, i_{sq}^* is zero. The reference value λ_r^* of the rotor flux at zero speed is calculated in the block diagram of Figure 8-16. The value of the rotor field angle θ_{rf} is assumed to be zero. The division in the block diagram of Figure 8-15 is prevented until λ_r takes on some finite (nonzero) value. This way, the three stator currents build up to their steady-state DC values. The rotor flux space vector will

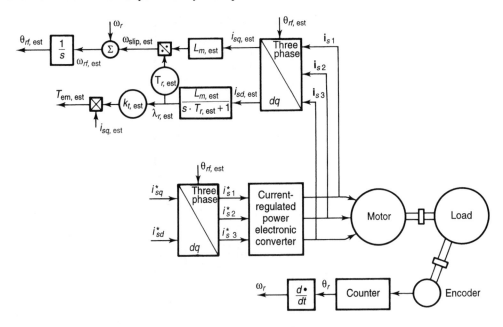

Figure 8.15. Block diagram of an induction motor-drive system, with reference currents, i_{sd}^* and i_{sq}^*, as inputs.

build up to be along the *d*-axis. Once the dynamics of the flux build-up is completed, the drive is ready to follow the speed and position commands.

Simulation of Rotor Flux–Oriented Induction Motor Models. The rotor flux–oriented vector-controlled induction motor drive presented in the previous section is simulated using MATLAB/Simulink, PSpice, and ATP. The control system and the power electronics converter are as shown in Figure 8-16. The purpose of this simulation is to look at the responses in the mechanical system, in the torque, and in the currents. The power electronics converter is modeled as a block where output currents are equal to the input current references. The motor parameters used in the simulations are $T_r = 0.05\,\text{s}$, $L_m = 60\,\text{mH}$, and $k_t = 0.95$ and correspond to a motor rating of 5 kW. The inertia of the motor and the constant of friction are $0.05\,\text{kgm}^2$ and $0.001\,\text{kgm}^2/\text{s}$, respectively. A fixed flux reference value of 1.77 Wb was used in the controller. A step change of 1 rad in the position reference was introduced at time $= 0.05\,\text{s}$, and a step load change of 10 Nm was given at time $= 0.3\,\text{s}$. The simulations were run with a maximum time of 0.5 s. Results of simulations when the estimated rotor time constant is 50% greater than the actual rotor time constant are also shown.

The states and parameters in the control system motor model are considered as estimated parameters and are distinguished from the actual motor parameters by an additional subscript, *est*. The measurements w_r and θ_r needed for the control are

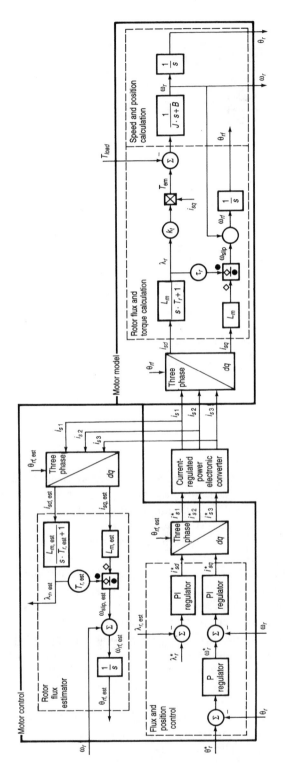

Figure 8-16. Block diagram of the motor-drive system with speed and position control loops.

taken from the states in the motor model. The motor is represented by the same equations in Figure 8-17 as for the rotor flux calculation in Figure 8-16.

Simulink System Representation. The motor drive system is represented using blocks in Simulink as shown in Figure 8-18. Such a representation allows the system structure to be maintained. The position reference and the load torque are given as inputs using the "Step Input" blocks. The controller, converter, and motor models are contained in the three subsystems. The multiplexer (mux) is used to pass the variables in the form of vectors from the motor model. The variables passed to the "To Workspace" blocks are available in the MATLAB environment for further analysis (if required). The "Scope" block is used to plot the variables passed to it.

Figure 8-19a shows the contents of the "Motor Model" block. All the blocks in Figure 8-19a, except the "three-phase to *dq*" block, are available in Simulink. The user needs only to select the required blocks from a large library of functions and assign the parameter values in the blocks. Simulink also allows the use of variable names in these blocks, which are passed through a block mask. The block mask is

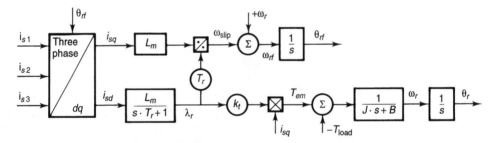

Figure 8-17. Rotor flux-oriented motor model and load model.

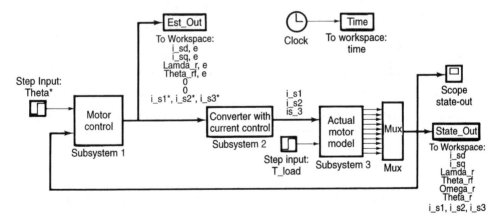

Figure 8-18. Topmost Simulink block diagram of a position-controlled induction motor drive.

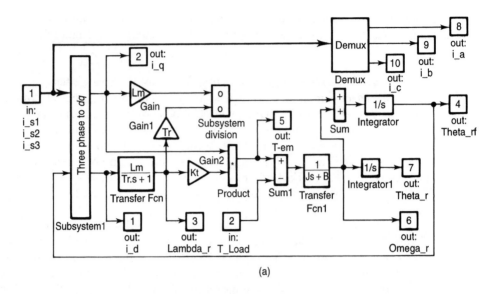

(a)

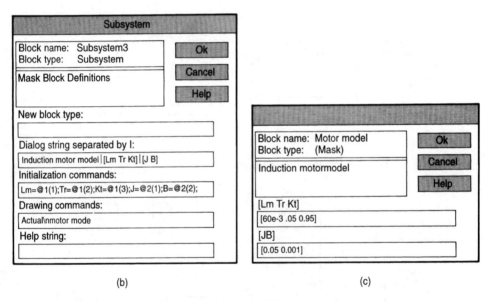

(b)					(c)

Figure 8-19. Simulink representation of (a) motor model; (b) programming
window for block mask; (c) block mask.

programmed as shown in Figure 8-19b, where the input fields, help string, and
drawing and initialization commands for the dialog box are defined. The resulting
user input window, which appears when the user double-clicks the masked motor
block, is shown in Figure 8-19c.

The block for transforming the three phase currents to the rotor flux-oriented
d–q axes currents is grouped as a subsystem. This block and the other subsystem

blocks in Figure 8-19 are implemented in a manner similar to the "Motor Model" block.

Simulation results with the estimated parameters equal to the actual parameters are shown in Figure 8-20. In Figure 8-21, the simulation results for a system where the estimated rotor time constant Tr_est is 50% larger than the actual rotor time constant Tr are shown.

PSpice Modeling of Indirect Field-Oriented Controlled Induction Motor Drive. The induction motor drive modeled in PSpice is based on the block diagram shown in Figure 8-22. The circuit file is listed in the appendix as A3. The equations presented earlier in this section (equations 8.10–8.23) are solved using the analog behavioral modeling feature of PSpice. As mentioned earlier, an ideal current-regulated converter is simulated. A small number (= 1E - 06 in the simulations) is added to the divisor in equation 8.16. This allows the simulation to proceed without any convergence problems during the initial bias point calculation. With the calculation of the bias point, the simulation begins from a steady-state condition.

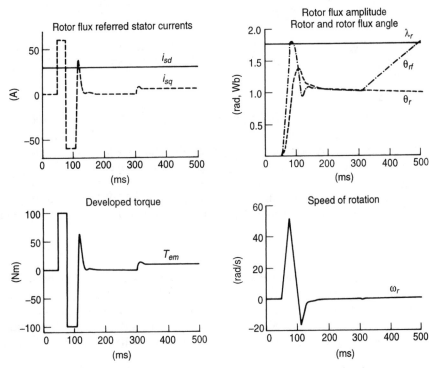

Figure 8-20. Simulation results with the same parameters in the motor model and the motor control block.

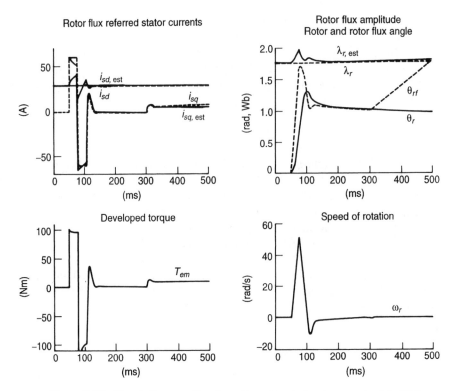

Figure 8-21. Simulation results where the estimated rotor constant is 50% larger than the actual rotor time constant.

ATP Modeling of Indirect Field-Oriented Controlled Induction Motor Drive. The motor drive simulated in ATP is also based on the block diagram in Figure 8-22. However, some of the variable names are different from those shown in the figure. The program listing is given in the appendix as A4. The motor drive was simulated with zero initial conditions. The zero initial conditions caused the rotor flux to increase to its steady-state value of 1.77 Wb in about 0.1 s. Hence, for the ATP simulations, a step change in the position reference was applied at time = 0.1 s.

Comparison of Results. All three programs gave similar results. Hence only results from Matlab/Simulink are presented here. In Matlab/Simulink and PSpice, where the maximum time step is specified, the error in the results increases when this parameter is set to a large value, for example, 1.0 ms. On the other hand, ATP does not show any significant variation in the results when the fixed time step is varied from 5 μs to 50 μs.

It is possible to include switches in the three programs and simulate the current-regulated converter with high-frequency switchings.

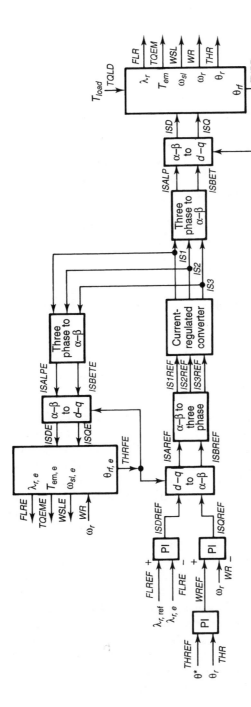

Figure 8-22. PSpice implementation of the induction motor drive. Node names are shown in italics.

8.9. POWER SEMICONDUCTOR DEVICE MODELS
FOR CIRCUIT SIMULATION

8.9.1. Introduction

When a simulation of a power electronic circuit is carried out over a large number of switching cycles, the goals of the simulation can be achieved by modeling the semiconductor power devices as ideal switches. However, some simulations are carried out for a duration of only a few switching cycles for such purposes as the design of snubber circuits or drive circuits, the calculation of the converter efficiency, or the estimation of stresses on the power devices and other components. The device models used in these latter simulations should be as complete and accurate as possible.

Unfortunately, the models currently available for such simulations are generally inadequate, especially in describing transient phenomena. This has led to the development of a large number of competing models for each of the widely used power devices, including pn junction diodes [43], Schottky diodes, BJTs [44, 45], MOSFETs [46–48], thyristors [49–51], GTOs [52, 53], IGBTs [54], and MCTs [55]. The person carrying out the simulation faces a bewildering and complex task of choosing the most appropriate device model to use.

8.9.2. Currently Available Models
and Their Shortcomings

The number of existing device models is too large to describe here. Two recent review articles [56, 57] provide some detail about many of the models. Some references to specific device models are given in the previous paragraph. Device models can be put under three classifications: generic, subcircuit, and mathematical. Generic models are the device models which are built into the generic versions of SPICE. These built-in models include pn junction diodes, Schottky diodes, BJTs, MOSFETs, and JFETs. There are no generic models for thyristors, GTOs, IGBTs, or MCTs. The generic models were intended to describe logic-level devices and do not accurately describe high-voltage device characteristics. The numerical values of the parameters of the built-in models can be adjusted to provide a better fit to the power device characteristics. A detailed description of all generic models is given in [58].

The inadequacies in this generic model has led to the use of the subcircuit feature of SPICE and its commercial derivatives in developing the most common type of power device model, the subcircuit model. The model is constructed by connecting the generic SPICE model along with passive components and controlled sources into a subcircuit. This type of model can be ported to nearly every type of simulator in use today. However, this class of device model suffers from several drawbacks [56]:

1. Complex and slow to simulate, often requiring several additional components to represent one equation.
2. Difficult parameter extraction procedure because each device characteristic may affect the values of several subcircuit components.
3. Limited flexibility of the subcircuit approach.

The mathematical model is the third type of device model. Mathematical equations representing some approximate description of the device behavior are inserted into the simulator. The mathematical model has the following advantages [56, 57]:

1. They are computationally more efficient: each device characteristic relates to a specific term in the equations.
2. Models can be created for any device. (A comprehensive mathematical model of the IGBT has been published recently [54].)

Implementing a mathematical model may be more difficult than implementing a subcircuit model depending upon the specific simulator being used. Some simulators do not allow the insertion of equations (e.g., generic SPICE) while others (SABER, KREAN) have a special high-level modeling language.

The procedure to obtain numerical values for the parameters of a device model (often termed parameter extraction) is a pervasive problem in simulation situations. Part of the problem lies in the complexity of the model, but often the problem is one of poor documentation. Models that appear to be in good argement with measurements are useless to the technical community if the parameter extraction procedure is usable only by the developer of the model or a specialist in the design of power devices.

8.9.3. Difficulties in Modeling Bipolar Devices

Bipolar devices including pn junction diodes, BJTs, IGBTs, thyristors, GTOs, and MCTs are difficult to model accurately, compared to majority carrier devices like MOSFETs. Carrier diffusion is important in the operation of these devices, and in principle, time-dependent partial differential equations should be used in the model. However, circuit simulators solve only systems of equations involving a single independent variable, time in the case of transient analysis or frequency in the case of frequency domain analysis. This requires that the power device models be lumped models that contain no dependence on spatial variables. Restricting the simulator to solving equations of a single independent variable is necessary in order that the simulator be able to solve circuits of reasonable complexity in a reasonably short period of time. Consequently, partial differential equations cannot be used in model descriptions.

Some special-purpose simulators such as PISCES solve partial differential equations involving both time and spatial variables. Such programs are finite element programs, and they can accurately and completely describe the device behavior.

However, such programs require substantially longer times to do a simulation than do widely used simulators such as SPICE.

Most device models avoid the difficulty of modeling spatially distributed effects such as carrier diffusion by using quasi-static stored charge approximations. Such approximations are used in all of the generic bipolar device models in SPICE. The quasi-static approximation is a source of significant error in describing the transient time-dependent behavior. Stored charge models of diodes do not produce soft reverse recovery or forward recovery in rectifiers. Stored charge models of BJTs, GTOs, and IGBTs fail to model correctly the current "tailing" behavior at device turn-off.

The quasi-static approximation assumes that during device turn-on and turn-off, the spatial dependence of the excess carrier distribution in a bipolar device retains the nonequilibrium steady-state functional dependence on position. The excess carrier density at every position is scaled up or down in proportion to the excess carrier density at the metallurgical junction and thus in proportion to the junction voltage since the excess carrier density at the junction is exponentially dependent on the junction voltage. (Since the relative spatial distribution of carriers $[n(x, t)/n(x = 0, t)]$ does not change in time in this approximation, it is termed quasi-static.) Integrating the excess carrier density over the spatial dimension yields a single quantity, the total excess charge Q, which is dependent on time only. The diode is now described by a lumped-element description which involves only time and requires only an ordinary differential equation.

When the quasi-static approximation is used to describe the turn-off of the power diode, there is no redistribution of excess carriers from one position to another, and the excess carrier density goes to zero everywhere in the drift region at the instant the current equals $-I_{rr}$. From this instant onward in time, the only mechanism for current flow is the displacement current through the space charge capacitance, and the current decays from $-I_{rr}$ toward zero as the space charge capacitance is charged up to the reverse bias voltage impressed by the external circuit. Since the capacitance is relatively small and the initial current (I_{rr}) is large, the charging time is quite short and the current thus appears to drop to zero nearly instantaneously, that is, to "snap off".

In the quasi-static approximation, there is no delay between a change in the carrier density at the metallurgical junction and a change in the carrier density some distance x into the drift region. This lack of delay is equivalent to the electrons and holes moving with infinite velocity which is physically unrealizable. Examination of the actual decay of the carrier distribution shown in Figure 8-23 for a diode clearly indicates a delay between changes in the carrier densities at different positions, indicating a finite carrier velocity.

8.9.4. Improvements in Bipolar Device Modeling

Two new approaches to modeling carrier diffusion effects based on spatial discretization show promise in reducing or removing the limitations of the quasi-static approximation. Spatial discretization converts spatially dependent partial dif-

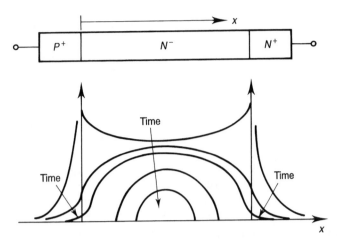

Figure 8-23. Decay of excess carrier distribution in the drift region of a power diode as the diode turns off. Substantial numbers of excess carriers still exist in the middle of the region even when the distribution has decayed to zero at the edges of the region.

ferential equations into finite difference equations of only one independent variable, that is, time. One approach to the discretization, termed the lumped-charge method [43, 49], divides the excess carrier distribution into regions such as shown in Figure 8-24. As time proceeds during the transient, the excess carriers in each region are treated as a single lump of charge which can change at different rates from the charge lumps in the neighboring regions. Even with only a small number of regions or lumps, the model predictions for the behavior of pn junction diodes [43]

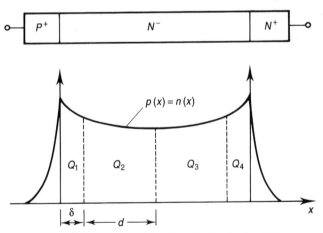

Figure 8-24. Partitioning of the excess carrier distribution in the drift region of a power diode into separate regions to approximately remove the quasi-static stored charge assumption used in generic SPICE diode models. Q1, Q2, and so on are the stored charges in each of the partitioned regions.

and thyristors [49] have proven to be reasonably accurate. The waveforms shown in Figure 8-25 present a comparison of the reverse recovery behavior of a BY329 fast-recovery diode as predicted by the built-in diode model in PSpice, the behavior predicted by an improved diode model [43], and the experimentally observed behavior. The two simulated waveforms were obtained from the same circuit (a simple step-down converter), the only changes being in the diode model. The simulations were constructed to match the experimental conditions as closely as possible. The improvement in simulation accuracy afforded by the improved model is evident in the figure.

In the second approach, a time-dependent spatial discretization scheme is used [57, 59]. The boundaries between regions of high and low excess carrier densities that change with time during the transients are the basis of the choice of discretization points. The partial differential equations again become finite difference equations of a single variable, that is, time, and standard methods are used for the time integration. This procedure is more complex than the lumped charge approach because at each time step, several iterations are required to find the location of the moving boundaries between low and high excess carrier densities. The entire procedure has been implemented as a subroutine and tested with diodes, GTOs, and IGBTs [57, 59]. Device models using this approach produce results in good agreement with experiment and are claimed to require no more CPU time than other device models.

8.9.5. Problems in Modeling Majority Carrier Devices

In principle, the modeling of majority carrier devices such as MOSFETs should be easier than bipolar devices because majority carrier devices do not involve diffusion effects. However, space charge effects such as depletion layers and associated depletion capacitance dominate the transient behavior of majority carrier devices. Accurate descriptions of such phenomena are needed for these devices. In constrast, these effects are not as important in minority carrier devices because most of the

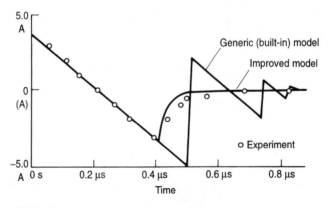

Figure 8-25. Comparison of simulations of the reverse recovery behavior of a BY329 diode with experimental measurements.

transient behavior is dominated by slower diffusion controlled effects. Historically, the description of space charge capacitance behavior in minority carrier devices has been relatively simplistic and not too accurate. Unfortunately, most of the descriptions of depletion layer capacitance behavior used in the built-in (generic) MOSFET models are based on these simplistic minority carrier device models. This lack of precision is further aggravated by the fact that the generic MOSFET models were developed for logic level operation in which the total change in drain source voltage from the off-state to on-state is only 5–10 volts. With such small changes in voltage, the depletion layer capacitances, which are a sublinear function of the drain source voltage, remain approximately constant. Hence, the developers of the generic MOSFET models have modeled the gate source and gate drain capacitances as being constant and independent of applied voltages.

However, power MOSFETs have large variations in gate drain and drain source voltages (10s to 100s of volts). This results in large variations in the effective gate drain capacitance (variations of 100:1 are common in high-voltage MOSFETs). The difference in the capacitance behavior between the generic logic-level model and a more accurate power MOSFET subcircuit model is illustrated by the plots of C_{gd} versus V_{DS} (with $V_{GS} = 0$) shown in Figure 8-26 for the MTP3055E (an n-channel MOSFET with a voltage rating of 60 V and a current rating of 12 A). The curve marked Motorola subcircuit model [46] (developed by Motorola, Inc., the manufacturer of the MTP3055E) accurately models (verified by comparison with experimental measurements) the nonlinear behavior of C_{gd}. The second curve, labeled SPICE model, shows a constant value of C_{gd} independent of V_{DS}. The PARTS parameter estimation program (MicroSim Corp.) in conjunction with a MPT3055 data sheet was used to find the parameter values for the built-in model.

The importance of accurate modeling of the capacitances in the MOSFET is illustrated by the behavior of the MOSFET drain source voltage at turn-on, as shown in Figure 8-27. Two separate curves are shown, one corresponding to the

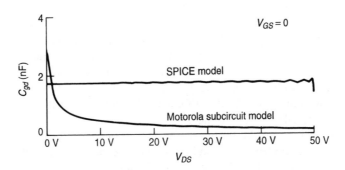

Figure 8-26. Gate drain capacitance of the MTP3055E power MOSFET as a function of the drain source voltage with the gate source voltage equal to zero. The curve-designated SPICE model is the gate drain capacitance variation of the generic (built-in) MOSFET model found in PSpice. The curve-designated Motorola subcircuit model is the capacitance variation of the subscircuit model for the MTP3055E developed by Motorola, Inc., and is available in its TMOS library.

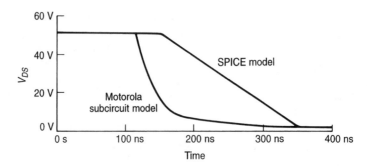

Figure 8-27. Simulation of the turn-on behavior of the drain source voltage of the MTP3055E embedded in a simple step-down converter circuit. The curve designated SPICE model used the generic (built-in) MOSFET model found in PSpice and the curve designated Motorola subcircuit model used a subcircuit model developed by Motorola, Inc.

built-in MOSFET model fitted by the PARTS program to the data sheet of a MTP3055E and the other corresponding to a subcircuit model developed by Motorola, Inc. for the MTP3055E. Both models were used in an identical step-down converter circuit. The two models produce significantly different behavior of the drain source voltage, with the subcircuit model behavior being the curve which most closely matches the experimental behavior.

8.9.6. Future Outlook

Several factors point to a significant improvement in the quality of power device models in the near (next few years) future. Methods for modeling carrier diffusion effects such as just described will be refined and applied to other devices such as MCTs [43, 49, 57, 59]. Modeling problems with majority carrier devices such as how to accurately describe nonlinear capacitances like the gate drain capacitance of a MOSFET have been recognized and improvements implemented [46–48]. New circuit simulators such as SABER and KREAN have been introduced which have a much more modular structure and provisions for adding new models relatively easily. This will make the development of new mathematical models much easier, and the portability of the models will be better.

The problems of better parameter extraction procedures and model verification are also being tackled on a broad front. The recent formation of the NIST (National Institure of Standards and Technology) Working Group on Model Validation is an example. The primary task of the group is to establish well-defined procedures for the comprehensive evaluation of circuit simulator models.

Future device models will include electrothermal effects in which electrical and thermal effects are coupled together. Nearly all existing models, if they contain any temperature-dependent parameters at all, have temperature as an independent input variable. However, realistic models should describe the physics which couple the electrical and thermal variables and device temperature should be an output of the simulation process just like voltages and currents. Some progress has been made recently in this area [54, 60].

8.10. CONCLUSION

In this section, the authors hazard a guess into the future of simulation based on the present trends. Driven by market pressures to reduce development cost and time, it is expected that simulation will become pervasive for its intrinsic advantages and educational benefits. User-friendliness will be absolutely essential. Most of the simulation programs already execute on personal computers. As the distinction between workstations and personal computers narrows, all programs will be easily accessible. If the present trend of declining computer costs coupled with rapidly increasing computational speed continues, the simulation packages that offer accuracy and robustness will excel over those which merely offer computational speed.

Multilevel simulators will allow the system to be modeled in modules that may be specified as interconnections of available element models or described by means of differential equations. This multilevel simulation will include digital controllers, for example, a DSP, whose code is written in C language and is interfaced with the rest of the system model within the same simulator in order to analyze its operation. Many of these simulators will operate in real time by means of CPUs operating in parallel, some of which are already available for testing of relays.

Models of semiconductor devices, magnetic elements, and parasitics will be refined to increase simulation accuracy. It will be possible to synthesize models from the waveforms either measured in the laboratory or given in the data sheets. Most simulators will have modules for neural network and fuzzy logic–based controllers. Expert systems will call up simulation programs as needed in the design and education processes.

There will be more application-specific programs that allow analysis and design of systems with a set of topologies and objectives. Many such programs are already available as modules within the general-purpose simulators. A good example of a stand-alone program is a tool for the analysis and design of switched-mode DC power supplies [14], where the speed of computation has been increased by approximately two orders of magnitude by recognizing the characteristics of this class of problems. The drawback of a large setup time with equation solvers will be overcome with the use of object-oriented programs, which will translate the user-specified interconnection of element models into a set of differential equations [61].

References

[1] Mohan, N., T. M. Undeland, and W. P. Robbins, *Power Electronics: Converters, Application and Design*, 2nd edition, John Wiley, New York, 1995.

[2] Bose, B., "Power electronics—recent advances and future perspectives," *IEEE Ind. Elec. Conf. Proc.*, pp. 14–15, 1993.

[3] Krauthamer, S., A. Bulawka, and R. Das, (eds.), *Proceedings of the 2nd Workshop on Smart Power/Power Integrated Circuits: Technology and Applications*, Pasadena, CA, December 8–9, 1993.

[4] Severns, R., "Efficient low voltage rectification techniques," Professional Education Seminar, *IEEE Applied Power Electronics Conf.*, Orlando, FL, 1994.

[5] Hingorani, N. G., "Flexible AC transmission," *IEEE Spectrum*, pp. 40–45, April 1993.

[6] Hingorani, N. G., and K. L. Stahlkopf, "High-power electronics," *Scientific American*, pp. 78–85, November 1993.

[7] Franz, G. A., "Multilevel simulation tools for power converters," *Records of the IEEE-Applied Power Electronics Conference*, pp. 629–633, March 11–16, 1990.

[8] Mohan, N., W. P. Robbins, T. M. Undeland, R. Nilssen, and O. Mo, "Simulation of power electronic and motion control systems—an overview," *Proc. IEEE*, Vol 82, no. 8, pp. 1287–1302, August 1994.

[9] Thottuvelil, V. J., "Challenges in computer-based analysis/simulation and design of power electronic circuits," *IEEE Workshop on Computers in Power Electronics*, pp. 1–8, 1988.

[10] Chandra, H. N., and V. J. Thottuvelil, "Modelling and analysis of computer power systems," *IEEE Power Electronics Specialists Conf. Rec.*, pp. 144–150, 1989.

[11] Thottuvelil, V. J., D. Chin, and G. C. Verghese, "Hierarchical approaches to modelling high-power-factor AC-DC converters," *IEEE Trans. Power Electronics*, pp. 179–187, April 1991.

[12] Vorperian, V., "Simplified analysis of PWN converters using PWM switch part 1 and part 2," *IEEE Trans. Aerospace and Electronic Systems*, Vol. 26, no. 3, pp. 490–505, May 1990.

[13] Mahabir, K., G. Verghese, J. Thottuvelil, and A. Heyman, "Linear averaged and sampled data models for large signal control of high power factor AC-DC converters," *IEEE PESC Rec.*, pp. 372–381, 1990.

[14] Ridley, R., "New simulation techniques for PWM converters," *Rec. IEEE-APEC*, pp. 517–523, 1993.

[15] Peterson, H. A., *Transients in Power Systems*, Dover, New York, 1966 (reprint of 1951 ed.).

[16] Vithayathil, J., C. Taylor, M. Klinger, and W. Mittelstadt, "Case studies of conventional and novel methods of reactive power control on an AC transmission system," *CIGRE SC 38-02*, Paris, 1988.

[17] Marti, J. R., and L. R. Linares, "Real-time EMTP-based transients simulation," paper presented at the IEEE-PES Summer Meeting, Vancouver, B.C., July 18–22, 1993.

[18] Lipo, T. A., "Analog computer simulation of a three-phase full wave controlled rectifier bridge," *Proc. IEEE*, Vol. 57, pp. 2137–2146, December 1969.

[19] Kassakian, J. G., "Simulating power electronic systems—a new approach," *Proc. IEEE*, Vol. 67, pp. 1428–1439, October 1979.

[20] *The Student Edition of MATLAB*, Prentice Hall, Englewood Cliffs, N.J., 1992.

[21] Elmquist, H., "SIMNON—an interactive simulation program for nonlinear systems—user's manual," Report 7502, Dept. of Automatic Control, Lund Institute of Technology, Lund, Sweden, April 1975.

[22] *ACSL Reference Manual*, 4th ed., Mitchell and Garthier Associates, Concord, MA, 1986.

[23] MATRIXx, Integrated Systems, Inc., 1993

[24] Bose, B. K., and P. M. Szczesny, "A microcomputer-based control and simulation of an advanced IPM synchronous machine drive system for electric vehicle propulsion," *IEEE Trans. Ind. Electr.*, Vol. 35, no. 4, pp. 547–559, November 1988.

[25] Round, S., and N. Mohan, "Comparison of frequency and time domain neural network controllers for an active filter," *IEEE Industrial Electronics Conf. Rec.*, pp. 1099–1104, 1993.

[26] Hingorani, N. G., J. L. Hays, and R. E. Crosbie, "Dynamic simulation of HVDC transmission systems on digital computers," *Proc. IEE*, Vol. 113, no. 5, pp. 793–802, 1966.

[27] Nagel, L. W., "SPICE2—a computer program to simulate semiconductor circuits," Memorandum No. ERL-M520, University of California, Berkeley, 1975.

[28] Harier, E., S. P. Norsett, and G. Wanner, *Solving Ordinary Differential Equations I*, Springer Verlag, 1987.

[29] Harier, E., and G. Wanner, *Solving Ordinary Differential Equations II*, Springer Verlag, Berlin, New York, 1991.

[30] McCalla, W. J., *Fundamentals of Computer-Aided Circuit Simulation*, Kluwer Academic Publishers, Boston, 1988.

[31] Dommel, H., *Electromagnetic Transients Program Reference Manual* (EMTP Theory Book), August 1986.

[32] Alverado, F., R. Lasseter, and J. J. Sanchez, "Testing of trapezoidal integration with damping for the solution of power transient problems," *IEEE Trans. PAS*, Vol. PAS-102, no. 12, pp. 3783–3790, December 1983.

[33] Dommel, H. W., and J. R. Cogo, "Simulation of transients in power systems with converters and static compensators," *Power Systems Computer Conference*, Austria, August 1990.

[34] Maguire, T. L., and A. M. Gole, "Digital simulation of flexible topology power electronic apparatus in power systems, *IEEE Trans. Power Delivery*, Vol. 6, no. 4, pp. 1831–1840, October 1991.

[35] Marti, J. R., and J. Lin, "Suppression of numerical oscillations in the EMTP," *IEEE Trans. Power Apparatus and Systems*, Vol. 4, pp. 739–747, May 1989.

[36] Blume, W., "Computer circuit simulation," *BYTE*, pp. 165–170, July 1986.

[37] Vladimirescu, A., *The SPICE BOOK*, John Wiley & Sons, Inc., New York, 1994.

[38] Meyer, W. S., and T.-H. Liu, *EMTP Rule Book*, Bonneville Power Administration, Portland, Oregon, 1982.

[39] Dommel, H. W., and W. S. Meyer, "Computation of electromagnetic transients," *IEEE Proc.*, Vol. 62, no. 7, pp. 983, 993, July 1974.

[40] *MATLAB*, The Math Works, Inc., Natick, MA, 1994.

[41] *PSpice*, MicroSim Corporation, Irvine, CA, 1995.

[42] *ATP Version of EMTP*, Canadian/American EMTP User Group, The Fontaine, Unit 6B, 1220 N.E., 17th Avenue, Portland OR.

[43] Lauritzen, Peter O., and Clif L. Ma, "A simple diode model with forward and reverse recovery," *IEEE Trans. Power Electronics*, Vol. 8, no. 4, pp. 342–346, October 1993.

[44] Kull, G. M., L. W. Nagel, S. Lee, P. Lloyd, J. Prendgast, and H. Dirks, "A unified circuit model for bipolar transistors including quasi-saturation effects," *IEEE Trans. Electron Devices*, Vol. ED-32, no. 6, pp. 1103–1113, June 1985.

[45] Xu, C. H., and D. Schroder, "A power bipolar junction transistor model describing the static and dynamic behavior," *Power Electronics Specialists Converence,* Milwaukee, WI, 1989.

[46] Cordonnier, C. E., "SPICE Model for TMOS," Motorola Application Note AN1043, Motorola, Inc.

[47] Scott, Robert, Gerhard A. Frantz, and Jennifer L. Johnson, "An accurate model for power DMOSFETs including interelectrode capacitances," *IEEE Trans. Power Electronics*, Vol. 6, no. 2, pp. 192–198, April 1991.

[48] Shenai, Krishna, "A circuit simulation model for high-frequency power MOSFETs," *IEEE Trans. Power Electronics*, Vol. 6, no. 3, pp. 539–547, July 1991.

[49] Ma, C. L., P. O. Lauritzen, P. Turkes, and H. J. Mattausch, "A physically-based lumped-charge SCR model," *Power Electronics Specialists Conf. Rec.*, pp. 53–59, 1993.

[50] Hu, C., and W. F. Ki, "Toward a practical computer aid for thyristor circuit design," *Power Electronics Specialists Conf.*, Atlanta, GA, 1980.

[51] Avant, R., and F. Lee, "A unified SCR model for continuous topology CDA," *IEEE Trans. Ind. Electronics*, Vol. IE-31, no. 4, pp. 352–361, November 1984.

[52] Tsay, C. L., R. Fischl, J. Schwartzenberg, H. Kan, and J. Barrow, "A high power circuit model for the gate turn off thyristor," *Power Electronics Specialists Conf. Rec.*, pp. 390–397, 1990.

[53] Apeldoorn, O., L. Schulting, and H.-Ch. Skudelny, *Power Electronics Specialists Conf. Rec.*, pp. 1074–1081, 1992.

[54] Hefner, Allen R., Jr., and Daniel M. Diebolt, "An experimentally verified IGBT model implemented in the SABER circuit simulator," *Power Electronics Specialists Conf. Rec.*, pp. 10–19, 1991.

[55] *MCT Handbook*, Harris Semiconductor Corp., Melbourne, FL., 1994.

[56] Lauritzen, P. O., "Power semiconductor device models for use in circuit simulators," *Conf. Rec. of IEEE Industry Applications Society*, pp. 1559–1563, 1990.

[57] Schroder, Dierk, "Modeling and CAE for power electronic topologies," *Proc. IECON*, pp. 626–636, 1993.

[58] Antognetti, R., and G. Massobrio, *Semiconductor Device Modeling with SPICE*, McGraw-Hill, New York, 1988.

[59] Metzner, D., T. Vogler, and D. Schroder, "A modular concept for the circuit simulation of bipolar power semiconductors," *European Power Electronics (EPE) Conf.*, Brighton, England, September 1993.

[60] Hefner, Allen R., and David L. Blackburn, "Simulating the dynamic electro-thermal behavior of power electronics circuits and systems," *IEEE Trans. Power Electronics*, Vol. 8, no. 4, pp. 376–385, October 1993.

[61] Cellier, F. E., B. P. Zeigler, and A. H. Cutler, "Object-oriented modeling: Tools and techniques for capturing properties of physical systems in computer code," *Proceedings CADCS '91—Computer-Aided Design in Control Systems*, Swansea, Wales, pp. 1–10, July 15–17, 1991.

Appendix

A1. PSpice Listing for Figure 8-8

```
BUCKCONV.CIR
* Buck (Step-Down) DC-DC Converter
.LIB PWR_ELEC.LIB
.PARAM RISE=9.8us, FALL=0.1us, PW=0.1us ,PERIOD=10us
*
VCONTL 50 0 0.75V

XLOGIC 50 0 52 PWM_RAMP
*
XSW      1      2 52 0 SWITCH
XD       0      2 SW_DIODE_WITH SNUB
*
L        2      3 5uH   IC=9.0A
C        3      0 100uF IC=5.5V
RLOAD    3      0 0.5
*
VD       1      0 8.0V
*
.TRAN  0.1us  100.0us 0s      0.1us uic
.PROBE
.END

.SUBCKT PWM_RAMP      150 100 152
RCNTL 150       100 1MEG
RRAMP 151       100 1MEG
VRAMP 151       100 PULSE(0 1V 0 {RISE} {FALL} {PW} {PERIOD})
EGATE 152       100 TABLE { V(150)    - V(151)  } = (-1.0, -1.0) (-0.005,-1.0)
+ (0.0,0.0) (0.005,1.0) (1.0,1.0)
RGATE 152       100 1MEG
.ENDS
```

A2. ATP/Models Program Listing for Figure 8-9

```
C     FILE NAME = THY1PH.DAT " THYRISTOR RECTIFIER 1-PHASE "
BEGIN NEW DATA CASE
C - - miscellaneous card
50.00E-6 75.0E-3
```

```
     1000      1
C - - BEGIN CONTROLLER DESCRIPTION
C - - use models for controller
MODELS
INPUT VSA {V(VSA)}
OUTPUT PLS1_2, PLS3_4
$INCLUDE,CONT.MOD
USE CONTL AS CONTL1
   INPUT VA:=VSA                          -- inputs from models to model
   OUTPUT PLS1_2:=PLS12, PLS3_4:=PLS34    -- outputs from model to models
ENDUSE
RECORD                                    -- to plot variables from models
   CONTL1.RAMP1 AS RAMP1
   CONTL1.COMP AS COMP
   CONTL1.DCMP AS DCMP
   CONTL1.PLS12 AS PLS12
   CONTL1.PLS34 AS PLS34
   CONTL1.RGEN1.MAGN AS MAGN
ENDRECORD
ENDMODELS
C - - BEGIN ELECTRICAL NETWORK DESCRIPTION
C - - branch data
C - - ac side of the rectifier
C - - Source inductance
00VSA   VMA                 0.2                                       1
C - - Rectifier ac-side inductance
00VMA   VA                  1.0
C - - dc side of the rectifier
00POSP  POS                 16.0                                     3
00POS   NEG           2.0                                            2
00POSP  NEG           1.0E+9        {for plotting only}              2
C - - snubbers (next 4 records)
00POSP  VA                  200.0     1.0
00POSP        POSP  VA
00VA    NEG   POSP  VA
00      NEG   POSP  VA
C - - Negligible resistances in series with thyristors 1 and 3 to avoid
C - - creating a loop of closed switches during current commutation
C - - from one thyristor pair to the next
00POSA  POSP              0.01
00POSG  POSP              0.01
BLANK CARD ENDING BRANCHES
C - - specify the four thyristors
11VA    POSA                                        PLS1_2         13
11      POSG                                        PLS3_4         13
11NEG   VA                                          PLS3_4         13
11NEG                                               PLS1_2         13
BLANK CARD ENDING SWITCHES
C - - ac source description
14VSA    169.7    60.0        -90.0
BLANK CARD ENDING SOURCES
C - - request for node voltage outputs
  VSA   VMA    VA
BLANK ENDING NODE VOLTAGE OUTPUT
BLANK
BLANK


C - - FILE NAME = CONT.MOD
MODEL CONTL                               -- begin model
INPUT VA                                  -- input from models
OUTPUT PLS12, PLS34                       -- output to models
```

```
VAR PLS12, PLS34, VCONTL, RAMP1, COMP, DCMP
CONST ALPHA {VAL: 45}, TENDEG {VAL: 500E-6}, dlay {VAL: 8.33e-03}
HISTORY COMP (DFLT: 0}, PLS12 {DFLT: 0}
DELAY CELLS DFLT :400
INIT                                     -- initialization
  VCONTL:=ALPHA/180
ENDINIT
$INCLUDE,RAMPGEN.MOD
EXEC
  IF (VA>=0) THEN
    USE RGEN AS RGEN1
      OUTPUT ramp1:=RAMP
      DATA FREQ:=120, AMPL:=1            -- data for use by rampgen.mod
    ENDUSE
  ELSE
    RAMP1:=0
  ENDIF
  COMP:=(RAMP1>=VCONTL)
  DCMP:=DELAY(COMP, TENDEG)
  PLS12:-(NOT DCMP AND COMP)
  PLS34:-DELAY(PLS12,dlay)               -- generation of pulses
ENDEXEC
ENDMODEL                                 -- end cont1
C

C - - FILE NAME = RAMPGEN.MOD
MODEL RGEN                               -- begin rgen
OUTPUT RAMP                              -- output to cont.mod
DATA    FREQ, AMPL
VAR     RAMP, MAGN, PERIODS
INIT
  MAGN:=AMP*(FREQ)                       -- initialization
  RAMP:=0
  PERIODS:=1
ENDINIT
EXEC
IF (T>0) THEN                            -- ramp generation
  IF (T<(PERIODS/FREQ)) THEN
    RAMP:=(RAMP+(TIMESTEP*MAGN))
  ELSE
    RAMP:=0
    PERIODS:=PERIODS+1
  ENDIF
ENDIF
ENDEXEC
ENDMODEL                                 -- end rgen
C
```

A3. PSpice Program Listing for Figure 8-22

```
Indirect Field-Oriented Control of an Induction Motor

.PARAM  LM=0.06,    TR=0.05,    KT=0.95; motor parameters
.PARAM  LMEST=0.06, TREST=0.05, KTEST=0.95; estimated motor parameters
.PARAM  J=0.05,     B=0.001;    mechanical constants
.PARAM  FLREF=1.77, TLD=10.0;   reference values for rotor flux, load torque
.PARAM  DEL=1.0E-04;            small number for use in the integrators

* ideal current regulated converter
```

```
E_IS1  IS1  0   VALUE = {V(IS1REF)}
E_IS2  IS2  0   VALUE = {V(IS2REF)}
E_IS3  IS3  0   VALUE = {V(IS3REF)}
*****************************************************************************
* MOTOR MODEL
*****************************************************************************
* 3 phase to alpha-beta
E_ISALP  ISALP  0  VALUE = {V(IS1) - 0.5*(V(IS2)+V(IS3))}
E_ISBET  ISBET  0  VALUE = {SQRT(3)*(V(IS2)-V(IS3))/2}

* alpha-beta to d-q
E_ISD ISD  0  VALUE = {V(ISALP)*COS(V(TH)) + V(ISBET)*SIN(V(TH))}
E_ISQ ISQ  0  VALUE = {-V(ISALP)*SIN(V(TH)) + V(ISBET)*COS(V(TH))}

* flux, torque, slip and speed calculation

* function "sdt" evaluates the time integral
E_FLR   FLR   0   VALUE = {sdt({LM}*V(ISD)-V(FLR))/{TR}}
E_TQEM  TQEM  0   VALUE = {{KT}*V(FLR)*V(ISQ)}
E_WSL   WSL   0   VALUE = {{LM}*V(ISQ)/((V(FLR)+1E-12)*{TR})}
E_WR    WR    0   VALUE = {sdt(V(TQEM)-V(TQLD)-{B}*V(WR))/{J}}
E_TH    TH    0   VALUE = {sdt(V(WSL)+V(WR))}
E_THR   THR   0   VALUE = {sdt(V(WR))}
*****************************************************************************
* ESTIMATION OF MOTOR VARIABLES
*****************************************************************************
* 3 phase to alpha-beta
E_ISALPE  ISALPE  0  VALUE = {V(IS1) - 0.5*(V(IS2)+V(IS3))}
E_ISBETE  ISBETE  0  VALUE = {SQRT(3)*(V(IS2)-V(IS3))/2}

* alpha-beta to d-q
E_ISDE  ISDE  0  VALUE = {V(ISALPE)*COS(V(THE)) + V(ISBETE)*SIN(V(THE))}
E_ISQE  ISQE  0  VALUE = {-V(ISALPE)*SIN(V(THE)) + V(ISBETE)*COS(V(THE))}

* flux, torque, slip and speed calculation
E_FLRE   FLRE   0  VALUE = {sdt({LMEST}*V(ISDE)-V(FLRE))/{TREST}}
E_TQEME  TQEME  0  VALUE = {{KTEST}*V(FLRE)*V(ISQE)}
E_WSLE   WSLE   0  VALUE = {{LM}*V(ISQE)/((V(FLRE)+1E-6)*{TREST})}
E_THE    THE    0  VALUE = {sdt(V(WSLE)+V(WR))}
*****************************************************************************
* CONTROLLER
*****************************************************************************
* regulators for generating flux ref., speed ref., d-q axes current references
E_FLREF  FLREF  0       VALUE = {FLREF}
X_WREF   THREF  THR  WREF    PI_AW_OL PARAMS: MAXOUT=377 PI_GAIN=100 PI_WO=10
X_ISDREF FLREF  FLRE ISDREF  PI_AW_OL PARAMS: MAXOUT=60  PI_GAIN=100 PI_WO=10
X_ISQREF WREF   WR   ISQREF  PI_AW_OL PARAMS: MAXOUT=60  PI_GAIN=10  PI_WO=100

* d-q to alpha-beta
E_ISAREF  ISAREF  0  VALUE = {V(ISDREF)*COS(V(THE)) - V(ISQREF)*SIN(V(THE))}
E_ISBREF  SBREF   0  VALUE = {V(ISDREF)*SIN(V(THE)) + V(ISQREF)*COS(V(THE))}

* alpha-beta to 3 phase
E_IS1REF  IS1REF  0  VALUE = (2*V(ISAREF)/3)

E_IS2REF  IS2REF  0  VALUE = {(-V(ISAREF)/3) + (V(ISBREF)/SQRT(3))}
E_IS3REF  IS3REF  0  VALUE = {(-V(ISAREF)/3) - (V(ISBREF)/SQRT(3))}
*****************************************************************************
* position reference
V_THREF  THREF  0  PULSE(0  1.0V  0.05s  0.1ms  0.1ms  1.0s  2.0s)
* load torque reference
```

```
V_TQLD    TQLD    0  PULSE(0  {TLD}  0.3s   1.0ms 0.1ms  1.0s  2.0s)
*************************************************************************
* subcircuit definition of PI regulator with anti-windup and output limiting
.SUBCKT  PI_AW_OL POSIN NEGIN OUTPUT PARAMS: MAXOUT=1 PI_GAIN=1 PI_W0=1
+        DEL=1E-04
*        name     in+   in-   out
* NOTE: All voltages are with respect to GND (0)
E_IN    SUM_IN   0  VALUE = {V(POSIN)-V(NEGIN)}
* switch at input of the integrator for anti-windup and its control voltage
S_INT   SUM_IN   INTEG_IN  SW_CTRL  0  SMOD
E_SW    SW_CTRL  0  VALUE = {MAXOUT-ABS(V(PIOUT))}
* PI with output limiting
E_INTOUT  INTEG_O  0  VALUE = {PI_GAIN*PI_W0*sdt(V(INTEG_IN))}
E_PIOUT   PIOUT    0  VALUE = {PI_GAIN*V(SUM_IN) + V(INTEG_O)}
E_LIMOUT  OUTPUT   0  VALUE = {LIMIT(V(PIOUT),-{MAXOUT},{MAXOUT})}
R_INTIN   INTEG_IN 0  1ohm

.MODEL  SMOD  VSWITCH [RON=0.001 ROFF=0.01MEG VON=0.02 VOFF=0.0]
.ENDS
*************************************************************************

.TRAN 0.5s 0.5s 0 0.1ms
.PROBE   V([ISDREF])  V([ISQREF])  V([THREF])  V([THR])  V([TH])  V([FLR])
+        V([FLRE])  V([TQEM])  V([WR])
.END
```

A4.   ATP Program Listing for Figure 8-22

```
C Indirect Field-Oriented Control of an Induction Motor
C Please note that the variable names used in this program may differ from the
C ones used in the text
BEGIN NEW DATA CASE
C 3456789012345678901234567890123456789012345678901234567890123456789012345678
C MISCELLANEOUS DATA CARDS (next two): Specify time step (50e-6s), final time
C (5e-1s), printout  frequency (once every 10000 points),  plot frequency (once
C every 3 points) and create a plot file ('1' in column 64 of second data card)
50.00E-6  5.0E-1
   10000        3                                                    1
C
TACS STAND ALONE
C - define load torque and position references
98THREF   = 1.0*(TIMEX .GE. 0.1) { step change at time t=0.1s
98TQLD    = 4.0*(TIMEX .GE. 0.3) { step change at time t=0.3s
C - constants for PI regulators; the letters A, B & C stand for the speed, ...
C - d-axis current & q-axis current regulators respectively
98MAXA    = 377
98GAINA   = 100
98W0A     = 10
98MAXB    = 60
98GAINB   = 100
98W0B     = 10
98MAXC    = 60
98GAINC   = 10
98W0C     = 100
C - ideal current regulated converter
98IS1     = IS1REF
98IS2     = IS2REF
98IS3     = IS3REF
C *********************************************************************
C MOTOR MODEL
C - 3 phase to alpha-beta
```

```
98ISALP   = IS1 - 0.5*(IS2+IS3)
98ISBET   = SQRT(3)*(IS2 - IS3)/2
C alpha-beta to d-q
98ISD     =  ISALP*COS(THRF) + ISBET*SIN(THRF)
98ISQ     = -ISALP*SIN(THRF) + ISBET*COS(THRF)
C - flux, torque, slip and speed calculation
 1FLR     +ISD                                       .060  { LM=0.060 }
 1.0
 1.0       0.05                                             { TR=0.050 }
98TQEM    = 0.95*FLR*ISQ                                    { KT=0.950 }
98WSL     = 0.06*ISQ/(FLR*0.05)
 1WR      +TQEM   -TQLD                               1.00
 1.0
 0.001     0.05                                             { B=0.001, J=0.050 }
 1THRF    +WR     +WSL                                1.00
 1.0
 0.0       1.0
 1THR     +WR                                        1.00
 1.0
 0.0       1.0
C
************************************************************************
C ESTIMATION OF MOTOR VARIABLES
C - 3 phase to alpha-beta
98ISALPE  = IS1 - 0.5*(IS2+IS3)
98ISBETE  = SQRT(3)*(IS2 - IS3)/2
C - alpha-beta to d-q
98ISDE    =  ISALPE*COS(THRFE) + ISBETE*SIN(THRFE)
98ISQE    = -ISALPE*SIN(THRFE) + ISBETE*COS(THRFE)
C flux, torque, slip and speed calculation
 1FLRE    +ISDE                                      .060  { LMEST=0.060 }
 1.0
 1.0       0.05                                             { TREST=0.050 }
98TQEME   = 0.95*FLRE*ISQE                                  { KTEST=0.950 }
98WSLE    = 0.06*ISQE/(FLRE*0.05)
 1THRFE   +WR     +WSLE                               1.00
 1.0
 0.0       1.0
C ************************************************************************
C - CONTROLLER; the PI regulators have anti-windup and output limiting
98FLREF   = 1.77
C - PI regulator for generating speed reference
98IINA    = (THREF-THR)*(ABS(PIOUTA) .LE. MAXA)
 1INTA    +IINA                                       1.00
 1.0
 0.0       1.0
98IOUTA   = GAINA*WOA*INTA
98PIOUTX  = GAINA*(THREF-THR) + IOUTA
 PIOUTA   +PIOUTX                                     1.00
98SIGNA   = SIGN(PIOUTA-MAXA)*(ABS(PIOUTA) .GT. MAXA)
98WREF 60-MAXA   +PIOUTA +MAXA                        0.0              SIGNA
C - PI regulator for generating Isdref
98IINB    = (FLREF-FLRE)*(ABS(PIOUTB) .LE. MAXB)
 1INTB    +IINB                                       1.00
 1.0
 0.0       1.0
98IOUTB   = GAINB*WOB*INTB
98PIOUTY  = GAINB*(FLREF-FLRE) + IOUTB
 PIOUTB   +PIOUTY                                     1.00
98SIGNB   = SIGN(PIOUTB-MAXB)*(ABS(PIOUTB) .GT. MAXB)
98ISDREF60-MAXB   +PIOUTB +MAXB                       0.0              SIGNB
```

```
C - PI regulator for generating Isqref
98IINC    = (WREF-WR)*(ABS(PIOUTC) .LE. MAXC)
 1INTC    +IINC                                        1.00
 1.0
 0.0       1.0
98IOUTC   = GAINC*WOC*INTC
98PIOUTZ  = GAINC*(WREF-WR)+IOUTC
  PIOUTC  +PIOUTZ                                       1.00
98SIGNC   = SIGN(PIOUTC-MAXC)*(ABS(PIOUTC) .GT. MAXC)
98ISQREF60-MAXC   +PIOUTC +MAXC                        0.0              SIGNC
C - d-q to alpha-beta
98ISAREF  = ISDREF*COS(THRFE) - ISQREF*SIN(THRFE)
98ISBREF  = ISDREF*SIN(THRFE) + ISQREF*COS(THRFE)
C - alpha-beta to 3 phase
98IS1REF  = 2*ISAREF/3
98IS2REF  = (-ISAREF/3)+(ISBREF/SQRT(3))
98IS3REF  = (-ISAREF/3)-(ISBREF/SQRT(3))
C REQUEST PLOTTING OF TACS VARIABLES
33THREF THR   THRF  THRFE WREF  WR    FLR   FLRE
33ISD   ISQ   ISDREFISQREFTQEM
BLANK ENDING TACS
BLANK DENOTING END OF FILE
```

Kouhei Ohnishi
Nobuyuki Matsui
Yoichi Hori

Chapter 9

Estimation, Identification, and Sensorless Control in AC Drives

9.1. INTRODUCTION

Since the early years of the twentieth century, electric motors for variable speed drives have been widely applied in large-capacity applications such as the steel industry and the automobile industry. In the early stage, DC motors were widely used for adjustable speed control. Since the late 1960s, however, AC motors have been replacing DC motors in a wide area of industry applications. Since AC drives required more complicated controllers in the beginning stage, they were not so economically feasible and did not meet with wide acceptance. However, allied to advances both in digital control technology and power semiconductor devices, AC drives became more and more economical and popular. In almost all areas, DC drives are now replaceable with AC drives.

However, there still exist some areas which are not suitable for AC drive applications. One of these is the area of applications, which requires precise torque control. For instance, injection machines need accurate torque control at a very low speed or in a standstill state. AC motors sometimes generate torque error or torque pulsation due to some parameter variations. To overcome such problems, more sophisticated techniques are necessary in the controller. These techniques employ the recent developments in digital control, including high-speed digital signal processors (DSPs) and parallel processing and are based on estimation or identification of motor parameters. A description of the recent advances in such areas with a focus on estimation and system identification is given in this chapter. The results are not only reflected in control design itself but also directly used to dispense with mechanical sensors. The chapter first introduces the electrical aspects of AC motor drives, emphasizing parameter estimation, flux identification, and speed estimation based on various methods, including self-tuning regulators, model reference adaptive systems, and so on. Important applications are the drives of induction motor and brushless motor without speed sensor. Theoretical analyses based on the physical viewpoint are presented, and the associated experimental results are shown. The chapter also describes the design of robust motion controllers which take mechanical aspects into account. By integrating these two aspects (electrical and mechanical), versatile applications will be possible. At the end of the chapter, a summary of the state of the art is given.

9.2. PARAMETER ESTIMATION IN AC DRIVES

9.2.1. Parameter Identification in Brushless Motors

Parameters of Brushless Motors. The control scheme of brushless motors with trapezoidal flux distributions (BLDM, brushless DC motor) is relatively simple. Usually it does not need parameter identification. Generally identification of the parameters is necessary for precise control of brushless motors with sinusoidal flux distribution (PMSM, permanent magnet synchronous motor), as less torque pulsation is required. Particularly such a method is employed for fine torque control. From the control viewpoint, the brushless motor has three electric parameters. The first is armature resistance, the second is armature inductance, and the third is EMF coefficient. They are significant parameters to be identified.

Two effective approaches are presented here. One is self-tuning regulator (STR), which has a tuning ability to make output-error zero inside the controller; the other is model reference adaptive system (MRAS), which has a referred model in the controller. It is important to note that direct applications of STR and MRAS to parameter identification do not always lead to successful results, because of the limitation of the processing time of the controller CPU. Since identification should be performed in parallel with current and speed control, it is essential to reduce the processing time for identification by a simple algorithm.

STR-Based Parameter Identifier. At steady-state, the PMSM has the simple equivalent circuit just like a DC motor. Terminal voltage, line current, and armature resistance are measured to identify the circuit parameters. Figure 9-1 shows an experimental evaluation of the influence of such parameter variations in armature current error at steady state. A current-regulated voltage source inverter supplies almost sinusoidal current. In the figure, the ordinate is the current control error due to the parameter variation and the parameter variation coefficient is defined as

$$K = \frac{\text{motor parameter}}{\text{controller parameter}} \tag{9.1}$$

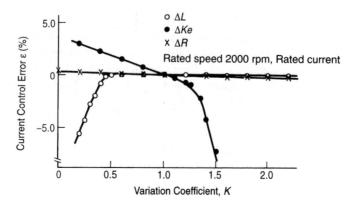

Figure 9-1. Experimental estimation of armature current error in steady state via parameter variations.

The results show that variations of the armature inductance and the EMF coefficient give rise to significant error in the performance, while the armature resistance variation does not. This means that the resistance drop is considerably smaller than both inductance drop and counter EMF. Figure 9-2 shows the STR-based identifier for armature inductance and EMF coefficient with associated experimental results. Here the inputs of the identifier are applied voltage and armature current. The armature current is obtained through a current sensor; however, the applied voltage is calculated using DC link voltage, PWM pattern, and dead-time information. To simplify the identification algorithm and save computation time, the following relations are used to identify the armature inductance and the EMF coefficient,

$$\hat{L} = \frac{v_d(n-2) - R_a i_d(n-2)}{i_d(n-1) - i_d(n-2) + \omega_r T i_q(n-2)} T \tag{9.2}$$

$$\hat{K}_e = \frac{v_q(n-2) - R i_q(n-2) - \dfrac{\hat{L}}{T}\{i_q(n-1) - i_q(n-2)\}}{\omega_r} + \hat{L} i_d(n-2) \tag{9.3}$$

where voltage and current transformed to the d–q axis are used and T and $\dot{\theta}(=\omega_r)$ are the current control period and motor speed, respectively. The estimation process is repeated 256 times, and the estimated values are averaged to avoid noises involved in calculations. Figure 9-3 shows the estimation errors against various load and speed conditions. From equation 9.2, since the estimation of the armature inductance uses division and the relatively small d-axis voltage, the error is a little larger compared to that of the EMF coefficient.

MRAS-Based Parameter Identifier. Figure 9-4 displays a MRAS-based identification approach, which includes a voltage-based motor model as a reference model. In addition, the figure shows the experimental results of identification of armature inductance and EMF coefficient for a tested motor. The input of the identifier is a current difference between model and actual motor. The current dif-

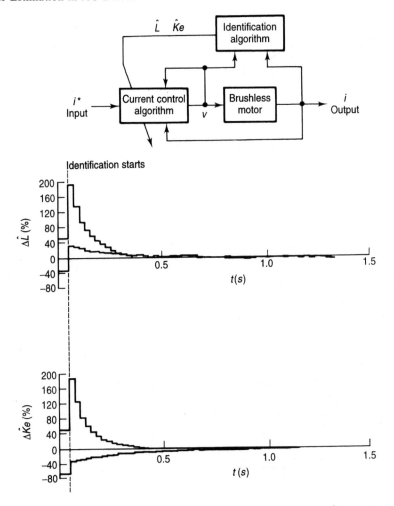

Figure 9-2. STR-based parameter identifier with experimental results. $\hat{L}$ is estimated armature inductance, $\hat{K}_e$ is estimated EMF constant.

ference is decomposed into two elements, from which the armature inductance and the EMF coefficient are identified. The identification algorithm is as follows. Rearranging the d–q axis voltage equation of the brushless motor, the reference model is given by the following equation.

$$\hat{i}(n-1) = \frac{T}{\hat{L}(n-2)}\left[v(n-2) - Ri(n-2)\right]$$

$$+ \; \omega_r(n-2)T \begin{bmatrix} -i_q(n-2) \\ i_d(n-2) - \hat{K}_e(n-2)/\hat{L}(n-2) \end{bmatrix} + i(n-2) \quad (9.4)$$

The current difference between the model and the actual motor is

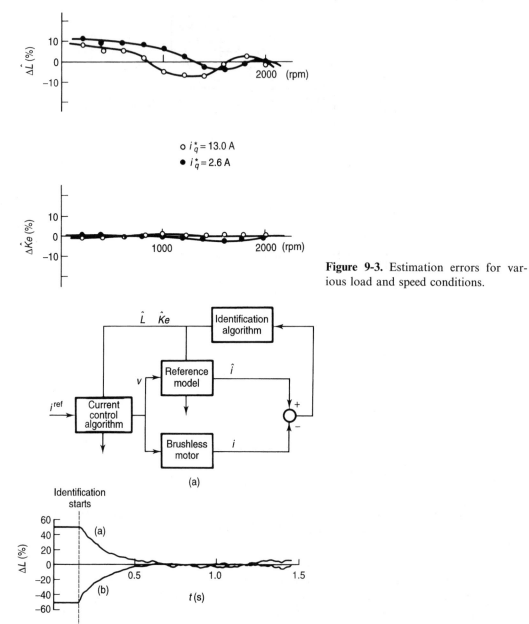

Figure 9-3. Estimation errors for various load and speed conditions.

Figure 9-4. MRAS-based parameter identifier with experimental results.

$$\Delta i(n-1) = \omega_r(n-2)T \begin{bmatrix} 0 \\ K_e/L - \hat{K}_e(n-2)/\hat{L}(n-2) \end{bmatrix}$$
$$+ T \left[\frac{1}{\hat{L}(n-2)} - \frac{1}{L} \right] [v(n-2) - Ri(n-2)] \qquad (9.5)$$

From the d-axis component of current, the d-axis current difference is

$$\Delta i_d(n-1) = T[v_d(n-2) - Ri_d(n-2)] \left[\frac{1}{\hat{L}(n-2)} - \frac{1}{L} \right] \qquad (9.6)$$

The current difference of the q-axis component is given under the assumption that the armature inductance could be identified as $L = \hat{L}(n-1)$ by using equation 9.5.

$$\Delta i_q(n-1) = \frac{T}{L}\dot{\theta} \, [K_e - \hat{K}_e(n-2)]. \qquad (9.7)$$

Using these two equations, the identification algorithm is summarized as

$$\begin{bmatrix} \hat{L}(n-1) \\ \hat{K}_e(n-1) \end{bmatrix} = K_P \, \mathrm{sgn} \, [A(n-2)]\Delta i(n-1) + K_I \sum_{k=1}^{n-1} \mathrm{sgn} \, [A(k-1)]\Delta i(k) \qquad (9.8)$$

where

$$A(k) = \begin{bmatrix} T\{v_d(n-2) - Ri_d(n-2)\} & 0 \\ 0 & \dot{\theta}(n-2) \end{bmatrix} \qquad (9.9)$$

and K_P and K_I are the gain matrices, respectively. Figure 9-5 shows the experimental estimation error under the various load and speed conditions. In this case, the estimation error of the EMF coefficient is larger compared to that of the armature inductance, particularly in low-speed range.

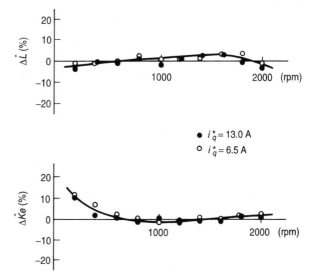

Figure 9-5. Experimental estimation error under various load and speed conditions.

Application of Parameter Identification to Torque Control. An interesting application example for the parameter identification of the brushless motor is a "torque sensorless" torque control. In Figure 9-6a, the conventional current-based torque control system is shown. The torque reference is divided by a torque constant

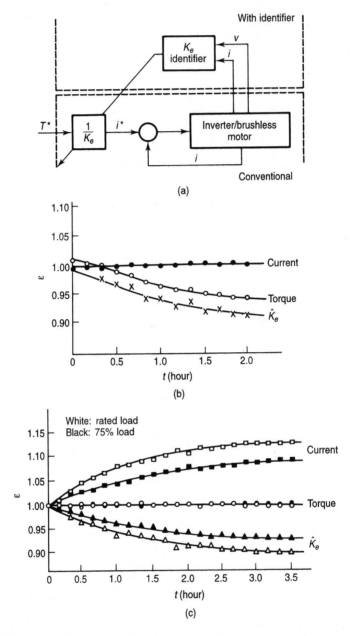

Figure 9-6. Torque sensorless torque control system and control characteristics.

(= EMF coefficient in SI unit) to generate a current reference. In Figure 9.6b, the abscissa is the operating time of motor with load which corresponds to temperature rise. With temperature rise, the torque constant decreases due to negative temperature coefficient of the permanent magnet. Since the armature current is controlled to be constant due to the current minor loop, the generated torque also decreases in proportion to torque constant as shown in the figure. However in Figure 9.6c, the identified torque constant is used to modify the current reference to compensate for decrease of the torque constant of the motor. As a result, the torque is maintained constant with an accuracy of less than 1% against temperature rise. To improve accuracy, the voltage calculation should be more precise by taking into account the turn-on and turn-off time of the switching devices and current dependency of the turn-off time, and so on.

9.2.2. Parameter Identification in Induction Motors

Parameters of Induction Motors. Basically induction motor in steady state is represented by the equivalent circuit in Figure 9-7. The classical no-load test, locked rotor test, and electrical quantity measurement test give identified parameters in Figure 9-7.

Recent computer technology makes it possible to carry out these tests on-line and in a real-time manner. In the process, as it is requisite to use fundamental components of voltage and current for the identification process, Fourier series expansion is usually used. Also some special methods have been proposed to measure electric quantities. For instance, to measure the stator resistance, the inverter is operated as a chopper mode. A sample of flowchart is shown in Figure 9-8 [4].

A direct application of this steady-state approach is an automatic boosting function at low-speed in V/f inverter-supplied induction motor drive. Figure 9-9 shows experimental results where the effectiveness of the boosting function in a low-speed range is observed [4]. The dead-time compensation of the inverter also prevents torque in the low-speed range from decreasing, since the dead time gives the effect in fundamental component.

Parameter Identification in Vector-Controlled Induction Motors. Vector-controlled induction motor is one of the promising driving actuators. There are two types of vector control in induction motors, and they can be specified as follows:

1. Field orientation control type
2. Slip frequency control type.

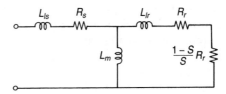

Figure 9-7. Equivalent circuit of caged induction motor without iron loss.

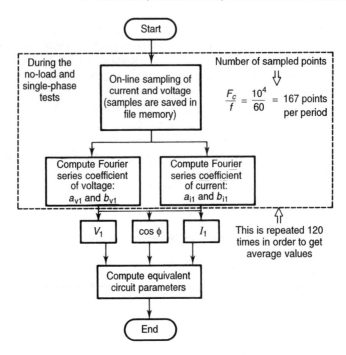

Figure 9-8. Flow chart of autonomous identification of induction motor parameters.

There are some direct orientation feedback loops in the former method, which is sometimes called direct vector control. A Hall device was used for the purpose in the first stage. On the contrary, the magnetic flux vector is not explicitly oriented in the latter method. The phase and the magnitude of magnetic flux are regulated through slip frequency control, and the latter is sometimes classified into indirect control. Since slip frequency control type is an inherently open loop control, the variation of electric parameters, particularly rotor resistance, gives significant effect to the performance. There have been many papers on the estimation of rotor resistance or rotor time constant for vector-controlled induction motors with shaft encoders. Most of the papers have been based on the LMS (least mean square) or similar approach. As shown in an early paper [15], PRBS (pseudo-random binary sequence) added to d-axis current reference is effective for the well convergence of the estimated rotor time constant. It will be shown later that such kind of persistently excited (PE) condition of d-axis current is indispensable for sensorless drive since rotor speed and rotor resistance cannot be estimated simultaneously.

Not only the secondary resistance but other parameters are also estimated. Holtz and coworkers proposed and realized a self-commissioning scheme for vector-controlled induction motor drive, where 80196 microcontrollers with ASIC are employed to identify the parameters, such as stator resistance, stator transient time constant, rotor time constant, rotor magnetizing current, and mechanical time constant [5]. Following his approach, several commercially available inverters have been suppled with such functions.

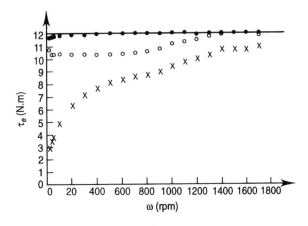

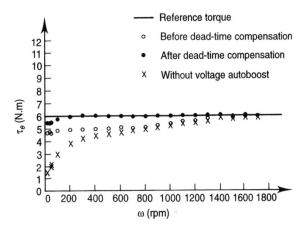

Figure 9-9. Experimental torque control characteristics (2.2 kW, four-pole induction motor).

Flux Estimation in Vector-Controlled Induction Motors. From the field orientation control view, many papers have been proposed to estimate the rotor flux. In general, the caged induction motor represented in Figure 9-7 has a dynamical system equation as

$$\begin{bmatrix} \dot{\boldsymbol{i}}_s \\ \dot{\boldsymbol{\Psi}}_r \end{bmatrix} = \begin{bmatrix} \boldsymbol{A}_{11} & \boldsymbol{A}_{12} \\ \boldsymbol{A}_{21} & \boldsymbol{A}_{22} \end{bmatrix} \begin{bmatrix} \boldsymbol{i}_s \\ \boldsymbol{\Psi}_r \end{bmatrix} + \begin{bmatrix} \boldsymbol{B}_1 \\ 0 \end{bmatrix} \boldsymbol{v}_s$$

$$\boldsymbol{i}_s = \begin{bmatrix} \boldsymbol{I} & 0 \end{bmatrix} \begin{bmatrix} \boldsymbol{i}_s \\ \boldsymbol{\Psi}_r \end{bmatrix} \tag{9.10}$$

where

$$\boldsymbol{A}_{11} = -\left\{ \frac{R_s}{\sigma L_s} + \frac{R_r(1-\sigma)}{\sigma L_r} \right\} \boldsymbol{I}$$

$$\boldsymbol{A}_{12} = \frac{L_m}{\sigma L_s L_r} \left\{ \frac{R_r}{L_r} \boldsymbol{I} - \omega_r \boldsymbol{J} \right\}$$

$$A_{21} = \frac{L_m R_r}{L_r} I$$

$$A_{22} = -\frac{R_r}{L_r} I + \omega_r J$$

$$B_1 = \frac{1}{\sigma L_s} I$$

$$\sigma = 1 - \frac{L_m^2}{L_s L_r}$$

$$I = \begin{bmatrix} 1 & 0 \\ 0 & 1 \end{bmatrix}, \quad J = \begin{bmatrix} 0 & -1 \\ 1 & 0 \end{bmatrix}$$

Since this equation is observable, it is possible to construct an observer to estimate the rotor flux. The minimum order observer is derived from the well-known Gopinath's method.

$$\dot{\hat{\Psi}}_r = A_{21} i_s + A_{22} \hat{\Psi}_r + G[\dot{i}_s - (A_{11} i_s + A_{12} \hat{\Psi}_r + B_1 v_s)] \tag{9.11}$$

The first two terms on the right-hand side shows the "simulation" of flux circuit, and the last on the right-hand side is a "correction" term. Figure 9-10 shows a realization.

Flux estimation error denoted by **e** is

$$\dot{e} = (A_{22} - GA_{12})e$$
$$= -He \tag{9.12}$$

Eigenvalues of **H** determine the error dynamics. The parameter variation of the induction motor gives transient error in flux observation. It is proven that sensitivity to the rotor resistance variation is minimized if **H** is skew symmetrical as

$$H = \alpha I - \beta J \tag{9.13}$$

Here $-\alpha \pm j\beta$ are the allocated poles of flux observer. α and β are

$$\alpha = \sqrt{\left(\frac{R_r}{L_r}\right)^2 + W^2 \left[\left(\frac{R_r}{L_r}\right)^2 + \omega_r^2\right]}$$

$$\beta = -\omega_r \tag{9.14}$$

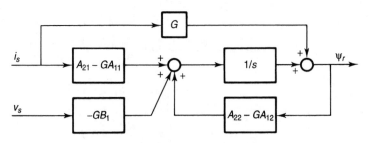

Figure 9-10. Flux observer realization.

W is a weighting coefficient to be determined by taking maximum variation ratio of rotor resistance ε, as

$$W \leq \sqrt{\left(1 + \frac{1}{\varepsilon}\right)^2 - 1} \tag{9.15}$$

For instance, for $\varepsilon = 0.3$, the recommended value of W is less than 3 or 4. Figure 9-11 is also effective in minimizing other parameter variations. When the estimated flux is fed back, this approach is a kind of an extension of original vector control by Blaschke [14]. Figure 9-11 is an example of a realization. The flux observer has a function similar to a flux detector.

9.3. SENSORLESS DRIVES OF AC MOTORS

Basically, the vector-controlled AC motors require speed or position sensors. However, these sensors bring several disadvantages from the standpoint of drive cost, reliability, machine size, and noise immunity. For these reasons, it is necessary to achieve the precise control of torque and speed without using position and speed sensors, that is, so-called sensorless drives of AC motors. In this chapter, first sensorless drive of brushless motors is described, then the induction motor is considered.

9.3.1. Sensorless Drives of Brushless Motors

As stated, there are two kinds of brushless motors: the motor with a trapezoidal flux distribution and that with a sinusoidal flux distribution. The approaches to sensorless drive of the brushless motor vary, depending on the rotor flux distribution. The brushless motor with a trapezoidal rotor flux distribution provides an attractive candidate, because two of the three stator windings are excited at a time. As a result, the unexcited winding can be used as a sensor [10, 11]; that is,

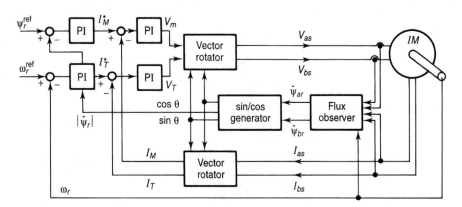

Figure 9-11. Flux observer–based field orientation.

the speed EMF induced in the unexcited winding is used to determine the rotor position and speed.

On the contrary, the brushless motor with a sinusoidal flux distribution excites three windings at a time and the sensorless control algorithm becomes complicated. Figure 9-12 shows an analytical model of a brushless motor where the d–q axis corresponds to an actual rotor position and the γ–δ axis is a fictitious rotor position. Since the actual rotor position is not known without a position sensor, the aim is to make the angular difference $\Delta\theta$ between the fictitious and actual rotor positions converge to zero.

Two approaches have been proposed. Both are the estimation of the angular difference by using the detected state variables and the estimated state variables which are obtained from a motor model in the controller. The approaches differ according to the motor model, that is,

- Voltage model-based drive [12]
- Current model-based drive [13]

These two are basically the model-based control and generally require on-line identification of the motor parameters if higher performance is required. However, it is interesting to note that the second method has robust control characteristics against the motor parameter variation.

In the voltage model-based sensorless drive, the voltage equation is given as follows, where P is the differential operator.

$$\begin{bmatrix} v_\gamma \\ v_\delta \end{bmatrix} = \begin{bmatrix} R + PL & -L\dot\theta_c \\ L\dot\theta_c & R + PL \end{bmatrix} \begin{bmatrix} i_\gamma \\ i_\delta \end{bmatrix} + K_E\dot\theta \begin{bmatrix} -\sin\Delta\theta \\ \cos\Delta\theta \end{bmatrix} \tag{9.16}$$

On the other hand, the voltage equation under the ideal condition that the fictitious and actual axes are coincident is

$$\begin{bmatrix} v_{\gamma M} \\ v_{\delta M} \end{bmatrix} = \begin{bmatrix} R + PL & -L\dot\theta \\ L\dot\theta & R + PL \end{bmatrix} \begin{bmatrix} i_\gamma \\ i_\delta \end{bmatrix} + K_E\dot\theta \begin{bmatrix} 0 \\ 1 \end{bmatrix} \tag{9.17}$$

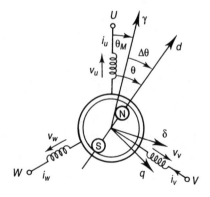

Figure 9-12. Analytical model of brushless motor.

Taking a difference between γ-axis voltage assuming $\Delta\theta$ is small, the following relation is obtained.

$$\Delta v_\gamma = v_\gamma - v_{\gamma M} = -K_E \dot{\theta} \sin \Delta\theta \simeq -K_E \dot{\theta} \cdot \Delta\theta \qquad (9.18)$$

Since the voltage difference can be calculated by the actual applied voltage in equation 9.16 and the model voltage calculated from equation 9.17, the angular difference can be made to converge to zero by the following rule:

if $\Delta v_r > 0$, then $\dot{\theta}_c$ decreases

if $\Delta v_r < 0$, then $\dot{\theta}_c$ increases $\qquad (9.19)$

(for clockwise rotation)

The current model is given in equation 9.20.

$$P\begin{bmatrix} i_\gamma \\ i_\delta \end{bmatrix} = \frac{1}{L}\left\{ \begin{bmatrix} v_\gamma \\ v_\delta \end{bmatrix} - \begin{bmatrix} R & -L\dot{\theta}_c \\ L\dot{\theta}_c & R \end{bmatrix}\begin{bmatrix} i_\gamma \\ i_\delta \end{bmatrix} - K_E\dot{\theta}\begin{bmatrix} -\sin \Delta\theta \\ \cos \Delta\theta \end{bmatrix} \right\} \qquad (9.20)$$

Similarly, the current difference is calculated and the result is as

$$\begin{bmatrix} \Delta i_\gamma \\ \Delta i_\delta \end{bmatrix} = \begin{bmatrix} i_\gamma - i_{\gamma M} \\ i_\delta - i_{\delta M} \end{bmatrix} = \frac{K_E T}{L}\begin{bmatrix} \dot{\theta}\sin \Delta\theta \\ -\dot{\theta}\cos \Delta\theta + \dot{\theta}_M \end{bmatrix} \simeq \frac{K_E T}{L}\begin{bmatrix} \dot{\theta}\Delta\theta \\ -\Delta\dot{\theta} \end{bmatrix} \qquad (9.21)$$

where T is a sampling period and $\dot{\theta}_M$ is the motor speed of the model. Equation 9.21 means that the current errors of each component of current correspond to position and speed errors, respectively. Therefore, the following algorithm is obtained.

$$\dot{\theta}_M = -K_\theta \int \Delta i_\delta \, dt \qquad (9.22)$$

$$\theta_c = \int (\dot{\theta}_M + K_\theta \Delta i_\gamma) dt \qquad (9.23)$$

Figure 9-13 shows torque-speed characteristics under a current model-based algorithm. The motor rating is 1.2 (kW), 6-poles, 1200 (rpm), 98 (kgf cm). The maximum speed is 1500 (rpm), the minimum speed is 60 (rpm), and a steady-state maximum speed error is within 0.4%.

9.3.2. Sensorless Drives of Vector Controlled Induction Motors

There have been many reports on the speed sensorless drive of the vector-controlled induction motors [6]. Various approaches have been proposed where the basic idea is estimation of speed by using applied voltage, line current, and frequency.

Slip frequency control approach is relatively simple, as shown in Figure 9-14. Here, the inverter frequency is controlled so that the vector control conditions are satisfied by estimating a slip frequency from the stator current transformed to the synchronously rotating coordinates (d-q axis) system. The slip frequency is given by

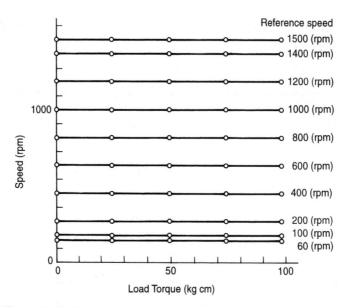

Figure 9-13. Torque-speed characteristics of sensorless brushless motor under current model-based control.

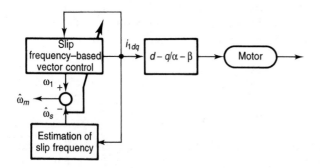

Figure 9-14. Basic schematic diagram of slip frequency–based sensorless control of induction motor.

$$\hat{\omega}_s = \frac{R_r}{L_r}\frac{i_{sq}}{i_{sd}} \tag{9.24}$$

and the motor speed is indirectly estimated by the inverter frequency and the estimated slip frequency as

$$\hat{\omega}_r = \omega - \hat{\omega}_s \tag{9.25}$$

In this approach, as the estimated slip frequency is directly fed back to the vector-controlled algorithm, the vector control and the speed estimation are coupled, and they should run simultaneously.

In the field orientation control approach, not only the speed but also the rotor flux are simultaneously estimated for the sensorless drive in a wide speed range. Schauder attacked this problem with MRAS [7]. His approach was based on the so-called voltage reference model where the speed terms are not explicitly included. The stator equation is used for correction of adjustable current reference model. The speed is estimated by a kind of error of flux components which is derived from Popov's stability criteria.

One modified implementation of this approach is shown in Figure 9-15, which is for application in the the low-speed range as well as mid- or high-speed ranges. The rotor flux observer based on the motor voltage model estimates the rotor flux by using the stator current transformed to the stationary coordinates (α–β axis) system, and the vector control is carried out on the basis of the estimated rotor flux while the motor speed is directly estimated through MRAS by using the flux and the stator current. Therefore, unlike the slip frequency–based approach, the estimated speed is used to adjust the motor model. The vector control can be decoupled with the speed estimation and be self-controlled. The speed is estimated according to the following relations.

$$\hat{\omega}_r = K_P \parallel \boldsymbol{e}_i \times \hat{\boldsymbol{\Psi}}_r \parallel + K_I \int \parallel \boldsymbol{e}_i \times \hat{\boldsymbol{\Psi}}_r \parallel dt \qquad (9.26)$$

$$\boldsymbol{e}_i = \boldsymbol{i}_s - \hat{\boldsymbol{i}}_s$$

It should be noted here that "$\times$" in equation 9.26 means the outer product. Since the speed estimation is based on the motor model, the parameter variation, especially the rotor resistance, has some effect on the speed estimation. With reference to this problem, Shin-naka clarified the impossibility of simultaneous estimation of both the speed and the rotor resistance theoretically [8].

From the voltage equation based on the stationary coordinates system, the relation

$$\frac{d}{dt} \parallel \boldsymbol{\Psi}_r \parallel^2 = 2\boldsymbol{\Psi}_r^T [-R_r \boldsymbol{i}_r + \omega_r \boldsymbol{J} \boldsymbol{\Psi}_r] \qquad (9.27)$$

$$= -2R_r \boldsymbol{\Psi}_r^T \boldsymbol{i}_r \qquad (9.28)$$

is obtained. Rearranging equations 9.27 and 9.28, equation 9.29 holds.

$$\begin{bmatrix} R_r \\ \omega_r \end{bmatrix} = [-\boldsymbol{i}_r \ \ \boldsymbol{J}\boldsymbol{\Psi}_r]^{-1} \frac{d}{dt} \boldsymbol{\Psi}_r \qquad (9.29)$$

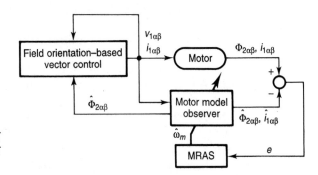

Figure 9-15. Basic schematic diagram of vector-controlled sensorless control of induction motor.

It is noted here that the following relation is obtained from basic equations of the vector-controlled induction motor.

$$\det\left[-i_r \quad J\Psi_r\right] = -\Psi_r^T i_r \tag{9.30}$$

By substituting equation 9.30 into equation 9.27, the following relation is derived.

$$\frac{d}{dt}\parallel \Psi_r \parallel^2 = 2R_r \det\left[-i_r \quad J\Psi_r\right] \tag{9.31}$$

Equation 9.31 and equation 9.29 mean that a simultaneous identification of the rotor resistance and the motor speed is possible only when the rotor flux is persistently time variant (PE condition). Under the vector control, the rotor flux is kept constant in principle for the orthogonality of rotor flux and rotor current. Then the simultaneous identification is theoretically impossible in the vector-controlled induction motor. This problem is overcome by adding small AC component to the d-axis current. The convergence is improved if such an AC component has a rich frequency spectrum like PRBS [15, 9].

Figure 9-16 shows an example of the sensorless control characteristics of 2.2 (kW), 4-poles induction motor. The maximum speed is 2400 (rpm) under the field-weakening control and the minimum speed is 15 (rpm). It is noted here that the speed control accuracy is 0.4% under the tuned condition; however, it is 1.4% under the detuned condition. Like the conventional vector control with sensors, the identification of the rotor resistance is important and difficult.

9.4. ROBUST MOTION CONTROL BY ESTIMATION OF MECHANICAL PARAMETERS

9.4.1. Estimation of Disturbance Torque

In general, the outputs of the motion system are position or force. There is a certain relational function between them as equation 9.32.

$$f = g(x) \tag{9.32}$$

In this equation, f is force applied to the mechanical system, and x is deviation by f. The control stiffness is defined as in equation 9.33.

$$\text{control stiffness} = \frac{\partial f}{\partial x} \tag{9.33}$$

Ideal force control has zero stiffness, and ideal position control has infinite stiffness. Any compliant or hybrid motion occupies the midway place between position and force control.

The robust controller should be both insensitive to external disturbances and parameter variations. In the motion system, the former characteristics correspond to a very high rejection capability against disturbance effects. As the external disturbance is the load, a robust motion controller should have an infinite control stiffness.

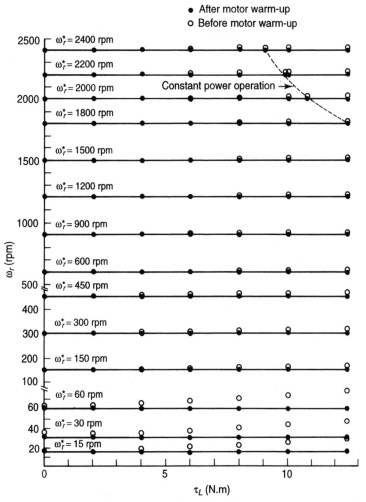

Figure 9-16. Example of control characteristics of stator flux–controlled sensorless vector control of induction motor.

To realize a versatile motion system whose control stiffness changes widely, the total motion system should have the double cascade structure of the acceleration reference generator to regulate total stiffness and the acceleration controller as shown in Figure 9-17.

To clarify the robust motion controller, at first, simple one-degree-of-freedom motion is analyzed. The dynamical equation is

$$J\frac{d\omega}{dt} + T_l = T_m \tag{9.34}$$

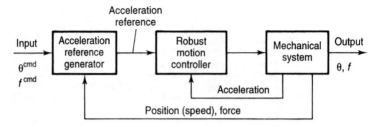

Figure 9-17. General structure of motion system based on robust control.

$$J = \text{inertia about motor shaft (kgm}^2)$$
$$T_l = \text{load torque (Nm)}$$
$$T_m = \text{motor torque (Nm)}$$

The load torque is the sum of inertial torque T_{int}, external torque T_{ext}, and friction torque T_{frc}. They are functions of position and/or time. The motor torque is the product of the generalized torque coefficient K_t corresponding with the magnetic flux by the generalized torque current I_a. Since the generalized torque current is assumed to be regulated by the high-gain current controller and the output current will completely coincide with its reference, the following equation holds.

$$T_m = K_t I_a = K_t I_a^{\text{ref}} \tag{9.35}$$

Combining equation 9.35 with equation 9.34, the following equation is obtained.

$$J\frac{d\omega}{dt} = K_t I_a^{\text{ref}} - (T_{\text{int}} + T_{\text{ext}} + T_{\text{frc}}) \tag{9.36}$$

In equation 9.36, the parameter variations denoted by Δ are shown in equation 9.37,

$$\begin{aligned} J &= J_n + \Delta J \\ K_t &= K_{tn} + \Delta K_t \end{aligned} \tag{9.37}$$

Using equation 9.37, the parameter variation and the load are treated in the torque dimension. The sum of both gives the disturbance torque.

$$\begin{aligned} T_{\text{dis}} &= T_l + \Delta Js\omega - \Delta K_t I_a^{\text{ref}} \\ &= T_{\text{int}} + T_{\text{ext}} + T_{\text{frc}} \\ &\quad + (J - J_n)s\omega + (K_{tn} - K_t)I_a^{\text{ref}} \end{aligned} \tag{9.38}$$

The basic dynamic equation 9.34 is transformed into equation 9.39 by equations 9.37 and 9.38.

$$T_{\text{dis}} = K_{tn}I_a^{\text{ref}} - J_n\frac{d\omega}{dt} \tag{9.39}$$

The left side of the equation is the sum of unknown factors, that is, the unpredictable load and the unknown parameter variation; however, the right side of the equation is

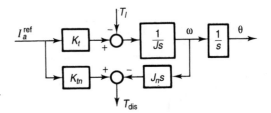

Figure 9-18. Calculation of disturbance torque based on acceleration.

known or detectable. Thus the disturbance torque can be calculated as shown in Figure 9-18.

As Figure 9-18 has a pure differentiation process, it is modified to be realizable as shown in Figure 9-19, where one low-pass filter (LPF) is inserted. Although any LPF is applicable, a simple first-order LPF is chosen here. In this case, the disturbance is estimated as

$$\hat{T}_{\text{dis}} = \frac{g}{s+g} T_{\text{dis}} \tag{9.40}$$

where g is a cutoff angular frequency of a first-order LPF. If g is large enough, the estimated disturbance torque is almost similar to the real one. Figure 9-19 is called a *disturbance observer*.

By direct feedback of the estimated disturbance torque shown in Figure 9-20, the modified diagram shown in Figure 9-21 is obtained. Figure 9-21 means that the disturbance has little effect on the motion system, since the feedback loop of disturbance is just the same as the feedforward effect of disturbance to cancel it. By attaching an auxiliary gain element J_n/K_{tn} in front of the current controller, it is clear that the physical meaning of the input of Figure 9-21 is acceleration. It is possible to extend such a robust motion control from one-degree-of-freedom systems to multi-degrees-of-freedom systems [17].

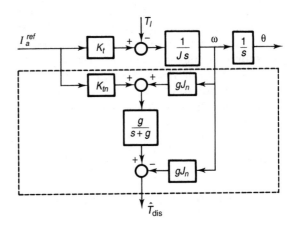

Figure 9-19. Calculation of disturbance torque by disturbance observer.

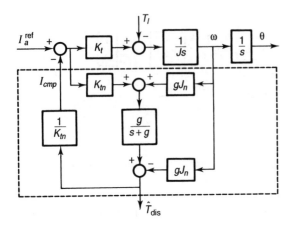

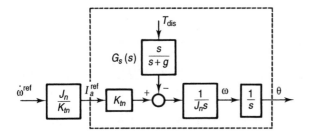

Figure 9-20. Feedback of estimated disturbance torque.

Figure 9-21. Acceleration controller by modifying Figure 9-20.

9.4.2. Estimation of Instantaneous Speed and Varied Inertia

For more accurate motion control, the instantaneous speed accuracy is very important. The incremental position encoder with a very short sampling time will lose resolution of the speed due to a small number of incremental pulses in a sampling period. On the contrary, in case of longer sampling time, resolution will be higher. However, the total motion system tends to be unstable. The instantaneous speed observer which is an expansion of the disturbance observer solves this antinomy; that is, accuracy is kept higher even in the case of a very short sampling time. Figure 9-22 shows a timing chart in such a case.

At the shorter sampling points of T_2 (represented by $k = 0, 1, 2, ..., K$), the total acceleration torque $T_{\text{mech}}[m, k]$ ($1 < k < K$) is given by the sum of the motor torque and the estimated disturbance torque as

$$\hat{T}_{\text{mech}}[m, k] = K_{tn}i[m, k] + \hat{T}_{\text{dis}}[m] \tag{9.41}$$

By integrating equation 9.41, the instantaneous speed at T_2 points can be estimated by

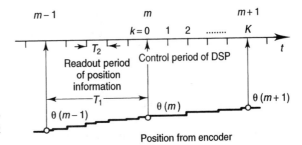

Figure 9-22. Timing chart of the instantaneous speed observer (position input type).

$$\hat{\omega}[m,k] = \hat{\omega}[m,k-1] + \frac{T_2}{2}\left(\frac{\hat{T}_{\text{mech}}[m,k] + \hat{T}_{\text{mech}}[m,k-1]}{J_n}\right) \tag{9.42}$$

The position is calculated by integrating equation 9.42 and the position error $\Delta\theta$ can be obtained at the next point, $k = K$, when reading out the counter.

$$\Delta\theta = \hat{\theta}[m,k] - \theta[m] \tag{9.43}$$

It is important to evaluate $\Delta\theta$ in the observer design. Suppose that $\gamma_1\Delta\theta$ $(0 < \gamma_1 < 1)$ is caused by $\Delta\omega_0$ (the error in the initial value of the estimated speed) and $\gamma_2\Delta\theta$ $(0 < \gamma_2 < 1)$ by $\Delta T_{\text{dis}\,0}$ (the error in the estimated disturbance in the section), namely

$$\gamma_1\Delta\theta = T_1\Delta\omega \tag{9.44}$$

$$\gamma_2\Delta\theta = \frac{T_1^2}{2J_n}\Delta T_{\text{dis}} \tag{9.45}$$

Based on equations 9.44 and 9.45, the values in the next section of $[m+1,k]$ are modified as follows before starting the speed estimation.

$$\hat{T}_{\text{dis}}[m+1] = \hat{T}_{\text{dis}}[m] - \Delta T_{\text{dis}} \tag{9.46}$$

$$\hat{\omega}[m+1,0] = \hat{\omega}[m,k] - \frac{T_1}{J_n}\Delta T_{\text{dis}} - \Delta\omega \tag{9.47}$$

The observer poles can be designed by γ_1 and γ_2 introduced in equations 9.44 and 9.45 [22]. The main advantage of the speed observer is the improvement of system stability by the equivalently reduced sampling period. By adding adaptive algorithm to instantaneous speed observer, the varied inertia is identified and the self-tuning regulator (STR) is realized.

At the mth sampling point of T_1 to read the encoder, the mechanical system's behavior is approximately given by

$$J\dot{\omega}[m] = K_{tn}i[m] + T_{\text{dis}}[m] \tag{9.48}$$

and

$$J\dot{\omega}[m-1] = K_{tn}i[m-1] + T_{\text{dis}}[m-1] \tag{9.49}$$

Since the variation in the torque coefficient K_t is included in the disturbance torque T_{dis}, a parameter K_{tn} is used in these equations. By subtracting equation 9.49 from equation 9.48, the following is obtained.

$$J\Delta\dot\omega[m] = K_{tn}\Delta i[m] + \Delta T_{\text{dis}}[m] \tag{9.50}$$

Equation 9.50 is the basic equation for inertia moment identification. It is important that $\Delta T_{\text{dis}}[m]$ can be assumed random Gaussian, if the disturbance torque is assumed to be constant in the neighboring sections of T_1. By summing up the equation error between the left and right terms of equation 9.50, the objective function to be minimized takes the form of

$$f(J) \equiv \sum_{m=1}^{M}(J\Delta\dot\omega[m] - K_{tn}\Delta i[m])^2 \to \min \tag{9.51}$$

By taking the partial differentiation of $f(J)$ by J, the estimation equation of the inertia moment is obtained by

$$\frac{\hat{J}}{K_{tn}} = \sum_{m=1}^{M}\frac{\Delta\dot\omega[m]\Delta i[m]}{\sum_{m=1}^{M}(\Delta\dot\omega[m])^2} \tag{9.52}$$

The actual calculation is performed by equation 9.53 replacing the oldest data with the newest data which comes into the rectangular window of equation 9.51.

$$\frac{\hat{J}}{K_{tn}}[m] = \sum_{j=m-M+1}^{m}\frac{\Delta\dot\omega[j]\Delta i[j]}{\sum_{j=m-M+1}^{m}(\Delta\dot\omega[j])^2} \tag{9.53}$$

The identified inertia moment is applied to the speed controller in a STR manner as is shown in Figure 9-23.

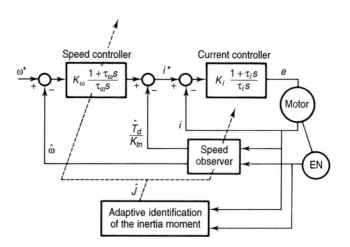

Figure 9-23. STR based on inertia identification.

Figure 9-24 shows the identification performance when the inertia moment varies from nominal value to three times of it. The varying inertia moment is estimated exactly with enough response time.

The systems described here led to improvement both in the robustness and preciseness in the motion systems. It is interesting that robust control is used for very fast improvement of control characteristics like disturbance rejection, and adaptive identification helps the robust control to increase the stability in a relatively slow mode [24].

9.5. CONCLUSION

A description of motion system is given in this chapter with an emphasis on both estimation and identification of parameters and control variables of AC motor–driven motion systems. In modern electrical drive systems, it is required to take not only the electrical aspect but also the mechanical ones into total system design.

Improvement in the electrical aspect needs various information pertaining to electrical machines and power electronic circuits of AC variable speed drives. Important techniques of identification or estimation of parameters and control variables in AC drives are explained. Such information includes machine parameters, flux, and so on. There is some theoretical limit of performance in the identification or estimation process. For the mechanical phase, the estimation of the disturbance torque, instantaneous speed, and varied inertia are described.

It is shown that the total robustness is attained by integrating the electrical improvement and the mechanical improvement. There are a wide variety of controllers based on combinations of these two aspects depending on applications. Further research is expected for total performance improvement.

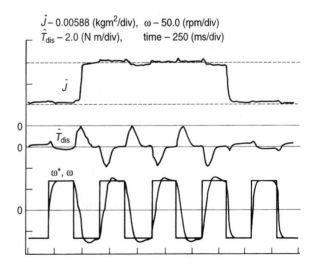

$\hat{J} - 0.00588$ (kgm^2/div), $\omega - 50.0$ (rpm/div)
$\hat{T}_{dis} - 2.0$ (N m/div),   time $- 250$ (ms/div)

$\hat{J}$

0

$\hat{T}_{dis}$

0

ω^*, ω

0

Figure 9-24. Identification performance of inertia moment.   ($T_2 = 100\,\mu s'$, $T_1 = 5\,ms'$, pole: $z = 0.6$, $M = 50$.) The inertia moment changes between J_n and $3J_n$.

Acknowledgment

The authors would like to express their thanks to B. K. Bose and A. Denker for their helpful suggestions in preparing this chapter.

References

[1] Bose, B. K., guest editor, "Power electronics and motion control—special issue," *Proc. IEEE*, Vol. 82, no. 8, 1994.

[2] Wang, C., D. W. Novotony, and T. Lipo, "An automated rotor time constant measurement system for indirect field-oriented drives," *IEEE Trans. Ind. Appl.*, Vol. 24, no. 1, 1988.

[3] Depenbrok, M., and N. R. Klaes, "Determination of induction machine parameters and their dependencies on saturation," *Conf. Rec. 1989 IEEE Ind. Appl. Soc.*, Part 1, 1989.

[4] Gastli, A., M. Iwasaki, and N. Matsui, "An automated equivalent circuit parameter measurements of an induction motor using a V/F PWM inverter," *Proc. IEEJ (Institute of Electrical Engineers of Japan), IPEC-Tokyo*, 1990.

[5] Khambadkone, A. M., and J. Holtz, "Vector-controlled induction motor drive with a self-commissioning scheme," *IEEE Trans. Ind. Elec.*, Vol. 38, no. 5, pp. 322–327, 1991.

[6] Nandam, P. K., G. F. Cummings, and W. G. Dunford, "Experimental study of an observer-based shaft sensorless variable speed drive," *Proc. 1991 IAS Annual Meeting*, Part 1, 1991.

[7] Schauder, C., "Adaptive speed identification for vector control of induction motors without rotational transducers," *Conf. Rec. 1989 IEEE Ind. Appl. Soc.*, Part 1, 1989.

[8] Shin-naka, S., "A unified analysis on simultaneous identification of velocity and rotor resistance of induction motor," *Trans. IEEJ*, Vol. 113-D, no. 12, 1993.

[9] Kubota, H., and K. Matsuse, "Simultaneous estimation of speed and rotor resistance of field oriented induction motor without rotational transducer," *Trans. IEEJ*, Vol. 112-D, no. 9, 1992.

[10] Iizuka, K., "Microcomputer control for sensorless brushless motor," *IEEE Trans. Ind. Appl.*, Vol. IA-21, 1985.

[11] Erdman, M. D., H. B. Harms, and J. L. Oldenkamp, "Electronically commutated DC motor for the appliance industry," *Conf. Rec. IEEE-IAS*, 1984.

[12] Matsui, N. , and M. Shigyo, "Brushless DC motor control without position and speed sensors," *Conf. Rec. IEEE-IAS*, 1990.

[13] Matsui, N., T. Takeshita, and K. Yasuda, "A new sensorless drive of brushless DC motor," *Conf. Rec. IEEE IECON '92*, 1992.

[14] Blaschke, F., "Das Prinzip der Feldorientierung, die Grundlage fur der Transvektor-Regulung von Drehfeldmaschinen," *Siemens-Z*, pp. 757–760, 1971.

[15] Leonhard, W., *Control of Electrical Drives*, Chap. 12, Springer-Verlag, Berlin, Germany, 1985.

[16] Ohishi, K., et al., "Torque-speed regulation of DC motor based on load torque estimation," *Int. Power Electr. Conf. IPEC-Tokyo*, Vol. 2, pp. 1209–1216, 1983.

[17] Nakao, M., et al., "A robust decentralized joint control based on interference estimation," *Proc. IEEE Int. Conf. Robotics and Automation*, Vol. 1, pp. 326–331, 1987.

[18] Murakami, T., et al., "Torque sensorless control in multidegrees-of-freedom manipulator," *IEEE Trans. Ind. Electr.*, Vol. 40, pp. 259–265, 1993.

[19] Ohnishi, K., et al., *Recent Advances in Motion Control*, Nikkan Kogyo Shimbun, Tokyo, 1990.

[20] Ohishi, K., et al., "Microprocessor-controlled DC motor for load-insensitive position servo system," *IEEE Trans. Ind. Electr.*, Vol. 34, pp. 44–49, 1987.

[21] Umeno, T., et al., "Two degrees of freedom controllers for robust servomechanism—Their application to robot manipulators without speed sensors," *IEEE 1st Int. Workshop Advanced Motion Control*, Yokohama, pp. 179–188, 1990.

[22] Hori, Y., et al., "An instantaneous speed observer for high performance control of DC servomotor using DSP and low precision shaft encoder," *4th European Conf. Power Electronics (EPE '91)*, Vol. 3, pp. 647–652, 1991.

[23] Awaya, I., et al., "New motion control with inertia identification function using disturbance observer," *Proc. IEEE IECON '92*, Vol. 1, pp. 77–80, 1992.

[24] Hori, Y., "Robust and adaptive control of a servomotor using low precision shaft encoder," *IEEE IECON '93*, Hawaii, 1993.

[25] Hori, Y., "Disturbance suppression on acceleration control type DC servo system," *Proc. IEEE PESC '88*, Vol. 1, pp. 222–229, 1988.

Chapter 10

Hoang Le-Huy

Microprocessors and Digital ICs for Control of Power Electronics and Drives

10.1. INTRODUCTION

The recent developments in electrical drive technology are motivated by the increasing requirements of industrial applications for higher performance, better reliability, and lower cost. They are due to the advances in several areas in particular power electronics, control theory, and microprocessor technology.

During the last two decades, power electronics has gained significant advances in several sectors. New power switches with better characteristics have been introduced (GTOs, MOSFETs, IGBTs), and new converter configurations as well as efficient commutation schemes (resonant converters) have been studied and experimented. Numerous advanced control algorithms for AC drives (self-controlled synchronous motor, field-oriented control, etc.) have been developed. Today, power electronic systems have attained an unusual high degree of complexity so that their control becomes more and more sophisticated.

The control of a power electronic system requires several functions of a different nature: signal filtering, regulation, drive signal generation, measurement, monitoring, protection, and so on. For a long time, the implementation of these functions has relied mainly on analog technology using a hardwired approach. Control circuits were built using operational amplifiers (op amps), nonlinear integrated circuits (ICs), and digital ICs. The last were used especially to implement sequential and combinatory logic functions in converter control circuits. The development of microprocessors has promoted the use of digital technology in the control of power electronic systems using a software approach that provides greater flexibility and better performance.

Microprocessors, introduced first in 1971, have evolved from simple 4-bit architecture with limited capabilities toward complex 64-bit architecture with tremendous processing power. The recent advances in microprocessor technology reside in several areas: processor, system architecture, and memory devices. Several advanced processors have been recently introduced such as reduced instruction set computing (RISC) processors, digital signal processors (DSPs), and transputers. High-performance computer architectures have been developed using cache memories, multiple bus architecture, pipeline structure, and multiprocessor structure. High-speed and high-capacity memories have been introduced in response to the requirements of high-speed processors. All these developments have pushed the performance of microprocessor-based systems to a unprecedented high level with ever-lower cost.

Conventional applications of microprocessors are mostly in computing and data processing areas such as computers, communication systems, image and speech processing, and so on. In real-time control of power electronic systems, although the requirements are different from those of data processing, the capabilities of microprocessors can be advantageously utilized. The requirements of control area are numerous and very demanding to the processor on many aspects. To obtain a large control bandwidth and to be able to implement complex control algorithms, processors with high execution speed are needed. To control large systems in which several microcomputers operate under the coordination of a central microcomputer, designers will need processors having high-speed communication capabilities.

In the control of power electronic systems, depending on the specifications, a wide range of processors can be used such as general-purpose microprocessors, microcontrollers, and advanced processors (DSPs, RISC processors, parallel processors). The advances in very-large-scale integration (VLSI) technology has permitted the implementation of specific devices using application-specific integrated circuit (ASIC) methodology to fulfill the requirements of particular applications.

The performance of digitally controlled drives is constantly improved by using faster microprocessors capable of complex computing. Since higher precision and higher computing speed can be achieved, the control bandwidth is pushed toward the analog limits.

Sophisticated algorithms from modern control theory, such as state feedback control, optimal control, and adaptive control, can be implemented in real-time using high-performance microprocessors and specialized ASICs. "Sensorless" electrical drives are now possible since the system state variables can be estimated by state observers implemented on microprocessors. Emerging technologies from the artificial intelligence field such as expert systems, neural networks, and fuzzy logic can be now applied to motion control systems owing to DSPs and ASIC chips. The high degree of integration results in reduced parts count and thus contributes to enhance the overall system reliability. The fault tolerance of control systems can be improved by implementing diagnosis functions in the control system. Also, self-testing and self-tuning capabilities can be added.

The development of microprocessor-based control systems becomes more and more complex so that sophisticated tools are required for the design, simulation, and testing of the target systems. To generate effective real-time code for complex microprocessor-based control systems, several programming languages have been developed, for example, C, C++, and Parallel C.

Real-time control of complex power electronic systems requires the concurrent execution of several tasks. Numerous real-time operating systems (RTOSs) have been developed for popular microprocessors permitting efficient scheduling and dispatching tasks in a multitasking environment.

With all these advances combined, today power electronic control engineers have in hand all the necessary tools to produce a "fully digital" design with high control bandwidth, sophisticated control algorithms, and multiprocessing capability. In this chapter, we will present the new developments in microprocessor technology and its application in the control of power electronic systems.

10.2. MICROCOMPUTER CONTROL OF POWER ELECTRONIC SYSTEMS

A power electronic system processes electric energy using static power converters. In general, it can consist of one or several power converters feeding passive loads (as in power supply systems) or active loads (as electric machines in variable speed drives). The proper operation of such a system requires a control system with functions determined by the system specifications.

10.2.1. Controlling Power Electronic Systems

Consider a typical power electronic system, an electric motor drive, which consists of three basic components—the electric motor, the power converter, and the control system—as shown in Figure 10-1. The electric motor drives the mechanical load directly or through reducing gears. The power converter controls the power flow from the source to the motor by activating the power switches according to the drive signals provided by the control system.

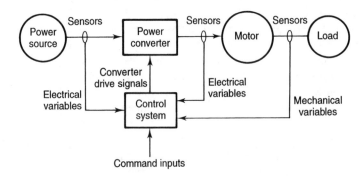

Figure 10-1. Control of an electric drive.

The system electrical and mechanical variables that are required for control and protection purposes are provided by various sensors. The number of sensors to be used will depend on the adopted control scheme. In "sensorless" motion control systems, the number of sensors are reduced to a minimum, since only electrical sensors are used. In this case, the control system has to estimate or reconstitute the required system variables from motor voltages and currents.

Depending on the specific motion control application, an electrical motor drive can be of one of the following types: torque (or force) control, acceleration control, speed control, or position control. The complexity of a control system depends on the control configuration and strategy that are to be implemented, and also on the performance specifications. The control configuration and strategy depend much on the type of motor. For example, the control system of a DC drive fed by AC-to-DC thyristor converter is less complex than that of an AC drive fed by pulse width modulated inverter. A simple proportional integral (PI) speed regulation loop is less complex than an adaptive vector-controlled system.

The basic function of the control system is to regulate the system variables (torque, speed, position, acceleration) according to input commands. In a general manner, the control system is responsible for system control and converter control. System control functions concern the functioning sequence, control algorithms, and regulation of the system variables. Thus, it has to accomplish a number of tasks such as acquisition and processing of system variables, acquisition of command inputs, and implementation of the system logic and control algorithms. Converter control functions deal with converter operating sequences and drive signal generation (pulse width modulation, for example).

The auxiliary control functions take care of the reliability of the system such as supervision (monitoring of the system variables), testing, and protection. For a long time, the implementation of control systems for power electronic systems has followed a hardwired approach in which analog and digital integrated circuits are used. During the last two decades, with the developments of microprocessors and peripheral circuits, digital technology has gradually replaced analog technology in conventional control functions and allowed the implementation of advanced control functions previously unattainable. Today, digital technology is considered as the sole and unrivaled approach for high-performance advanced motion control systems because of the unique possibilities it can offer.

Hardwired Implementation. In a hardwired approach, analog and digital components (integrated circuits) with "embedded" functions are connected following a specific topology to provide the desired function:

- Analog devices such as operational amplifiers combined with resistors and capacitors are used to implement signal processing and regulation circuits.
- Nonlinear devices (comparator, function generator, etc.) are used in converter drive signals generation, time delay, and so on.

- Digital integrated circuits (logic gates, flip-flops, counters, registers, etc.) are used to implement logic functions such as sequence generation, pulse distributing circuit, and so on.

The main advantage of the hardwired implementation is the "parallel" operation of the function blocks, thus resulting in high execution speed.

A major drawback of hardwired approach is the "rigidity" of the design, since any modification will require changes in hardware. In addition, the complexity of the hardware increases with the complexity of the control algorithms. It is difficult or impossible to implement advanced algorithms such as state feedback or adaptive control using hardwired approach.

Microcomputer-Based Implementation. In this approach, control functions are defined and realized by the software (or code) contained in the system memory. A major advantage of the software approach is the flexibility of the design due to the programmability of the microprocessors. The functions implemented by software can be readily modified without modifying the system hardware. Customization, modification, adaptation, and upgrades of the control system to a specific drive can be readily done, even after the hardware has been installed.

It has been pointed out that most of the control functions in an electric drive can be implemented by software. The sole limitation is imposed by the execution speed of the processor. The logic and computing capabilities of microprocessors are exploited in the implementation of the system logic (operating system), control algorithms, regulation schemes, signal filtering, and the like. With their increasing capabilities, microprocessors can be used to implement advanced control algorithms in real time. The communication capability of microprocessors permits the transfer of a large amount of data between the controllers in a large system where several drives operate in a networked configuration. This situation becomes more and more common in a high-productivity industrial context.

The control performance can be greatly enhanced by using multiprocessor architectures. In such configurations, high-speed communication units are required to relieve the central processing unit (CPU) of this time-consuming task. Another advantage of microprocessor-based control systems is the multitasking capability permitting several tasks to run concurrently in real time. This feature is essential in most power electronic systems where real-time tasks are multiple.

Hardware/Software Trade-offs. In a real-time context, the execution speed of the control functions is of primary importance. With hardware implementation, the execution speed is very fast, but the hardware complexity increases as the functions are added. With software implementation, the hardware complexity is fixed, but the execution time increases as the functions are added. Very often, the design engineer has to consider this question and to make some trade-offs between hardware and software. The decision is usually influenced by the desired sampling time and the processor execution speed. Some functions are more suited for hardware implementation, such as converter drive signals generation. Some others are better realized by software, such as regulation and adaptive algorithms.

With the increasing improvements in microprocessor performance, control functions are gradually migrating from hardware to software. In a "fully digital" control system, all the functions are implemented by software with a minimum of hardware used as interface between the control system and the power circuit.

10.2.2. Microcomputer Control of Power Electronic Systems

Digital control of a power electronic system requires a microcomputer-based control system (or digital controller), built around one or several microprocessors, that processes data and implements control algorithms under digital form. The structure of a microcomputer-based control system can be presented under two aspects: hardware and software configurations. The hardware configuration depends on the processors and the bus system used, and also on the interface circuits. The software configuration depends mainly on the functions performed by the control system.

The hardware configuration of a microcomputer-based control system will be discussed in detail in Section 10.4. Here, we consider the software configuration (organization of the control functions) and discuss issues concerning digital control such as sampling rate, word length and data types, and multitasking control.

The software configuration of a typical microprocessor-based control system for power electronic systems is shown in Figure 10-2. This diagram illustrates the basic functions that can be classified as five groups as follows:

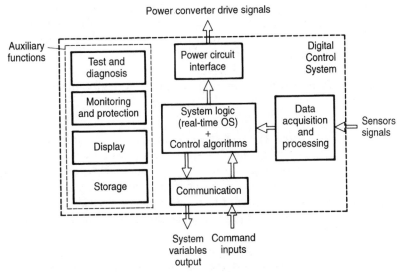

Figure 10-2. Basic functions of a digital motion control system.

- *Data Acquisition and Processing.* The system electrical and mechanical variables (voltage, current, position, speed, acceleration, torque, force) provided by various sensors are sampled and converted into numerical data by data converters (analog-to-digital converter, resolver-to-digital converter, etc.). The acquired data are then processed (filtered, scaled, etc.) to be usable by the control functions.
- *Communication.* The communication function is necessary to receive the command inputs from the operator and/or from the coordinating computer. It is also required for transmitting different system variables necessary for the high-level control to the coordinating computer.
- *System Logic and Control Algorithms.* This group performs the main functions of the control system. It implements the system logic and control algorithms that govern the system operation. The system logic typically includes a real-time operating system that concurrently schedules and executes multiple tasks. The control algorithms include the identification, estimation, regulation, and control schemes. Depending on the commands, operating conditions, and strategy implemented, the RTOS selects the sequences and the control algorithms to be executed by the processor.
- *Power Circuit Interface.* This function is required to adapt the control software and hardware to the power converter inputs. It generates the required drive signals for the power converter.
- *Auxiliary Functions*:
    - *Display.* System variables and/or alarm signals can be displayed on light emitting diode (LED), liquid crystal display (LCD), or cathode ray tube (CRT) displays.
    - *Storage.* System variables can be stored in memory or in mass storage systems (diskette, hard disk, optical disk, magnetic tape).
    - *Monitoring and Protection.* Important system variables can be continuously monitored. Any malfunction can produce alarm signals, and the system can be forced to perform emergency sequences to protect the system components.
    - *Test and Diagnosis.* Tests can be performed to adjust the system parameters and to detect malfunctions and their sources.

Sampling Rate. The variables in power electronic systems are continuous functions of time (analog signals), so conversion is needed to transform them into numerical data that can be processed by the digital control system. This is done by sampling first the continuous time signal to obtain a discrete time signal. An analog-to-digital converter is then used to produce the numerical equivalent of the discrete samples. The operation of sampling and conversion of an analog signal is illustrated in Figure 10-3.

In a digital control system, the sampling rate may affect the control performance, particularly the disturbance rejection limits, the smoothness of the response, and the sensitivity to parameter variations. The sampling rate of a digital control

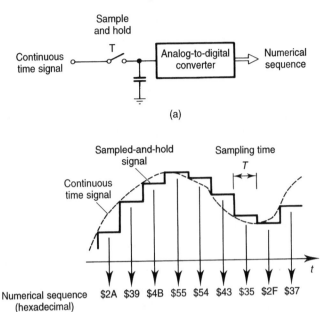

Figure 10-3. Sampling and conversion of an analog signal (a) functional diagram; (b) waveforms.

system depends on the system dynamics and the processor speed. The selection of the sampling rate involves a compromise between performance and cost, and the best choice is the slowest sampling rate that meets all performance specifications. The sampling theorem applied to a closed loop system gives an absolute lower bound for the sampling rate that is twice the required closed loop bandwidth of the system. However, in practice, to ensure some smoothness in the response and to limit the magnitude of the control steps, a sampling rate between 10 and 50 times the desired closed loop bandwidth is typically used [5].

In motion control systems where the electrical and mechanical time constants may be significantly different, improvements in performance can be realized by using different sampling rates on the electrical and mechanical variables. For example, in a conventional motor control configuration where the current, speed, and position are controlled by three cascaded loops, the bandwidth of the inner loop is typically many times higher than that of the outer loop so that three sampling rates can be used, one for each control loop.

Word Length and Data Types. The variables in a microprocessor-based control system can be represented using one of two number formats: fixed point and floating point. Microprocessors are also divided into two types, fixed point and floating point processors, depending on the capability of their arithmetic-logic unit (ALU) to handle a specific format. While fixed point processors are designed to handle fixed point numbers, floating point processors operate directly on data using floating point format. A fixed point processor can also handle floating point numbers by software but at the expense of increased processing time. Figure 10-4 shows the two number formats represented by 32-bit words.

The word length of a processor is determined by its data bus width. In a 16-bit processor, for instance, the data path for operands is 16 bits wide. The word length of a processor also represents the size of the operands upon which the processor can operate. It affects directly the resolution of the variables and the software execution speed. In motion control systems, the most utilized word lengths are 8 bit and 16 bit. With the introduction of high-performance microprocessors, digital controllers are moving toward 32-bit word length.

The processor word length determines the resolution and dynamic range of the variables. With fixed point processors, there is a linear relationship between word length and dynamic range because each additional bit adds a higher power to the base 2. Resolution in such processors is fixed by the value of the least significant bit (LSB). However, to take full advantage of the resolution, variables in fixed point processors must be properly scaled.

On the other hand, dynamic range of floating point processors increases much more drastically with word length since each additional bit allows the exponential term to grow by a factor of 2. For a given word length, however, floating point resolution is not as good as fixed point resolution because fewer digits are assigned to the mantissa. One of the advantages of floating point processors is that scaling of the variables is not necessary which can reduce significantly the execution time of real-time control algorithms. The resolution and dynamic range of fixed and floating point formats are shown in Table 10-1.

Multitasking Control. Multitasking control is of first importance in a power electronic system because of the multitude of time-dependent tasks to be executed by the control system. Even in a simple DC drive, for example, real-time tasks are numerous and demanding such as converter drive signals generation, current regulation, speed regulation.

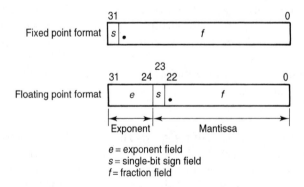

Figure 10-4. Fixed point and floating point number formats (32 bits).

TABLE 10-1 RESOLUTION AND DYNAMIC RANGE OF VARIABLES VERSUS FORMAT
AND WORD LENGTH

Format	Fixed point			Floating point	
Word length	8-bit	16-bit	32-bit	16-bit*	32-bit†
Resolution	3.922×10^{-3}	1.526×10^{-5}	2.328×10^{-10}	3.8147×10^{-6} (min) 0.0625 (max)	7.0065×10^{-46} (min) 2.0282×10^{31} (max)
Dynamic range	255	65,535	4.2950×10^{9}	32762	5.7896×10^{76}

*(4-bit exponent + 12-bit mantissa)

†(8-bit exponent + 24-bit mantissa)

The functions required to control a power electronic system can be partitioned into several tasks (or processes) that are executed according to a control scheme or strategy. A priority must be assigned to each task depending on its importance for the operation of the system. In general, the control tasks can be classified into two groups:

1. *Real-time tasks* are usually interrupt-driven processes that are synchronized to the events in the power system. High priorities are generally assigned to these tasks since they are vital for the proper operation of the power system. Every time the CPU receives an interrupt signal, the real-time task assigned to this interrupt must be executed within a very short delay. An example of real-time task is the regulation of the currents in an AC drive. This task must be executed at regular intervals (or sampling periods) to ensure that the motor currents follow the references.

2. *Nonreal-time tasks* are usually low-priority ones that are also necessary for the system operation but are not time critical. They can be executed in the background when the microprocessor is not busy executing real-time tasks. An example of nonreal-time (or background) task is the display of system variables.

This foreground/background system represents the most commonly used task scheduling scheme for power electronic systems because the number of foreground tasks is fixed and known a priori. In a multitasking control system, a real-time operating system is required to schedule and dispatch tasks and to assume inter-task communication. Features of real-time operating systems are discussed in Section 10.8.

10.2.3. Processor Selection

Because of the multiple factors to be considered in the design of a micropro-cessor-based control system, it is difficult to establish rules to apply for processor selection. The designer can make a judicious choice of processor only after having defined all the control functions to execute as well as the desired performance. This can be done during the first stage of the development cycle discussed in Section 10.8. In this section, we examine in a general manner the different factors that may influence the choice of processors for the control of power electronic systems.

The choice of processors is critical and many factors have to be considered such as word length and data type (fixed point or floating point), architecture, processing speed, mathematics capability, time processing capability, and interrupt handling capability. For a given control system, the selection of an appropriate processor will depend also on the criteria that the designer establishes for the application:

- Desired performance: sampling time, control accuracy, response time and so on.
- Hardware: bus system, availability and so on.
- Software: availability and performance of development and debugging tools
- System cost.

Processor Architecture. The range of processors to handle complex real-time control problems such as control of power electronic systems can fall into the following categories:

1. *General-purpose microprocessors* are central processing units that require external memory and support chips to complete the control system.
2. *Microcontrollers* are control-oriented devices that offer a high level of integration on a single chip. They typically incorporate a central processing unit and peripheral units required for real-time operation such as programmable timer system, A/D converter, interrupt controller, pulse width modulator, and digital input-output (I/O).
3. *Application-specific integrated circuits* are custom, semicustom, or standard ICs that may integrate on-chip several analog and digital components necessary for a specific application: microprocessor, A/D converter, pulse width modulator (PWM), I/O interface, and so on.
4. *Digital signal processors* are high-speed processors with Harvard architecture designed for efficient execution of signal processing algorithms. In these processors, the multiply-accumulate (MAC) operation is optimized and repeat instruction is implemented.
5. *Reduced-instruction-set-computing microprocessors* are high-speed processors with pipeline architecture and simple instructions. The complex operations are executed by software.
6. *Parallel processors* are high-performance microprocessors equipped with communication links that can be connected in network to operate concurrently to increase processing speed and/or to improve control flexibility and performance.

The microcomputer architecture can significantly influence the system performance. It can limit the attainable control performance and restrict the features that the designer can put into his system.

Mathematics Capability. In control systems, mathematical operations are required to implement filters and control and regulation algorithms. The most utilized operations are multiplication, addition, square root, trigonometric functions (sine, cosine, tangent, etc.). The implementation of filters and controllers, from conventional proportional integral derivative (PID) to complex controllers, as well as identification and estimation schemes, necessitates intensively the form of multiply-accumulate operation (sum of products). This operation is effectively handled by digital signal processors. The implementation of advanced control algorithms such as state controllers, state observers, adaptive controllers, and so on requires complex matrix operations.

Processing Speed. The processing speed is an important factor but not the sole one that determines the performance of a microprocessor. We have to consider at the same time the processing speed and the processor capabilities.

The processing speed of a microprocessor is usually measured in MIPS (millions of instructions per second). The floating point processing speed is measured in MFLOPS (millions of floating point operations per second). These measures appear to be inaccurate in comparing different architectures, in particular complex instruction set computing (CISC) versus RISC. Therefore, system designers have to rely on benchmarks to evaluate the performance of microprocessors.

Benchmarks are short programs developed in an attempt to mimic the behavior of actual applications permitting the measure of microprocessors performance. Whetstone and Dhrystone benchmarks have been the most utilized for a long time. Recently, System Performance Evaluation Corp. (SPEC) has introduced a benchmark set, known as SPEC92, including one suite of 6 integer operation programs and one suite of 14 floating point operation programs. These suites contain widely used operations and constitute an effective and useful measure of microprocessors performance for most data processing applications. For control applications, SPEC92 benchmark results can provide an accurate base of performance comparison. If a more precise measure of control-specific performance is needed, the design engineer will have to write his own benchmark programs taking into account the actual operations required by his application.

Real-Time Operation Capability. Real-time control of power electronic systems requires the interrupt capability, time processing capability, and high-speed I/O. The interrupt capability is required to synchronize the control software to internal or external events. The *interrupt latency* is defined as the time delay between an interrupt request and the start of the service routine. It must be as small as possible compared to the sampling period to ensure the determinism of the control. *Interrupt priority* management is necessary for handling interrupts from different sources.

Context switching is an important operation in multitasking control where the operating system is often required to change the tasks according to the operation conditions and the control strategy. The context of a task (or process) may consist of a program counter, stack pointer, register set, address translation tables, path and I/O descriptors, and amounts of private code and data. The processor must be able to handle context switching with minimum delay to avoid performance degradation.

Time processing units are needed for different time-related operations such as power converter drive signals generation, period measurement, periodic interrupt generation, pulse width modulation, delay generation, baud rate generation, and the like. High-speed parallel I/O ports are required for communication with sensors, displays, and power converter.

Communication Capability. Communication capability is essential to most power electronic systems to operate in a networked environment in which several processors operate under the coordination of one central processor. The communication between microprocessors can be serial or parallel. Serial communication is typically used for low- and medium-speed data transfer. The transfer rate is an important parameter which is measured in bauds (bits per second) or Kbytes/s (kilobytes per second). Parallel communication is used essentially for high-speed transfer of large amounts of data in a processor network. The transfer rate in this case is measured in Mbytes/s (megabytes per second).

A communication protocol is necessary to ensure minimum transmission error. Communication protocols can be implemented by hardware in specialized communication units or by software.

10.2.4. Digital Versus Analog Control

For many years, analog controllers have dominated the control of power electronic systems. But the new requirements from industrial applications are such that they cannot adequately respond from lack of processing capabilities. Even in small systems where the performance requirements are relatively modest, digital technology can be used with advantages over analog approach because of the additional features they can provide. In complex motion controllers such as field-oriented or adaptive controllers, where computationally intensive algorithms are to be implemented, digital technology is the sole solution.

The advantages of microprocessor-based control systems are mainly related to the programmable nature of microprocessors and their computing and communication capabilities. We can cite the most important ones:

- Microprocessor-based control systems are programmable. The control strategy and configuration can be modified at software level providing a high operation and adaptation flexibility.
- Owing to their computing capabilities, microprocessors can be used to implement complex advanced control algorithms such as observer-based state

controllers, adaptive controllers, Kalman filtering, and nonlinear control. Also self-tuning and adaptive tuning can be implemented in real time.

- In microprocessor-based motion control systems, "sensorless" operation is possible since the system state variables (position, speed, flux) usually provided by sensors can be estimated by the microprocessor from the motor voltages and currents. The number of sensors (position, speed, flux) can be thus reduced.

- The communication capabilities of microprocessors make it possible to operate the electric drives in a networked configuration where the coordination must be done by a central computer, or in an automated system containing several intelligent drives.

The disadvantages and limitations of digital motion control systems are due mostly to the inherent properties of discrete systems that result in sampling, quantization, and truncation errors. These errors can affect seriously the load disturbance rejection limits of the controller. Also, the computation delay limits the system bandwidth and can affect the control stability when too large.

Table 10-2 summarizes the advantages and limitations of analog and digital technologies in control of power electronic systems. One can note that the advantages of digital technology outnumber its disadvantages. Furthermore, the recent developments in VLSI, processor technology, and control theory tend to minimize the disadvantages shown. Data converters can be now incorporated on the same chip as the CPU to reduce parts count. The execution speed of new-generation processors is such that the bandwidth of digital motion control systems can now attain values comparable to that of analog ones. Self-tuning systems can be implemented to provide optimum control performance without adjustment. Simulation and computer-aided design (CAD) software are now available to help the design engineer to reduce the analysis and design time of digital control systems.

10.3. MICROCOMPUTER BASICS

A microcomputer is a programmable digital system that processes data in executing stored instructions. A microcomputer consists basically of three components: microprocessor, memory, and input-output (I/O) unit. The communication between these blocks is done over a bus set.

10.3.1. Microcomputer Architecture

Basic Architecture. The architecture of a microcomputer is mainly determined by the architecture and operation of the microprocessor itself. Figure 10-5 shows a block diagram of a basic microcomputer. This architecture, also known as von Neumann architecture, is the most utilized up to the present. It consists essentially of three components:

TABLE 10-2   COMPARISON OF ANALOG AND DIGITAL TECHNOLOGIES
FOR CONTROL OF POWER ELECTRONICS SYSTEMS

Control technology	Analog	Digital
Advantages and capabilities	• High bandwidth. • High resolution. • No data conversion is required. • Specific control functions are available as off-the-shelf ICs. • Analysis and design methods are well known. • Adjustment by potentiometers or variable capacitors is easy and fast.	• Programmable solution: modifications, upgrades, or adaptations are done by software. • Less sensitive to environment. • Can implement advanced control algorithms. • Capable of self-tuning, adaptive control, and nonlinear control functions. • Capable of "sensorless" operation. • Capable of additional functions (monitoring, diagnosis, protection, test, etc.) • Communication capability: can be incorporated in a networked control system. • Flexible storage capability
Disadvantages and limitations	• Temperature drift: control performance depends on passive and active components characteristics that change with temperature. • Component aging: periodic adjustments are required to maintain good performance. • Hardware design: modifications, upgrades, or adaptations must be done at hardware level. • Can implement only simple designs (PID, lead-lag). • Sensitive to noise. • No communication capability. • No effective storage capability.	• Data converters are required. • Analysis and design methods are more complex. • Sampling and resolution can affect the load disturbance rejection limits. • Computation delay limits the system bandwidth and can affect stability. • Quantization and truncation errors can affect the control precision. • During adjustment phase, access to intermediate variables is difficult.

1. *Microprocessor* is the central processing unit that controls the computer operation and executes the instructions.

2. *Memory* stores the program and data. The program consists of the instructions to be executed by the microprocessor. Since the instructions do not change during the operation of the system, the program is usually stored in read-only memory (ROM). The data are the information that the program needs for its execution and that the program produces during its execution.

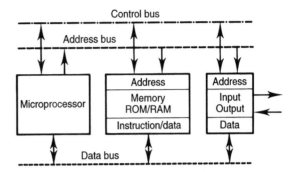

Figure 10-5. Basic microcomputer architecture.

Constant parameters are usually stored in ROM while transient data are stored in random access memory (RAM) or read-write memory.

3. *I/O unit* is the interface by which the microprocessor communicates with the outside world. It can read from or write to peripheral units through the I/O unit.

The bus set consists of three buses which transport information between the system blocks:

1. Address bus is used for memory and I/O addressing.
2. Data bus transports instructions and data from and to the memory.
3. Control bus transports control signals between the CPU and the memory and I/O unit.

Instruction Cycle. The operation of a microcomputer is characterized by its instruction cycle. A typical microprocessor basic instruction cycle consists of four operations:

1. Fetch: the CPU gets the instruction from the memory.
2. Decode: the CPU decodes the instruction.
3. Read: the CPU reads the operands from the memory.
4. Execute: the CPU executes the instruction.

In the microcomputer basic architecture, the instructions are executed in a sequential manner as shown in Figure 10-6.

Advanced Architectures. In the basic architecture where both instructions and data are stored in the same memory and travel over the same bus set, the microcomputer throughput is limited by the bus bandwidth and the memory access time. Advanced microcomputer architectures make use of registers, pipeline, cache, and multiple bus to accelerate the data transfer and reduce the memory access time so as to enhance the overall execution speed.

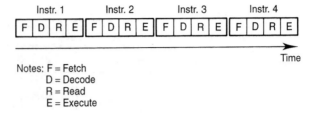

Notes: F = Fetch
D = Decode
R = Read
E = Execute

Figure 10-6. Instruction cycles of a basic microcomputer.

- *Cache memory* is an auxiliary high-speed memory in which some active portion of the slower-speed main memory is duplicated. When a memory request is generated, the request is first presented to the cache memory, and if the cache cannot respond, the request is then presented to main memory. With cache memory, the memory access time is reduced. In several advanced processors, cache memory resides on the same chip as the CPU to minimize propagation delay.

- *Virtual memory* systems attempt to make optimum use of the microcomputer main memory, while using an auxiliary memory, usually a hard disk, for backup. The main memory is used only to store active data. As they become inactive, they are moved back to the hard disk. With virtual memory systems, a microcomputer can address a memory larger than physical memory with high performance and low cost.

- In *pipeline architectures*, the basic operations (fetch, decode, read, execute) are assumed by separate units and the successive of an instruction stream are allowed to overlap so the CPU can handle several instructions at the same time. The execution time can be thus reduced as illustrated by the example shown in Figure 10-7.

- Figure 10-8 shows the *multiple bus architecture* (*Harvard architecture*) in which two separate bus sets are used to transport independently instructions and data. The *program bus set* consists of address and data buses connected to the *program memory*. The *data bus set* consists of address and data buses connected to the *data memory*. With this dual-bus system, the CPU can access at the same time the program memory and the data memory.

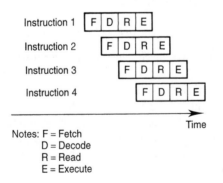

Instruction 1
Instruction 2
Instruction 3
Instruction 4

Time

Notes: F = Fetch
D = Decode
R = Read
E = Execute

Figure 10-7. Pipeline execution of instructions.

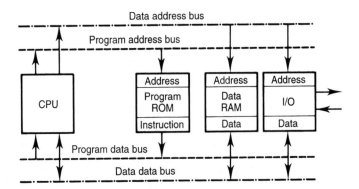

Figure 10-8. Harvard architecture.

Pipeline operation is thus possible and the processor throughput can be enhanced. Harvard architecture is commonly used in digital signal processors and in recent 32-bit and 64-bit microprocessors. In some high-performance microprocessors, three separate bus sets are used to ensure the data transfer between the CPU and program and data memories, and between data memory and I/O units by direct memory access (DMA).

10.3.2. Microprocessors

Architecture. A microprocessor is characterized by its architecture and instruction set. Most general-purpose microprocessors use the von Neumann architecture in which data and instructions are stored in the same memory and travel over the same bus set. A typical 8-bit microprocessor contains a program counter (PC), an internal CPU memory (scratch pad memory and micromemory), general registers, an instruction register, and a control unit, as shown in Figure 10-9. This architecture represents the basic structure of a "conventional" CPU.

The architecture of 16-bit and 32-bit microprocessors is more complex and may contain additional units for advanced functions such as register file, cache memory, pipeline, direct memory access (DMA) controller, and so on. Advanced microprocessor architectures are presented in Section 10.6.

Instruction set. The instruction set of a microprocessor is closely related to its internal architecture. There are two large classes of microprocessors: complex instruction set computing and reduced instruction set computing processors.

CISC microprocessors are so called because of the large number of instructions they can execute. The instruction set of CISC microprocessors can handle both basic operations and complex functions. Typically, the instructions are microcoded and can take many clock cycles to complete.

A typical 8-bit microprocessor instruction set includes the following basic instruction groups:

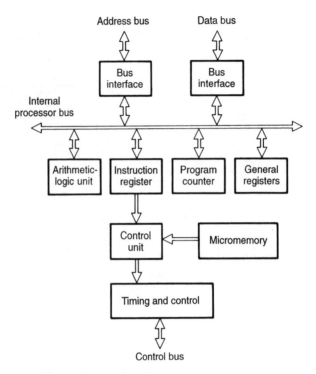

Figure 10-9. Basic microprocessor architecture.

- Accumulator and memory instructions (loads, stores, and transfers; arithmetic operations; multiply and divide; logical operations; data testing and bit manipulation; shifts and rotates)
- Stack and index register instructions
- Condition code register instructions
- Program control instructions (branches, jumps, subroutine calls and returns, interrupt handling)
- Miscellaneous (NOP, STOP, TEST)

The addressing modes in a typical 8-bit microprocessor are numerous and include immediate, direct, indirect, register, indexed, and inherent addressing.

The instruction sets of 16-bit and 32-bit microprocessors and microcontrollers are even more complex because of the additional sophisticated operations they can handle. For example, Motorola's 16-bit microcontrollers (M68HC16 family) have a number of special instructions specifically designed for digital signal processing.

RISC architecture is used in recent 32-bit and 64-bit microprocessors. These processors are characterized by a large register set and a small instruction set containing frequent simple instructions that can be executed in one clock cycle. Complex operations must be implemented in software by using sequences of simple instructions. More details on RISC processors are presented in Section 10.6.

10.3.3. Memory

The memory IC technology has evolved rapidly since the last decade due to the increasing demands for performance. There are now several types of memory devices which can be classified as two large families: volatile and nonvolatile memories. *Volatile memories* retain valid data only when they are powered. Data will be lost when the power is shut off. *Nonvolatile memories* retain valid data indefinitely or during a very long time after the power is shut off.

The memory chips presently available are usually classified following their function as detailed next.

- *Random-access memory* (RAM) is volatile memory which is used for storage of variables and temporary data. RAM chips are read-write memories and are classified as static RAM and dynamic RAM, depending on the storage method used.
  - *Static RAM* (SRAM) chips store data in register arrays built with flip-flops. A static RAM maintains the data as long as the chip is supplied. The density of static RAM is relatively low so the capacity of available chips is small. On the other hand, the access time of static RAMs is very short. Typical access time of static RAMs attains now 8 nanoseconds (ns). Main applications of static RAMs are cache memory and video RAM.
  - *Dynamic RAM* (DRAM) chips store data as electric charges in capacitors so periodic refresh is required to keep data valid. The density of dynamic RAM is very high because of the small number of devices per memory cell. Dynamic RAM chips are available with high-capacity (up to 16 Mbits at the present time). The access time of dynamic RAMs is longer than that of static RAMs. Typical values are in the 20–100 ns range. Principal application of dynamic RAMs is microcomputer main memory.
- *Read-only memory* are mask-programmed by the chip manufacturer. ROMs are used for storage of application program instructions in a definitive version. They are also used for storage of permanent data such as look-up tables, code converters, and the like.
- *Programmable read-only memory* (PROM) is a type of ROM that can be programmed by the user. A PROM chip contains an array of fusible links connected with logic gates. Data are written into a PROM by selectively "blowing" the links. Programming of PROMs is done off-board using a PROM programmer.
- *Erasable programmable read-only memory* (EPROM) is a type of PROM that can be reprogrammed a limited number of times. EPROMs store data as electric charges in an array of floating gate devices. The entire stored data can be erased by exposing the chip to ultraviolet light through a transparent window on the package. EPROMs must be removed from system and erased before reprogramming. EPROMs are typically used for program and data storage during the development phase of microprocessor-based systems.

- *Flash EPROM* is a special type of EPROM with fast-erasing mechanism. The entire stored data can be erased on-board by using a short electric pulse. Flash EPROMs must be reprogrammed a block at a time. They are typically used as mass storage devices (solid-state disk drive).
- *Electrically erasable programmable read-only memory* (EEPROM) is a type of EPROM in which each a byte can be erased and reprogrammed by electric pulses. Thus EEPROMs can be reprogrammed on-board under software control permitting on-site customization and adaptation of the control software.

10.3.4. Input-Output

Depending on the processor architecture and the application requirements, the transfer of data between the microcomputer and the outside world can be done by one of three methods:

1. *Programmed I/O.* In programmed I/O, input-output operations are accomplished by special instructions using an addressing space separate from the memory addressing space. Special I/O control signals are required. The special instructions are typically OUT and IN which transfer data from the CPU to the I/O device and vice versa. Figure 10-10a shows a typical programmed I/O circuitry.

2. *Memory-mapped I/O.* In memory-mapped I/O, the I/O devices "reside" in the same address space as the memory and appear to the CPU as memory locations. Input-output operations are accomplished using all memory-related instructions. Figure 10-10b shows a typical memory-mapped I/O circuitry.

3. *Direct Memory Access (DMA).* In normal I/O operations, the data transfer is done under the control of the CPU as illustrated in Figure 10-11a. This mode occupies the CPU during the transfer and is suited for transfer of small amounts of data. In the case where a large amount of data is to be transferred, the overheads can be reduced by giving to I/O devices access to the microcomputer's memory. Since the transfer is done directly between the memory and the I/O devices, the transfer rate can be increased. A special device called *DMA controller* is thus required for the transfer control. During the transfer, the CPU can execute other tasks that do not require access to the memory. The DMA operation principle is illustrated in Figure 10-11b.

10.4. REAL-TIME CONTROL USING MICROCOMPUTERS

In power electronic systems, a real-time control system is required to acquire data, emit control signals, and interact with the power system at precise times to ensure proper operation of the system with specified performance. A real-time control

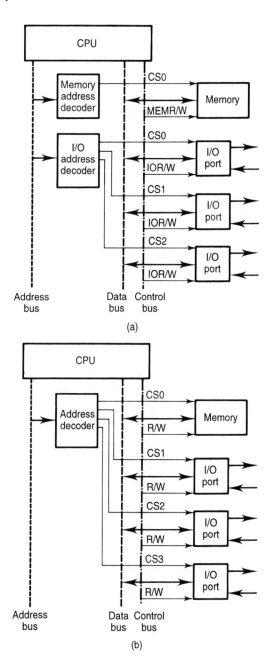

Figure 10-10. Two input-output schemes: (a) programmed I/O; (b) memory-mapped I/O.

system has to execute time-dependent tasks and the response time to an event must be small compared to sampling time.

General-purpose microprocessors are processing units with arithmetic and logic capabilities that are designed with data processing applications in mind. In order to use microprocessors in real-time control applications, additional resources for con-

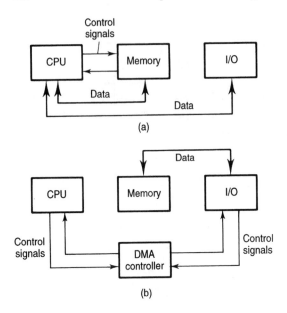

Figure 10-11. Data path in two I/O modes: (a) CPU-controlled transfer; (b) DMA transfer.

trol functions are required. A typical configuration of a control system using a single microprocessor with von Neumann architecture is shown in Figure 10-12. It consists of the microprocessing unit (MPU) and peripheral devices usually needed in embedded applications such as coprocessor, RAM, ROM, EPROM, EEPROM, I/O ports, A/D and D/A converters, timers, pulse width modulator, and communication interface. In this configuration, the MPU and peripheral devices reside on several chips. The MPU communicates with the memory and peripherals over the on-board data, address, and control buses.

In advanced architectures where multiple bus sets are used, the system configuration may be different because the communication between the CPU and the peripheral devices are done over separate buses. The same peripheral devices are still required, though.

10.4.1. Digital Input-Output

Parallel input-output operations in a microprocessor-based motion control system are of first importance since the controller has to interact as fast as possible with the power system. The parallel I/O ports can be classified as general-purpose ports and handshake ports as shown in Figure 10-13.

General-purpose I/O ports are usually bit programmable; that is, the direction of each line can be programmed by writing a corresponding bit in a data direction register. This feature is important since in motion control, bit-manipulation operations are often required to control the I/O lines individually.

Handshake I/O ports are usually used for data transmission under the control of software. Typically, two additional lines (READY and ACKNOWLEDGE) are

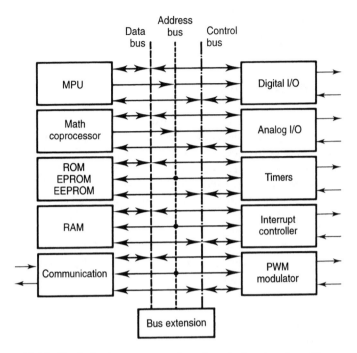

Figure 10-12. Typical microprocessor-based control system (von Neumann architecture).

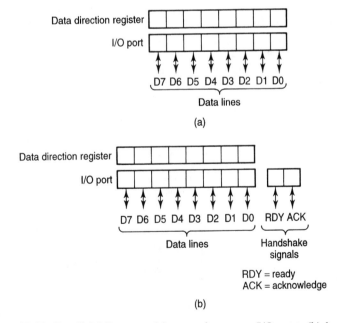

Figure 10-13. Parallel I/O ports: (a) general-purpose I/O port; (b) handshake I/O port.

used to implement a handshake protocol. This mode requires more overheads than general-purpose I/O.

10.4.2. Analog Input-Output

Since the CPU processes data under digital form, data converters are required for interfacing with the power system. The digital control signals from the microprocessor are converted into analog voltages by digital-to-analog (D/A) converters. The analog signals provided by different sensors (voltage, current, torque, speed, position, etc.) are converted into digital form by appropriate data converters such as analog-to-digital (A/D) and resolver-to-digital (R/D) converters.

Digital-to-Analog (D/A) Converters. D/A converters are needed to transform the numerical outputs from control algorithms into analog control signals for the power system. A functional diagram illustrating the operation principle of a D/A converter is shown in Figure 10-14. The most important characteristics of D/A converters for control systems are resolution, accuracy, linearity, and settling time. Complete D/A converters, with data latch and control logic, ready for microprocessor interface are available as monolithic or hybrid function blocks .

Analog-to-Digital (A/D) Converters. A/D converters perform the conversion of the analog signals from different sensors into digital words readable by the CPU. The resolution and conversion speed of the A/D converters are the most important features to be considered. The resolution of the A/D converters affects directly the precision of the control system because it determines the resolution of the feedback signals. The A/D conversion speed determines the admissible sampling interval for the highest-dynamics variable, usually the motor currents.

There are three main types of A/D converters:

1. *Integrating* A/D converters are relative slow devices so their use in real-time control system is not desirable.

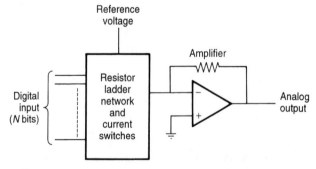

Figure 10-14. Digital-to-analog converter functional diagram.

2. *Successive-approximation* A/D converters are high-speed devices suited for use in real-time control systems. Their operation principle is illustrated by the functional diagram shown in Figure 10-15a. The conversion time depends on the resolution and the internal clock frequency. Typical conversion time for a 12-bit converter ranges from 1 μs to 10 μs.

3. *Flash* A/D converters are very-high-speed devices and are usually used for conversion of high-frequency signals. The fast conversion speed is attained by using a very large number of comparators as shown in Figure 10-15b. Conversion rate of typical 8-bit flash converters can attain 250 MSPS (megasamples per second). High-resolution flash converters can be obtained by using two or more steps of lower-resolution flash conversion.

If several analog signals have to be acquired and converted, an analog acquisition system can be used which typically consists of a multiplexer, a sample-and-hold amplifier, and an A/D converter. Complete analog acquisition systems are available as monolithic and thick film hybrid devices. In some microcontrollers, an entire analog acquisition system is included on chip which reduces considerably the parts count. Figure 10-16 shows a block diagram of a typical analog acquisition system. In this system, the analog channels are sampled and converted sequentially. The total conversion time is thus proportional to the number of channels. In systems where the conversion time is critical, one A/D converter can be used for each channel so the analog signals are converted in parallel. This approach is increasingly adopted because of the decreasing price of A/D converters.

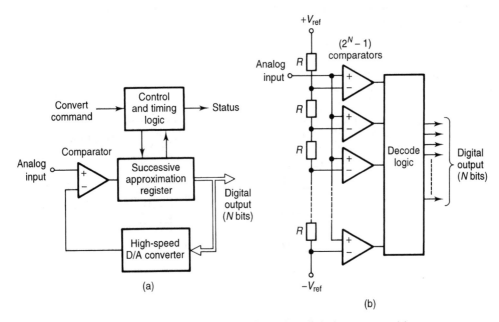

Figure 10-15. Operation principle of analog-to-digital converters: (a) successive-approximation A/D converter; (b) flash A/D converter.

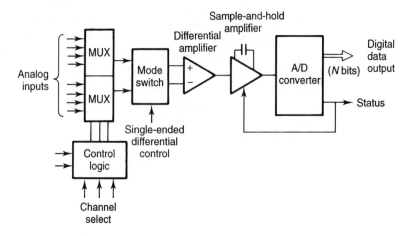

Figure 10-16. Analog data acquisition system.

Resolver-to-Digital (R/D) Converters. Resolvers are rugged position sensors used in many industrial robotics systems. An R/D converter converts the resolver output signals ($\sin\theta$, $\cos\theta$) into digital position data readable by the microprocessor. Most R/D converters operate on tracking closed loop principle which is illustrated by a functional block diagram shown in Figure 10-17. The most important characteristics of a R/D converter are the resolution (number of bits used to represent the angular position) and the maximum tracking rate (in revolution per second).

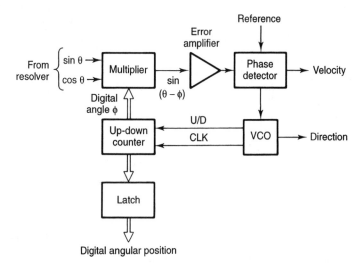

Figure 10-17. Functional block diagram of a R/D converter.

10.4.3. Interrupt Controller

In motion control systems, time-dependent tasks are required to synchronize with internal or external events. This can be accomplished by exploiting the interrupt capability of the microprocessors. In response to an interrupt request, the CPU suspends temporarily the current program and jumps to a service routine. At the end of the service routine, the CPU returns to the suspended program. Figure 10-18 illustrates the interrupt process in microcomputers.

Interrupts can be generated internally by exceptional conditions (overflow, software interrupt, etc.) or externally by peripheral devices (timers, I/O devices, etc.). Upon reception of a valid interrupt, the CPU will finish the current instruction and enter an interrupt sequence which consists, in general, of the following operations:

- Identify the source of the interrupt.
- Save the program counter and other CPU registers in the stack.
- Branch to the service routine assigned to the interrupt.

At the end of the interrupt service routine, the CPU executes a "return from interrupt" instruction that recovers the program counter and other CPU registers from the stack. Then, the CPU can resume the program from where it has left off.

An important parameter of an interrupt system is the *latency* which is defined as the time delay between the reception of an interrupt request and the starting of the service routine. An effective interrupt management is necessary to provide a minimum latency to optimize the control performance.

The identification and dispatching of interrupts can be accomplished by software or by a specialized circuit called *interrupt controller*. The two commonly used techniques are polling and vectoring. In *polling systems*, the CPU identifies the source of interrupt by polling so the response time is variable and unpredictable. In *vectoring systems*, the interrupting device identifies itself to the CPU by using its specifically assigned flag bit or its own interrupt request (IRQ) line. The execution branches directly to the service routine associated to the identified interrupt. The

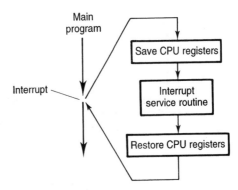

Figure 10-18. Interrupt operation.

response time in this system is constant, which is a desirable feature for real-time control.

In many systems, it is required that priority is attributed to interrupts. This can be done internally by the CPU or externally by a priority encoder circuit as shown in Figure 10-19. The priority attribution scheme can be static (fixed priority) or dynamic (priority can be changed during program execution). Interrupts play an important role in electric motor control systems where they are generally used for scheduling real-time tasks. Periodic interrupt signals with different sampling rates required by the control system are usually generated by programmable timers (presented in the next section).

10.4.4. Time Processing Devices

Time processing devices are used for the control of power electronic systems that require several time-related functions such as time delay, event counting, pulse width, period, and frequency measurement, power converter drive signal generation (pulse width modulation), real-time interrupt, and watchdog function. Time processing devices are typically built around programmable timers.

Programmable Timers. A programmable timer consists generally of a counter associated with logic control circuits. Programmable timers are controlled by software that can execute various operations such as load a count, read the contents, change counting mode, change clock rate, detect special conditions, and the like. Additional logic circuits are usually used to execute complex functions such as input-capture, output-compare, watchdog, real-time interrupt, and so on. A typical time processing unit included in microcontrollers is relatively complex and may contain several function blocks as shown in Figure 10-20.

Input Capture and Output Compare Operations. Two important operations of a time processing unit are the measurement of the elapsed time between two external events and generation of precise time delay controlled by software. These two operations require special functions called *input capture* and *output compare*.

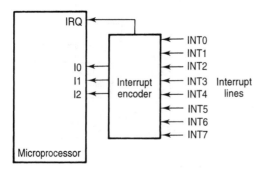

Figure 10-19. Interrupt priority encoding scheme.

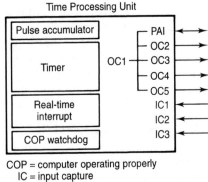

Figure 10-20. Time processing unit contained in Motorola 68HC11 microcontroller.

COP = computer operating properly
IC = input capture
OC = output compare
PA = pulse accumulator

Input capture function permits one to record the time at which a specific external event occurs. This is accomplished by latching the contents of a free-running counter when a rising or falling edge is detected at the input. The time at which the event occurs is saved in a register. The functional block diagram of an input capture circuit and waveforms are shown in Figure 10-21. By recording the times of successive edges on an input signal, the software can determine its period and/or pulse width.

Output compare function is used to program an action to occur at a specific time which is when the counter contents reach a value stored in a register. The

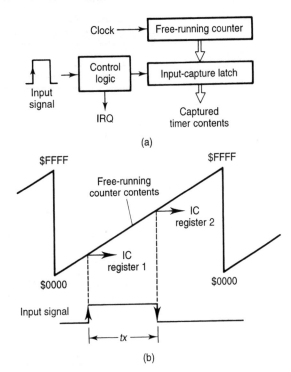

Figure 10-21. Input capture function: (a) functional diagram; (b) waveforms.

functional block diagram of an output compare circuit and waveforms are shown in Figure 10-22. Output compare function can be used to generate a pulse or a pulse train with a specific duration or to produce a precise time delay. By controlling sequentially the values stored in output compare registers, the software can generate a pulse width modulated signal for driving DC choppers or PWM inverters in electrical drives.

10.4.5. Communication Interface

Transfers of data between a microcomputer and other microcomputers or peripherals can be accomplished by serial or parallel transmission. Serial transmission is usually used because it is less expensive than parallel transmission. Serial transmission can be synchronous or asynchronous depending on the required transfer speed and the amount of data.

Synchronous Serial Communication. In synchronous communication, clock pulses are present in the data stream and are used to synchronize the transmission. The clock can be sent on a separate line or interleaved with data and sent on the same line.

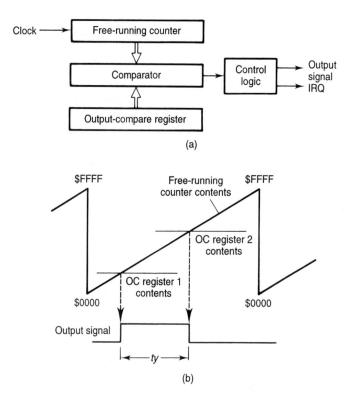

Figure 10-22. Output compare function: (a) functional diagram; (b) waveforms.

Synchronous serial peripheral interface (SPI) is a specialized data communication unit that is needed to interface the microcomputer and the communication lines. Figure 10-23a shows the waveforms of a typical synchronous peripheral interface for data transmission. Because of its high efficiency, synchronous transmission is suited to high-speed transfer of large amounts of data between microcomputers where channels may be noisy or of great length.

Asynchronous Serial Communication. In asynchronous communication, there is no clock contained within the data stream. The transmitter sends data at a programmed frequency and the receiver operates at the same nominal frequency. The receiver clock is required to resynchronize on each character. Specialized data communication units are needed to interface the microcomputers and the communication channel. They are commonly known as universal asynchronous receiver transmitter (UART) or asynchronous serial communication interface (ASCI). Figure 10-23b shows the waveforms of a typical asynchronous serial communication interface.

The efficiency of asynchronous transmission is lower than that of synchronous transmission because of the controls bits required for each data character. Asynchronous communication is used typically to connect the microcomputer to a CRT terminal or personal computer. Also, several widely distributed microcomputers can use their UARTs to form a serial communication network.

Parallel Communication. Parallel communication is used when high-speed data transfer is required. For the same clock rate, parallel communication is faster than serial communication because the bits are transmitted in packages on several lines. Connection of parallel communication ports requires multiple-conductor

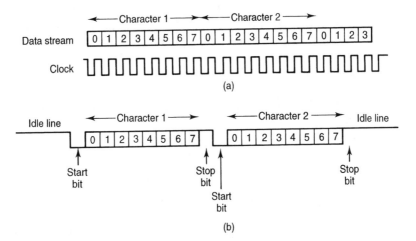

Figure 10-23. Serial communication: (a) synchronous mode; (b) asynchronous mode.

cables and connectors. Parallel communication is typically used to connect microprocessors in a multiprocessor architecture. Details on parallel communication ports in transputers and parallel DSPs are given in Section 10.6.

10.5. MICROCONTROLLERS

Microcontrollers are control-oriented devices that contain the microprocessor and several peripheral devices on the same chip. The data, address, and control buses are thus implemented on chip. For embedded control applications, microcontrollers are preferred because of the reduced parts count and enhanced reliability and performance.

The evolution of microcontrollers has followed that of microprocessors. As a general rule, the computing performance of microcontrollers increases with the data bus width. At the present time, control engineers have a broad choice of microcontrollers available on the market with various features. It is impossible to describe here all of them. We will examine some microcontroller families designed and produced by Intel Corp. and Motorola, Inc., which are among the most popular microcontrollers used in embedded control applications. Their structures and characteristics can give us a good idea on the features and performance of typical microcontrollers presently available that can be used for the control of power electronic systems.

10.5.1. Intel Microcontrollers

The microcontrollers produced by Intel Corporation for control applications consist of three main families: MCS-51, MCS-96, and i960. Each family consists of several models (or versions) having different peripheral devices and available in various package formats to better fit specific application needs.

The MCS-51 family is based on an 8-bit CISC microprocessor. On-chip peripheral devices include RAM, ROM, serial port (asynchronous), and timers. Figure 10-24 shows a block diagram of the MCS-51 microcontroller.

The MCS-96 family is based on a 16-bit CISC microprocessor with a large register file to improve the execution speed. This microcontroller contains several peripheral devices: ROM (or EPROM), serial port (UART), high-speed I/O, timers, and 8-bit pulse width modulator. Figure 10-25 shows a block diagram of the MCS-96 microcontroller.

The i960 family implements 32-bit RISC architecture with on-chip floating point coprocessor. The i960 microprocessor uses a highly parallel superscalar core with an on-chip 4 Kbyte instruction cache for frequently used instructions and a 1 Kbyte data cache for constants and other frequently used data. Table 10-3 shows the key features of the MCS-51, MCS-96, and i960 microcontroller families offered by Intel Corporation.

TABLE 10-3 KEY FEATURES OF INTEL'S MICROCONTROLLER FAMILIES

Microprocessor	MCS-51 Family	MCS-96 Family	i960 Family
Architecture	CISC	CISC Register-to-register	RISC Load/store
Data bus width	8 bits	16 bits	32 bits
Address space	64 Kbytes	64 Kbytes	4 gigabytes, linear
On-chip memory	128–256 bytes data RAM 4–16 Kbytes ROM 4–16 Kbytes EPROM	Up to 8 Kbytes EPROM	
Registers	26 general-purpose registers	232-byte register file	• 16 global, 32-bit registers • 16 local, 32-bit registers • 4 local register sets stored on-chip • register scoreboarding
Floating point capability			On-chip 80-bit FPU • IEEE 754 format • 4 80-bit registers • 4 millions Whetstones/s
Interrupts	Interrupt controller • 6–19 interrupt sources • 5–11 vectors • 2 priority levels	Interrupt controller • 28 interrupt sources • 16 vectors	Built-in interrupt controller • 32 priority levels • 256 vectors • supports 8259A
Bus speed	12 MHz	12 MHz	20 MHz
Timers	2 to 3, 16-bit timer/counters	2 16-bit timers 16-bit watchdog timer 16-bit up/down counter with capture	
A/D	8-channel, 8-bit A/D converter (87C51GA only)	8-channel, 10-bit A/D converter with S/H	
Pulse width modulated output		PWM output	
I/O ports	4 to 7 8-bit I/O ports	5 8-bit I/O ports	
Serial ports	Full-duplex serial port	Full-duplex serial port	

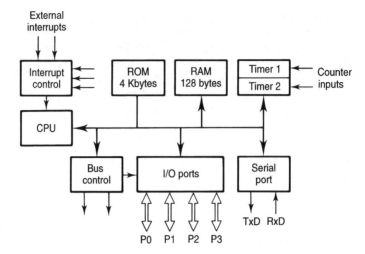

Figure 10-24. Block diagram of the Intel MCS-51 microcontroller.

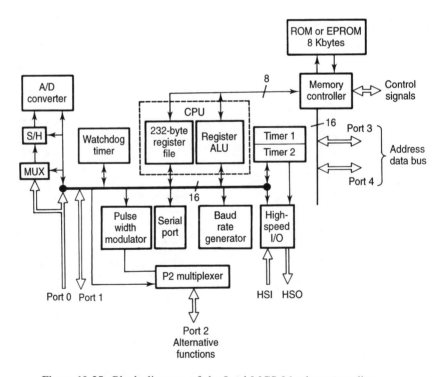

Figure 10-25. Block diagram of the Intel MCS-96 microcontroller.

10.5.2. Motorola Microcontrollers

The microcontrollers produced by Motorola, Inc., for control applications are also divided into three main families: M68HC11, M68HC16, and M68830. Each family consists of several models (or versions) having different peripheral devices and in various package formats to suited a wide range of applications.

The M68HC11 family is based on an 8-bit CISC microprocessor (the MC6801 core). This microcontroller contains a wide variety of peripheral devices needed for embedded control: RAM, ROM, EPROM, EEPROM, time processing unit, communication interfaces (synchronous and asynchronous), and analog data acquisition system. Figure 10-26 shows a block diagram of the M68HC11 microcontroller.

The M68HC16 family is based on a 16-bit CISC microprocessor with DSP instructions. The M68HC16 architecture uses modules including a CPU core,

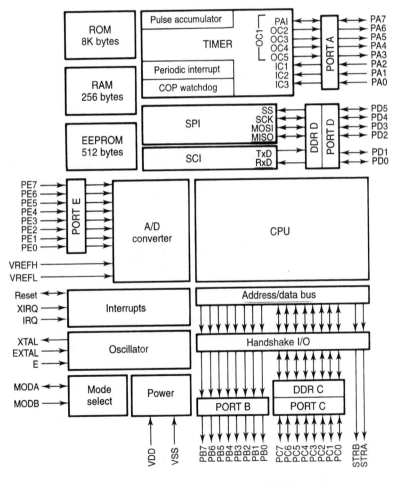

Figure 10-26. Block diagram of the Motorola M68HC11 microcontroller.

RAM, 10-bit A/D converter, general-purpose timer (GPT), queued serial module (QSM), system integration module (SIM), clock control, and port or chip selects. Figure 10-27 shows a block diagram of the M68HC16 microcontroller.

The M68830 family is based on the MC68000 microprocessor.

Table 10-4 shows the key features of the M68HC11, M68HC16, and M68332 microcontroller families offered by Motorola, Inc.

10.6. ADVANCED MICROPROCESSORS FOR CONTROL OF POWER ELECTRONIC SYSTEMS

The performance of a digital controller for power electronic systems depends not only on the processor structure, computing capabilities and execution speed but also on the implemented control algorithms. These factors mutually interact so that they must be considered together when the control performance is to be improved. Advanced microprocessors with higher computing capabilities and execution speed can help motion control engineers to increase the sampling rate that results in lower quantization noise and computational delays. The control bandwidth, stability, and

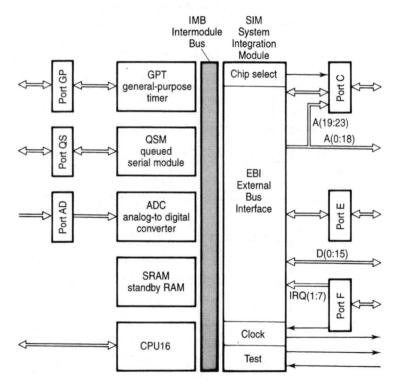

Figure 10-27. Block diagram of the Motorola M68HC16 microcontroller.

TABLE 10-4   KEY FEATURES OF MOTOROLA'S MICROCONTROLLER FAMILIES

Microprocessor	68HC11 Family	68HC16 Family	68830 Family
Architecture	CISC	CISC with DSP capability	CISC based on MC68000 core
Data bus width	8 bits	16 bits	32 bits
Address space	64 Kbytes	1 Mbytes program 1 Mbytes data	16 Mbytes
On-chip memory	256–1 Kbytes RAM 4–12 Kbytes ROM 4 Kbytes EPROM 512–2 Kbytes EEPROM	1–2 Kbytes static RAM 8–48 Kbytes ROM	2–4 Kbyres static RAM
Registers			• 8 32-bit data registers • 7 32-bit address registers
DSP capability		• Multiply-accumulate (MAC) • Repeatable operation (RMAC)	
Floating point capability			
Interrupts	Interrupt controller • 18 interrupt sources • 20 vectors • fixed hardware priority (17 levels)	Exception processing • 52 predefined vectors • 200 user-definable vectors • 8 interrupt priority levels	Exception processing • 64 predefined vectors • 192 user definable vectors • 8 interrupt priority levels
Bus speed	2 MHz	16.8 MHz	16.8 MHz
Timers	One 16-bit free-running counter with one four-stage prescaler • 3–4 input captures • 4–5 output compares Watchdog timer 8-bit pulse accumulator	2 16-bit free-running counters with one seven-stage prescaler • 3–4 input captures • 4–5 output compares Watchdog timer	2 16-bit free-running counters with one seven-stage prescaler • 3–4 input captures • 4–5 output compares Watchdog timer
A/D	8-channel, 8-bit A/D converter with S/H	8-channel, 10-bit A/D converter with S/H	8-channel, 10-bit A/D converter with S/H (68F333 and 68334)

TABLE 10-4 (CONT'D)

Microprocessor	68HC11 Family	68HC16 Family	68830 Family
Pulse width modulated ouput	Software controlled	2 PWM outputs	2 PWM outputs
I/O ports	Five 8-bit I/O ports	46–95 I/O lines	16–64 I/O lines
Serial ports	• serial communication interface • synchronous serial peripheral interface	• serial communication interface • queued serial peripheral interface	• serial communication interface • queued serial peripheral interface

load disturbance rejection can be thus enhanced. Also, advanced control algorithms such as state feedback, adaptive, fuzzy, and neural control can be implemented.

During recent years, a number of advanced microprocessors have been developed including digital signal processors, reduced instruction set computing processors, and parallel processors. The performance of the new-generation processors is significantly enhanced compared to conventional microprocessors. Their main characteristics are (1) multiple bus sets for increased transfer rate, (2) instruction and data caches for pipelined operations, and (3) simple instruction sets for efficient compilation. These characteristics result in higher execution speed and efficient compilation of high-level languages. Also, the new processors are capable of superscalar and multiprocessing operations. Superscalar operation is the execution of two or more instructions per cycle by two or more execution units. Multiprocessing operation is the concurrent execution of several codes in several processors connected in network.

Most of these high-performance advanced microprocessors are designed with high-speed computing and data processing in mind. Their main application domains include personal computers, image processors, graphics workstations, and massively parallel computers. Their application in control of power electronic systems will require the development of appropriate peripheral devices to take full advantage of the available processing power. This section gives an overview of the main features of these advanced processors and discusses their application in the control of power electronic systems.

10.6.1. Digital Signal Processors (DSPs)

DSPs began to appear roughly about 1979. Since then, several DSP generations with increasing performance have been introduced by many manufacturers. These chips have been developed specifically for real-time computing in digital signal processing applications. Many DSPs function as embedded real-time processors in special-purpose hardware such as modems, speech coders, speech synthesizers, speech recognition systems, and image processing systems.

Table 10-5 shows popular recent DSP families offered by different manufacturers.

DSP Internal Structure and Features.

DSP Internal Structure and Features. Most DSPs are built with a Harvard architecture, where data and instructions occupy separate memories and travel over separate buses as shown in Figure 10-8. Because of this dual bus structure, the processor can fetch simultaneously an instruction and a data operand. Pipelined operation of instructions and data transfer is thus possible, resulting in a higher instruction throughput rate. The pipeline can be from two to four levels deep, depending on the architecture. To optimize the processing speed, important operations such as multiplication and shift are implemented in hardware instead of using software or microcode. In recent DSPs, the execution speed is further enhanced by using several independent units, multiple bus sets, and additional units such as instruction cache, register file, and dual-access memories.

The operation of DSPs is optimized so that most of the instructions are executed in a single cycle. Third-generation DSPs can even perform parallel multiply and ALU operations on integer or floating point data in a single cycle. The multiply/ accumulate operation is the basic operation which is optimized in DSPs. This operation is used in most signal processing and control algorithms (digital filters, FFT, PID controllers, etc.) that can be expressed as a sum of products:

$$y_n = a_1 y_{n-1} + a_2 y_{n-2} + \cdots + b_0 x_n + b_1 x_{n-1} + \cdots$$

Special instructions are also used to enhance the execution speed of signal processing and control algorithms. An example is the block repeat capability of DSPs that permits the reduction of the number of instruction cycles. Key features of the TMS320 DSP family of Texas Instruments, Inc., are examined next.

Texas Instruments DSPs : The TMS320 Family.

Texas Instruments DSPs : The TMS320 Family. The TMS320 family of 16- to 32-bit single-chip digital signal processors consists of five generations: three fixed point and two floating point devices. The 16-bit fixed point devices are TMS320C1x, TMS320C2x, and TMS320C5x. The 32-bit floating point devices are TMS320C3x and TMS320C4x.

The 16-bit fixed point devices implement Harvard-type architecture. The 32-bit floating point devices use a register-based architecture with multiple bus sets with program cache. Typical architectures of TMS320 devices are illustrated in Figures 10-28 and 10-29 showing the block diagrams of the TMS320C50 (16-bit fixed point DSP) and TMS320C30 (32-bit floating point DSP). Tables 10-6 and 10-7 summarize the key features of the Texas Instruments TMS320 family.

DSPs in Control of Power Electronic Systems.

DSPs in Control of Power Electronic Systems. In control of power electronic systems, the high computation capability of DSPs has been exploited to increase the sampling rate and to implement complex signal processing and control algo-

TABLE 10-5   FEATURES OF POPULAR RECENT DIGITAL SIGNAL PROCESSOR FAMILIES

Manufacturer	Model	Date	Description	Instruction cycle time
Analog devices	ADSP-2100	1986	16/40-bit fixed point	125 ns
	ADSP-2100A	1988	16/40-bit fixed point 16 × 24 program cache	80 ns
	ADSP-2201/2	1988	A 2100A with on-chip RAM, ROM, peripherals	60 ns
	ADSP-2105	1990	A low-cost 2101	100 ns
	ADSP-2111	1990	A 2101 with host interface port	60 ns
	ADSP-21020	1991	32/40-bit floating point, 32 × 48 program cache	40 ns
	ADSP-21060	1994	32-bit floating point, 2 serial ports, 128K × 32 RAM	25 ns
AT&T	DSP16/16A	1987/88	16-bit fixed point	55/25 ns
	DSP32/32C	1984/88	32-bit floating point	160/80 ns
Motorola	DSP56000	1987	24-bit fixed point mask-programmed	75–97.5 ns
	DSP56001	1987	24-bit fixed point	75–97.5 ns
	DSP56116	1990	16-bit fixed point 2K × 16 program RAM, 2K × data RAM, 16-bit timer, 24 I/O pins	50 ns
	DSP96001	1990	32-bit IEEE floating point, one external bus set	50 ns
	DSP96002	1990	32-bit IEEE floating point, two external bus sets	50 ns
NEC	μPD77230	1985	32-bit floating point	150 ns
	μPD77220	1986	24/48-bit fixed point	100 ns
	μPD77C25	1988	Updated version of 7720A (early NEC DSP)	122 ns
	μPD6380/ 6381	1990	Updated version of 77C25, digital audio	122 ns
	μPD77810	1990	Combined 77C25 and 8-bit microprocessor	
	μPD77240	1990	Updated version of 77230	90 ns
Texas Instruments	TMS320C1x	1986	16-bit fixed point	160–200 ns
	TMS320C2x	1987	16-bit fixed point	80–100 ns
	TMS320C3x	1988	32-bit floating point	50–74 ns
	TMS320C4x	1990	32-bit floating point, 6 communication ports	40–50 ns
	TMS320C5x	1990	16-bit fixed point	35–50 ns

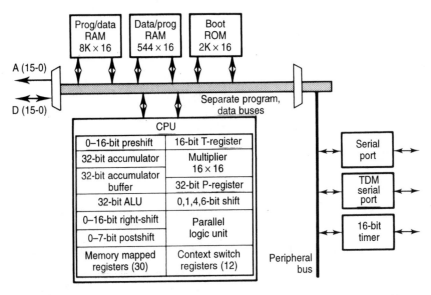

Figure 10-28. Block diagram of the Texas Instruments TMS320C50 16-bit fixed point DSP.

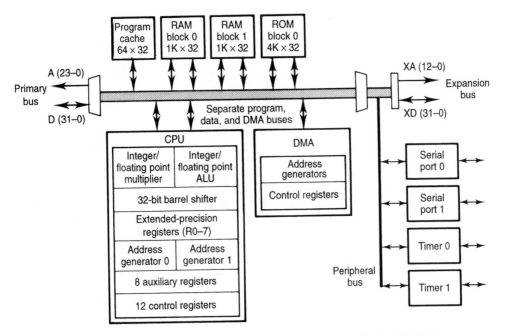

Figure 10-29. Block diagram of the Texas Instruments TMS320C30 32-bit floating point DSP.

TABLE 10-6   KEY FEATURES OF TEXAS INSTRUMENTS 16-BIT FIXED POINT DSPS

Features	TMS320C1x	TMS320C2x	TMS320C5x
Architecture	Harvard architecture	Harvard architecture	Advanced Harvard architecture
On-chip program/data RAM		1568 words	9K × 16-bit
On-chip data RAM	144 to 256 words	544 words	1056 × 16-bit
On-chip program ROM	1.5K to 4K words	4K words	8K × 16-bit
On-chip boot ROM			2K × 16-bit
External memory space	4K words	128K words	224K × 16-bit
Arithmetic-logic unit	32-bit ALU	32-bit ALU	32-bit ALU
Accumulator	32-bit accumulator	32-bit accumulator	• 32-bit accumulator • 32-bit accumulator buffer
Registers	2 auxiliary registers	Up to 8 auxiliary registers	• 8 auxiliary registers • 11 context-switch registers
Parallel logic unit			16-bit parallel logic unit
Shifters	0- to 16-bit barrel shifter	16-bit parallel shifter	• 0- to 16-bit left and right barrel shifters • 64-bit incremental data shifter
Multiplier	16 × 16 multiplier with a 32-bit product	16 × 16 parallel multiplier with a 32-bit product	16 × 16 parallel multiplier with a 32-bit product
Serial ports	1 full-duplex serial port (TMS320C14)	1 full-duplex serial port	• 1 synchronous full duplex • 1 TDM (time-division multiple-access)
Direct memory acccess		Concurrent DMA using an external hold	Concurrent DMA using an external hold
Timers	4 independent timers (TMS320C14)	16-bit timer	16-bit timer

TABLE 10-6　(CONT'D)

Features	TMS320C1x	TMS320C2x	TMS320C5x
Parallel I/O ports	16-pin bit-selectable port (TMS320C14)	16 parallel I/O ports	64K parallel I/O ports (16 memory mapped)
JTAG interface*			JTAG boundary scan logic
Interrupts	15 external/internal interrupts (TMS320C14)	3 external interrupt lines	4 external interrupt lines
Instruction cycle time	160–200 ns	80–100 ns	35–50 ns
Execution speed	6.25 MIPS	12.8 MIPS	6.8–20 MIPS
Multiply/accumulate	Single-cycle MAC	Single-cycle MAC	Single-cycle MAC

*Joint test action group interface = IEEE 1149.1 Standard for boundary scan testing.

rithms. The implementation of conventional algorithms (PID, feedforward, etc.) and advanced algorithms (Kalman filtering, state observer, adaptive control, etc.) utilizes the same basic operations as the signal processing algorithms and can thus benefit from the DSPs capabilities.

The signal processing capability of DSPs can also be used to reduce the number of sensors (in particular, position, speed, and flux sensors) in motion control systems. "Sensorless" operation is thus possible because the system variables usually provided by sensors can be estimated from the electrical variables. In adaptive control, the system parameters and/or state variables can be estimated using a state observer that can be effectively implemented using a DSP. Owing to their enhanced execution speed, DSPs can also be used with advantages to implement neural network and fuzzy logic-based motion control systems.

A major drawback of DSPs in motion control is the lack of on-chip resources to support real-time operation, such as complex timer unit, high-speed I/O ports, and complex interrupt controller. Additional chips are therefore necessary resulting in an increase of system hardware complexity.

To remedy this drawback, two approaches have been adopted by chip manu-facturers: (1) adding real-time control resources to an existing DSP and (2) adding DSP capabilities to an existing microcontroller.

Following the first approach, Texas Instruments, Inc., offers a special device, the TMS320C14 DSP microcontroller, specifically developed for motion control applications. This device contains on a single chip a TMS320C15 CPU and basic peripherals needed in controllers and typically found in 16-bit microcontrollers. These peripherals include 16 pins of bit I/O, 4 timers, 6 channels of PWM , 4 capture inputs for optical encoder interface, a serial port with UART mode, and 15 internal-

TABLE 10-7 KEY FEATURES OF TEXAS INSTRUMENTS 32-BIT FLOATING POINT DSPS

Features	TMS320C3x	TMS320C4x
Architecture	Register-based CPU architecture	Register-based CPU architecture
On-chip program/ data RAM	Block 0 1K × 32 Block 1 1K × 32	Block 0 1K × 32 Block 1 1K × 32
Cache	Instruction cache 64 × 32	Instruction cache 128 × 32
On-chip program ROM	4K × 32 (TMS320C30)	
On-chip boot ROM	Boot loader (TMS320C31)	Boot loader
External memory space	8K words on local bus (13 bits) 2M words on global bus (24 bits)	2G words on local bus (31 bits) 2G words on global bus (31 bits)
Arithmetic-logic unit	32-bit integer/32-bit logical/40 bit floating point	32-bit interger/32-bit logical/40-bit floating point ALU
Accumulators	8 extended-precision registers (accumulators) (32-bit integer/40-bit floating point)	12 extended-precision registers (accumulators) (32-bit integer/40-bit floating point)
Registers	• 2 auxiliary register arithmetic units • primary register file (28 registers)	• 2 auxiliary register arithmetic units • primary register file (32 registers)
Shifters	32-bit barrel shifter	32-bit barrel shifters
Multiplier	24-bit integer/32-bit floating point single-cycle multiplier	32-bit integer/40-bit floating point single-cycle multiplier
Serial ports	2 bidirectional serial ports supporting 8/16/24/32-bit transfers	
Direct memory access	On-chip DMA controller for concurrent I/O and CPU operation	DMA coprocessor 6 DMA channels
Communication ports		6 high-speed ports (20 Mbytes/s)
Timers	2 32-bit timers	2 32-bit time/event counters
Parallel I/O ports		
JTAG interface		
Interrupts	4 external interrupt lines DMA interrupts Internal interrupts	4 external interrupt lines DMA interrupts Internal interrupts
Instruction cycle time	50–74 ns	40–50 ns
Execution speed	13.5–20 MIPS 27–40 MFLOPS	275 MOPS 50 MFLOPS
Multiply/accumulate	Single-cycle MAC	Single-cycle MAC

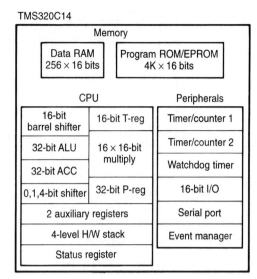

TMS320C14

Figure 10-30. Texas Instruments TMS320C14/E14 DSP microcontroller key features.

external interrupts. Figure 10-30 shows the key features of the TMS320C14 DSP microcontroller.

In the second approach, Motorola, Inc., adds DSP capabilities to the new 16-bit microcontrollers, the M68HC16 family, to enhance the execution speed of signal processing and control algorithms. Another drawback of DSPs is the high price of DSP-based systems. Many DSP commercial VME boards (boards using the VME bus structure) are available, but the price is very high compared to microcontroller boards. Also, the development of DSP-based real-time controllers requires very sophisticated development and debugging tools.

10.6.2. Reduced Instruction Set Computing Processors

RISC is a style of computer architecture that emphasizes the processor simplicity and efficiency. RISC processors have been developed to enhance the execution speed by using a pipelined architecture and a reduced instructions set containing a few simple instructions and by migrating complex operations to software. The term RISC is used by contrast to CISC which is usually associated with conventional microprocessors. In general, RISC architecture is characterized by a large register file and instruction cache and absence of data cache. To illustrate typical RISC architecture, block diagrams of two RISC processors are shown in Figures 10-31 and 10-32. They are, respectively, the Motorola MC88100 and the Integrated Device Technologies 79R3000A (this processor is based on the MIPS Technologies R3000 architecture).

Typical characteristics of current RISC processors are listed as follows:

- Reduced instruction set (50 to 75 instructions)
- Single-cycle execution

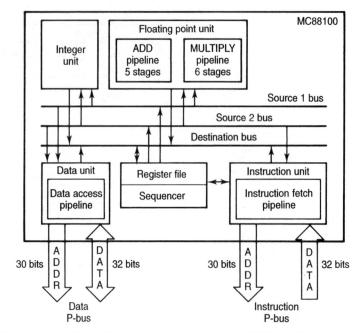

Figure 10-31. Block diagram of the Motorola MC88100 RISC processor.

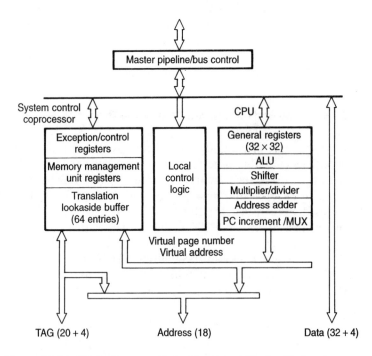

Figure 10-32. Block diagram of the Integrated Device Technology
IDT79R3000A RISC processor.

TABLE 10-8 CHARACTERISTICS OF POPULAR 32-BIT RISC PROCESSORS [9]

Microprocessors	i860	M88000	SPARC	R3000
Company	Intel Corp.	Motorola, Inc.	Sun Micro-systems, Inc.	MIPS Techno-logies, Inc.
Instruction size	32 bits	32 bits	32 bits	32 bits
Address space (size, model)	32 bits, flat	32 bits, flat	32 bits, flat	32 bits, flat
Data alignment	Aligned	Aligned	Aligned	Aligned
Data addressing modes	2	3	2	1
Protection	Page	Page	Page	Page
Page size	4 Kbytes	4 Kbytes	4–64 Kbytes	4 Kbytes
I/O	Memory mapped	Memory mapped	Memory mapped	Memory mapped
Integer registers (size, model, number)	31 GPR × 32 bits	31 GPR × 32 bits	31 GPR × 32 bits	31 GPR × 32 bits
Separate floating point registers	30 × 32 or 15 × 64 bits	0	32 × 32 or 16 × 64 bits	16 × 32 or 16 × 64 bits
Floating point format	IEEE 754 single, double	IEEE 754 single, double	IEEE 754 single, double	IEEE 754 single, double

- Instructions are implemented directly in hardware, precluding the need for microcoded operations
- Simple fixed format instructions (32-bit opcodes, 2 formats maximum)
- Simplified addressing modes (3 modes maximum)
- Register-to-register operation for data manipulation instructions
- Memory access by load-store operations
- Large register file (greater than 32 registers)
- Simple efficient instruction pipeline visible to compilers

Table 10-8 shows the main characteristics of four popular 32-bit RISC processors.

Chip manufacturers recently have introduced several RISC processors with performance significantly enhanced compared to the former 32-bit RISC processors. Main characteristics of these new-generation microprocessors, developed mostly for computing and data processing applications, are summarized in Table 10-9. In this

TABLE 10-9   CHARACTERISTICS OF CURRENT HIGH-PERFORMANCE RISC MICROPROCESSORS [60]

Microprocessor	Alpha 21064	MIPS R4400SC	PA7100	Power PC 601	Super Sparc	Pentium
Company	Digital Equipment Corp.	MIPS Technologies Inc.	Hewlett-Packard Co.	Apple Comp., Inc., IBM Corp. Motorola Inc.	Sun Micro-systems, Inc.	Intel Corp.
Date	1992	1992	1992	1993	1992	1993
Type	RISC	RISC	RISC	RISC	RISC	CISC
Width, bits	64	64	32	32	32	32
On-chip cache, KB (instruction/data)	8/8	16/16	None	32 unified	20/16	8/8
Off-chip cache, MB (instruction/data)	16	4	1/2	External controller	External controller	External controller
No. of registers (general-purpose/floating point)	32/32	32/32	32/32	32/32	136/32	8/8
Superscalar operation (instruction issue rate per cycle)	2	1	2	3	3	1
No. of independent units	4	NA	3	3	5	3
No. of pipeline stages (integer/floating point)	7/10	7/10	5/6	4/6	4/5	5/8
Multiprocessing support	Yes	Yes	Yes	Yes	Yes	Yes
Technology	0.68 μm CMOS	0.6 μm CMOS	0.8 μm CMOS	0.65 μm CMOS	0.7 μm CMOS	0.8 μm BiCMOS
Die size, mm	15.3 × 12.7	12 × 15.5	14.2 × 14.2	11 × 11	16 × 16	17.2 × 17.2
Transistors, millions	1.68	2.3	0.85	2.8	3.1	3.1
Clock, MHz	200	150	100	80	60	66

NA – Not available.

table, the Intel's Pentium microprocessor (a high-performance CISC processor) is shown for comparison purpose.

RISC Processors in Control of Power Electronic Systems. RISC architecture has been designed to optimize the execution speed of continuous streams of instructions by using caches, pipelines, registers, and memory management units (MMUs). Such architecture can adversely affect the determinism, latency, and response time in real-time control systems. RISC processors have not been widely accepted for real-time control systems where the requirements are different from those of data processing, because of some problems related to the architecture.

In control of power electronic systems, an efficient interrupt control is required. RISC processors generally handle interrupts completely in software so that the efficiency depends largely on the software. Many RISC processors have only a single interrupt vector with no hardware support for multiple interrupts. RISC processors typically have large register sets which directly affect context switching times and interrupt handling times. In a context switch, when the processor must change the control task in response to an event or because the task scheduler has switched to another task, the contents of all the registers has to be saved and then restored. This often results in relatively long context switching times compared to CISC processors. To reduce context switching and interrupt handling times, RISC systems often take advantage of compiler optimization and do not save or restore a full context on task switches or interrupts. The program cache in RISC processors is also a problem in real-time control. During context switches and other real-time events, the large cache has to be flushed and rewritten. This can have undesirable effects on the determinism of the real-time control system.

The situation is changing with the introduction of special RISC processors dedicated for control in which RISC cores are combined with other functions to address the specialized needs of real-time embedded applications. Examples of such RISC processors include the Advanced Micro Device AMD29000, the Intel i960, and the Advanced RISC Machines ARM60 and ARM600 processors. To answer the needs of real-time control, Integrated Device Technology, Inc., has developed a family of derivative MIPS processors specially addressing embedded applications that are called *RISControllers.* Examples include the R3001, R3041, R3051, and R3081 families. These devices integrate on the same chip a RISC core and functional units such as clock generator unit, system control coprocessor, instruction cache, data cache, bus interface unit, and floating point coprocessor in order to reduce the total system chips count. Figure 10-33 shows a block diagram of the IDT R3051 family of RISControllers.

10.6.3. Parallel Processors: Transputers and Parallel DSPs

Parallel computing concepts have been introduced almost 25 years ago, but they have really become reality owing to recent advances in VLSI and processor technologies permitting the construction of multiprocessor structures in which several pro-

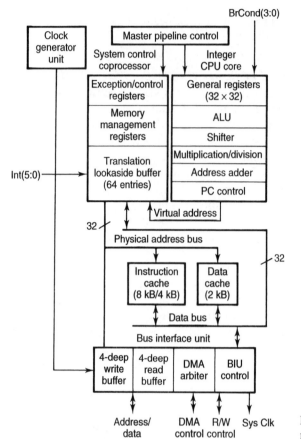

Figure 10-33. Integrated Device Technology's R3051 family block diagram.

cessors run concurrently. Multiprocessor architecture requires microprocessors with high communication capability as building blocks. The transputer, introduced about ten years ago by INMOS, Inc., is a device specially designed for parallel processing which is basically a microprocessor with on-chip memory and communication links. Texas Instruments, Inc., has recently introduced the TMS320C40, a parallel 32-bit floating point DSP with six high-speed communication links, designed for parallel processing.

Depending on the nature of the data processing algorithms, multiprocessor architecture can take several forms: linear arrays, two-dimensional arrays, hypercube, and the like. It has been shown that the distributed memory multiple instruction multiple data (MIMD) structure is well suited for the control of electrical drive systems because the control functions can be partitioned into modules operating in parallel. Since the interprocessor communication is generally very intensive in such configurations, the processors must be equipped with several high-speed communication links over which data are exchanged.

The major difficulty in multiprocessor real-time control systems concerns the partition of tasks between the processors in order to balance the computing load and to optimize the speeding factor (ratio of the execution times on a single processor and on a multiprocessor system). The interprocessor communication must be minimized (in time and in number) in order to reduce the overheads. The synchronization between the different tasks executed concurrently on several processors is also a critical point to be considered.

Transputers. The IMS T800 is a member of the INMOS transputer family. It integrates a 32-bit 10 MIPS processor, four serial communication links, 4 Kbytes of RAM, and a floating point unit on a single chip. An external memory interface allows access to a total memory of 4 gigabytes. A block diagram of the IMS T800 transputer is shown in Figure 10-34.

The IMS T9000, the latest transputer offered by INMOS, integrates a 32-bit integer processor, a 64-bit floating point processor, 16 Kbytes of cache memory, a communication processor, and 4 high-speed communication links. The T9000 is capable of a peak performance of 200 MIPS and 25 MFLOPS. It has been designed for multiprocessing and real-time applications. The communication system includes four serial links with a speed of 20 Mbytes/s each and a communication processor. This latter manages all link communications, operating concurrently with the main CPU so that data transfers do not adversely affect CPU operation. Figure 10-35 shows a block diagram of the IMS T9000 transputer.

The transputer instruction set has been designed for efficient implementation of high-level language compilers. The instructions are of the same format and chosen to give a compact representation of the operations of most frequent occurrence in

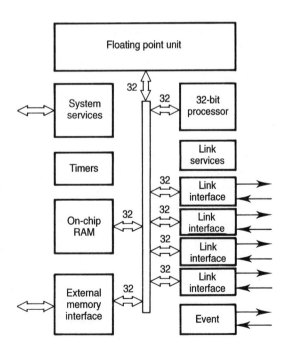

Figure 10-34. The Inmos IMS T800 32-bit transputer block diagram.

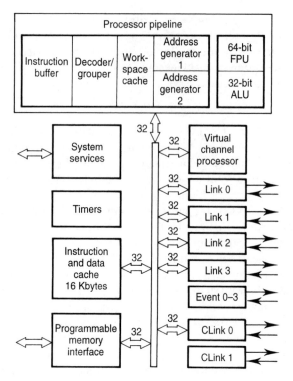

Figure 10-35. The Inmos IMS T9000 64-bit transputer block diagram.

programs. Transputers can be programmed in sequential languages such as C, Pascal, Ada, and FORTRAN. However, the facilities for concurrence and communication provided by the transputer architecture can be fully exploited by using the parallel Occam language specifically designed for transputers. This language enables a system to be described as a collection of concurrent processes which communicate one with another, and with the outside world, via communication channels.

Parallel DSPs: Texas Instruments' TMS320C40. The TMS320C40 is a 32-bit floating point DSP designed for multiprocessor systems. The CPU consists of 40-bit floating point/integer multiplier, ALU, 32-bit barrel shifter, 32-word primary register file, expansion register file, and two auxiliary register arithmetic units. Optimized for mathematically intensive applications, the CPU architecture and instruction set use high-level languages to achieve high performance and small code size. On-chip hardware supports IEEE format conversion, division, square root functions, and byte and half-word accessibility. Key features of the TMS320C40 parallel DSP are illustrated by the block diagram shown in Figure 10-36.

For direct processor-to-processor communication, the C40 contains six asynchronous, high-speed parallel communication ports with a maximum transfer rate of

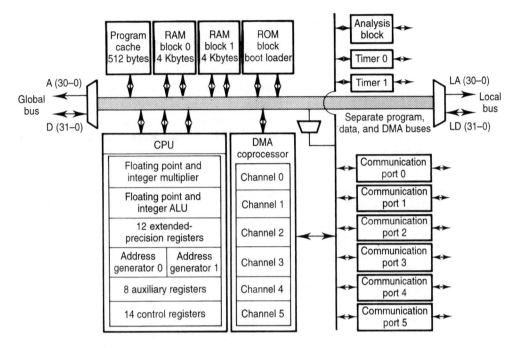

Figure 10-36. The Texas Instruments TMS320C40 parallel DSP architecture.

20 Mbytes/s each. Figure 10-37 illustrates the structure of one communication port of the C40. Each port independently buffers all input and output data transfers (by using two FIFOs), provides automatic arbitration and handshaking, and supports synchronization for the CPU or DMA.

The C40 features on-chip DMA coprocessor to support interprocessor communications concurrently with calculations being executed in the CPU.

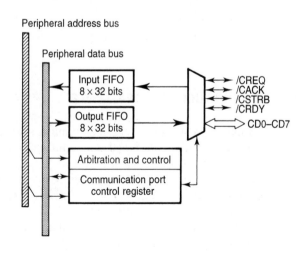

Figure 10-37. One parallel communication port in the TMS320C40.

The C40 computing capacity is adequate for scientific and motion control calculations (50 MFLOPS). The six parallel high-speed communication links permit building large processor networks with various topologies: tree structure, two-dimensional array, hypercube, and the like. The operation of the communication ports, by DMA or by CPU, is almost independent of the CPU so that communication overheads are minimum. Theoretical communication speed is 20 Mbytes/s which is acceptable for real-time control requirements. Several commercial VME boards using the C40s are now available with communication links ready for network connection.

Parallel Processors and Multiprocessor Systems in Control of Power Electronic Systems. Multiprocessor systems can be used with advantages in control of power electronic systems because of the performance they can provide. By applying parallel computing design, the control functions can be partitioned into several processors that operate in concurrence resulting in very fast computing time, and high sampling rates are thus possible. Parallel processors facilitate the implementation of multitasking motion control in which multiple programs must run simultaneously.

One major drawback of multiprocessor systems for motion control is the system price. Also, the development tools for parallel real-time processing are not fully matured so that the development cycle may be longer than with single processor systems.

Transputers have also some drawbacks when used in motion control systems. Because the transputer and Occam language have been designed for data processing applications, some important features required for real-time control systems are not present. The main drawbacks from real-time point of view are summarized as follows [107]:

- The task scheduling scheme does not permit an effective priority service.
- The priority of different services cannot be dynamically changed.
- Multiple events cannot be directly handled.
- Formal error handling mechanisms are not provided.

10.7. ASICS FOR CONTROL OF POWER ELECTRONIC SYSTEMS

Application-specific integrated circuit is a generic term that is used to designate any integrated circuit designed and built specifically for a particular application. The ASIC concept has been introduced with the advances of VLSI technology which permits the user to tailor his or her design during the development stages of an IC to suit his or her needs. The complexity of an ASIC can be in a wide range from simple interface logic to complete DSP, RISC processor, neural network, or fuzzy logic controller. The ASIC design method and the availability of DSP and RISC cores will give the motion control engineer the ability to integrate complete system solutions in a few ASICs. In this section, we will examine the capabilities of ASICs and their application in control of power electronic systems.

10.7.1. ASIC Technology

The advancement of large-scale integration process has resulted in two major ASIC technologies, CMOS and BiCMOS, that have attained feature sizes of 0.5 μm. With CMOS process, it is possible to manufacture ASIC devices with 250,000 gates or higher (one gate is generally defined as a single NAND gate). On the other hand, BiCMOS gate arrays (containing bipolar and CMOS devices) will offer greater operating speed at the expense of a more complex process and lower densities.

- *CMOS ASICs* are offered as standard cells and gate arrays technologies. With standard cells, processor cores can be integrated with different memory blocks and logic modules, providing great flexibility. However, the prototyping cost is much higher. On the other hand, with CMOS gate arrays (sea of gates technology), memory blocks and logic functions can be designed. Several CMOS gate arrays are offered with a fixed number of available gates and I/O buffers and processor cores. A 0.8 μm CMOS ASIC can contain up to 250,000 gates. With a 0.5 μm CMOS process, it is now possible to pack up to 600,000 usable gates in a single device.
- *BiCMOS ASICs* combine CMOS transistors and bipolar transistors using sea of gates technology. The operating frequency of BiCMOS devices is relatively high (100 MHz) because of the drive capacity of bipolar transistors. However, the density is lower, for example, a 0.8 μm BiCMOS ASIC can contain only up to 150,000 gates. With 0.5 μm BiCMOS technology, it is possible to obtain ICs having up to 300,000 usable gates.
- *Mixed-signal ASICs* (containing both digital and analog components on the same chip) are recently offered by several chip suppliers providing more possibilities for integration of complex systems. These chip-level systems can implement combined analog-digital designs that formerly required board-level solutions. Analog cells include operational amplifiers, comparators, D/A and A/D converters, sample-and-hold, voltage references, and RC active filters. Logic cells include gates, counters, registers, microsequencer, programmable logic array (PLA), RAM, and ROM. Interface cells include 8-bit and 16-bit parallel I/O ports as well as synchronous serial ports and UARTs.
- *RISC and DSP cores* are now offered as megacells by several chip suppliers permitting the design of customized advanced processors using an ASIC design methodology. Building blocks such as DSP cores, RISC cores, memory, and logic modules can be integrated on a single chip by the user using advanced Computer-Aided Design (CAD) tools. As an example, Texas Instruments, Inc., offers DSP cores in the C1x, C2x, C3x, and C5x families as ASIC core cells. Each core is a library cell, including a schematic symbol, a

timing simulation model for the simulation engine, chip layout files, and a set of test patterns.

10.7.2. ASIC Design

The design process of an ASIC consists of three main stages:

1. Logic design and simulation
2. Placement, routing, and connectivity check
3. Mask layout and prototype production.

The end user can enter the design process following the semistandard, semicustom, and full-custom paths, depending on the specific requirements of her application. In semistandard design path, the end user submits her design proposal under the form of high-level specifications. The ASIC supplier tailors the IC design in accordance with mutually negotiable specifications. With semistandard ASICs, cost is highly negotiable if predicted volume is sufficient and trustworthy, and the IC manufacturer might retain some rights to resell the chip or parts of its design to others. In semicustom design path, the end user establishes the specifications, performs the logic design (schematic capture and design verification) and simulation using CAD tools usually provided by the ASIC supplier. He then submits a CAD netlist (a list of simulated network connections) and the performance specifications. The chip supplier then performs the placement, routing, connectivity check, and mask layout merging precharacterized physical blocks into a mosaic with its own unique customized metallization and builds the prototype chip. In full-custom design path, in addition to the semicustom design stages, the end user also goes through placement, routing, and connectivity check of his design. The chip supplier takes care only of mask layout and prototype production.

The design of semicustom ASICs can be performed using gate arrays or standard cells technologies.

A gate array is a CMOS LSI chip consisting of p devices, n devices, and tunnels in a repetitive, ordered structure on either a silicon or a sapphire substrate. All device nodes (gates, drains, and sources) are accessible. Gate arrays are available for both single-layer and multi-layer metallization. To design his ASIC using a gate array, the end-user defines the connections of the individual devices to realize the desired functions. At the fabrication stage, only metallization layers are deposited on the silicon. Signal routing over the gates makes the gates beneath unusable. In this approach, gate utilization factor is usually about 70-90 percent. Macros such as RAM and ROM are very inefficient for implementation. However, lower cost and quicker production time are expected for this technology.

In cell-based approach, no fixed positions for gates and routing channels are predefined. The integrated circuit is designed using libraries of building blocks with specific logic functions. The chip supplier provides in general extensive libraries of well characterized and verified standard cells, supercells, and megacells. To design his ASIC, the end user combines the library cells into the configuration that performs the functions required by his specific application. The fabrication process involves the etching of the required gates as well as the deposition metallization of layers. Standard-cell technology offers a better utilization factor for silicon. Dedicated macros for RAM and ROM ensure reduced gates count and minimum silicon area. A longer fabrication time is expected since more steps are required.

The design of ASICs is performed usually in open architecture CAD systems on graphics engineering workstations using different software tools: schematic capture, simulation and fault simulation, logic optimization and synthesis, placement and routing, layout versus schematic, design rule check, and functions compiler.

The design of large ASICs typically uses a high-level design language (HDL, hardware description language) to help designers to document designs and to simulate large systems. The most common hardware description languages are Verilog and VHDL (the latter conforms to IEEE 1076 standard).

The design of a high-performance mixed-signal IC is inherently more difficult than the design of a logic IC. The variety of analog and digital functions requires a cell-based approach. Thorough simulation and layout verification are necessary to ensure the functionality of the prototype ASIC.

10.7.3. Field-Programmable Gate Arrays and Programmable Logic Devices

Field-programmable gate arrays (FPGAs) are a special class of ASICs that differs from mask-programmed gate arrays in that their programming is done by end users at their site with no IC masking steps.

An FPGA consists of an array of logic blocks that can be programmably connected to realize different designs. Current commercial FPGAs utilize logic blocks that are based on one of the following: transistor pairs, basic small gates (two-input NANDs and exclusive-ORs), multiplexers, look-up tables, and wide fan-in AND-OR structures.

The programming of FPGAs is via electrically programmable switches that are implemented by one of three main technologies:

1. *Static RAM technology.* The switch is a pass transistor that is controlled by the state of a static RAM bit. A SRAM-based FPGA is programmed by writing data in the static RAM.

2. *Antifuse technology.* An antifuse is a two-terminal device that irreversibly changes from a high-resistance to a low-resistance link when electrically programmed by a high voltage.

3. *Floating gate technology.* The switch is a floating gate transistor that can be turned off by injecting a charge on the floating gate. The charge can be removed by exposing the floating gate to ultraviolet (UV) light (EPROM technology) or by using an electric voltage (EEPROM technology).

The design process of an FPGA consists of three main stages:

1. Logic design and simulation
2. Placement, routing, and connectivity check
3. Programming.

This process is the same as that used for a semicustom ASIC gate array, except for the last stage, and uses mostly the same software tools.

Current FPGAs offer complexity equivalent to an 20,000-gate conventional gate array and typical system clock speeds of 40–60 MHz. This size is much smaller than mask-programmed gate arrays but large enough to implement relatively complex functions on a single chip.

The main advantage of FPGAs over mask-programmed ASICs is the fast turn-around that can significantly reduce design risk because a design error can be corrected quickly and inexpensively by reprogramming the FPGA.

Programmable logic devices (PLDs) are uncommitted arrays of AND and OR logic gates that can be organized to perform dedicated functions by selectively making the interconnections between the gates. Recent PLDs have additional elements (output logic macro cell, clock, security fuse, tristate output buffers, and programmable output feedback) that make them more adaptable for digital implementations. The most popular PLDs are PALs (programmable array logic) and GALs (generic array logic). Programming of PLDs can be done by blowing fuses (in PALs) or by EEPROM or SRAM technologies which provide reprogrammability.

The main advantages of PLDs compared to FPGAs are the speed and ease of use without nonrecurring engineering cost. The size of PLDs is, on the other hand, smaller than that of FPGAs. Current PLDs offer complexity equivalent to 8000 gates and speed up to 100 MHz.

10.7.4. Examples of ASICs for Control of Power Electronic Systems

In power electronics control systems, ASIC technology permits the design engineer to tailor the processor and the peripheral devices to obtain the desired specifications for his application. Using ASIC methodology, a motion control engineer can design her own control system on one or several chips using building blocks such as DSP or RISC cores, memory, analog, and logic modules. Optimized integration level and performance can be thus achieved. The high integration level results in a reduced chips count that can lower significantly the fabrication cost and improve the system reliability.

A disadvantage of ASICs in motion control systems is the lack of flexibility to modify or to adapt the design to different types of motor drives, once the chip is built. To change the design, even a small detail, it is necessary to go back to the initial design stages. The high development and fabrication cost for an ASIC can be thus justified only in large volume production.

In small-volume production and in prototyping stages, FPGAs offer a realistic alternative to full gate arrays design to implement specific motion control functions of medium complexity requiring less than 20,000 gates.

Chip manufacturers are now offering a number of standard ASICs that perform complex functions in drive control systems such as coordinates conversion (a–b–c/d–q conversion), pulse width modulation, PID controllers, fuzzy controllers, neural networks, and so on. Such devices can be used with advantages in motion control designs allowing reduction of processor computing load and increase of the sampling rate. In the following, some examples of commercial ASICs designed for motion control are presented.

- Analog Devices AD2S100/AD2S110 AC vector controller performs the Clarke and Park transformations usually required for implementing field-oriented control of AC motors. The Clarke transform converts a three-phase signal (a–b–c coordinates) into an equivalent two-phase signal (α–β coordinates). The Park transform rotates the resulted vector at the update position information applied to the input (α–β to d–q coordinates).
- Hewlett-Packard HCTL-1000 is a general-purpose digital motion control IC which provides position and velocity control for DC, DC brushless, and stepper motors. The HCTL-1000 executes any one of four control algorithms selected by the user: position control, proportional velocity control, trapezoidal profile control for point to point moves, and integral velocity control.
- The Signetics HEF4752V AC motor control circuit is an ASIC designed for the control of three-phase pulse width modulated inverters in AC motor speed control systems. A pure digital waveform generation is used for synthesizing three 120° out-of-phase signals, the average voltage of which varies sinusoidally with time in the frequency range 0 to 200 Hz.
- American Neuralogix NLX230 Fuzzy microcontroller is a fully configurable fuzzy logic engine containing an 1-of-8 input selector, 16 fuzzifiers, a minimum comparator, a maximum comparator, and a rule memory. Up to 64 rules can be stored in the on-chip, 24-bit-wide rule memory. The NLX230 can perform 30 millions rules per second.
- Intel 80170X ETANN (Electrically Trainable Analog Neural Network) simulates the data processing functions of 64 neurones, each of which is influenced by up to 128 weighted synapse inputs. The chip has 64 analog inputs and outputs. Its control functions for setting and reading synapse weights are digital. The 80170X is capable of 2 billions multiply-accumulate operations (connections) per second.

10.8. DESIGN OF MICROPROCESSOR-BASED CONTROL SYSTEMS

The development of a microprocessor-based control system for power electronics is a complex task consisting of several stages usually completed by several engineers. It involves the design of both hardware and software components and their integration in considering various factors such as system performance specifications, processor computing capacities, hardware availability, software development and debugging tools, and system cost. This development can follow the same guidelines as that adopted for any real-time control system. However, the motion control designer has to pay particular attention to the constraints imposed by the control configuration and strategy since the final design can be greatly affected.

10.8.1. Development Cycle

Figure 10-38 shows a general flow diagram illustrating the different stages involved in the development cycle of a real-time microprocessor-based motion control system. The main design stages will be discussed shortly.

10.8.2. System Requirements and Preliminary Design

During the first development stage, the system requirements have to be established under the form of specifications. The system specifications must be as detailed as possible since the subsequent stages will depend on them. The specifications concern not only the functions that the motion control system has to accomplish

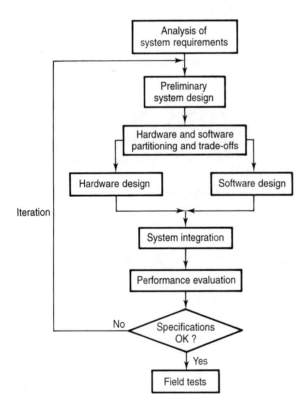

Figure 10-38. Development cycle of microprocessor-based real-time motion control systems.

but also the performance it must provide. The functional specifications must detail the control strategy and configuration as well as the different control and regulation tasks that the control system has to accomplish. The major performance specifications of a motion control system concern the response time, the response accuracy, and the communication interface.

The system specifications will permit the engineer to achieve his preliminary design and proceed to the selection of one or many appropriate microprocessors capable to accomplish the required tasks. Processor selection is a major task that requires a good knowledge of the desired functions and the performance of the final system. If an ASIC or an ASIC chips set is selected instead of commercially available processors, it will be necessary to design the ASICs in considering the desired system functions.

10.8.3. Hardware and Software Partitioning and Trade-Offs

The analysis of the system functions and performance and the processor capabilities will permit the designer to partition the functions into modules. It has been proven that the modular design approach provides in general a high flexibility concerning analysis, design, construction, test, and debugging of the system.

The trade-offs between hardware and software approaches to implement the different functions in a motion control system are done on the basis of performance-cost compromise. Most of the functions can be implemented by software or hardware. By software, the processing time allowed to a given function will increase the sampling period resulting in performance decrease. By hardware, additional devices required by the functions will increase necessarily the system complexity and cost. Depending on the complexity of a specific function, the advantages of hardware approach versus software approach are not always evident. In general, hardware implementation is used to reduce the computing load of the CPU. In motion control, several time-consuming complex functions such as coordinates conversion (a–b–c/d–q conversion) or pulse width modulation can be advantageously implemented using ASICs.

10.8.4. Hardware Design

The hardware development will consist in selecting appropriate components to build the control system hardware according to the specifications established. Depending on the complexity and the desired performance of the control system, the design engineer can decide one of two possible hardware configurations: single-board or multiboard microcomputer.

A single-board microcomputer contains the microprocessor and all peripheral devices (including interface) on the same board. A major advantage of single-board microcomputer is that no system bus is required and the interconnection is simple. The control board can be connected directly to the power system through interface circuits that reside on-board.

A multiboard microcomputer consists typically of a card cage with a power supply and a backplane using a specific bus system. Several boards are connected to the backplane: processor board which may include several microprocessors and peripheral devices, digital I/O board, analog I/O board, and special power system interface board. The main advantage of this approach is the flexibility of hardware configuration. On the other hand, the interconnection is more complex, and the hardware cost is usually higher.

The motion control engineer may have to choose between two alternatives: (1) to build the processor board and the I/O boards to suit her application and (2) to use commercial processor and I/O boards that are available for different bus systems. Building her own processor board will ensure optimum utilization of the hardware, but the development time may be long. The use of commercial boards reduces the hardware development and debugging time. However, these boards are not always optimized for the specific application. If the design is based on an ASIC or an ASIC chip set, the processor board must be custom designed using a standard bus system.

We consider here the case where commercially available boards are used to implement the control system, so the hardware design will take a minor role compared to the software design.

The bus system used will depend on the selected processor and the operation environment decided for the motion control system. Most 8-bit and 16-bit microcontroller boards are available with simple bus systems such as EISA and STD buses. Advanced microprocessor boards are available with high-speed, high-performance bus systems such as VMEbus, S-Bus, and FutureBus+. Some bus systems widely used by board manufacturers are described in the paragraphs that follow.

ISA Bus. The ISA bus is used in IBM-PC and compatibles and consists of several versions.

- The 8-bit ISA bus (62-conductor connector) supports 8-bit data and 20-bit address.
- The 16-bit ISA bus (62-conductor connector + 36-conductor connector) supports 16-bit data and 24-bit address.
- The EISA (extended ISA) bus (62-conductor connector + 36-conductor connector) is a 32-bit bus that support 32-bit data and address.

The data transfer on ISA bus is synchronous with a maximum rate of 1 Mbytes/s. Boards design for ISA bus can be full size (10.6 cm × 33.5 cm), half-size, or quarter-size.

STD Bus (IEEE P961 Standard). The STD bus is based on a backplane having 56 conductors that supports 8-bit data and 16-bit address. The data transfer on STD bus is synchronous with a maximum rate of 1 Mbytes/s. STD bus boards are of 4.5 in. × 6.5 in. size. Recently, the STD32 bus has been introduced to support 32-bit data and address.

VME Bus (IEEE P1014 Standard). The VME bus is based on a backplane consisting of two sets of 96 conductors. VME boards are available as (1) single wide (16 cm × 10 cm) for 16-bit data bus and 24-bit address bus and (2) double width (16 cm × 23.4 cm) for 32-bit data bus and 32-bit address bus. The data transfer on VME bus is asynchronous with a maximum rate of 24 Mbytes/s.

10.8.5. Software Design

The development of real-time control software can be done following three main stages: simulation, off-line development, and real-time integration. In the first stage, the control algorithms are developed using a simulation environment such as MATLAB or Xmath. The algorithms and control configurations are tested and debugged using a high-level language that facilitates the development task. In the second stage, the control tasks are written as several modules which are tested individually first. Then, the modules are combined and tested in an off-line context. Finally in the third stage, the control modules are integrated with a real-time operating system, and the whole is tested and debugged in real-time conditions.

The software developed for real-time control typically contains a real-time operating system that schedules and manages different specific functions under the form of modules (tasks or processes). With the exception of very simple systems where the RTOS kernel can be written by the control engineer, the control software is generally built around an acquired real-time kernel. Figure 10-39 shows a typical structure of software for real-time control of power electronic systems.

Real-time operating systems with various features are commercially available for different microprocessors. Some examples of commercial RTOS are VxWorks from Wind River Systems, Inc., SPOX from Spectron, Inc., and pSOS+ from Integrated Systems, Inc. The important features of a RTOS are real-time multitasking, multiprocessing, and source and symbolic debugging capabilities. Real-time multitasking features include event-driven scheduling, dynamically prioritized

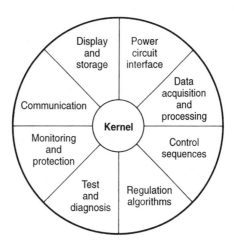

Figure 10-39. Typical structure of real-time control software.

tasks, synchronization and communication facilities (between tasks), timer services, and handling of device interrupts. A multiprocessing feature provides facilities for programmers to employ a variety of schemes for multiprocessor structures (shared memory, communication links, private bus). Source and symbolic debugging facilitates the debugging tasks in multitasking software.

The control software can be written by using assembler language or high-level language. Assembler has always been recognized as an effective programming language for real-time control systems because it gives access to the processor internal structure. The codes can be optimized to use efficiently the available memory space and to optimize the execution speed. A major drawback of assembler programming resides in the processor dependence of the developed software. At present, C language is widely accepted as a programming language for real-time control systems because of its portability and effectiveness in manipulating hardware resources. The developed codes using C language can be brought to another processor generation or to another microprocessor with minimum modifications. However, the generated code is less compact than those produced by assembler language. Several optimized C compilers are now available for microprocessors, microcontrollers, DSPs, and RISC processors, capable of generating very compact codes. In general, C or C++ language is used for software coding except for time critical functions that are better implemented in assembler language. For multiprocessor systems, Parallel C appears to be accepted as an appropriate high-level language for real-time parallel software development.

10.8.6. System Integration and Performance Evaluation

The design of the microprocessor-based control system is completed by integrating hardware and software together. The interaction between the two parts can be studied in real time by running the developed software on actual hardware or by using an in-circuit emulator. The performance of the control system can be thus evaluated under real operating conditions and compared to the specifications established during the first design step.

To evaluate the performance of the designed system, effective tools are needed. The evaluation tools, which may include a test system and a hardware simulator, have to be developed at the same time as the control system.

The test system includes hardware and software facilities which are added to the target system to collect and analyze data. In general, data acquisition capability must be included to acquire system variables and internal variables must be made available. Also, test procedures must be established to systematically evaluate the control performance.

A hardware simulator is needed when the controlled power electronic system is of very high power or difficult to operate in real conditions. The hardware simulator allows the design engineer to reproduce realistic operating conditions in test laboratory. The iteration through all the design steps will be necessary to ensure that the system functions conform to the specifications and the performance criteria are met.

10.9. DEVELOPMENT TOOLS

The development of a microprocessor-based control system necessitates effective tools to design, implement, integrate, and test hardware and software. Since the microprocessors become more and more complex, the development tools must be sufficiently sophisticated to fully exploit the capabilities of the processors.

10.9.1. Development System

A development system is required at every stage of the development cycle for a microprocessor-based control system. The role of a development system is to provide the designer an environment and necessary tools to edit, compile, test, and debug the control software in operating conditions similar to that of the actual system. Depending on the hardware and software decided, the development system may be very simple or very complex. In general, the complexity of a development system increases with that of the microprocessor.

Integrated development systems for a specific microprocessor family are available. But their price is high, and their application is limited to this family only. We can obtain the same features with higher flexibility in using development systems based on personal computers (IBM-PC, Macintosh, etc.) or UNIX workstations (Sun, DEC, etc.).

Low-Cost Development Setup. For the development of 8-bit and 16-bit microcontroller applications, the manufacturers offer a wide variety of evaluation boards (EVBs) and evaluation modules (EVMs) which can be connected to a personal computer (PC) to form a complete development system.

Figure 10-40 shows a typical development system based on an EVB. This latter is connected to the PC by a serial link (RS-232). The PC is equipped with text editor, assembler, C compiler, and simulator for code development. The executable code is downloaded to the EVB for execution and debugging in real time with the aid of a monitor program residing in the EVB's memory (EPROM). The capabilities of this monitor are usually limited but sufficient for developing small systems.

In the production phase, the EVB itself can be used as control system hardware. The debugged software can be then programmed in EPROM and replaces the EVB monitor program. The hardware development cost in this case is minimum. The

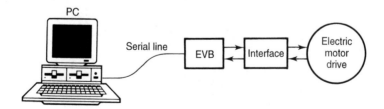

Figure 10-40. Low-cost development system based on an EVB.

designer can also decide to design his own microcontroller board (based on the EVB design) to suit specific applications.

PC-Based Development Setup. If the control system is designed with add-on boards plugged in a PC, then a PC-based development system would be appropriate. Figure 10-41 shows a typical development setup based on a PC and plug-in boards. The PC bus is used for communication between the PC itself and the control processor board. As in the previous setup, control software is developed in the PC, which is equipped with text editor, assembler, C compiler, simulator, and debugger. The executable code is downloaded to the control processor memory for execution and debugging in real-time with the aid of a debugger in the PC.

In the production phase, the same configuration can be used as control system hardware. The debugged software will be programmed in EPROM, which replaces the RAM on the processor board. The hardware development cost is also minimum.

Advanced Development Setup. The development of real-time control systems based on advanced processors (DSPs, RISC processors, transputers) requires sophisticated development systems with advanced software tools. The commercially available boards for these processors are most of the time VME boards. A typical development setup for advanced processors based on a workstation and VME boards is shown in Figure 10-42.

The control hardware resides in a VME card cage, including processor board, digital I/O board, analog I/O board, and power system interface board. Since the workstation bus is not VME compatible, a bus adaptor is required to connect the workstation bus to the VME bus (for example, S-bus to VME-bus).

The control software is developed on the workstation which can be connected to a Ethernet network. Development software, including text editor, assembler/linker, C compiler, simulator, and debugger, resides in the workstation or in the network.

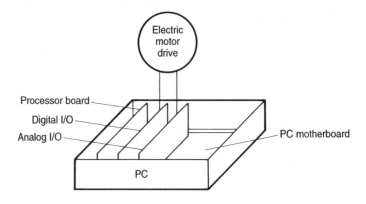

Figure 10-41. Development system based on a PC and add-on boards.

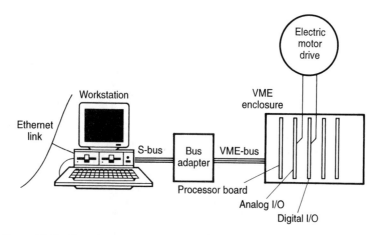

Figure 10-42. Development system based on a workstation and VME boards.

The executable code is downloaded to the processor board memory for execution and debugging in real-time with the aid of a debugger on the board itself.

10.9.2. Software Development Tools

The development and debugging of control software for a microprocessor-based control system will require different tools depending on the selected processors, system complexity, and development approach. The degree of sophistication of the development tools will affect the development time and the performance.

The software development tools for real-time control systems include typically text editors, C compilers, macro assemblers, and linkers. Debugging tools include debuggers, simulators, and in-circuit emulators. The software development and debugging tools can run on different computer platforms, the most popular ones are XWindows on Sun computers, and Windows on PCs.

Typical software development process is illustrated by the flow diagram shown in Figure 10-43. Various development tools are the following:

- *Text editor* allows one to create and edit source files (C language, Assembly language, macro).
- *C compiler* translates C source code into microprocessor Assembly language source code.
- *Assembler* translates Assembly language source files into machine language object files. Assembler source files can contain microprocessor instructions, assembler directives, and macro directives.
- *Linker* combines object files into a single executable object module in performing relocation and resolving external references.

Various debugging tools are the following:

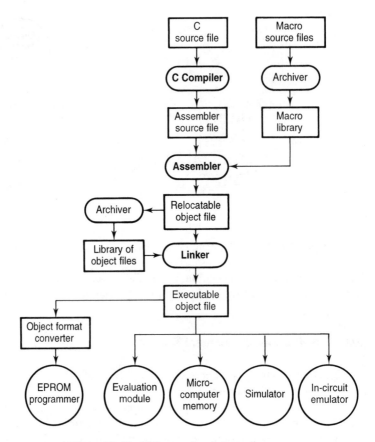

Figure 10-43. Software development process.

- *Simulator* is a software program running on a host computer that simulates the microprocessor instructions.
- *In-circuit emulator (ICE)* is a special tool connected to the target system microprocessor socket through a probe to control the target system. An ICE contains usually a processor of same type and permits symbolic debugging in the target system.
- *Evaluation module* is a development board built around the specific microprocessor intended for full-speed emulation and hardware debugging.

10.10. APPLICATION EXAMPLES

The application of microprocessors and digital ICs in the control of power electronic systems is a very broad issue and cannot be adequately covered without going into a very detailed description of every work. Nevertheless, we can have a general idea of the current development trends in examining typical recent research and development works on the subject. This section describes two digital control systems for power electronic systems in which microcontrollers and advanced microprocessors are used to implement control functions.

EXAMPLE 1   Digital Control of Permanent Magnet (PM) Synchronous Motor Drive for Electric Vehicle Propulsion [46]

In this work, a multiprocessor system based on a 16-bit microcontroller (Intel 8097) and two DSPs (Texas Instruments TMS32010) for the control of a PM synchronous motor drive is described. Figures 10-44 and 10-45 show, respectively, a simplified diagram of the controller hardware and the control configuration in constant-torque region.

The input signal processor (ISP) is responsible for the acquisition and processing of the motor currents while the output signal processor (OSP) produces reference current waveforms for the current regulators. Peripheral devices (A/D, D/A, and R/D converters) are used to interface the DSPs with the power system.

The 8097 microcontroller is primarily responsible for estimating and regulating the torque and flux of the PM motor. Inputs to the estimators are provided by the input signal processor and outputs from the regulators are transmitted to the output signal processor which transforms them into three-phase current references. Interprocessor communication is accomplished by using first-in, first-out (FIFO) registers.

EXAMPLE 2   Digital Control of Induction Motor Drive [80]

In this work, an induction motor drive system using a multitransputer network to handle real-time control, signal processing, housekeeping, and diagnosis functions is described. Figure 10-46 shows the configuration of the transputer system in which the vector control algorithm is implemented.

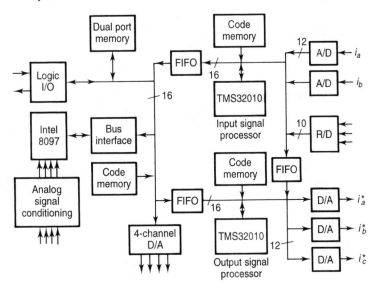

Figure 10-44. Diagram of the digital controller for an interior permanent magnet synchronous motor drive for electric vehicle propulsion [46].

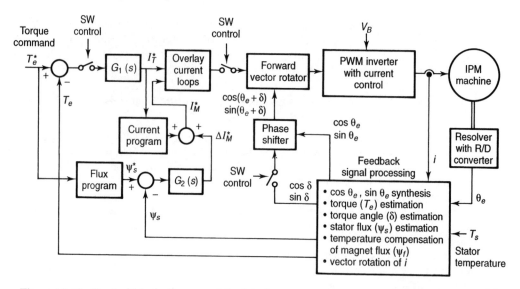

Figure 10-45. Control block diagram of the interior permanent magnet synchronous motor drive system for electric vehicle propulsion in constant torque region [46].

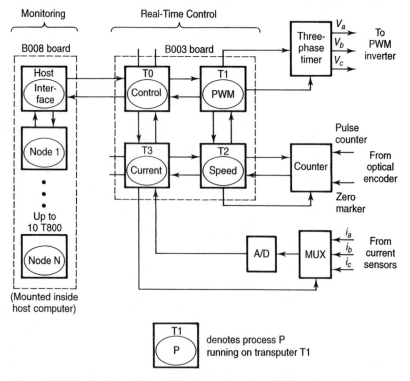

Figure 10-46. Configuration of the transputer-based induction motor control system [80].

The transputer system consists of a standard B008 board containing up to ten T800 transputers, hosted by a personal computer, and a B003 board containing four T414 transputers. The B008 board runs the transputer development system (TDS) and also assumes monitoring functions. Real-time control of the drive system is achieved by the B003 board. The interfaces with the induction motor drive include a three-phase timer for PWM waveform generation, a counter for position and speed measurement, and an A/D converter for current sensing.

The real-time control software is partitioned into modules (processes): PWM waveform generation, current control, speed control, system control, and interface. These processes are allocated to different transputers that operate concurrently.

10.11. CONCLUSION

During the last two decades, microprocessor technology has advanced with a continuous rhythm and several processor generations have been developed with increasing computing capabilities. Today's microprocessors provide ever more computing power for most demanding applications. This advancement is expected to continue during the next decade at an accelerating rate.

Recently introduced microprocessors can be classified into three large categories according to their structure: CISC processors, RISC processors, and digital signal processors. The computing capabilities of these devices, tremendously enhanced as compared to the last generation of microprocessors, include superscalar operation and multiprocessing operation which result in higher execution speed and more complex processing.

The new advanced microprocessors are designed primarily for computing and data processing applications, but control systems can also benefit from their enhanced performance. The potentials of new microprocessors in control of power electronic systems are numerous. The high execution speed of new microprocessors can help to increase the sampling rate. The bandwidth can thus attain values comparable to that of analog controllers, and the load disturbance rejection can be enhanced. The high computing capabilities of today's processors allow one to implement in real-time complex control algorithms such as state feedback control, optimal control, state observers, Kalman filters, and adaptive control with high sampling rate. Next-generation intelligent controllers using expert systems, neural networks, and fuzzy logic can be implemented using advanced processors and ASICs. The multiprocessing capability of new generation processors will permit developing multiprocessor motion control systems where parallel operation can solve present processing speed problems. It will also facilitate the implementation of multitasking motion control systems.

With the present development rate, it is difficult to project with accuracy the future developments in microprocessor technology in the next decade. Nevertheless, by observing the microprocessor industry behavior in the last decade, one can expect that during the next three years significant developments will be achieved in microprocessors addressed to real-time motion control systems. The developments will not be in the CPU computing performance but in the integration level that will permit the incorporation of more complex devices on the same chip. Also, the semicustom ASIC approach will be privileged for the design of microcontrollers for motion control systems.

In the next five years, trends for future developments in microprocessors and ICs for control of power electronic systems can be outlined as follows:

1. Development of new devices for real-time control that incorporate on the same chip advanced processors (DSP, RISC) and all required peripheral devices.

2. Development of semicustom ASIC technology permitting a motion control engineer to design a complete single-chip microcontroller containing advanced processors (RISC, DSP) and peripheral devices (digital and analog) suited for his application.

3. Development of cost-effective control-oriented parallel processors suitable for building multiprocessor motion control systems.

4. Development of neural networks and fuzzy logic chips suitable for implementing neural or fuzzy motion control systems.

The development of sophisticated real-time control systems requires effective hardware and software tools that are being constantly improved to respond to new requirements created by the new microprocessors.

Today's VLSI technology is sufficiently advanced to permit an end user to participate in the development stages of an ASIC specifically suited for his application. Although motion control engineers will continue to use off-the-shelf microprocessors, microcontrollers, DSPs, and RISC processors in their designs, new trends will lend to the semicustom ASIC design path because of the performance and possibilities it can offer. The evolution of VLSI and microprocessor technologies is expected to continue with an accelerating pace during the next decade. In a very near future, it will be possible for a motion control engineer to design a complete custom RISC-based or DSP-based controller and have it built into an ASIC.

References

Books

[1] Bose, B. K., *Microcomputer Control of Power Electronics and Drives*, IEEE Press, New York, 1987.

[2] Bose, B. K., *Power Electronics and AC Drives*, Prentice Hall, Englewood Cliffs, NJ, 1986.

[3] de Carlini, U., and U. Villano, *Transputers and Parallel Architectures: Message-Passing Distributed Systems*, Ellis Horwood, Chichester, West Sussex, U.K., 1991.

[4] Dote, Y., *Servo Motor and Motion Control Using Digital Signal Processors*, Prentice Hall, Englewood Cliffs, NJ, 1990.

[5] Franklin, G. F., J. D. Powell, and M. L. Workman, *Digital Control of Dynamics Systems*, Addison-Wesley, Reading, MA, 1990.

[6] Graham, I., and T. King, *The Transputer Handbook*, Prentice Hall, Englewood Cliffs, NJ, 1990.

[7] Lawrence, P. D., and K. Mauch, *Real-Time Microcomputer System Design*, McGraw-Hill, New York, 1987.

[8] Leonhard, W., *Control of Electrical Drives*, Springer-Verlag, New York, 1985.

[9] Patterson, D. A., and J. L. Hennessy, *Computer Architecture: A Quantitative Approach*, Morgan Kaufmann, San Mateo, 1990.

[10] Peatman, J. B., *Design with Microcontrollers*, McGraw-Hill, New York, 1988.

[11] Phillips, C. L., and H. T. Nagle, Jr., *Digital Control System Analysis and Design*, Prentice Hall, Englewood Cliffs, NJ, 1984.

[12] Laplante, Phillip A., *Real-Time Systems Design and Analysis*, IEEE Press, New York, 1992.

Databooks

[13] *The T9000 Transputer: Products Overview Manual*, INMOS Limited, Almondsbury, Bristol, U.K., 1991.

[14] *The Transputer Databook*, INMOS Limited, Almondsbury, Bristol, U.K., 1989.

[15] *Transputer Technical Notes*, Prentice Hall, Englewood Cliffs, NJ, 1989.

[16] *The Transputer Applications Notebook: Architecture and Software*, INMOS Limited, Almondsbury, Bristol, U.K., 1989.

[17] *The Transputer Applications Notebook: Systems and Performance*, INMOS Limited, Almondsbury, Bristol, U.K., 1989.

[18] *RISC Microprocessor Components & Subsystems Data Book,* Integrated Device Technology, Inc., Santa Clara, CA, 1992.

[19] *8-Bit Embedded Controller Handbook*, Intel Corporation, Santa Clara, CA, 1989.

[20] *16-Bit Embedded Controller Handbook*, Intel Corporation, Santa Clara, CA, 1989.

[21] *32-Bit Embedded Controller Handbook*, Intel Corporation, Santa Clara, CA, 1989.

[22] Motorola. *M68HC11 Reference Manual*, Prentice Hall, Englewood Cliffs, NJ, 1989.

[23] *MC68HC16Z1 16-Bit Modular Microcontroller*, Motorola, Inc., Phoenix, AZ, 1990.

[24] *MC88100 RISC Microprocessor User's Manual*, Motorola Inc., Phoenix, AZ, 1988.

[25] *MC88200 Cache/Memory Management Unit User's Manual*, Motorola, Inc., Phoenix, AZ, 1988.

[26] *DSP56000/DSP56001 Digital Signal Processor User's Manual*, Motorola, Inc., Phoenix, AZ, 1990.

[27] *DSP96002 IEEE Floating-Point Dual-Port Processor User's Manual*, Motorola, Inc., Phoenix, AZ, 1989.

[28] *TMS320C1x User's Guide*, Texas Instruments Inc., Houston, TX, 1989.

[29] *TMS320C14/E14 User's Guide*, Texas Instruments Inc., Houston, TX, 1988.

[30] *TMS320C2x User's Guide*, Texas Instruments Inc., Houston, TX, 1990.

[31] *TMS320C3x User's Guide*, Texas Instruments Inc., Houston, TX, 1991.

[32] *TMS320C4x User's Guide*, Texas Instruments Inc., Houston, TX, 1991.

[33] *TMS320C5x User's Guide*, Texas Instruments Inc., Houston, TX, 1990.

[34] *Digital Signal Processing Applications with the TMS320 Family*, Texas Instruments Inc., Houston, TX, 1986.

[35] *Digital Signal Processing Applications with the TMS320 Family*, Vol. 2, Texas Instruments Inc., Houston, TX, 1990.

[36] *Digital Signal Processing Applications with the TMS320 Family*, Vol. 3, Texas Instruments Inc., Houston, TX, 1990.

[37] *Digital Control Applications with the TMS320 Family*, Texas Instruments Inc., Houston, TX, 1991.

Papers

[38] Andrews, W. "RISC finding a place in real-life, realtime applications," *Computer Design*, Vol. 32, no. 8, pp. 67–70 and 86–88, August 1993.

[39] Asher, G. M., and M. Summer, "Parallelism and the transputer for real-time high-performance control of AC induction motors," *IEE Proc.*, Vol. 137, Part D, no. 4, pp. 179–188, July 1987.

[40] Ben-Brahim, L., and A. Kawamura, "Digital current regulation of field-oriented controlled induction motor based on predictive flux observer," *IEEE Trans. Ind. Appl.*, Vol. 27, no. 5, pp. 956–961, September/October 1991.

[41] Best, J., J. M. Pacas, and K. Peters, "New generation of intelligent drives with SERCOS interface," *Proc. Intelligent Motion '92 Conf.*, Nürnberg, pp. 289–302, April 1992.

[42] Booker, A., and J. McKeeman, "Design considerations for RISC microprocessor in realtime embedded systems," *Computer Design*, Vol. 32, no. 8, pp. 71–72, August 1993.

[43] Bose, B. K., "Motion control technology—present and future," *IEEE Trans. Ind. Appl.*, Vol. IA-21, no. 6, pp. 1337–1342, November/December 1985.

[44] Bose, B. K., "Power electronics and motion control: Technology status and recent trends," *IEEE Trans. Ind. Appl.*, Vol. 29, no. 5, pp. 902–909, September/October 1993.

[45] Bose, B.K., T. J. Miller, P. M. Szczesny, and W. H. Bicknell, "Microcomputer control of switched reluctance motor," *IEEE Trans. Ind. Appl.*, Vol. IA-22, no. 4, July/August 1986.

[46] Bose, B. K., and P. M. Szczesny, "A microcomputer-based control and simulation of an advanced IPM synchronous machine drive system for electric vehicle propulsion," *IEEE Trans. Ind. Elect.*, Vol. 35, no. 4, pp. 547–559, November 1988.

[47] Bowes, S. R., and P. R. Clark, "Simple microprocessor implementation of new regular-sampled harmonic elimination PWM techniques," *IEEE Trans. Ind. Appl.*, Vol. 28, no. 1, pp. 89–95, January/February 1992.

[48] Bowes, S. R., and P. R. Clark, "Transputer-based harmonic-elimination PWM control of inverter drives," *IEEE Trans. Ind. Appl.*, Vol. 28, no. 1, pp. 72–80, January/February 1992.

[49] Bowes, S. R., and P. R. Clark, "Transputer-based optimal PWM control of inverter drives," *IEEE Trans. Ind. Appl.*, Vol. 28, no. 1, pp. 81–88, January/February 1992.

[50] Brickwedde, A., "Microprocessor-based adaptive speed and position control for electrical drives," *IEEE Trans. Ind. Appl.*, Vol. IA-21, no. 5, pp. 1154–1161, September/October 1985.

[51] Cecati, C., and D. Q. Zhang, "A general-purpose, low cost multiple microprocessor system for electrical drives control," *Proc. IECON '91 Conf.*, pp. 647–652, November 1991.

[52] Cecati, C., and M. Tursini, "Vector control algorithms implementation for inverter-fed permanent magnet synchronous motor using tranputer," *Proc. IECON '91 Conf.*, p. 171–176, November 1991.

[53] Cecati, C., F. Parasiliti, and M. Tursini, "Microcomputer-based speed control of permanent-magnet synchronous motor drives," *Proc. IECON '92 Conf.*, pp. 101–106, November 1992.

[54] Cerruto, E., A. Consoli, A. Raciti, and A. Testa, "Adaptive fuzzy control of high performance motion systems," *Proc. IECON '92 Conf.*, pp. 88–94, November 1992.

[55] Chance, R. J., and J. A. Taufiq, "A TMS32010 based near optimized pulse width modulated waveform generator," *IEEE-IAS '88 Conf. Rec.*, pp. 903–908, 1988.

[56] Chung-Ming Young, T. L., and C. H. Liu, "Microprocessor-based controller design and simulation for a permanent magnet synchronous motor drive," *IEEE Trans. Ind. Elect..* Vol. IE-35, no. 4, pp. 516–523, 1988.

[57] T. D. Collings, and W. J. Wilson, "A fast-response current controller for microprocessor-based SCR-DC motor drives," *IEEE Trans. Ind. Appl.*, Vol. 27, no. 5, pp. 921–927, September/October 1991.

[58] Dhaouadi, R., and N. Mohan, "DSP-based control of a permanent magnet synchronous motor with estimated speed and rotor position," *EPE '91*

European Conf. Power Electronics and Applications Proc., pp. 1.596–1.602, 1991.

[59] Dote, Y., and K. Kano, "DSP-based neuro-fuzzy position controller for servomotor," *Proc. IECON '92 Conf.*, pp. 986–989, November 1992.

[60] Geppert, L., "Not your father's CPU," *IEEE Spectrum*, Vol. 30, no. 12, pp. 20–23, December 1993.

[61] Greene, J., E. Hamdy, and S. Beal, "Antifuse field programmable gate arrays," *Proc. IEEE*, Vol. 81, no. 7, pp. 1042–1056, July 1993.

[62] Guo, Y., H. C. Lee, and B. T. Ooi, "Extensible digital-signal-processing modules for real-time control and simulation," *Proc. IECON '93 Conf.*, pp. 2229–2234, November 1993.

[63] Hanselmann, H., "Implementation of digital controllers—a survey," *Automatica*, Vol. 23, no. 1, pp. 7–32, 1987.

[64] Harashima, F., S. Konda, and K. Ohnishi, "Multimicroprocessor-based control system for quick response induction motor drive," *IEEE Trans. Ind. Appl.*, Vol. IA-21, no. 3, pp. 602–609, May/June 1985.

[65] Harley, G. R., et al., "Transputer based digital controller with high performance I/O and SVM ASIC for AC drives," *EPE '91 European Conf. Power Electronics and Applications Proc.*, pp. 1.603–1.606, 1991.

[66] Hossain, A., and S. Suyut, "A new method of state machine controller design and implementation using programmable logic devices for industrial applications," *IEEE-IAS '93 Conf. Rec.*, pp. 2077–2083, 1993.

[67] Iizuka, K., et al., "Microcomputer control for sensorless brushless motor," *IEEE Trans. Ind. Appl.*, Vol. IA-21, no. 3, pp. 595–601, May/June 1985.

[68] Iwasaki, M., and N. Matsui, "DSP-based high performance speed control system of vector control IM with load torque observer," *Proc. Int. Power Electronics Conf.*, pp. 436–441, 1990.

[69] Iwasaki, M., and N. Matsui, "Robust speed control of IM with torque feedforward control," *Proc. IECON '91 Conf.*, pp. 627–632, November 1991.

[70] Ji, J. K., and S. K. Sul, "DSP-based self-tuning IP speed controller for rolling mill DC drive," *Proc. IECON '93 Conf.*, pp. 2276–2281, November 1993.

[71] Ko, J. S., J. H. Lee, S. K. Chung, and M. J. Youn, "A robust digital position control of brushless DC motor with dead beat load torque observer," *IEEE Trans. Ind. Elect.*, Vol. 40, pp. 512–520, October 1993.

[72] Kubo, K., M. Watanabe, and T. Ohmae, "A fully digitalized speed regulator using multimicroprocessor system for induction motor drives," *IEEE Trans. Ind. Appl.*, Vol. IA-21, no. 4, pp. 1001–1008, July/August 1985.

[73] Kubota, H., K. Matsuse, and T. Nakano, "DSP based adaptive flux observer of induction motor," *IEEE-IAS '91 Annual Meeting Conf. Rec.*, pp. 380–384, 1991.

[74] Kumar, P. P., R. Parimelalagan, and B. Ramaswami, "A microprocessor-based DC drive control scheme using predictive synchronization," *IEEE Trans. Ind. Elect.*, Vol. 40, pp. 445–452, August 1993.

[75] Kwon, B. H., and B. D. Min, "A fully software-controlled PWM rectifier with current link," *IEEE Trans. Ind. Elect.*, Vol. 40, pp. 355–363, June 1993.

[76] Leonhard, W., "Microcomputer control of high dynamic performance AC drives—a survey," *Automatica*, Vol. 22, no. 1, pp. 1–19, 1986.

[77] Li, W., and R. Venkatesan, "A highly reliable parallel processing controller for vector control of AC induction motor," *Proc. IECON '92 Conf.*, pp. 43–48, November 1992.

[78] Low, T. S., T. H. Lee, and K. T. Chang, "A nonlinear speed observer for permanent-magnet synchronous motors," *IEEE Trans. Ind. Elect.*, Vol. 40, pp. 307–316, June 1993.

[79] Luk, P. C. K., M. G. Jayne, and D. Rees, "The transputer control of variable speed induction motor drives," *EPE '91 European Conf. Power Electronics and Applications Proc.*, pp. 1.574–1.579, 1991.

[80] Luk, P. C. K., "On applying parallel processing to a versatile induction motor drive system," *Proc. IECON '93 Conf.*, pp. 907–912, November 1993.

[81] Marchesoni, M., G. Rossi, A. Scaglia, and P. Segarich, "Development and test of a new advanced DSP based architecture for robotics drives control," *Proc. IECON '93 Conf.*, pp. 1848–1853, November 1993.

[82] Matsui, N., "Recent trends in AC motion control," *Proc. IECON '92 Conf.*, pp. 25–30, November 1992.

[83] Matsui, N., and H. Ohasi, "DSP-based adaptive control of a brushless motor," *IEEE-IAS '88 Conf. Rec.*, pp. 375–380, 1988.

[84] Matsui, N., and M. Shigyo, "Brushless DC motor control without position and speed sensors," *IEEE Trans. Ind. Appl.*, Vol. 28, no. 1, pp. 120–127, January/February 1992.

[85] Naitoh, H., and S. Tadakuma, "Microprocessor-based adjustable-speed DC motor drives using model reference adaptive control," *IEEE Trans. Ind. Appl.*, Vol. IA-23, no. 2, pp. 313-318, March/April 1987.

[86] Naunin, D., and H. C. Reuss, "Synchronous servo-drive: A compact solution of control problems by means of a single-chip microcomputer," *IEEE Trans. Ind. Appl.*, Vol. 26, no. 3, pp. 408–414, May/June 1990.

[87] Naunin, D., S. Beierke, and P. Heidrich, "Transputers control asynchronous servodrives," *EPE '91 European Conf. on Power Electronics and Applications Proc.*, pp. 1.584–1.589, 1991.

[88] Norum, L., A. K. Adnanes, W. Sulkowski, and L. A. Aga, "The realization of a permanent magnet synchronous motor drive with digital voltage vector selection current controller," *Proc. IECON '91 Conf.*, pp. 182–187, November 1991.

[89] Ohishi, K., et al., "Microprocessor controlled DC motor for load insensitive position servo system," *IEEE Trans. Ind. Elect.*, Vol. IE-34, no. 1, pp. 44–49, 1987.

[90] Ostiguy, D., "Implementing DSP solutions in ASICs," *ICSPAT, The International Conference on Signal Processing Applications and Technology*, pp. 416–421, November 1992.

[91] Pillay, P., C. R. Alle, and R. Budhabhathi, "DSP-based vector and current controllers for a permanent magnet synchronous motor drive," *IEEE-IAS '90 Conf. Rec.*, pp. 539–544, 1990.

[92] Pollmann, A. J., "Software pulse width modulation for microprocessor control of AC drives," *IEEE Trans. Ind. Appl.*, Vol. IA-22, no. 4, July/August 1986.

[93] Retif, J. M., B. Allard, X. Jorda, and A. Perez, "Use of ASIC's in PWM techniques for power converters," *Proc. IECON '93 Conf.*, pp. 683–688, November 1993.

[94] Rose, J., A. El Gamal, and A. Sangiovanni-Vincentelli, "Architecture of field-programmable gate arrays," *Proc. IEEE*, Vol. 81, no. 7, pp. 1013–1029, July 1993.

[95] Simar, R., Jr., et al., "Floating-point processor join forces in parallel processing architectures," *IEEE Micro Mag.*, pp. 60–69, August 1992.

[96] Suemitsu, W., M. Ghribi, P. Viarouge, and H. Le-Huy, "Current regulation of a permanent-magnet synchronous motor using the TMS320C30 DSP," *Proc. IECON '92 Conf.*, pp. 1412–1416, November 1992.

[97] Summer, M., and G. M. Asher, "PWM induction motor drive using the INMOS transputer parallel processor," *IEEE-APEC '88 Conf. Rec.*, pp. 121–129, 1988.

[98] Suyitno, A., J. Fujikawa, H. Kobayashi, and Y. Dote, "Variable-structured robust controller by fuzzy logic for servomotors," *IEEE Trans. Ind. Elect.*, Vol. 40, pp. 80–88, February 1993.

[99] Tagawa, K., et al., "Design of a multi-DSP system using TMS320C25 and optimal scheduling for digital controllers," *Proc. IECON '92 Conf.*, pp. 1391–1396, November 1992.

[100] Tamaki, K., et al., "Microprocessor-based robust control of a DC servo motor," *IEEE Control Systems Mag.*, Vol. 5, pp. 30–35, October 1986.

[101] Trimberger, S., "A reprogrammable gate array and applications," *Proc. IEEE,*, Vol. 81, no. 7, pp. 1030–1041, July 1993.

[102] Tzou, Y. Y., and H. J. Wu, "Multimicroprocessor-based robust control of an AC induction servo motor," *IEEE Trans. Ind. Appl.*, Vol. 26, no. 3, pp. 441–449, May/June 1990.

[103] Vukosavic, S. N., and M. R. Stojic, "On-line tuning of the rotor time constant for vector-controlled induction motor in position control application," *IEEE Trans. Ind. Elect.*, Vol. 40, pp. 130–138, February 1993.

[104] Webster, M. R., et al., "A development system for high performance controllers for AC drives using transputers and parallel processing," *EPE '91 European Conf. Power Electronics and Applications Proc.*, pp. 1.580–1.583, 1991.

[105] Whitby-Strevens, C., "Transputers—past, present, and future," *IEEE Micro Mag.*, pp. 16–19 and 76–82, December 1990.

[106] Xu, X., and D. Novotny, "Implementation of direct stator flux orientation control on a versatile DSP based system," *IEEE-IAS '90 Conf. Rec.*, pp. 404–409, 1990.

[107] Zhang, D. Q., C. Cecati, and E. Chiricozzi, "Some practical issues of the transputer based real-time systems," *Proc. IECON '92 Conf.*, pp. 1403–1407, November 1992.

Chapter 11

Bimal K. Bose

Expert System, Fuzzy Logic, and Neural Networks in Power Electronics and Drives

11.1. INTRODUCTION

Expert system, fuzzy logic, and neural network belong to the area of artificial intelligence (AI), a major discipline in computer science. The AI techniques are recently finding widespread applications in science and engineering. The computing can be classified as "hard computing," or precise computation, and "soft computing," or approximate computation. The area of expert system belongs to hard computing, whereas soft computing encompasses fuzzy logic, neural network, and probablistic reasoning techniques, such as genetic algorithms, chaos theory, belief networks, and parts of learning theory.

The research in AI is very fascinating and challenging, and a large segment of the scientific and engineering community is devoting effort in this area. What is AI? AI is basically embedding human intelligence in a machine so that a machine can think intelligently like a human being. The term was used systematically since the Dartmouth College conference in 1956 when "artificial intelligence" was defined as "computer processes that attempt to emulate the human thought processes that are associated with activities that require the use of intelligence." Can a computer really think and make intelligent decisions? Or is it as good as the program implanted in it by a human programmer? The computer intelligence has been debated since its invention and will possibly continue so forever. There is no denying the fact that the human brain is the most complex machine on earth, and we hardly understand it and its behavior in thinking, learning, and reasoning for complex problems. Neurobiologists have tried to understand the microstructure of the brain nervous system and analyze how it functions. On the other hand, behavioral scientists like psychologists and psychiatrists have tried to analyze our thought process with a top-down approach. Possibly, the intricacies of the brain—a carefully guarded secret of God—will be ever shrouded with mystery in spite of our painstaking research for hundreds of years. However complex is the human thought process, there is no denying that computers can have adequate intelligence to help solve our problems that are difficult to solve in traditional ways. Therefore, it is true that today "artificial intelligence" techniques are finding widespread applications in practically all segments of our society. With the AI tools, a system is often defined as "intelligent," "learning," or having "self-organizing" or "self-adaptation" capability. However limited computer intelligence is, it has at least superiority over human intelligence in several aspects. The computer can process a problem extremely fast compared to a human being; it can work relentlessly without being tired and fatigued, and its problem-solving capability is not impaired with anger, emotion, boredom, and other frailties of a human being.

The goal of this chapter is to discuss the application of expert system, fuzzy logic, and neural network in power electronics and drives. In spite of widespread applications, these AI tools have hardly penetrated the power electronics area. In the author's opinion, AI technology will have major impact on power electronics in the coming decades. In this chapter, we will review the principles of expert system, fuzzy logic, and neural network in simple language and then supplement the understanding with adequate number of examples.

11.2. EXPERT SYSTEM

Expert system has traditionally been the most important branch of artificial intelligence. It is the forerunner among all the AI techniques, and from the beginning (1960s) to 1980s, both terms (expert system and artificial intelligence) have been used synonymously in the literature. Expert system has recently found wide applications in industrial process control, medicine, geology, agriculture, information management, military science, and space technology, just to name a few. However, its applications in power electronics and drives area are relatively few.

Historically, it was perceived in early times that the human brain makes decisions on the basis of "yes"–"no" or "true"–"false" reasoning. In 1854, George Boole first published his article "Investigations on the Laws of Thought," which gave birth to Boolean algebra and set theory. Gradually, the advent of electronic circuits and solid-state integrated circuits ushered the modern era of von Neumann–type sequential digital computation. Digital computers were known as "intelligent" machines because of their capability to process human thought—like yes (or logic 1) no (or logic 0) reasoning. Of course, using the same binary logic, computers can also solve complex scientific, engineering, and other data processing problems. Since the 1960s and in the early 1970s, it was believed that computers had severe limitations to solve only the algorithmic-type problems. An entirely new way of programming a computer which closely matches the human logical thinking process, called "expert system," was born. In the last two decades, the technology has practically reached the stage of maturity. It is true that in many applications, expert system programs outperform human experts. However, exaggerated and unrealistic performance goals claimed by the proponents of the technology have shadowed it with a cloud of controversy in the professional community.

11.2.1. Expert System Principles

What is an expert system? An expert system is basically an "intelligent" digital computer program that is designed to embed the expertise of a human being in a certain domain. Human thinking, learning, and reasoning processes are very complex as mentioned before, and a human being may have multiple areas of expertise. Consider a power electronics technician who has a special or domain expertise for fault diagnosis of a converter system. The technician can apply power to the converter and measure voltages and currents at appropriate points with the help of an oscilloscope and other measuring instruments. Then, based on his knowledge or expertise, he can conclude which devices have become defective or which control elements are giving faulty performance. He has learned or acquired this knowledge by education and experience over a prolonged period of time. The question is: Is it possible to implant the same knowledge in a computer program so that it can replace the human expert? The answer is a qualified "yes," because we need to understand that no computer program, however sophisticated, can ever replace human thinking. A conventional CAD program is characterized by heavy numeric computation and light logical computation, and its algorithm can be described by a flow chart. The program can have knowledge embedded in it, and the computation flow path can be altered by the logical signals. If the embedded knowledge requires change, time-consuming alteration of the program will be required. An expert system program's features, on the other hand, are entirely different because of its different structure. Figure 11-1 shows the basic elements of an expert system and its organization. The core of the expert system is the knowledge base, and for this reason, an expert system is also called a *knowledge-based system*. This domain knowledge is acquired from the domain expert for implanting in the computer program. The domain expert may be the power electronics engineer or technician who may not have the requisite software expertise to write the expert system program. Knowledge engineering is defined as the branch of computer science that deals with the acquisition of knowledge and representing this knowledge in efficient computer software. A knowledge engineer is basically a software specialist who gets the knowledge from the domain expert and translates it into software. Often, the domain expert (that is,, the power electronics engineer) can become educated in knowledge engineering and then he and the knowledge engineer are the same person. The knowledge base, as shown, is classified into two parts:

1. Expert knowledge
2. The data, facts, and statements that support the expert knowledge.

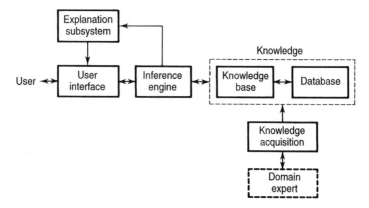

Figure 11-1. Basic elements of an expert system.

A limited amount of data can be directly implanted in the knowledge. For large data, a separate data base is used to support the knowledge. This is analogous to a human expert using the help of a catalog. The knowledge is basically organized in the form of a set of IF - THEN production rules, and for this reason, expert system is often called *a rule-based system*. A typical rule, as shown in Figure 11-2, can read as

```
Rule 1:
    IF : DC LINK VOLTAGE <200 V AND
         AC LINE VOLTAGE IS ZERO AND
         MACHINE SPEED >50% RATED SPEED
 THEN : REDUCE MACHINE SPEED BY 20%
```

In this rule, a diagnostic expert system prevents shut-down of an AC drive by pumping up the DC link voltage with regenerative braking in case of short-time line power outage. The IF statement of a rule is defined as premise (or antecedent or condition), and the THEN statement is defined as consequent (or conclusion or action). The segments of IF and THEN statements are normally connected by logical AND, OR, and NOT operations for drawing the conclusion. Within each segment again, there may be logical or arithmetic operation or execution of a command, which will be discussed later. Each rule has a number of parameters that may have numerical, logical, or textual values. In the example rule, DC link voltage,

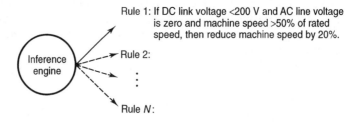

Figure 11-2. A set of production rules interfacing the inference engine.

AC line voltage, and machine speed are the parameters that are usually represented by symbols in the software. A rule is "fired" or executed if the premise part is true, and then the consequent THEN statement is executed. The rules can be designed to handle probability through certainty factors and probability-based models, such as Bayesian approach. Rules can also be hybrid by mixing with fuzzy logic. One outstanding feature of the knowledge base is that it is structured so that the knowledge and data can be easily altered or updated. This off-line alteration may be necessary as know-how on the problem changes or technology advance mandates this alteration. Sometimes, on-line alteration of the knowledge base is possible based on "machine learning." The knowledge can sometimes be defined as "shallow" or "deep." The shallow knowledge corresponds to that directly derived from the domain expert (see Figure 11-1), whereas the deep knowledge-based rules can be derived from the system model, which is based on designer's or researcher's knowledge.

The inference engine, as the name indicates, is basically the control software of an expert system that tests the rules in systematic order and attempts to draw inference or conclusion. This means that the rules whose premise part is true will be fired, and control actions determined by the THEN statements will be taken. The inference engine also controls the user interface, as shown. The rules are tested by inference engine by using either forward chaining or backward chaining method. A forward chaining or antecedent rule works on the principle of deductive logic; that is, the premise part is tested first, and if it is true, then the rule is fired. A backward chaining or consequent rule, on the other hand, works on the principle of inductive logic, that is, the inference engine hypothesizes the inference or consequent part of the rule and then tests backward for the premise part to be true for the rule's validity. This is like a medical doctor assuming that a patient has a certain disease and then trying to match the symptoms with it. It is interesting to note that one of the most successful expert system programs is Mycin, which was developed by Stanford University in the early 1970s for medical diagnosis of infectious diseases. In an expert system, both forward and backward chaining rules are mixed strategically for best performance. The user interface of expert system, either directly or through the explanation subsystem, is very important because often the user is an unskilled or semiskilled person trying to consult the expert system. The user must communicate in natural language in a user-friendly manner because he is usually unfamiliar with the intricacies of expert system software and its language. The expert system seeks parameter values from the user through the user interface, and based on these values, the knowledge is searched and the relevant rules are fired. The solution then appears on the monitor screen. If the expert system is designed for real time control, the sensors will supply the signal values and the resulting control action may be like tripping a circuit breaker, sounding an alarm or printing a message.

The user interface with the explanation subsystem helps the user to communicate with the expert system in simple English or other natural languages. This part makes the expert system distinctive from the algorithmic-type programs. An efficient user interface design is as important as the effective knowledge base design because the program is consulted by a skilled or semiskilled user. The user interface has essentially two parts: one is getting data or parameter values from the user and the other is providing consultation results and explanation information to the user. In the former case, the information may be asked with a menu, or direct data value may be inputted. For example, the consultation screens for application of an AC drive may be as follows:

What type of motor are you using in your drive?

IMCR Induction motor cage type
IMWR Induction motor wound rotor type
BLDM Brushless DC motor
PMSM Permanent magnet synchronous motor
WRSM Wound rotor synchronous motor

or

What is the machine horsepower rating?

In the previous screen, the user responds by highlighting a particular item in the menu, and in the later case, the numerical value is typed. In the user dialog, a series of such questions will be asked, and based on the user's response, the knowledge base will be searched, the appropriate rules will be fired, and finally the solution of the problem will be given on the screen. Many questions asked to the user may appear meaningless because the technical expertise of the user in the particular domain may be low. Therefore, a HELP message can be provided on the screen for the questions that are not straightforward. This message gives technical explanation as the background of the question. One of the key objectives in writing the expert system programs is heavy operator or user training in a particular technical area with these HELP messages. In addition to HELP messages, the program has the capability to explain the questions asked by responding to the user's WHY and HOW commands. The WHY command explains why the value of the particular parameter is needed when the user is prompted for a parameter value. This clarifies any confusion or ambiguity in the user in spite of his clear technical knowledge and helps to answer the question that can be deciphered by the expert system. The HOW command explains to the user how the expert system came to such a conclusion. This explanation convinces the user that a sensible conclusion has been reached. Note that in the consultation, HELP, and solution screens, pictures can be drawn to supplement the text with the help of graphics programs.

11.2.2. Knowledge Base

An expert system knowledge is usually structured in the form of a tree that consists of a root frame and a number of subframes as indicated in Figure 11-3. A simple knowledge base can have only one frame, that is, the root frame, whereas a large and complex knowledge base may be structured on the basis of multiple frames, as indicated in the figure. The root frame and each subframe are organized so that each has its respective rules and parameters and characteristics. The frame-based architecture gives modular hierarchical organization of a large knowledge

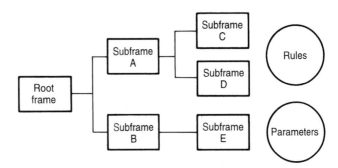

Figure 11-3. Frame structure of a knowledge base.

base. The root frame is the primary core and may have child subframes (A and B) and grandchild subframes (C, D, and E), as indicated in the figure. All the subframes have the root frame as an ancestor. A subframe's ancestors include all the frames that are in direct line to the root frame. A frame's descendants include all the frames that are in a direct line away from the root frame. A frame can have many ancestors, but it can have only one parent. A parent frame has no limit to the number of child frames. Generally, a subframe has access to the parameters of its ancestors but not to those of its descendants. However, a frame can have access to the rules of its descendants, but not to those of the ancestors.

Each subframe can be considered as a subdomain of expert knowledge. Consider, for example, the problem of power supply product selection from a large company for a certain application. The root frame may correspond to the expertise a of general sales manager, and subframes A and B correspond to application engineer's expertise in PWM mode and resonant link power supplies, respectively. The user interfaces the root frame in the beginning and holds dialog regarding his application needs and characteristics. Based on this conversation, if PWM mode power supply appears to be the choice, the subframe A will be activated and the user dialog will start with it. Or else, subframe B will be activated for choice of resonant link power supplies. A characteristic antecedent rule in the root frame will cause this activation or "instantiation." The parameters in the respective subframe will get values (such as type of load, number of line phases, supply voltages, etc.) from the user, make appropriate calculations, and then determine the catalog item satisfactory for the user. The subframes C, D, and so on may correspond to auxiliary features, such as price, delivery, and installation conditions. A frame or subframe may access external routines to derive conclusions that will be discussed later.

In a large and complex knowledge base, meta-rules and other forms of meta-knowledge can increase the efficiency with which the expert system reaches a conclusion. Unlike conventional programs, an expert system is said to have learning capability because of meta-knowledge. *Meta-knowledge* is knowledge about the operation of a knowledge base, and *meta-rules* are rules about operation of the rules. Practically, the knowledge base has two levels of operation: the usual domain level and the meta-level. Meta-level knowledge determines the most efficient strategy of operation that the domain level can take. The knowledge base can be made to learn from experience which rules are most useful; that is, most likely to be fired. Then, the expert system can follow to test these rules first during consultation.

Avoidance of test for the unlikely rules will enhance the speed of execution; that is, improve efficiency of knowledge base search. An example of a meta-rule to guide the order of rule search within a frame is

```
MRule 1:
IF RULES 1, 3, 9 and 10 are skipped 12 times
consecutively,
THEN test these rules at the end
```

The knowledge in a knowledge base can be categorized as declarative or factlike knowledge and procedural or methodlike knowledge. The *declarative knowledge* is basically what to do (i.e., kernel of the knowledge base) whereas the *procedural knowledge* is concerned with how to do (i.e., knowledge base organization with frames, rules, and parameters).

Finally, what language should be used in the software of an expert system? In principle, any language can be used in the development of an expert system, but some are more efficient than others. Since the major part of an expert system program processing is symbolic or nonnumeric, a symbolic processing language, such as PROLOG or LISP, has traditionally been popular. Of course, these languages have limited numeric computation capability. The LISP or its dialects are particularly strong candidates because of their power and flexibility. The numeric computation intensive high level languages, such as BASIC, FORTRAN, PASCAL, and the like have also been used, but they have limited symbolic processing capability. The inherent drawbacks of high level languages are that they are slow but user-friendly and can be used mainly for off-line problem solving. For on-line control applications, particularly for power electronic systems, a fast, lower-level language, such as C or even Assembly language, may be essential.

11.2.3. Expert System Shell

An *expert system shell* is basically a development system for designing an expert system. It provides an efficient and user-friendly software environment to the knowledge engineer for building an expert program. The developed program for a client's use can be consulted within the shell or exported to a compatible computer. Today, there are so many good commercial expert system shells available that hardly anyone builds an expert system from ground zero. A good shell should have the following features:

- A highly interactive environment for program development and testing.
- Good debugging and value checking aids.
- Easy access to external data bases and problem-solving aids.
- Ability to incorporate graphics.
- A mechanism for handling uncertainty both from the developer and from the end user.

- Ability to explain in English (or other natural language) why the system is asking for information and how it has reached conclusion.
- A window-oriented interface with extensive on-line help.
- A comprehensive rule-entry language that is similar to English or other natural language.
- A full-screen editor.
- Meta-rules for sophisticated rule control.
- Ability to extend knowledge base through the use of user-defined LISP (or other language) functions.
- A frame capability that allows the knowledge base structure to be divided into logically different but related segments.
- The ability to trigger actions, such as updating displays.
- Additional means (other than inferencing and prompting) to evaluate and set parameter values.

A large number of expert system shells [1, 4] based in mainframe, mini, and personal computers are available for different applications. The features of a few commercial shells based on a personal computer are summarized as follows:

- **1ST CLASS**—Ist-Class Expert Systems, Inc.
  IBM PC
  Microsoft Pascal and Assembler
- **ART**—Inference Corporation
  Sun, VAX, Lisp Machine, IBM PC, TI Explorer
  LISP C
- **PC PLUS**—Texas Instruments
  IBM PC, TI Business Pro, Compaq Desk-Pro
  PC Scheme
- **GURU**—Micro Data Base Systems, Inc.
  IBM PC
  C
- **EXSYS**—Exsys, Inc.
  IBM PC, SUN, Apple Macintosh, VAX
  C
- **KES**—Software Architecture and Engineering, Inc.
  IBM PC, SUN, VAX, Apollo
  C

The discussion on detailed features of these shells is beyond the scope of this chapter. Since the features of these shells have a lot of common elements, Texas Instruments' PC PLUS will be briefly reviewed here as an example for completeness.

The PC PLUS (Personal Consultant PLUS) expert system shell [5, 6] has all the essential features for development of a good expert system discussed so far and is well suited for practically all the applications in power electronics systems, except possibly the real-time control. The shell operates in DOS environment of an IBM compatible personal computer. The shell uses PC SCHEME language which is a dialect of LISP. The program developer is required to have some familiarity with PC SCHEME. On a full-screen editor, the knowledge base development starts by defining the basic elements, such as DOMAIN (the heading that appears on the knowledge base screen), frame name, TRANSLATION (text that describes the purpose of the frame), and GOALS (parameters for which PC PLUS tries to determine a value). Then, after adding the frame properties, the parameters and their properties are to be entered. The IF ... THEN ... statements of the rules are written on a template setting with English-like Abbreviated Rule Language (ARL). Of course, PC SCHEME can also be used for rule development. The example rule given before can be written in ARL as

```
Rule 001 :
IF :: DCVL < 200 AND ACLV = NO AND MC-SPD > 0.5
THEN :: MC-SPD = MC-SPD * 0.8
```

where DCVL, ACVL, and MC-SPD are the corresponding parameter names. The property of every rule is to be defined. Once the knowledge base is created by systematically defining the frames, rules, and parameters and their properties, the program is compiled and ready to execute with the built-in inference engine and user-supplied parameter values. After compiling, every rule is translated into English by the PC PLUS for checking its correctness. After the program development, a run-time version or client program is created to operate stand-alone in a DOS environment. The inference engine in the PC PLUS is automatically loaded with the knowledge base and user interface to the client program. The user dialog occurs in pure English, as mentioned before. The client program can be generated either in LISP for nontime-critical application or in C for less memory and time critical application. When the program is resident in the shell, the developer can easily alter or update it, but no program modification is possible in the client environment. The knowledge is organized in hierarchical frame-based architecture with HELP, WHY, and HOW messages, as discussed before. A new frame can be instantiated by using the CONSIDERFRAME function as follows:

```
IF :: MOTOR-TYPE = INDUCTION-MOTOR
THEN :: CONSIDERFRAME IM
```

where the antecedent rule transfers the operation from the root frame to the IM subframe. The inference engine defaults to backward chaining rule test unless forward chaining is specifically instructed by antecedent property of the rule.

During consultation, the shell can access external files and programs as indicated in Figure 11-4. A limited amount of data can be directly embedded in the program, but for a larger size of data, such as product catalog consultation, dBASE database files can be consulted. With the help of dBASE functions, the shell can obtain (DBASE-RETRIEVE), change information (DBASE-REPLACE), add

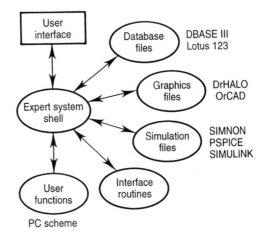

Figure 11-4. External interface of an expert system shell.

information (DBASE-APPEND), remove information (DBASE DELETE), or print information (DBASE-REPORT) from the database. The LOTUS 1-2-3 spreadsheets can also be consulted from the shell. The rules in the knowledge base can process simple logic and arithmetic computations with the help of PC SCHEME, but for complex calculations, such as solving a differential equation, it can access an external DOS program, as shown in Figure 11-5. The function WRITE-DOS-FILE loads the parameter values to an input data file. The DOS-CALL function halts the shell, temporarily returns control to DOS, executes the DOS program, and then restarts the shell. Finally, READ-DOS-FILE function receives the data from output data file. The shell can control and transfer data to a simulation program which will be discussed later. Expert system–aided simulation of a power electronic system is very important before building a breadboard or prototype. A powerful feature of the shell is that it can integrate pictures in the knowledge base, as mentioned before. The picture is first created with a third-party graphics editor, such as DR HALO (Media Cybernetics) or ORCAD (Orcad Systems). The shell graphics utility SNAPSHOT captures the picture and then compresses it into a file. As the knowledge base needs the picture during consultation, the SNAPSHOT expansion tool restores the picture

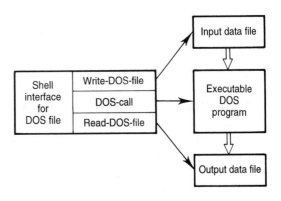

Figure 11-5. Expert system shell interface to a DOS program.

by expanding it and makes it available. Finally, user functions either in PC SCHEME or other languages can be used with the shell. The functions can again be either user defined or user written.

11.2.4. Design Methodology

A systematic procedure for building an expert system with the help of PC PLUS shell is described in [5, 6]. In fact, all the discussions so far on the expert system principle and shell are applicable for development of an expert system. The application examples in the next section will give insight to the design methodology. In summary, the general procedure can be defined as follows:

1. Analyze whether the problem has sufficient elements such that it deserves to be an expert system program. Otherwise, an algorthmic program may be adequate. Fuzzy logic or neural network described in later sections can also be considered for solving the problem, if necessary.
2. Get all the information or knowledge related to the problem and organize it in the form of matrices, charts, and so on. The domain expert and knowledge engineer are considered to be the same person.
3. Select a suitable language for the program. Real-time control may be C based, but the off-line program may be LISP based. The program is considered off-line and based on a personal computer.
4. Select a shell. Consider that it is PC PLUS.
5. Analyze the knowledge and partition it into a number of frames. For a small problem, one frame is adequate. Define the properties of the frames.
6. Determine the parameters of each frame and define their properties.
7. Determine the rules in each frame and define their properties.
8. Plan all the pictures and write graphic files with DR HALO or ORCAD.
9. Develop the knowledge step by step with the help of the shell and debug.
10. Design the user dialog for getting consultation data.
11. Design the HELP routines.
12. Develop DOS programs, dBASE programs, and other interface as necessary and integrate with the knowledge base.
13. Debug and test the complete program.
14. Generate the client program diskette and deliver. The inference engine is automatically loaded from the shell.

11.2.5. Application in Power Electronics and Drives

Expert system techniques have practically reached the mature state of evolution at the present time. However, their application in power electronics and drives are only few. They can be applied in practically all aspects, such as model generation from test data, selection of appropriate vendor product based on application specifications, automated circuit or system design, performance optimization in simulation, on-line or off-line fault diagnostics, fault-tolerant control, automated performance tests, performance-optimized real-time and the like. In this section, a few applications will be briefly reviewed.

Selection of AC Drive Product. With the help of an expert system, a semi-skilled user can select a drive product best suited for his application [8]. Normally, the user determines the application needs and then consults a company application engineer extensively to decide the drive system to be purchased. The application engineer with his knowledge of drive technology and company products makes some calculations, consults the catalog, and then recommends the appropriate drive. For a large company, there may be a general application engineer, who after preliminary consultation, may direct the user to the respective application engineer with expertise in induction motor drives or synchronous motor drives. In expert system, the application engineer's expertise and the product catalogs may be implanted in the knowledge base. The user holds conversation with the expert system and supplies all the relevant application information. Based on this dialog, the knowledge and database will be searched, and the appropriate product will be recommended. The user himself may have power electronics expertise and give a choice on induction or synchronous machine, voltage-fed, or current-fed inverter; select power device; and have preference on a particular company product. A typical rule in PC PLUS ARL is given as

```
IF  :: MOTOR = CAGE-ROTOR-INDUCTION-MOTOR AND
          APPLICATION = FAN AND
          LINE VOLTAGE = 220-V AND
          PHASE = THREE AND
          POWER = 20-HP AND
          SPEED REVERSAL = NO

THEN :: PRODUCT = TOSHIBA-VT130H0U-2220
```

Figure 11-6 shows the simplified flow diagram for consultation. The drive oversizing is determined by the acceleration-deceleration needs and correspondingly the inverter oversizing factor is calculated with the help of a DOS file. Figure 11-7 shows the frame structure of the knowledge base. The root frame includes the "goal" parameters of the program, the parameters to get "initial data" from the user, and rules to control instantiation of subframes. The subframes IMCR, IMWR, and PMM relate to knowledge of cage rotor induction motor, wound rotor induction motor and permanent magnet motor, respectively. The subframes with dashed outline indicate that these have provision for future expansion. The database stores the product catalog including the load type information and can be accessed from any subframe. The particular expert system can help in the selection of induction motor drive from product catalogs of three different companies.

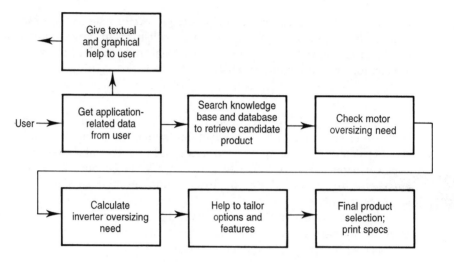

Figure 11-6. Flow chart for drive product selection.

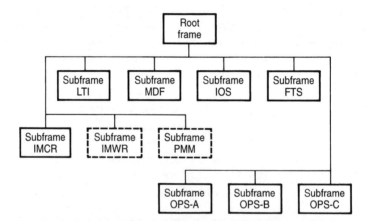

LTI = load-type information, MDF = motor derating factor, IOS = inverter oversizing factor
FTS = features of the valid candidate drives, IMCR = induction motor (cage rotor)
IMWR = induction motor (wound rotor), PMM = permanent magnet motor
OPS-A/OPS-B/OPS-C = various options the manufacturer provides for selection by the user

Figure 11-7. Frame structure for drive product selection.

Monitoring and Diagnostics.   Industrial plant fault diagnostics is one of the most popular and early applications of expert system [12, 13]. The diagnostics may be based on off-line or on-line. In off-line diagnostics (or trouble shooting), the plant is shut down and the expert system is used to identify the fault. In this case, the expertise of a diagnostic technician is embedded in the knowledge base. The procedure may be static or dynamic. A static expert system advises the trouble-shooting procedure step-by-step to the operator, and the observed symptoms are fed as input in the form of a dialog. The expert system then searches the knowledge base and gives the conclusion on the screen. In a dynamic system, all the foregoing procedures are automated, and the conclusions are reached without user intervention.

The on-line diagnostics based on expert system may be quite complex. Here, the main objective is to maintain reliability and safety of the operating plant and avoid unnecessary shut-down. For example, in a voltage-fed inverter AC drive, the AC line voltages and currents, DC link voltage and current, IGBT gate drive logic signals, machine stator voltages and currents, and stator winding temperature signals can be sensed and fed to a microcomputer that embeds the diagnostic program. The monitoring, alarm processing, and protection functions can easily be integrated with the diagnostic system. An example rule of on-line diagnostics was given before. The expert system may be designed to monitor the general health of a drive, avoid preventable shut-down, and give fault-tolerant control of the system.

Design, Simulation, and Control of Drive. In the traditional approach for AC drive design, the power electronics engineer with the expertise of drive technology designs the system using paper, pencil, calculator, and CAD programs. He then simulates the system and optimizes the performance with iteration of the parameters. A laboratory breadboard is then built to verify the performance, before finally building the prototype. In this expert system application, the expertise for the design of indirect vector-controlled induction motor (IVC-IM) drive (Figure 11-8) is embedded in the knowledge base to help a semiskilled designer to automate the drive design, simulate, observe, and optimize the performance. The implementation stages are as follows:

1. Design the converter system [9].

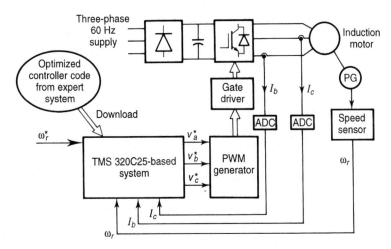

Figure 11-8. Indirect vector-controlled induction motor drive with TI–TMS320C25-based controller.

2. Simulate the converter-machine system and optimize the power circuit design [10].

3. Design the IVC-IM drive, make hybrid simulation, optimize the control, and download the real time controller routine to the experimental drive [11].

The expertise related to the voltage-fed PWM converter system, shown in Figure 11-8, is stored in the knowledge base with the help of frames, rules, and parameters. The data related to the power semiconductor devices are stored in the database. The graphical data in the specification sheets are stored in the form of equations using the software program TableCurve (Jandel Scientific). The user supplies the details of load condition, line power supply, and other specifications during consultation. Based on this information, the expert system calculates the ratings of power semiconductors and retrieves the designed components from the database. The ratings of snubber components and filter capacitor are also determined. The conduction and switching losses of power devices are calculated, and based on the thermal model, the heatsink sizes are also determined. Finally, the circuit of the complete converter system with a table of designed components is displayed on the screen. An example of a simple design rule in ARL is

```
IF :: ID-AV IS KNOWN AND
      FORM-FACTOR IS KNOWN AND
      NO-OF-PHASES = THREE-PHASE
THEN :: IS-RMS = FORM-FACTOR * ID-AV
```

where the precomputed line current form factor is supplied to the knowledge base in the form of segmented linear equations. This rule is used to calculate the rms current rating of the rectifier diode.

After completion of the design, the expert system passes parameter values of the design to a converter simulation program in PC-SIMNON (Engineering Software Concepts) or other languages. The simulation of the designed system, controlled by the expert system according to the user's choice, then passes the results back to the expert system. The knowledge base contains the expertise for guiding simulation, makes intelligent observation, and performs optimization. The iterated simulation will optimize the snubber design and verify that the worst case voltages and currents are within the safe device rating. The simulation results can be passed to PC-MATLAB (Math Works) for further signal processing, such as plotting the fast Fourier transform (FFT) of a wave. The expert system software is highly modular, and can easily be upgraded to include other converter topologies.

The interface between the expert system based in PC PLUS shell and the SIMNON simulation is somewhat involved and is shown in Figure 11-9. The complexity arises because both accept the input data and save the output data in different specific formats. As shown, the expert system writes and reads data in ASCII format, whereas SIMNON reads and writes in special labelled format. Also, for the simulation process to be completely automated and controllable by the user from the expert system, the simulation command files (or macros) have to be modified from the

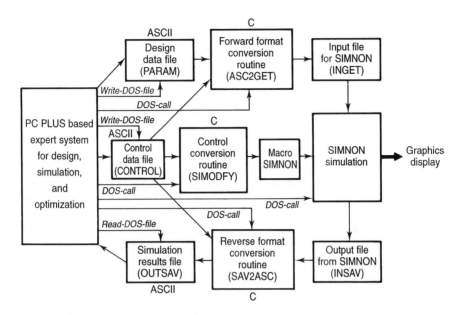

Figure 11-9. Expert system interface with simulation program in SIMNON showing all the interface routines.

expert system. The control conversion routine performs this function. During a typical consultation session (upon user request), a rule such as the following is executed:

```
IF :: DESIGN-DONE AND
      USER-CHOICE = SIMULATION

THEN :: WRITE-DOS-FILE ''PARAM.DOT'' We INPUT_VOLT
        SOURCE_L
        FILT_CAP VDO K MFs POWER_IGBT_IC T_RISE SNUB_R
        SNUB_C
        LEAD_L T_FALL
        AND WRITE-DOS-FILE ''CONTROL.DAT'' CHOICE NO-OF-
        PHASES
        AND DOS-CALL ''ASC2GET.EXE''
        AND DOS-CALL ''SIMODIFY.EXE''
        AND DOS-CALL ''SIMON.EXE''
        AND DOS-CALL ''SAV2ASC.EXE''
        AND READ-DOS-FILE ''OUTSAV.DAT''IMAX
```

This rule exports results of the design from the expert system in the form of a parameter list to the file PARAM and the control parameters for simulation to the file CONTROL. The expert system then calls and executes the C-based interface routine ASC2GET to achieve the forward format conversion of the data available from the expert system. The output of this routine, stored in the file INGET, is readable by SIMNON. The routine ASC2GET receives the complete list of parameters from the expert system, but the file INGET receives only the parameters necessary for the subsystem selected by the user, as indicated in the file CONTROL. Then, the expert system calls another C-based control conversion routine SIMODFY to generate the self-executing macro file SIMNON for controlling the simulation as chosen by the user. When SIMNON is invoked, this macro automatically directs SIMNON to read the data available in INGET, perform simulation, store and display values of the critical quantities, and save and export values of interest to the user to be used by the expert system in the output file INSAV. The expert system then executes the reverse data format conversion routine SAV2ASC to convert the data file INSAV to the ASCII file OUTSAV so that it is readable by the expert system.

In the final phase of the project, the complete drive system was designed and studied by hybrid simulation. Then, after optimization of the controller performance, the controller codes in TMS320C25 digital signal processor are downloaded to the experimental drive, as indicated in Figure 11-8. Figure 11-10 shows the structure of of expert system–based programs and their interface. The user, in the beginning, supplies the detailed performance specifications and power supply conditions. Based on these, the machine, converter (as discussed before), and control are designed. The speed loop and the local current controllers (synchronous current control) have proportional and integral gains that are initialized conservatively based on prior analysis. These gains are then optimized by expert system based on the simulation results. The current loops are tuned before the speed loop.

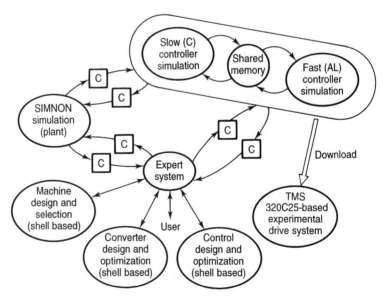

Figure 11-10. Expert system–based design and simulation of vector-con - trolled drive showing the simulation interface routines.

An important feature of the simulation is that it is hybrid; that is, the converter and machine are simulated in SIMNON (nonreal time), whereas the controller is simulated in real time with the help of TMS320C25 DSP simulator. The speed loop is simulated in C, but the fast-current loops are simulated in Assembly language. The objective of real-time controller simulation is that the control software after proper tuning can be directly downloaded to the experimental system. The advantage of this procedure is obvious. The plant and controller simulation occurs continuously in sequential manner with a certain sampling time (200 μs, in this case) with the help of a DOS batch file, as shown in Figure 11-11. Note that SIMNON simulation is very slow compared to controller simulation and that, at the end of each simulation period, it halts and passes the state values to the controller. Similarly, the controller halts after its simulation interval and passes the state values to SIMNON. The SIMNON output states that are in decimal format are converted to HEX format for the controller, and vice versa. Similar conversion is required for controller-to-expert system interface. At any time, if the user requests display of response, all the simulation programs go to halt mode. Instead of SIMNON, another simulation language such as SIMULINK (Math Works) can be used.

11.3. FUZZY LOGIC

Fuzzy logic is another form of artificial intelligence, but its history and applications are more recent than expert system. It is argued that human thinking does not always follow crispy "yes"-"no" logic, but is often vague, uncertain, indecisive, or fuzzy. Based on this, Lofty Zadeh, a computer scientist at the University of California, Berkeley, originated the "fuzzy logic" or fuzzy set theory in 1965 [16] that gradually

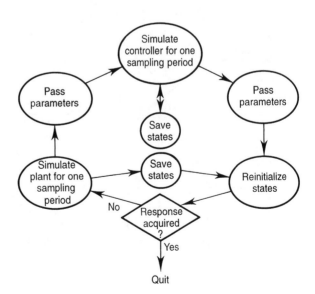

Figure 11-11. Continuous simulation of plant (in SIMNON) and controller (in C and Assembly).

emerged as a discipline in AI. The general methodology of reasoning in fuzzy logic and expert system are the same, and therefore, it is often defined as "fuzzy expert system." Fuzzy logic helps to supplement the expert system, and it is sometimes hybrided with the latter to solve complex problems. Fuzzy logic has recently been applied in process control, modeling, estimation, identification, diagnostics, stock market prediction, agriculture, military science, and so on. It is interesting to note that there is a fuzzy logic version of Mycin, the medical diagnostic program in expert system.

11.3.1. Fuzzy Logic Principles

What is fuzzy logic? Fuzzy logic, unlike Boolean or crispy logic, deals with problems that have vagueness, uncertainty, imprecision, or qualitativeness, as mentioned before. It tends to mimic human thinking, which is often fuzzy in nature. In conventional set theory based on Boolean logic, a particular object or variable is either a member (logic 1) of a given set or it is not (logic 0). On the other hand, in fuzzy set theory based on fuzzy logic, a particular object has a degree of membership in a given set that may be anywhere in the range of 0 (completely not in the set) to 1 (completely in the set). This property helps fuzzy logic to deal with nonstatistical uncertain situations in a fairly natural way. It may be mentioned that although fuzzy logic deals with imprecise information, it is based on sound quantitative mathematical theory that has been advanced in recent years. In this section, the simple fuzzy logic theory that is applicable to process control, modeling, and estimation will be discussed.

A fuzzy variable has values that are expressed by natural English language. For example, as shown in Figure 11-12a, the stator temperature of a motor can be

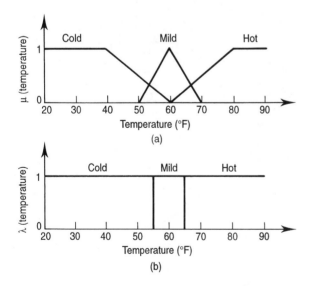

Figure 11-12. Representation of temperature using (a) fuzzy sets, (b) crispy sets.

defined by qualifying linguistic variables (also defined as *fuzzy sets* or *fuzzy subsets*) COLD, MILD, or HOT, where each is represented by triangular or straight-line segment membership functions (MFs). The fuzzy sets can have more subdivision, such as ZERO, VERY COLD, MEDIUM COLD, MEDIUM HOT, VERY HOT, and so on, for precision description of the variable. Although the triangular membership function is most commonly used, the shape may be trapezoidal or Gaussian (bellshaped). In Figure 11-12a, if the temperature is below 40°F, it belongs completely to the set COLD; that is, the membership function value is 1, whereas for 55°F, it is in the set COLD by 30% (MF = 0.3) and to the set MILD by 50% (MF = 0.5). For the temperature above 80°F, it belongs completely to the set HOT (MF = 1). The membership functions can either be defined by mathematical equation or look-up table. In Figure 11-12b, the corresponding crispy or Boolean classification is provided for comparison; that is, the temperature range below 55°F belongs to the COLD set only (MF = 1), 55°F to 65°F belongs to MILD set only (MF = 1), and above 65°F belongs to HOT set only (MF = 1). The sets are not member (MF = 0) beyond the defined ranges. The numerical interval that is relevant for the description of a fuzzy variable (temperature) is defined as *universe of discourse* (20°F to 90°F in Figure 11-12).

The basic properties of Boolean logic are also valid for fuzzy logic, and these are described in the paragraphs that follow. Let $\mu A(x)$ denote the degree of membership of a given element x in the fuzzy set A.

Union: Given two fuzzy sets A and B, defined on a universe of discourse X, the union $(A \cup B)$ is also a fuzzy set of X, with the membership function given as

$$\mu_{A \cup B}(x) = \max[\mu_A(x), \ \mu_B(x)] \tag{11.1}$$

where x is any element of X. This is equivalent to Boolean OR logic.

Intersection: The intersection of two fuzzy sets A and B of the universe of discourse X, denoted by $A \cap B$, has the membership function given by

$$\mu_{A \cap B}(x) = \min [\mu_A(x), \ \mu_B(x)] \tag{11.2}$$

This is equivalent to Boolean AND logic.

Complement or Negation. The complement of a given set A of the universe of discourse X is denoted by $\neg A$ and has the membership function

$$\mu_{\neg A} = 1 - \mu_A(x) \tag{11.3}$$

This is equivalent to Boolean NOT operation. Figure 11-13 illustrates the foregoing properties with triangular membership functions.

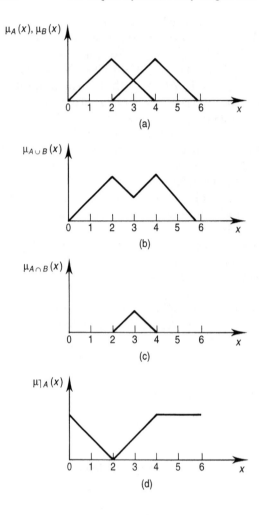

Figure 11-13. Basic operation involving fuzzy sets: (a) fuzzy sets A and B; (b) union $A \cup B$; (c) intersection $A \cap B$; (d) negation $\neg A$.

11.3.2. Fuzzy Control

The control algorithm of a process that is based on fuzzy logic is defined as *fuzzy control*. A fuzzy controller essentially embeds the experience and intuition of a human plant operator, and sometimes those of designer and researcher of the plant. The design of a conventional control is normally based on the mathematical model of a plant. If an accurate mathematical model is available with known parameters, it can be analyzed, for example, by a Bode or Nyquist plot, and a controller can be designed for specific performance. Such a procedure is tedious and time consuming, although CAD programs are available for such design. Unfortunately, for complex processes, such as cement plant, nuclear reactor, and the like, a reasonably good mathematical model is difficult to find. On the other hand, the plant operator may have good experience for controlling the process. A power electronics system model is often ill defined. Even if the plant model is known, there may be parameter variation problems. Sometimes, the model is multivariable, complex, and nonlinear, such as dynamic d–q model of an AC machine. Vector or field-oriented control of a drive can overcome this problem, but accurate vector control is nearly impossible, and wide parameter variation problems may remain in the system. To combat these problems, various adaptive and robust control techniques based on self-tuning regulation (STR), model referencing adaptive control (MRAC), and sliding mode or variable structure control have been developed. Fuzzy control, on the other hand, does not need any model of the plant. It is based on plant operator experience and heuristics, as mentioned earlier, and it is very easy to apply. Fuzzy control is basically adaptive and gives robust performance for plant parameter variation. In fact, the fuzzy control performance is better than that of the methods already mentioned.

Since the development of fuzzy logic theory by Zadeh, its first application to the control of dynamic process was reported by Mamdani [18] and Mamdani and Assilian [19]. This is an extremely significant contribution because it stirred widespread interest by later workers in the field. They were concerned with the control of a small laboratory steam engine. The control problem was to regulate the engine speed and boiler steam pressure by means of the heat applied to the boiler and the throttle setting of the engine. The difficulties with the process were that it was nonlinear, noisy, and strongly coupled, and no mathematical model was available. The fuzzy control designed purely from the operator's experience was found to perform well and was better than manual control. Nearly at the same time, Kickert and Lemke [20, 21] examined the fuzzy control performance of an experimental warm water plant where the problem was to regulate the temperature of water leaving a tank at a constant flow rate by altering the flow of hot water in a heat exchanger contained in the tank. The success of Mamdani and Assilian's work led King and Mamdani [22] to attempt to control the temperature in a pilot scale batch chemical reactor by fuzzy algorithm. Later, Rutherford and Carter [23] and Ostergaard [24] successfully applied fuzzy control to a sinter stand and heat exchanger, respectively. Then, Tong [25] applied it to the control of a pressurized tank containing liquid. These results indicated that fuzzy control is very useful for complex processes and gave superior performance than the conventional proportional-integral-derivative (PID) control. These successes later attracted many workers in the field.

Let us now review the principles of fuzzy control. Fuzzy control, similar to expert system-based control, is described by IF-THEN production rules, as discussed before. An example of fuzzy rule (also called *fuzzy implication*) in a process control is

```
IF TEMPERATURE IS LOW AND
PRESSURE IS HIGH
THEN SET THE FUEL VALVE TO MEDIUM
```

The heuristic rule for a boiler control is derived from the operator's experience. Here the temperature, pressure, and fuel valve setting are the fuzzy variables that are described by the corresponding fuzzy sets LOW, MEDIUM, HIGH, and so on. Each set has the characteristic membership function, as shown in Figure 11-

12a. In comparison, a corresponding structure of an expert system rule may appear as

```
IF TEMPERATURE < 150°F AND PRESSURE > 50 PSI
THEN SET THE FUEL VALVE 50%
```

where temperature, pressure, and fuel valve setting are the parameters. The rule processes the logic variables generated by the parameter conditions. The fuzzy expert system rules contain fuzzy knowledge. It can be shown that fuzzy control requires much fewer rules compared to that of expert system.

To design a fuzzy controller, a fuzzy rule base consisting of a set of rules (see, for example, Table 11-2) must be constructed. Consider a hypothetical fuzzy speed control system for a DC motor (described later), where the speed loop error (E) and rate of error change, (i.e., the error change (CE) within a sampling time of digital control) are used to determine the rate of change in control (i.e., change in control DU), which in this case, is the increment of command armature current Ia^*. A part of the rule base, as indicated in Figure 11-14, would be

```
Rule 1 :
IF error (E) is zero (Z) AND change in error (CE)
is negative small (NS) THEN change in control (DU) is
negative small (NS)
```

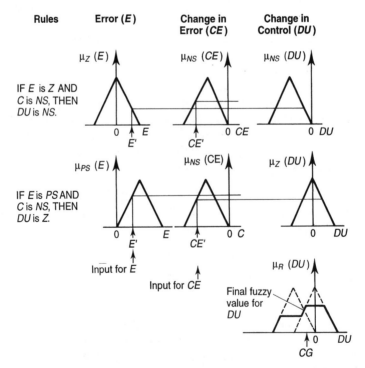

Figure 11-14. Fuzzy rules for control and composition by SUP-MIN principle.

```
Rule 2 :
IF error (E) is positive small (PS) AND change in error
(CE) is negative small (NS) THEN change in control (DU)
is zero (Z)
```

Here, error (E), change in error (CE), and change in control (DU) are the fuzzy variables, with possible values given by fuzzy sets such as positive small (PS), negative small (NS), and so on. As illustrated in the figure, a given numerical value of a fuzzy variable can be a member of more than one fuzzy set. This means that, for a particular input pair of values (E and CE), more than one rule could be activated or "fired." Therefore, there should be a method to combine the individual control actions of the fired rules, such that a single, meaningful action is taken. In fuzzy logic terms, the composition or inference operation is the mechanism by which such a task is performed. Several composition methods, such as MAX-MIN (or SUP-MIN) and MAX-DOT, have been proposed in the literature, but the SUP-MIN method is most popularly used. Figure 11-14 illustrates the fuzzy composition by SUP-MIN principle for the two stated rules. Note that the output membership function of each rule is given by MIN operator, whereas the combined fuzzy output is given by the SUP operator.

The general structure of fuzzy feedback control system is given in Figure 11-15. The final control signal U is inferred from the two-state variables error (e) and change in error (ce). The e and ce signals are per unit (p.u.) signals derived from the actual E and CE signals by dividing with the respective gain factors GE and GC. Similarly, the output control U is derived by multiplying the per unit output by the scale factor GU and then summing. The advantage of fuzzy control in terms of per unit values is that the same controller can be applied to all the plants of the same family. The gain factors can be fixed or programmable: programmable gain factors can control sensitivity of operation in different regions of control or the same control strategy can be applied in similar loops. The fuzzy controller is designed to process fuzzy quantities only. Therefore, all crispy input values must be converted to fuzzy sets before being used. The process is called *fuzzification operation*. This is performed by considering the crispy input values (see Figure 11-14) as "singletons" (defined by

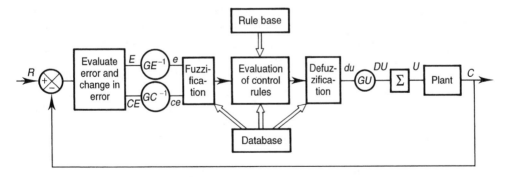

Figure 11-15. Structure of fuzzy control in feedback system.

fuzzy sets that have membership value of 1 for the given input value and 0 for all other points in the universe of discourse). If, on the other hand, the input data values are fuzzy, the degree of membership is determined as the maximum degree of membership value for the intersection of the membership functions for the fuzzy data values and the corresponding fuzzy sets. The final fuzzy output after evaluation of the valid control rules, as indicated in Figure 11-14, is to be converted to a crispy value required by the plant. This is called *defuzzification operation* and can be performed by a number of methods of which the center of gravity (also known as *centroid*) and height methods are common. A centroid defuzzification method calculates the output crispy value corresponding to the center of gravity of the output membership function, which is given by the general expression

$$U_0 = \frac{\int \mu(u)du}{\int \mu(u)du} \tag{11.4}$$

In the height method, the centroid of each output membership function for each rule is first evaluated. The final output is then calculated as the average of the individual centroids, weighted by their heights (degree of membership) as

$$U_0 = \frac{\sum_{i=1}^{n} u_i \mu(u_i)}{\sum_{i=1}^{n} \mu(u_i)} \tag{11.5}$$

Finally, the database provides the operational definitions of the fuzzy sets used in the control rules, fuzzification, and defuzzification operations. The defuzzifier output can be mathematically expressed as

$$DU = K_p \frac{dE}{dt} + K_i E \tag{11.6}$$

considering DU, E, and dE/dt as continuous functions and K_i and K_p are equivalent fuzzy controller gains. Integrating equation 11.6,

$$U = (K_p + \frac{K_i}{s})E \tag{11.7}$$

Equation 11.7 indicates that fuzzy feedback control as explained in Figures 11-14 and 11-15 is equivalent to proportional integral (PI) control with programmable proportional and integral gains K_p and K_i, respectively. There are two methods of implementation of fuzzy control: (1) the rigorous mathematical computation for fuzzification, evaluation of control rules, and defuzzification in real time and (2) the look-up table method where the output is precomputed from the input data and stored in computer in the form of matrix look-up table. A feedforward neural network that will be described later can also be trained to implement a fuzzy controller. It should be evident by now that in fuzzy control the structured fuzzy knowledge is dependent on heavy numeric processing compared to dominant logical processing of crispy structured knowledge in expert system. A numerical example is given later to clear the concept.

In spite of the advantages in fuzzy control, the main limitation is the lack of a systematic procedure for design, analysis, and calibration of the control system. The heuristic and iterative approach to fine-tune the rule base and membership functions can be very time consuming. If an approximate mathematical model of the system is available, it can be simulated and the controller fine-tuning can be done on simulation. Two other difficulties in fuzzy control are (1) lack of completeness of the rule base (the controller must be able to give a meaningful control action for every condition of the process) and (2) lack of definite criteria for selection of the shape of membership functions, their degree of overlapping, and the levels of data quantization. Recently, a fuzzy neural network (FNN) and a genetic algorithm techniques have been proposed to solve some of these problems.

11.3.3. Modeling and Estimation

Fuzzy logic principles can also be applied to modeling and estimation. A process such as a nuclear reaction is difficult to describe by a reasonably good mathematical model, but its operational characteristics can be described by a set of fuzzy rules. Such a fuzzy logic–based model can help to enhance the performance of fuzzy control in the same way as a mathematical model–based conventional control gives superior performance. Similarly, fuzzy estimation technique can be applied to a process where a mathematical estimation model is not known, ill defined, or has parameter variation problem. The fuzzy modeling and estimation can use rule-based method, as described before, or relational method described by Takagi and Sugeno [27], which is also known as *Sugeno's method*. Figure 11-16 shows the principle of relational estimation of rms line current (I_s) for a triac-controlled line current (described later), where W = current pulse width and H = pulse height. The idea behind the fuzzy relational approach is to define the regions where the output can be expressed as linear functions of the inputs. Basically, it is a hybrid method that combines the fuzzy and mathematical methods. As shown in Figure 11-16, the premise part of the rule is identical with that in the rule-based approach, but the consequents are described by equations. Rule 1 in the figure can be stated as

```
IF W is MEDIUM AND H is MEDIUM
THEN I_s = A01 + A11.W + A21.H
```

The consequents are linear functions of W and H, and the parameters A_{ij} are constant coefficients. The A_{ij} can be determined by multiregression linear analysis and then fine-tuned by observation or simulation study. The linear equation outputs are then defuzzified; that is, the weighted average of the consequents is evaluated by the respective membership values to determine the crispy output. It can be shown [40] that the relational method of estimation requires less number of rules, and gives better accuracy, and the algorithm development time is less time consuming than the rule-based approach.

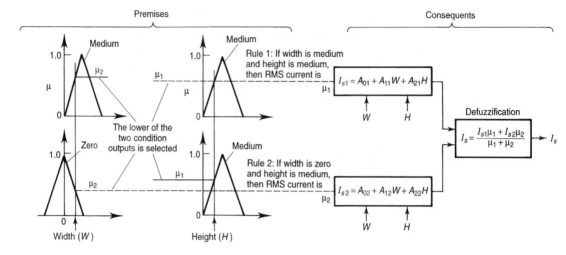

Figure 11-16. Principle of relational estimation (Sugeno's method)

11.3.4. Design Methodology and Control Implementation

The preceding discussion will give adequate guidance for fuzzy control design methodology and practical implementation. The step-by-step numerical example given in the next section will make it more clear. In summary, the general design procedure can be given as follows:

1. First, analyze whether the problem has sufficient elements such that it deserves to be a fuzzy control system. Otherwise, consider conventional control. Expert system or neural network-based control can also be considered, if necessary.

2. Get all the information from the operator of the plant to be controlled. Get information about the design and operational characteristics of the plant from the plant designer, if possible.

3. If a plant model is available, develop a simulation program with conventional control and study the performance characteristics.

4. Identify the control elements where fuzzy logic is to be applied.

5. Identify the input and output variables for the fuzzy controller.

6. Define universe of discourse of each.

7. Determine the fuzzy sets and the corresponding membership function shape of each. For a sensitive variable, the number of fuzzy sets should be more.

8. Determine the rule table.

9. Determine scale factor of the variables.

10. If the model is available, simulate the system with the defined fuzzy controller and observe performance under different conditions of operation. Iterate the rule table and membership functions until the performance is

satisfactory. For a plant without model, fuzzy control is to be designed very conservatively and fine-tuned by testing on the operating plant.

11. Implement the controller in real time and further iterate for optimum performance.

The fuzzy controller can be implemented either by dedicated hardware or by microcontroller/DSP software. The dedicated hardware can again be analog or digital. General-purpose dedicated hardware chips have recently been available from various vendors [28]. ASIC control chips can be developed that can coexist with other ASIC chips or software controller. With software control, the program can coexist with other control routines of the system. The features of a few fuzzy logic tools and products [28] are summarized in Table 11-1.

11.3.5. Application in Power Electronics and Drives

Fuzzy logic application in power electronics and drives is relatively new. Li and Lau [29] applied fuzzy logic to microprocessor-based servo motor controller, assuming a linear power amplifier. They compared the fuzzy-controlled system performance with that of PID control and model referencing adaptive control (MRAC) and demonstrated the superiority of the former. Da Silva et al [30] developed a fuzzy adaptive controller and applied it to a four-quadrant power converter for the first time. Gradually, fuzzy control gathered momentum for other applications. In this section, a few example applications from the literature will be discussed.

A Phase-Controlled Converter DC Motor Drive. Fuzzy logic can be applied in feedback control of a drive system [31]. With nonlinearity, parameter variation, and load disturbance effects, it can provide fast and robust control. The drive may be based on DC or AC machine. Since vector-controlled AC drive and DC drive have identical dynamic models, the same fuzzy control principle is applicable in either case. The methodology in other feedback control systems essentially remains the same.

Figure 11-17 shows the DC motor speed control system under consideration where the power circuit consists of a phase-controlled bridge converter that drives a separately excited DC motor. The speed control loop has inner current control loop, as shown. The current loop output V_s' is added with the feedforward counter EMF signal V_c to generate the control signal V_s, which then generates the firing angle α by cosine wave crossing method. In the figure, as shown, fuzzy control is applied to speed loop and current loop as well as for converter transfer function linearization.

The converter may operate in either continuous or discontinuous conduction mode. At low speed when the counter EMF is small, the conduction will be continuous, but at high speed the conduction will tend to be discontinuous. Figure 11-18a gives the plot of V_d (p.u.) - I_a (p.u.) relation for different firing angle, where V_d (p.u.) and I_a (p.u.) are the per unit armature voltage and armature current, respectively. The boundary between continuous and discontinuous conduction is shown in the figure. The gain of the converter at continuous conduction is constant,

TABLE 11-1   FUZZY LOGIC TOOLS AND PRODUCTS

Company	Product	Description
American Neuralogix, Inc.	NLX 230 fuzzy microcontroller	Has 8 digital inputs, 8 digital outputs, 16 fuzzifiers; holds 64 rules. Evaluates 30M rules/sec.
	ADS 230 fuzzy microcontroller development system	PC-compatible system uses NLX 230 with analog and digital I/O.
	NLX 110 fuzzy pattern correlator	Correlates eight 1-Mbit patterns; expandable to 256 n-bit patterns.
	NLX 112 fuzzy data correlator	Performs pattern matching on serial data streams.
Hyperlogic Corp.	Cubicalc	Software for developing fuzzy logic applications. Runs under MS Windows with 286 or higher processor. Simulates fuzzy and non-fuzzy systems.
	Cubicalc-RTC	A superset of Cubicalc. Provides runtime compiler support and libraries for linking. Compatible with Microsoft C and Borland C.
	Cubicalc run-time source code	Generates C source code for use in compiling to a specific processor.
	Cubicard	Includes Cubicalc-RTC and PC-based hardware for analog and digital I/O.
Motorola, Inc.	Fuzzy logic kernel for microcontrollers	Fuzzy processing kernels for 68HC05 and 68HC11 microcontrollers. Includes fuzzy knowledge-base generator to create code for kernel.
	Fuzzy logic educational kit	Interactive training tool provides good introduction for understanding and using fuzzy logic. Runs under MS Windows. Includes demonstration version of Fide (from Aptronix).
Togal Infralogic, Inc.	TILShell + fuzzy C development system	Complete fuzzy development system generates C code and includes debug, fuzzy simulation, and graphical analysis tools. Tutorial included.
	Microcontroller evaluation packages	Fuzzy development systems for Hitachi H8/300, H8/500, and HMCS400; Intel 8051; and Mitsubishi 37450.
	Microcontroller production licenses	Unlimited production license.
	FC110	Digital fuzzy logic processor (IC).
	FC110 development system	Hardware and software development system for FC110. Versions support IBM PC/AT bus, Sbus, and VMEbus.

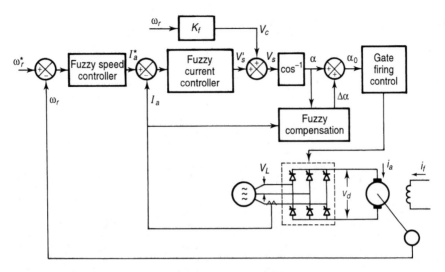

Figure 11-17. Phase-controlled converter DC drive with fuzzy control

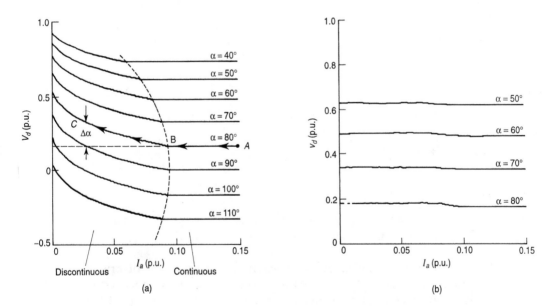

Figure 11-18. Theoritical V_d (p.u.) - I_a (p.u.) characteristics: (a) without compensation; (b) with compensation.

but it varies at discontinuous conduction at different operating points. The objective of fuzzy compensation is to generate a compensating angle $\Delta\alpha$ as a function of firing angle α and armature current I_a in order to linearize the converter transfer characteristics. A constant converter gain helps to obtain improved transient response of the drive at all operating conditions. The rule base designed for fuzzy compensation is given in matrix form in Table 11-2. A typical rule has the following structure:

TABLE 11-2   RULE BASE FOR NONLINEARITY COMPENSATION

I_a \ α	NB	NS	Z	PS	PB
NVB	NVB	PB	PB	PB	PB
NB	NVB	Z	Z	Z	Z
NM	NVB	NS	NVS	NVS	NVS
NS	NVB	NM	NS	NS	NS
Z	NVB	NB	NM	NM	NS
PS	NVB	NVB	NB	NM	NM
PM	NVB	NVB	NB	NB	NB
PB	NVB	NVB	NVB	NB	NB
PVB	NVB	NVB	NVB	NVB	NB

NB = Negative big

NM = Negative medium

NS = Negative small

NVB = Negative very big

NVS = Negative very small

PB = Positive big

PM = Positive medium

PVB = Positive very big

Z   = Zero.

```
IF I_a is small negative (NS) AND α is small positive (PS)
THEN Δα is small negative (NS)
```

The rule base is developed by heuristics from the viewpoint of practical system operation. The current I_a is treated as normalized value. Figure 11-19 shows the membership function plots of the variables α, I_a (p.u.), and $\Delta\alpha$. The universe of discourse of the variables covers the whole discontinuous conduction region. The sensitivity of a variable determines the number of fuzzy sets, as mentioned before. The universe of discourse of α is described by 5 fuzzy sets, whereas I_a (p.u.) and $\Delta\alpha$ are described by 9 and 11 sets, respectively. The linguistic terms used for the sets are for convenience only and must be interpreted in "context-free" grammar. The triangular membership functions are symmetrical, and 50% overlap has been considered. Therefore, at any point of the universe of discourse, no more than two fuzzy sets will have nonzero degree of membership. It is evident that for any input data of I_a (p.u.) and α, only four rules (at the most) will be valid in the entire rule base given in Table

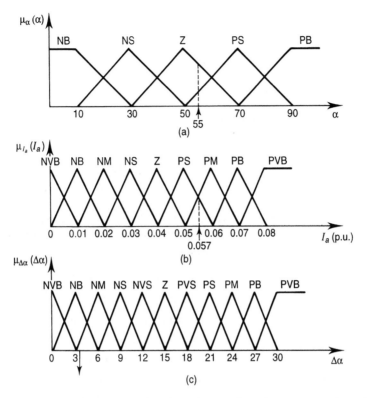

Figure 11-19. Membership functions for nonlinearity compensation: (a) firing angle; (b) armature current; (c) compensating angle.

11-2. The algorithm for fuzzy linearization can be summarized as follows. A numerical example is included in each for clarity.

1. Sample the DC current I_a and firing angle α from the current control loop. Convert I_a to I_a (p.u.) by dividing with the base current. Assume that: I_a = 5.6 A, α = 55°, I_{base} = 98.5 A, and I_a (p.u.) = 5.6/98.5 = 0.057.

2. Calculate the interval indices I and J (that identify the interval number in the fuzzy sets) for α and I_a (p.u.), respectively as

$$I = \text{INT}\,\frac{\alpha + 10}{20}$$

$$J = \text{INT}\,\frac{(I_a\,(\text{p.u.}) + 0.01)}{0.01}$$

$$I = 3, \; J = 6.$$

3. Calculate the degree of membership of α and I_a (p.u.) for the leftmost fuzzy set using I and J, respectively,

$$\mu_z(\alpha) = \frac{20I + 10 - \alpha}{20}$$

$$\mu_{ps}(I_a\,(\text{p.u.})) = \frac{0.01I - I_a\,(\text{p.u.})}{0.01}$$

$$\mu_z(55°) = 0.75, \quad \mu_{ps}(0.057) = 0.3.$$

4. Evaluate degree of membership for the other sets by complement relation

$$\mu_{ps}(\alpha) = 1 - \mu_z(\alpha)$$

$$\mu_{pm}(I_a\,(\text{p.u.})) = 1 - \mu_{ps}(I_a\,(\text{p.u.}))$$

$$\mu_{ps}(55°) = 0.25, \quad \mu_{pm}(0.057) = 0.7.$$

5. Identify the four valid rules in Table 11-2 (stored as look-up table) and calculate the degree of membership μR_i contributed by each rule R_i (i = 1, 2, 3, 4], using MIN operator

$$\mu_{R1} = \min\{\mu_z(\alpha),\ \mu_{PS}(I_a\,(\text{p.u.}))\}$$
$$= \min\{0.75,\ 0.3\} = 0.3$$
$$\mu_{R2} = \min\{\mu_{PS}(\alpha),\ \mu_{PS}(I_a\,(\text{p.u.}))\}$$
$$= \min\{0.25,\ 0.3\} = 0.25$$
$$\mu_{R3} = \min\{\mu_z(\alpha),\ \mu_{PM}(I_a\,(\text{p.u.}))\}$$
$$= \min\{0.75,\ 0.7\} = 0.7$$
$$\mu_{R4} = \min\{\mu_{PS}(\alpha),\ \mu_{PM}(I_a\,(\text{p.u.}))\}$$
$$= \min\{0.25,\ 0.7\} = 0.25$$

6. Retrieve the amount of correction $\Delta\alpha_i$, i = 1, 2, 3, 4 corresponding to each rule, from Table 11-2.

$$\Delta\alpha_1 = (\alpha = Z,\ I_a\,(\text{p.u.}) = PS) \rightarrow \Delta\alpha_1 = NB = 3°$$
$$\Delta\alpha_2 = (\alpha = PS,\ I_a\,(\text{p.u.}) = PS) \rightarrow \Delta\alpha_2 = NM = 6°$$
$$\Delta\alpha_3 = (\alpha = Z,\ I_a\,(\text{p.u.}) = PM) \rightarrow \Delta\alpha_3 = NB = 3°$$
$$\Delta\alpha_4 = (\alpha = PS,\ I_a\,(\text{p.u.}) = PM) \rightarrow \Delta\alpha_4 = NB = 3°$$

7. Calculate the crispy value $\Delta\alpha$ by the height defuzzification method as

$$\Delta\alpha = \frac{0.30 \cdot 3 + 0.25 \cdot 6 + 0.70 \cdot 3 + 0.25 \cdot 3}{0.30 + 0.25 + 0.70 + 0.25}$$

Figure 11-18b shows the V_d (p.u.) - I_a (p.u.) plot after phase compensation.

Next, fuzzy control was applied to speed and current control loops. Since both the loops have essentially first order characteristics (after applying fuzzy compensation), intuitively the same fuzzy controller should be valid for both the loops except that the scale factors are different. The general input variables in the fuzzy rule base are

$$E(k) = R(k) - C(k) \tag{11.8}$$
$$CE(k) = E(k) - E(k-1) \tag{11.9}$$

where
$$E(k) = \text{loop error}$$
$$CE(k) = \text{change in loop error}$$
$$R(k) = \text{reference signal}$$
$$C(k) = \text{output signal}$$
$$k = \text{sampling instant}$$

The variables can be expressed as per unit quantities as

$$e \text{ (p.u.)} = \frac{E(k)}{GE}$$
$$ce \text{ (p.u.)} = \frac{CE(k)}{GC}$$
$$du \text{ (p.u.)} = \frac{DU(k)}{GU}$$

where GE, GC, and GU are the respective scale factors. In the rule base, a typical rule structure is

```
IF e(k) is x AND ce(k) is y THEN du(k) is z
```

After implementing the current and speed loops, performances were evaluated on the system. Figure 11-20a shows the speed and current response that covers both continuous and discontinuous regions. Figure 11-20b shows the similar response but with four times the shaft inertia load. The fuzzy control verifies the robust deadbeat performance with inertia variation and load torque disturbance.

Induction Motor Drive Efficiency Optimization Control. In variable frequency drives, machines are normally operated at rated flux to give optimum transient response. However, at light load, this causes excessive core loss impairing efficiency of the drive. The flux can be programmed at a light-load, steady-state condition to improve the drive efficiency [34]. Figure 11-21 explains the on-line search method of efficiency optimization by flux programming in an indirect vector-controlled induction motor drive. Assume that the machine operates initially at rated flux in steady-state with the load torque and speed, as indicated. The rotor flux is decremented in steps by reducing the magnetizing component (i_{ds}) of stator current. This results in increase of the torque component of current i_{qs} (normally by the speed loop) so that the developed torque remains the same. As the core loss decreases with the decrease of flux, the copper loss increases, but the system (converter and machine) loss decreases improving the overall efficiency. This is reflected in the decrease of DC link power P_d, as shown. The search is continued until the system settles at the minimum input power point A.

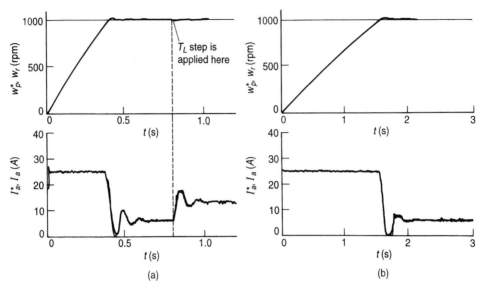

Figure 11-20. Fuzzy speed control loop response with stepped speed and torque load: (a) nominal inertia; (b) four times nominal inertia.

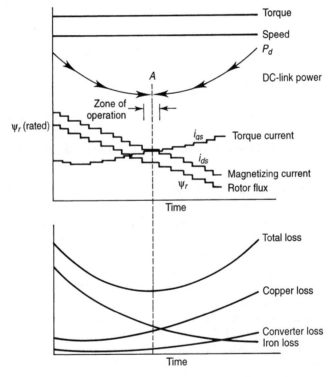

Figure 11-21. On-line search method of flux programming efficiency improvement control for vector-controlled induction motor drive.

Figure 11-22 shows the block diagram of a vector-controlled induction motor drive incorporating the fuzzy logic–based efficiency controller [35], as previously discussed. The fuzzy controller has the advantage that it operates in noisy environment with parameter variation and adaptively decrements the step size of excitation current so that fast convergence is attained. In Figure 11-22, the speed control loop generates the torque component of current i_{qs}, as indicated. The vector rotator receives the torque and excitation current commands i_{qs} and i_{ds}, respectively, from two positions of a switch: the transient position (1), where the excitation current is established to the rated value (i_{dsr}) and the speed loop feeds the torque current; and the steady-state position (2), where the excitation and torque currents are generated by the fuzzy efficiency controller and feedforward torque compensator which will be explained later. The fuzzy controller becomes effective at steady-state condition; that is, when the speed loop error $\Delta \omega_r$ approaches zero. Note that minimization of DC link power also minimizes the system input power. Figure 11-23 explains the fuzzy efficiency controller operation. The DC link power P_d is sampled and compared with the previous value to determine the increment $\Delta P_d(k)$. In addition, the last excitation current (Δi_{ds}^{*} (p.u.) $(k-1)$) polarity is reviewed. With these input signals, the decrement step of Δi_{ds}^{*} (p.u.) is generated from fuzzy rules through fuzzy inference and defuzzification, as discussed before. The scale factors are programmable and are given by the expressions

$$P_b = a \omega_r + b \tag{11.10}$$

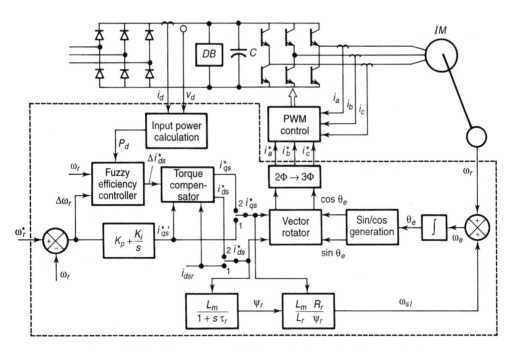

Figure 11-22. Vector-controlled drive with fuzzy efficiency optimizer.

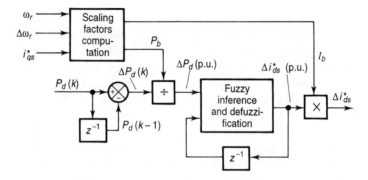

Figure 11-23. Efficiency optimizer control block diagram.

$$I_b = C_1\omega_r - C_2\hat{T}_e + C_3 \tag{11.11}$$

The variables ΔP_d (p.u.) and Δi_{ds}^* (p.u.) are each described by seven asymmetric triangular membership functions, whereas Δi_{ds}^* (p.u.) $(k-1)$ has only two (positive and negative) membership functions. An example rule can be given as

```
IF the power increment (ΔPd) is negative small (NS)
     AND the last ids (Δi*ds (p.u.) (k-1)) is negative (N)
THEN the new excitation current increment (Δi*ds (p.u.))
     is negative small (NS)
```

Figure 11-24 explains the principle of feedforward pulsating torque compensation. This compensator functions to compensate the loss of torque due to decrementation of i_{ds} by injecting an equivalent Δi_{qs}^* so that the developed torque remains the same. Otherwise, slow-speed loop compensation creates large pulsating torque at low frequency, which may be harmful for the drive. The compensating current $\Delta i6_{qs}^*(k)$ can be computed as

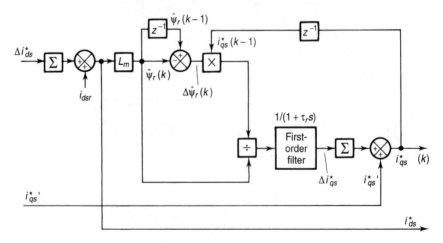

Figure 11-24. Feedforward pulsating torque compensation block diagram.

$$\Delta i^*_{qs}(k) = \frac{\psi_r(k-1) - \psi_r(k)}{\psi_r(k)} \, i^*_{qs}(k-1) \tag{11.12}$$

Since the response of Δi^*_{qs} is instantaneous, but Δi^*_{ds} responds with the rotor time constant, for correct torque matching Δi^*_{qs} is to be delayed by the same rotor time constant, as indicated in the figure. Figure 11-25 shows performance of fuzzy efficiency controller in conjunction with the feedforward torque compensator. The system abandons the efficiency optimization mode and establishes the rated flux as speed command or load torque changes.

Slip Gain Tuning Control. On-line slip gain tuning of an indirect vector-controlled induction motor drive has been a challenging research topic in the recent years [37, 38]. A detuned slip gain, caused by variation of machine parameters, particularly the rotor resistance, gives undesirable transfer characteristics for torque and flux, and therefore, the system may fall into instability in extreme condition. Fuzzy logic principle can be effective in solving this problem. Figure 11-26 shows the block diagram of an indirect vector-controlled drive where the fuzzy tuning controller has been incorporated, and Figure 11-27 shows the details of fuzzy MRAC–based tuning controller. The scheme depends on reference model computation of reactive power (Q^*) and d-axis voltage (v^*_{ds}) at the machine terminal for ideally tuned condition as

$$Q^* = \omega_e(L_s i^{*\,2}_{ds} + L_\sigma i^{*\,2}_{qs}) \tag{11.13}$$
$$v^*_{ds} = R_s i^*_{ds} - \omega_e L_\sigma i^*_{qs} \tag{11.14}$$

where all the parameters and variables are in standard notation. The foregoing reference models are then compared with estimation of the quantities given by

$$Q = v_{qs} i_{ds} - v_{ds} i_{qs} \tag{11.15}$$
$$v_{ds} = v^s_{qs} \, \sin\theta_e + v^s_{ds} \, \cos\theta_e \tag{11.16}$$

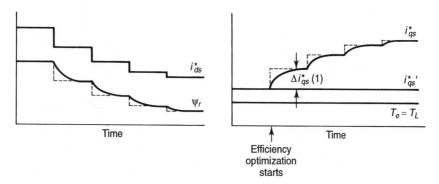

Figure 11-25. Waveforms explaining efficiency optimization and torque compensator control.

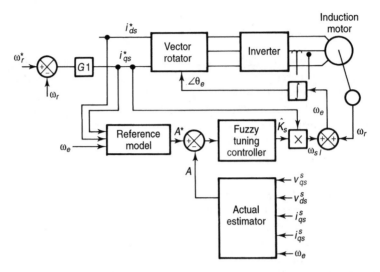

Figure 11-26. Indirect vector-controlled induction motor drive showing the proposed fuzzy slip gain tuning controller.

where $\cos\theta_e$ and $\sin\theta_e$ are the unit vectors. The loop errors are divided by the respective scale factors to derive the per unit variables for manipulation in the fuzzy controller. There are, in fact, two fuzzy controllers in Figure 11-27, and each is designed with the respective rule base and fuzzy sets. The controller FLC-1 generates a weighting factor K_f that permits appropriate distribution of Q control and v_{ds}-control on the torque-speed (i.e., i_{qs} - ω_e) plane. The objective is to assign high sensitivity to detuning control by dominant use of Q control in low-speed, high-

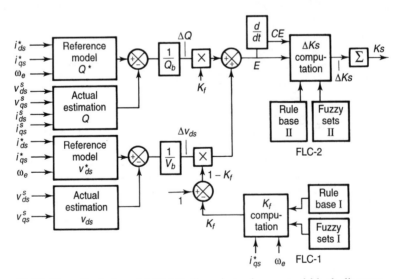

Figure 11-27. Fuzzy logic–based MRAC slip gain tuning control block diagram.

torque region and ν_{ds} control in high-speed, low-torque region. An example rule from the 2 × 2 rule matrix can be given as

```
IF speed (ωₑ) is high (H) and torque (iqs) is low (L)
THEN weighting factor (Kf) is low (L)
```

The combined error signal for both the loops in Figure 11-27 is given as

$$E = \Delta Q K_f + \Delta \nu_{ds}(1 - K_f) \tag{11.17}$$

The fuzzy controller FLC-2 generates the corrective incremental slip gain K_s based on the combined detuning error (E) and its derivative (CE). Basically, it is an adaptive feedback controller for fast convergence at any operating point irrespective of the strength of E and CE signals. In summary, at ideally tuned condition of the system, both Q and ν_{ds}, and correspondingly the E and CE signals, will be zero and the slip gain K_s will be set to the correct value (K_{so}). If the system is detuned with parameter variation, the actual Q and ν_{ds} variables will deviate from the respective reference variables and the resulting error will alter the K_s value until the system becomes tuned; that is, $E = 0$. The effects of detuning on torque and rotor flux transients are illustrated in Figure 11-28. In Figure 11-28a, the slip gain was set to twice the correct value ($K_s/K_{so} = 2$), resulting in higher order dynamics for torque

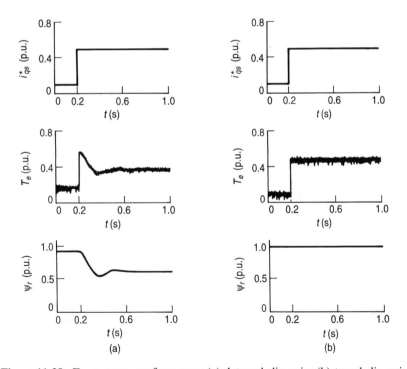

Figure 11-28. Fuzzy tuner performance: (a) detuned slip gain; (b) tuned slip gain.

and flux transients as well as underfluxing condition. In Figure 11-28b, on the other hand, the tuned slip gain condition gives ideal transient response.

Estimation of Distorted Waves. Power converters usually generate distorted voltage and current waves, and it is often necessary to calculate parameters, such as total rms value, fundamental rms value, displacement power factor, power factor, and the like for monitoring and control [40]. A fuzzy logic–based waveform estimation has the advantage of fast response, multiple outputs from a single premise of a rule, and immunity of noise and drift from the sensors. Basically, it is a pattern recognition process; that is, a certain wave pattern dictates the foregoing parameter values. Here, fuzzy logic is essentially used as a pattern recognizer, similar to that of a neural network, which will be discussed later.

Both rule-based and relational methods, as discussed before, can be used for the estimation. Figure 11-29 illustrates the rule-based estimation for a triac-controlled line current with resistive load (such as light dimmer). The current wave pattern is characterized by the width (W) and the height (H) variables, and a fuzzy estimator is programmed (or trained) to estimate the total rms current (I_s), fundamental rms current (I_f), displacement power factor (DPF), and power factor (PF) parameters. The input variable W and the output variables I_s (p.u.), I_f (p.u.), DPF, and PF of the fuzzy estimator, as shown in the figure, are each defined by 12 asymmetrical membership functions. Note that MFs are nonidentical because each

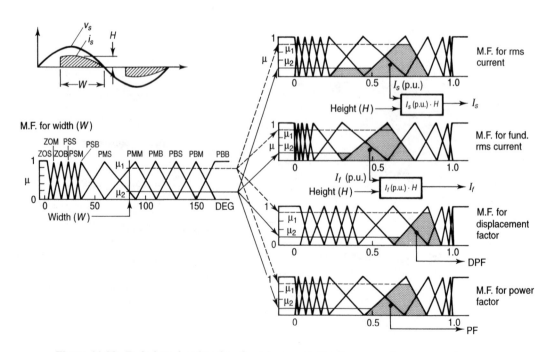

Figure 11-29. Rule-based estimation for triac controller line current.

output has different degree of nonlinearity. The iteration of MFs, their shape, and the rule base for the best estimation accuracy are based on simulation results. Obviously, there are altogether 12 rules each with multiple consequents. The two rules for the example case are

```
Rule 1:
IF width (W) is PMS
THEN I_s(p.u.) = PMS, I_f(p.u.) = PMS, DPF = PMS, PF = PMS


Rule 2:
IF width (W) is PMM
THEN I_s(p.u.) = PMM, I_f(p.u.) = PMM, DPF = PMM, PF = PMM
```

Note that although the consequents of each rule are identical, the actual output will be different as shown by the shaded areas because of nonidentical shape of the MFs. The outputs are defuzzified by height method to convert to the crispy values. The I_s (p.u.) and I_f (p.u.) are then multiplied by the input variable H to determine the actual currents whereas DPF and PF are obtained directly. Figure 11-30 explains the relational estimation method for the triac controller current wave. The method needs example data to fine-tune the coefficients of linear equations for correct estimation. The simulation results were used in the regions of width W defined by the 12

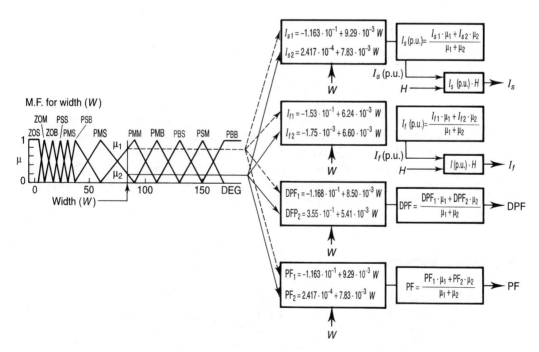

Figure 11-30. Relational estimation for triac controller line current.

fuzzy sets to generate these data. In each region, multiregression analysis was performed to define the parameters Aij for all the estimated outputs. Note that the estimation equations that give solution in terms of per unit values are a function of W only. The actual currents are obtained by multiplying with H after defuzzification, whereas DPF and PF are obtained directly after defuzzification. There are altogether 12 rules as in the rule-based method, and as shown in Figure 11-30, the two example rules fired are given as

```
Rule 1:
IF W is PMM
```
$$\text{THEN}\quad I_s\ (\text{p.u.}) = -1.163 \cdot 10^{-1} + 9.29 \cdot 10^{-3}\,W$$

$$I_f\ (\text{p.u.}) = -1.53 \cdot 10^{-1} + 6.24 \cdot 10^{-3}\,W$$

$$\text{DPF} = 1.168 \cdot 10^{-1} + 8.50 \cdot 10^{-3}\,W$$

$$\text{PF} = -1.163 \cdot 10^{-1} + 9.296 \cdot 10^{-3}\,W$$

```
Rule 2:

IF W is PMS
```

$$\text{THEN}\quad I_s(\text{p.u.}) = 2.417 \cdot 10^{-4} + 7.83 \cdot 10^{-3}\,W$$

$$I_f(\text{p.u.}) = -1.75 \cdot 10^{-3} + 6.60 \cdot 10^{-3}\,W$$

$$\text{DPF} = 3.55 \cdot 10^{-1} + 5.41 \cdot 10^{-3}\,W$$

$$\text{PF} = 2.417 \cdot 10^{-4} + 7.837 \cdot 10^{-1}\,W$$

The estimation results for both the approaches are shown in Figure 11-31 for the entire width W and compared with the actual values. The estimation accuracy in the relational method was found to be better than the rule-based approach. It is not surprising since the former hybrids the fuzzy and mathematical methods. The accuracy, of course, of a rule-based method can be improved by using a larger number of fuzzy sets for input and output variables.

11.4. NEURAL NETWORK

Artificial neural network (ANN) or neural network (NN) is the most generic form of AI for emulation of human thinking compared to expert system and fuzzy logic. The conventional digital computer is very good in solving expert system problems and somewhat less efficient in solving fuzzy logic problems, but its limitations in solving pattern recognition and image processing–type problems have been seriously felt

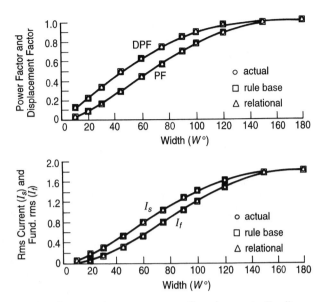

Figure 11-31. Fuzzy estimator accuracy for triac controller line current by rule-based and relational method.

since the late 1980s and early 1990s. As a result, people's attention was gradually focused to ANN technology, which could solve such problems very efficiently. Fundamentally, the human brain constitutes billions of nerve cells, called *neurons*, and these neurons are interconnected to form the biological neural network. Our thinking process is generated by the action of this neural network. The ANN tends to simulate the biological neural network with the help of dedicated electronic computational circuits or computer software. The technology has recently been applied in process control, identification, diagnostics, character recognition, robot vision, and financial forecasting, just to name a few.

The history of ANN technology is old and fascinating. It predates the advent of expert system and fuzzy logic technologies. In 1943, McCulloch and Pitts first proposed a network composed of binary-valued artificial neurons that was capable of performing simple threshold logic computations. In 1949, Hebb proposed a network learning rule that was called *Hebb's rule*. Most modern network learning rules have their origin in Hebb's rule or a variation on it. In the 1950s, the dominant figure in neural network research was psychologist Rosenblatt at Cornell Aeronautical Lab who invented the Perceptron, which represents a biological sensory model, such as eye. Widrow and Hoff proposed ADALINE and MADALINE (many ADALINES) and trained the network by their delta rule, which is the forerunner of the modern back-propagation training method. The lack of expected performance of these networks coupled with the glamor of the von Neumann digital computer in the late 1960s and 1970s practically camouflaged the neural network evolution. The modern era of neural network with rejuvenated research practically started in 1982 when Hopfield, a professor of chemistry and biology at Cal Tech, presented his invention at the National Academy of Science. Since then, many network models and learning rules have been introduced. Since the beginning of the 1990s, the neural network as AI tool has captivated the attention of a large segment in the scientific community. The technolgy is predicted to have a significant impact on our society in the next century.

11.4.1. Neural Network Principles

Artificial neural network (or neural network), as the name indicates, is the interconnection of artificial neurons that tend to simulate the nervous system of the human brain. It is also defined as *neurocomputer* or *connectionist system* in the literature. When a person is born, the cerebral cortex of the brain contains roughly 100 billion neurons or nerve cells, and each neuron output is interconnected from 1000 to 10,000 other neurons. A biological neuron is a processing element that receives and combines signals from other neurons through input paths called *dendrites*. If the combined signal is strong enough, the neuron "fires," producing an output signal along the axon that connects to dendrites of many other neurons. Each signal coming into a neuron along a dendrite passes through a synaptic junction. This junction is an infinitesimal gap in the dendrite that is filled with neurotransmitter fluid that either accelerates or retards the flow of electrical charges. The fundamental actions of the neuron are chemical, and this neurotransmitter fluid produces electrical signals that go to the nucleus or soma of the neuron. The adjustment of the impedance or conductance of the synaptic gap leads to "memory" or "learning process" of the brain. According to this theory, we are led to believe that the brain has distributed memory or intelligence characteristics giving it the property of *associative memory*, but not as digital computerlike central storage memory addressed by CPU. Otherwise, a patient when recovering from anesthesia will forget everything.

An artificial neuron is a concept whose components have direct analogy with the biological neuron. Figure 11-32 shows the structure of an artificial neuron reminding us of analog summer-like computation. It is also called neuron, processing element (PE), neurode, node, or cell. The input signals X_1, X_2, X_3, $\ldots, X_n$ are normally continuous variables but can also be discrete pulses that occur in the brain. Each of the input signals flows through a gain or weight, called *synaptic weight* or *connection strength* whose function is analogous to that of the synaptic junction in a biological neuron. The weights can be positive (excitory) or negative (inhibitory), corresponding to acceleration or inhibition, respectively, of the flow of electrical signals. The summing node accumulates all the input weighted signals (activation signal) and then passes to the output through the transfer function, which is usually nonlinear. The transfer function can be a step or threshold type (that passes logical 1 if the input exceeds a threshold, or else 0), signum type (output is $+1$ if the input exceeds a threshold, or else -1), or linear threshold type (with the output clamped to $+1$). The transfer function can also be a non-linear continuously varying type, such as sigmoid (shown in Figure 11-32), hyper-

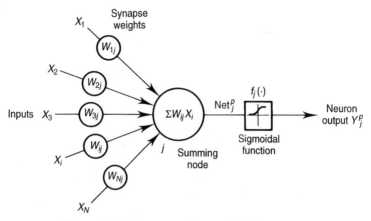

Figure 11-32. Structure of an artificial neuron.

bolic-tan, inverse-tan, or Gaussian type. The sigmoidal transfer function is most commonly used and is given by

$$Y = \frac{1}{1 + e^{-\alpha X}} \qquad (11.18)$$

where α is the coefficient or gain that determines slope of the function that changes between the two asymptotic values (0 and $+1$). With high gain, it approaches as a step function. The sigmoidal function is nonlinear, monotonic, and differentiable and has the largest incremental gain at zero input signal. These properties have special importance in neural network. These transfer functions are characterized as *squashing functions*, because they squash or limit the output values between the two asymptotes. It is important to mention here that a nonlinear transfer function gives nonlinear mapping property of neural network; otherwise, the network property will be linear.

The actual interconnection of biological neurons is not well known, but scientists have evolved more than 60 neural network models and many more are yet to come. Whether the models match well with those in the brain is not very important, but it is important that the models help in solving our scientific, engineering, and many other real-life problems.

Neural networks can be classified as *feedforward* (or layered) and *feedback* (or recurrent) types, depending on the interconnections of the neurons. It can be shown that whatever problems can be solved by feedback network can also be solved by the equivalent feedforward network with proper external connection. A network can also be defined as *static* or *dynamic*, depending on whether it is simulating static or dynamical systems. At present, the majority of applications (roughly 90%) use feedforward architecture, and it is of particular relevance to power electronic applications. Figure 11-33 shows the structure of a feedforward multilayer network with three input and two output signals. The circles represent the neurons, and the dots in the connections represent the weights. The transfer functions are not shown in the

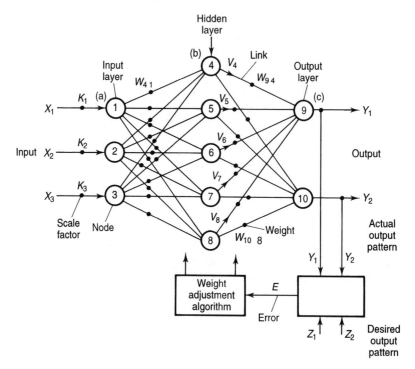

Figure 11-33. Structure of feedforward neural network showing back propagation training.

figure. The back-propagation training principle, as indicated, will be discussed later. The network has three layers:

1. Input layer
2. Hidden layer
3. Output layer.

With the number of neurons in each layer, as shown, it is defined as 3-5-2 network. The hidden layer functions to associate the input and output layers. The input and output layers (defined as *buffers*) have neurons equal to the respective number of signals. The input layer neurons do not have transfer function, but there is scale factor in each input to normalize the input signals. There may be more than one hidden layer. The number of hidden layers and the number of neurons in each hidden layer depend on the complexity of the problem being solved. The input layer transmits the computed signals to the hidden layer, which in turn, transmits to the output layer, as shown. The signals always flow in the forward direction. There is no self, lateral, or feedback connection of neurons. The network is defined as "fully connected" when each of the neurons in a given layer is connected with each of the neurons in the next layer (as indicated in Figure 11-33) or "partially connected" when some of these connections are missing. Network input and output signals may be logical (0, 1), discrete bidirectional (±1) or continuous variables. The sigmoid output signals can be clamped to convert to logical variables. The architecture of neural network makes it obvious that basically it is a parallel input–parallel output multidimensional computing system where computation is done in a distributed manner, compared to sequential computation in a conventional computer that takes help of centralized CPU and storage memory. It is definitely closer to analog computation.

11.4.2. Training of Feedforward Neural Network

How does a feedforward neural network described earlier perform useful computation? Basically, it performs the function of nonlinear mapping or pattern recognition. This is defined as *associative memory*. With this associative memory property of the human brain, when we see a person's face, we remember his name. With a set of input data that correspond to a definite signal pattern, the network can be "trained" to give a correspondingly desired pattern at the output. A trained neural network, like a human brain, can associate a large number of output patterns corresponding to each input pattern. The network has the capability to "learn" because of the distributed intelligence contributed by the weights. The input-output pattern matching is possible if appropriate weights are selected. In Figure 11-33, there are altogether 25 weights, and by altering these weights, we can get 25 degrees of freedom at the output for a fixed input signal pattern. The network will be initially "untrained" if the weights are selected at random, and the output pattern will then totally mismatch the desired pattern. The actual output pattern can be compared with the desired output pattern, and the weights can be adjusted by an algorithm until the pattern matching occurs; that is, the error becomes acceptably small. Such training should be continued with a large number of input-output example patterns. At the completion of training, the network should be capable not only of recalling all the example output patterns but also of interpolating and extrapolating them. This tests the learning capability of the network instead of a simple look-up table function. This type of learning is called *supervised learning* (learning by a teacher, which is similar to alphabet learning by children). There are other types of learning described in the literature, such as unsupervised or self-learning where the network is simply exposed to a number of inputs, and it organizes itself in such a way as to come up with its own classifications for inputs (stimulation-reaction mechanisms, similar to the way people initially learn language) and reinforced learning where the learning performance is verified by a critic. With a learning procedure, as described, a neural network can solve a problem satisfactorily, but compared to human learning or expert system knowledge, it cannot explain how it generated a particular output. Again, for continuous output signals, the solution is only approximately similar to fuzzy logic computation.

The learning for pattern recognition in a feedforward network will be illustrated by the optical character recognition (OCR) problem [43], as shown in Figure 11-34a. The problem here is to convert the English alphabet characters into a 5-bit code (which can be considered as data compression) so that altogether $2^5 = 32$ different characters can be coded. The letter "A" is represented by a 5×7 matrix of inputs

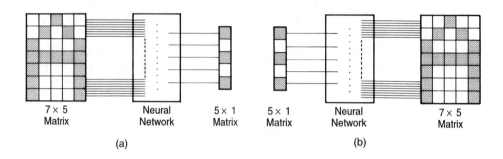

(a)   (b)

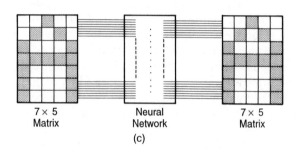

7 × 5
Matrix

Neural
Network

7 × 5
Matrix

(c)

Figure 11-34. Neural network mapping for the letter "A": (a) input-output mapping; (b) inverse mapping; (c) input-output mapping in an autoassociative network.

consisting of logical 0's and 1's. The input vector of 35 signals is connected to the respective 35 neurons at the input layer. The three-layer network has 11 neurons in the hidden layer and 5 neurons in the output layer corresponding to the 5 bits (in this case, 10101), as indicated. The network uses sigmoidal transfer function that is clamped to logical output. The input-output pattern mapping is performed by supervised learning; that is, altering the network weights (800 altogether) to the desired values. If now the letter "B" is impressed at the input and the desired output map is 10001, the output will be totally distorted with the previous training weights. The network undergoes another round of training until the desired output pattern is satisfied. This is likely to deviate the desired output for "A." The back-and-forth training will satisfy output patterns for both "A" and "B." In this way, a large number of training exercises will eventually train the network for all the 32 characters. Evidently, the nonlinearity of the network with logical clamping at the output makes such pattern recognition possible. It is also possible to train a network for inverse mapping; that is, with the input vector 10101, the output vector maps the letter "A," as shown in Figure 11-34b. The procedure for training is the same as before. This case is like data expansion instead of compression. It is possible to cascade Figures 11-34a and b so that the same letter is reproduced. This arrangement has the advantage of character transmission through a narrow-band channel. Again, it is possible to train a network such that the output pattern is the same as the input pattern that is indicated in Figure 11-34c. This is called an *auto-associative network* compared to the *heteroassociative networks* discussed earlier. The benefit for auto-associative mapping is that if the input pattern is distorted, the output mapping will be clean and crispy because the network is trained to reproduce the nearest crispy output. The same benefit is also valid for Figure 11-34a. This inherent noise or distortion-filtering property of a neural network is very important in many applications. A neural network is often characterized as fault tolerant. This means that if a few weights become erroneous or several connections are destroyed, the output remains virtually unaffected. This is because the knowledge is distributed throughout the network. At the most, the output will degrade gracefully for larger defects in the network compared to catastrophic failure that is the characteristic of conventional computers.

Back Propagation Training of Feedforward Network. The back propagation is the most popular training method for a multilayer feedforward neural network, and it was first proposed by Rumelhart, Hinton, and Williams in 1986. Basically, it is the generalization of the delta rule proposed by Widrow and Hoff. Because of this method of training, the feedforward network is often defined as the *back-prop network*. The network (see Figure 11-33) is assigned random positive and negative weights in the beginning. For a given input signal pattern, step-by-step calculations are made in the forward direction to derive the output pattern. A cost functional given by the squared difference between the net output and the desired net output for the set of input patterns is generated, and this is minimized by a gradient descent method altering the weights one at a time starting from the output layer. The equations for the output of a single processing unit, shown in Figure 11-32, are given as

$$\text{net}_j^p = \sum_{i=1}^{N} W_{ij} X_i \tag{11.19}$$

$$Y_j^p = f_j \left(\text{net}_j^p \right) \tag{11.20}$$

where j = the processing unit under consideration

p = input pattern number

X_i = output of the ith neuron connected to the jth neuron

net_j^p = output of the summing node, that is, jth neuron activation signal

N = number of neurons feeding the jth neuron

f_j = nonlinear differentiable transfer function (usually a sigmoid)

Y_j^p = output of the corresponding neuron.

For the input pattern p, the squared output error for all the output layer neurons of the network is given as

$$E_p = \frac{1}{2}(d^p - y^p)^2 = \frac{1}{2}\sum_{j=1}^{S}(d_j^p - y_j^p)^2 \qquad (11.21)$$

where d_j^p = desired output of jth neuron in the output layer

y_j^p = corresponding actual output

S = dimension of the output vector

y = actual net output vector

d^p = corresponding desired output vector.

The total squared error E for the set of P patterns is then given by

$$E = \sum_{p=1}^{P} E_p = \frac{1}{2}\sum_{p=1}^{P}\sum_{j=1}^{S}(d_j^p - y_j^p)^2 \qquad (11.22)$$

The weights are changed to reduce the cost functional E to a minimum acceptable value by gradient descent method, as mentioned before. For perfect matching of all the input-output patterns, $E = 0$. The weight update equation is then given as

$$W_{ij}(k+1) = W_{ij}(k) + \eta \left(\frac{\delta E_p}{\delta W_{ij}(k)}\right) \qquad (11.23)$$

where η = learning rate

$W_{ij}(k+1)$ = new weight

$W_{ji}(k)$ = old weight.

The weights are iteratively updated for all the P training patterns. Sufficient learning is achieved when the total error E summed over the P patterns falls below a prescribed threshold value. The iterative process propagates the error backward in the network and is therefore called *error back propagation algorithm*. To be sure that the error converges to a global minimum but does not get locked up in a local minimum, a momentum term is added to the right of equation 11.23. Further improvement of the back-propagation algorithm is possible by making the learning rate gradually small (adaptive); that is,

$$\eta(k+1) = u\eta(k) \qquad \text{with } u < 1.0 \qquad (11.24)$$

so that oscillation becomes minimal as it converges to the global optimum point.

From the foregoing discussion, it is evident that neural network training is very time consuming, and this time will increase very fast if the number of neurons in the hidden layer or the number of hidden layers is increased. Normally, the training is done off-line with the help of a computer simulation program, which will be discussed later.

4.4.3. Fuzzy Neural Network

Fuzzy neural networks are systems that apply neural network principles to fuzzy reasoning [44, 45]. Basically, it emulates a fuzzy logic controller. This type of fuzzy controller emulation has the advantages that it can automatically identify fuzzy rules and tune membership functions for a problem. FNNs can also identify nonlinear systems to the same extent of precision as conventional fuzzy modeling methods do. The FNN topology can be either on rule-based approach or relational (Sugeno's) approach which was discussed before. A topology for closed loop adaptive control is shown in Figure 11-35. The network has two premises (loop error E and change in error CE) and one output (control signal U or DU). Each premise has three membership functions—SMALL(SM), MEDIUM(ME), and BIG—which are synthesized with the help of sigmoidal functions (f) giving each a Gaussian-type shape. The weight W_c controls the spacing whereas W_g controls the slope of the membership

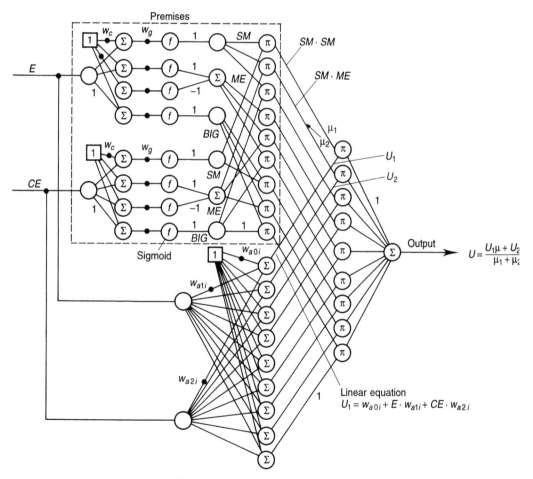

Figure 11-35. Fuzzy neural network.

functions. The weights are determined by back-propagation training algorithm. The premises are identical for both rule-based and relational topologies. The nine outputs of the premises after product (π) operation indicate that there are nine rules. The inferred value of the FNN is obtained as sum of the products of the truth values in the premises and linear equations in the consequences, as shown. A typical rule in Figure 11-35 can be read as

$$\text{IF } E \text{ is SM AND } CE \text{ is ME THEN } U = \frac{U_1 \mu_1 + U_2 \mu_2}{\mu_1 + \mu_2}$$

where

$$U_1 = W_{a01} + W_{a11} \cdot E + W_{a21} \cdot CE$$
$$U_2 = W_{a02} + W_{a12} \cdot E + W_{a22} \cdot CE$$

$$u_1 = \text{SM} \cdot \text{SM}$$

and

$$u_2 = \text{SM} \cdot \text{ME}$$

The generation of linear equations is shown in the lower part of the figure where the weights W_{a0i}, W_{a1i}, and W_{a2i} are the trained parameters. If necessary, the network can be expanded with larger number of membership functions, and more I/O signals can be added.

11.4.4. Design Methodology and Implementation

The general methodology for designing a neural network can be summarized as follows:

1. Analyze whether the problem has sufficient elements such that it deserves neural network solution. Consider alternative approach, such as expert system, fuzzy logic, or plain DSP-based solution.
2. Select feedforward network, if possible.
3. Select input nodes equal to the number of input signals and output nodes equal to the number of output signals.
4. Select appropriate input scale factors for normalization of input signals and output scale factors for denormalization of output signals.
5. Create input-output training data table. Capture the data from an experimental plant. If a model is available, make a simulation and generate data from the simulation result.
6. Select a development system. Assume that it is NeuralWorks Professional II/Plus.

7. Set up a network topology in the development system. Assume that it is a three-layer network. Select hidden layer nodes average of input and output layer nodes. Select transfer function. The training procedure is highly automated in the development system. The steps are given below.

8. Select an acceptable network training error E. Initialize the network with random positive and negative weights.

9. Select an input-output data pattern from the training data file. Change weights of the network by back propagation training principle.

10. After the acceptable error is reached, select another pattern and repeat the procedure. Complete training for all the patterns.

11. If a network does not converge to acceptable error, increase the hidden layer neurons or increase number of hidden layers (most problems can be solved by three layers), as necessary. A too high number of hidden layer neurons or number of hidden layers will increase the training time, and the network will tend to have memorizing property.

12. After successful training of the network, test the network performance with some intermediate data input. The weights are then ready for downloading.

13. Select appropriate hardware or software for implementation. Download the weights.

A flow chart for training is given in Figure 11-36. There is a large number of neural network development tools and products available in the market of which a few are shown in Table 11-3.

11.4.5. Application in Power Electronics and Drives

A neural network can be used for various control and signal processing applications in power electronics. Considering the simple input-output mapping property of a feedforward network, it can be used in one or multidimensional function generation. For example, $Y = \sin X$ function can be generated by a neural network [48]. In this case, the training is carried out with large Y versus X precomputed example data table for the full cycle. Although it appears like look-up table implementation, the trained network can interpolate between the example data values. Another example of similar application is the selected harmonic elimination method of PWM control, where the notch angles of a wave can be generated by a neural network for a given modulation index [60]. As mentioned before, a neural network can be trained to emulate a fuzzy controller.

A network can be trained for on-line or off-line diagnostics of a power electronics system. Consider, for example, a large drive installation where the essential sensor signals relating to the state of the system are fed to a neural network. The network output can interpret the "health" of the system for monitoring purposes. The drive may not be permitted to start if the health is not good. The diagnostic information can be used for remedial control, such as shut-down or fault-tolerant control of the system. Similarly, a network can receive an FFT pattern of a complex signal and be trained to derive important conclusions from it. A neural network can

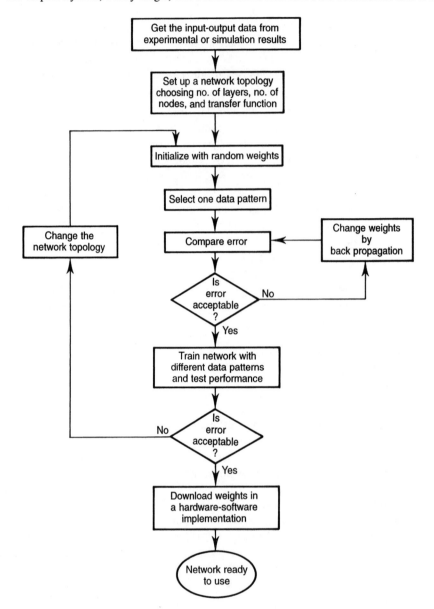

Figure 11-36. Flowchart for training a neural network.

receive time-delayed inputs of a distorted wave and perform harmonic filtering without any phase shift [48]. Although a feedforward network cannot incorporate any dynamics within it, a nonlinear dynamical system can be simulated by time-delayed input, output, and feedback signals. Consider, a nonlinear single-input, single-output (SISO) dynamical system which can be expressed mathematically by a nonlinear differential equation. The differential equation can then be expressed in finite difference form and simulated by a neural network. Narendra and Parthasarathy [49]

TABLE 11-3 NEURAL NETWORK TOOLS AND PRODUCTS

Company	Product	Description
Intel	80170NX*	Sixty-four neurons, 10,240 synaptic weights; can use third-party simulation software for training; iNNTS downloads weights to the chip, nonvolatile weight storage, and electrical programmability
NeuralWare/ NeuralWorks	Professional II/Plus	Development tool on AT or other platforms; use for BP, ART, LVQ2, PNN, RBF, SOM
California Scientific Software	BrainMaker	Development tool (Brainmaker, Network Toolkit, Brainmaker Prof.) for BP (DOS, MAC); Accelerator/Professional Accelerator (DSP-based AT cards)
MathWorks	MATLAB Toolbox	Collection of MATLAB functions for designing and simulating NN. Includes learning rules, transfer functions, and training and design procedures for BP, fully integrated with MATLAB
NeuroDynamX	DynaMind	Development tool for BP, Madaline, etc. neural accelerators (DSP-based AT cards)

ART = adaptive resonance theory, BP = back propagation, LVQ2 = learning vector quantization, PPN = probablistic neural network, RBF = radial basis feedback, SOM = self-organizing map.
*Electrically trainable IC analog FF neural netowrk.

identified four different types of SISO nonlinear models for neural network implementation as follows:

Model I:

$$x(k+1) = \sum_{i=0}^{n-1} \alpha_i x(k-i) + f[u(k), u(k-1), \ldots, u(k-m+1)] \quad (11.25)$$

Model II:

$$x(k+1) = f[x(k), x(k-1), \ldots, x(k-n+1)] \quad (11.26)$$
$$+ \sum_{i=0}^{m-1} \beta_i u(k-i)$$

Model III:

$$x(k+1) = f[x(k), x(k-1), \ldots, x(k-n+1)] \quad (11.27)$$
$$+ g[u(k), u(k-1), \ldots, u(k-m+1)]$$

Model IV:

$$x(k+1) = f[x(k), \ x(k-1), \ \ldots, \ x(k-n+1); \tag{11.28}$$
$$u(k), \ u(k-1), \ \ldots, \ u(k-m+1)]$$

where $[u(k), x(k)]$ represents the input-output pair of the plant at time k. The simulation structure for model I is shown in Figure 11-37, where the nonlinear static function $f(\cdot)$ is simulated by a feedforward neural network. Instead of forward model of a dynamical plant, a neural network can also be trained to identify an inverse dynamic model. In fact, inverse model simulation is simpler, which will be discussed later. Both forward and inverse models can be used for adaptive control of a plant. A few more neural network application examples for estimation and control are given in the paragraphs that follow.

Estimation of Distorted Wave. The problem here is to calculate accurately the rms current (I_s), fundamental rms current (I_f), displacement power factor (DPF), and power factor (PF) for a single-phase, antiparallel thyristor controller with R-L load, where the firing angle (α), load impedance (Z), and impedance angle (ϕ) are varying. We discussed similar estimation with fuzzy logic from the wave pattern in Section 11.3.5. The supply voltage is assumed to be sine wave and constant at 220 V. Figure 11-38 shows the network structure for estimation where it uses two hidden layers with 16 neurons in each hidden layer. The training data for the network were derived from the circuit simulation for different values of α, ϕ, and Z_b/Z, and the ouput parameters were calculated from the simulation waves with the help of MATLAB program. Figs. 11-39a, b, c, and d show the estimator [53] performance for the four output parameters where each estimated curve is compared with the corresponding calculated curve. In each case, the minimum α angle is restricted to be equal to α_{min}, that is, at the verge of continuous conduction. The I_s (p.u.) and I_f (p.u.) are converted to actual values by multiplying with the scale factor Z_b/Z for the input condition $Z_b/Z = 1.0$. The network was trained with the help of NeuralWorks

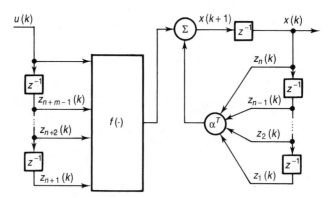

Figure 11-37. Model of a nonlinear dynamical system with a feedforward neural network.

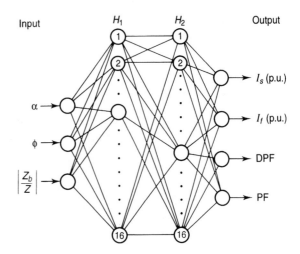

Figure 11-38. Neural network for estimation for triac controller line current.

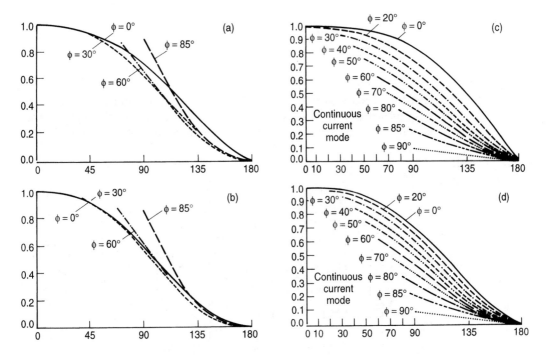

Figure 11-39. Neural network based estimation of a thyristor-controlled current wave: (a) rms current (I_s); (b) fundamental rms current (I_f); (c) displacement power factor (DPF); (d) power factor (PF).

Professional II/Plus. It required 14.3 million training steps, and the error was found to converge below 0.2%. A numerical example will make the estimation steps clear from Figure 11-39.

Example:

Input parameters

$$\text{Firing angle } (\alpha) = 60°$$
$$\text{Impedance angle } (\phi) = 30°$$
$$\text{Supply voltage } (V_s) = 220\,\text{V}$$
$$\text{Base impedance } (Z_b) = 220\sqrt{2}\ \Omega$$
$$\text{Load impedance } Z = 50\ \Omega$$

Estimated outputs

From Fig. 11-39a, I_s (p.u.) $= 0.79$; i.e., $I_s = 0.79 . 220\ \sqrt{2}/50 = 4.915$ A

From Fig. 11-39b, I_f (p.u.) $= 0.78$; i.e., $I_f = 0.78 . 220\ \sqrt{2}/50 = 4.853$ A

From Fig. 11-39c, $DPF = 0.84$

From Fig. 11-39d, $PF = 0.82$

Pulsewidth Modulation. Figure 11-40 shows a current control PWM scheme with the help of a neural network [54, 55]. The network receives the phase current error signals through the scaling gain K and generates the PWM logic signals for

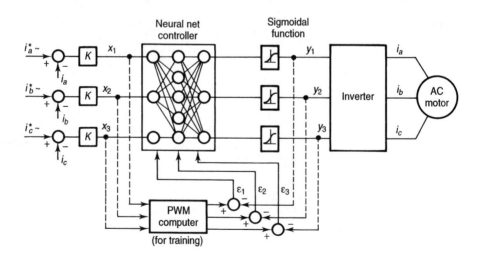

Figure 11-40. Neural network–based PWM controller.

driving the inverter devices. The sigmoidal function is clamped to 0 or 1 when the threshold value is reached. The output signals (each with 0 and 1) have eight possible states corresponding to the eight states of the inverter. If, for example, the current in a phase reaches the threshold value $+0.01$, the respective output should be 1, which will turn on the upper device of the leg. If, on the other hand, the error reaches -0.01, the output should be 0, and the lower device will be switched on. The network was trained with eight such input-output example patterns.

In another PWM scheme [55], the network was trained to generate the optimum PWM pattern for a prescribed set of current errors. The desired pattern was generated separately by a PWM computer, as shown. The desired pattern and the actual output pattern can be compared, and the resulting errors can train the network by back propagation algorithm. In a somewhat similar scheme, the network was trained to minimize the current errors within the constraint of switching frequency.

Drive Feedback Signal Estimation. Figure 11-41 shows the block diagram of a direct vector-controlled induction motor drive where a neural network-based estimator estimates the rotor flux (Ψ_r), unit vectors ($\cos\theta_e$ and θ_e) and torque (T_e) [56]. A DSP-based estimator is also shown in the figure for comparison. Since the feed-

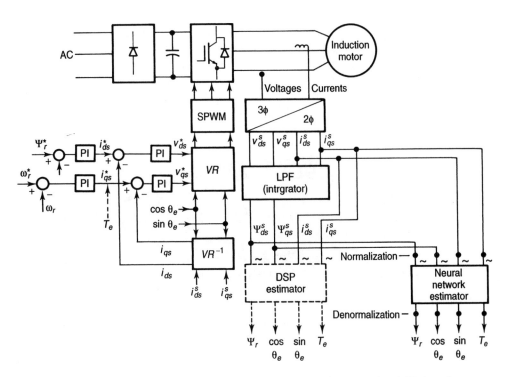

Figure 11-41. Vector-controlled induction motor drive showing DSP-based and neural network–based estimation.

forward network cannot incorporate any dynamics, the machine terminal voltages are integrated by a hardwared low-pass filter (LPF) to generate the stator flux signals, as shown. The variable frequency variable magnitude sinusoidal signals are then used to calculate the output parameters. Figure 11-42 shows the topology of the network where there are three layers and the hidden layer contains 20 neurons. A bias signal is coupled to all the neurons of the hidden and output layers (output layer connections are not shown) through a weight. The input layer neurons have linear transfer characteristics, but the hidden and the output layers have hyperbolic-tan type of transfer function to produce bipolar outputs. Figures 11-43a–d show, respectively, the torque, flux, $\cos \theta_e$, and $\sin \theta_e$ output of the estimator after successful training of the network with a very large number of simulation data sets. The estimator outputs are compared with the DSP-based estimator and shows good accuracy. With a switching frequency of 2 kHz (instead of 15 kHz), the estimator network was found to have relatively harmonic immune performance. The drive system was operated in the wide torque and speed regions independently with DSP-based estimator and neural network-based estimator and was found to have comparable performance.

Identification and Control of DC Drive. Figure 11-44 shows an inverse dynamic model-based indirect model referencing adaptive control scheme of a DC drive where it is desirable that the motor speed follows an arbitrary command speed trajectory [57]. The motor model with the load is nonlinear and time invariant, and the model is completely known. However, the reference model that the motor is to follow is given. The unknown nonlinear dynamics of the motor and the load are

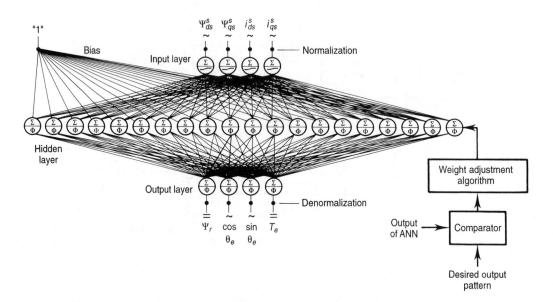

Figure 11-42. Topology of the estimator network.

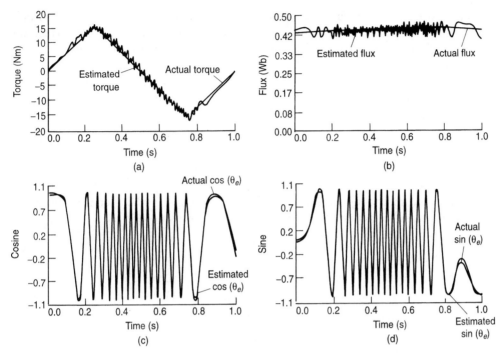

Figure 11-43. Estimator accuracy at 15 kHz inverter switching frequency:
(a) torque; (b) flux; (c) cosine wave; (d) sine wave.

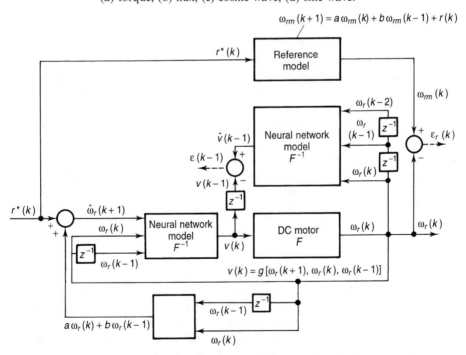

Figure 11-44. Model identification and adaptive control of a DC motor
using a neural network.

captured by a feedforward neural network. The trained network identifier is then combined with the reference model to achieve trajectory control of speed. Here, the motor electrical and mechanical dynamics can be given by the following set of equations,

$$K_v \omega_r(t) = v(t) - R_a i_a(t) - L_a \frac{di_a}{dt} \tag{11.29}$$

$$K_t i_a(t) = j \frac{d\omega_r}{dt} + B\omega_r(t) + T_L(t) \tag{11.30}$$

$$T_L(t) = K\omega_r^2(t) \, [\text{sign}(\omega_r(t))] \tag{11.31}$$

where the common square law torque characteristics have been considered. These equations can be combined and converted to the discrete form as

$$\omega_r(k+1) = \alpha \omega_r(k) + \beta \omega_r(k+1) + \gamma[\text{sign}(\omega_r(k))]\omega_r^2(k)] \tag{11.32}$$
$$+ \, \delta[\text{sign}(\omega_r(k))]\omega_r^2(k-1) + \zeta v(k)$$

$$\text{or } v(k) = g[\omega_r(k+1), \, \omega_r(k), \, \omega_r(k-1)] \tag{11.33}$$

where

$$g[\cdot] = \frac{1}{\xi} \begin{bmatrix} \omega_r(k+1) - \alpha\omega_r(k) - \beta\omega_r(k-1) - \gamma\,[\text{sign}(\omega_r(k)]\,\omega_r^2(k) \\ -\delta\,[\text{sign}(\omega_r(k)\,\omega_r^2(k-1) \end{bmatrix} \tag{11.34}$$

Equation 11.33 gives the discrete model of the machine. A three-layer network with five hidden-layer neurons was trained off-line to emulate the unknown nonlinear function g[.], which is basically the inverse model of the machine. The signals $\omega_r(k+1)$, $\omega_r(k)$, and $\omega_r(k-1)$ are the network inputs, and the corresponding output is g[.] or $v(k)$. The training data can be obtained from an operating plant, or else from simulation data if the model is available. The signal $\epsilon(k-1)$ is the identification error, which should approach zero after successful training. The network is then placed in the forward path, as shown, to cancel the motor dynamics. Since the reference model is asymptotically stable, and assuming that the tracking error $\epsilon_r(k)$ tends to be zero, the speed at $(k+1)$th time step can be predicted from the expression

$$\hat{\omega}_r(k+1) = 0.6\,\omega_r(k) + 0.2\omega_r(k+1) + r^*(k) \tag{11.35}$$

Therefore, for a command trajectory of $\omega_r(k)$, $r(k)$ can be solved from the reference model, and the corresponding $\omega_r(k+1)$, $\omega_r(k)$, and $\omega_r(k-1)$ signals can be impressed on the neural network controller to generate the estimated $v(k)$ signal for the motor, as shown. The parameter variation problem cannot be incorporated in the network with off-line training. The model emulation and adaptive control, as described, can also be extended to AC drives [58].

Flux Observer for Induction Motor Drive. Figure 11-45 shows a neural network–based adaptive flux observer for vector-controlled induction motor drive [59]. The neural observer, as indicated, receives the synchronously rotating frame stator voltage (ν_{ds}) and currents (i_{ds}, i_{qs}) and estimates the rotor flux magnitude and unit vectors for feedback control and vector transformation, respectively. In addition, there is a rotor time constant (T_r) identification unit that helps to fine-tune the estimator. The observer consists of two emulator units, that is, neural flux emulator and neural stator emulator. The flux emulator receives the delayed rotor flux, i_{ds} and i_{qs} at the input and calculates slip frequency and rotor flux at the output, as shown. The feedforward network is trained off-line first with machine model equations at decoupled condition using the nominal machine parameters. The neural stator emulator receives the flux, frequency, and stator currents at the input and generates the stator ν_{ds} signal at the output. This feedforward network is also trained off-line with the standard machine model equations using the nominal parameters. Once the off-line training for both the emulators is complete with nominal parameters, the fine-tuning for estimation is done on-line with the estimated T_r, as indicated. The on-line training is done in three stages:

1. Retrain the flux emulator with T_r-varying model equations
2. Retrain the stator emulator with T_r-varying stator model equations

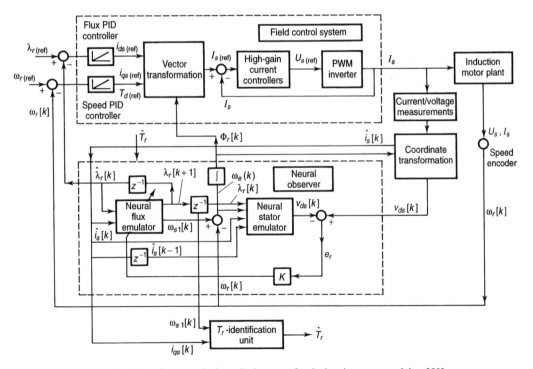

Figure 11-45. Neural network–based observer for induction motor drive [59].

3. Retrain the flux emulator with the on-line ν_{ds} loop error signal e_r. The complex observer is implemented with a sampling time of 1.0 ms.

11.5. SUMMARY

A brief but comprehensive review of expert system, fuzzy logic, and neural network is given in this chapter. The principles of each that are relevant to power electronics and drive control applications are described in a simple manner to make them easily understandable to readers with a power electronics background. Several applications in each area are then described that supplement theoretical principles. Finally, a glossary has been added at the end as a convenience to readers. Expert system, fuzzy logic, and neural network technologies have advanced significantly in recent years and have found widespread applications, but they have hardly penetrated the power electronics area. The frontier of power electronics that is already so complex and interdisciplinary will have a wide impact by these emerging technologies in the coming decades and will offer a greater challenge to the community of power electronic engineers.

11.6. GLOSSARY

Expert System

Artificial intelligence.   A branch of computer science where computers are used to model or simulate some portion of human reasoning or brain activity.

Backward chaining.   An inference method where the system starts from a desired conclusion and then finds rules that could have caused that conclusion.

Boolean logic.   A logic or reasoning method using Boolean variables (0 and 1).

Database.   A set of data, facts, and conclusions used by rules in expert system.

Declarative knowledge.   An actual content or kernel of knowledge.

Domain expert.   A person who has extreme proficiency at problem solving in a particular domain or area.

Forward chaining.   An inference method where the IF portion of a rule is tested first to arrive at conclusion.

Frame.   A treelike structuring of a knowledge base.

Inference.   The process of reaching a conclusion using logical operations and facts.

Inference engine.   The part of an expert system that tests rules and tends to reach conclusion.

Knowledge acquisition.   The process of transferring knowledge from domain expert by the knowledge engineer.

Knowledge base.   The portion of an expert system that contains rules and data or facts.

Knowledge engineering. The process of translating domain knowledge from an expert into the knowledge base software.

Knowledge representation. The process of structuring knowledge about a problem into appropriate software.

Meta-rule. A rule that describes how other rules should be used.

Procedural knowledge. Software structuring for representation of knowledge.

Rule. A statement in an expert system, usually in an IF-THEN format, which is used to arrive at a conclusion.

Shell. A software development system that is used to develop an expert system.

Symbolic processing. Problem solving based on the application of strategies and heuristics to manipulate symbols standing for problem concepts.

Fuzzy Logic

Centroid defuzzification. A method of calculating crispy output from center of gravity of the output membership function.

Degree of membership. A number between 0 and 1 that expresses the confidence that a given element belongs to a fuzzy set.

Defuzzification. The process of determining the best numerical value to represent a given fuzzy set.

Fuzzification. The process of converting nonfuzzy input variables into fuzzy variables.

Fuzzy composition. A method of deriving fuzzy control output from given fuzzy control inputs.

Fuzzy control. A process control that is based on fuzzy logic and is normally characterized by "IF ... THEN ..." rules.

Fuzzy expert system. An expert system composed of fuzzy IF-THEN rules.

Fuzzy implication. Same as fuzzy rule.

Fuzzy rule. IF - THEN rule relating input (conditions) fuzzy variables to output (actions) fuzzy variables.

Fuzzy set (or fuzzy subset). A set consisting of elements having degrees of membership varying between 0 (nonmember) to 1 (full member). It is usually characterized by a membership function, and associated with linguistic values, such as SMALL, MEDIUM, etc.

Fuzzy set theory. A set theory that is based on fuzzy logic.

Fuzzy variable. A variable that can be defined by fuzzy sets.

Height defuzzification. A method of calculating a crispy output from a composed fuzzy value by performing a weighted average of individual fuzzy sets. The heights of each fuzzy set are used as weighting factors in the procedure.

Linguistic variable. A variable (such as temperature, speed, etc.) whose values are defined by language, such as LARGE, SMALL, etc.

Membership function. A function that defines a fuzzy set by associating every element in the set with a number between 0 and 1.

Singleton.   A special fuzzy set that has the membership value of 1 at particular point and 0 elsewhere.

SUP-MIN composition.   A composition (or inference) method for constructing the output membership function by using maximum and minimum principle.

Universe of discourse.   The range of values associated with a fuzzy variable.

Neural Network

ANN (artificial neural network).   A model made to simulate biological nervous system of human brain.

Associative memory.   A type of memory where an input (which may be partial) serves to retrieve an entire memory that is the closest match of input information.

Autoassociative memory.   A memory in which entering incomplete information causes the output of complete memory, that is, the part that was entered plus the part that was missing (same as content addressable memory).

Back propagation.   A supervised learning method in which an output error signal is fed back through the network altering connection weights so as to minimize that error.

Back propagation network.   A feedforward network that uses back-propagation training method.

Computational energy.   A mathematical function defining the stable states of a network and the paths leading to them.

Connection strength.   Gain or weight in the connection (or link) between nodes through which data pass from one node to another.

Content addressable memory (CAM).   A memory that is addressed by the partial content (unlike address addressable memory in digital computer) (same as auto-associative memory).

Dendrite.   The input channel of a biological neuron.

Distributed intelligence.   A feature in neural network in which the intelligence is not located at a single location, but is spread throughout the network.

Gradient descent method.   A learning process that changes a neural network's weights to follow the steepest path toward the point of minimum error.

Heteroassociative memory.   A network where one pattern input generates a different pattern at the output.

Learning.   The process by which a network modifies its weights in response to external input.

Learning rate.   A factor that determines the speed of convergence of the network.

Neuron.   A nerve cell in a biological nervous system; a processing element (or node or cell or neurode) in a neural network.

Pattern recognition.   The ability to recognize a set of input data instantaneously and without conscious thought, such as recognizing a face. The ability of a neural network to identify a set of previously learned data, even in the presence of noise and distortion in the input pattern.

Perceptron. A neural network designed to resemble a biological sensory model.

Recurrent network. Same as feedback network.

Sigmoid function. A nonlinear transfer function of a neuron that saturates to 1 for high positive input and 0 for high negative input.

Squashing function. A transfer function where the output squashes or limits the values of the output of a neuron to values between two asymptotes.

Supervised learning. A learning method in which an external influence helps to correct its output.

Synapse. The area of electrochemical contact between two neurons.

Training. A process during which a neural network changes the weights in orderly fashion in order to improve its performance.

Transfer function. A function that defines how the neuron's activation value is transferred to the output.

Unsupervised learning. A method of learning in which no external influence is present to tell the network whether its output is correct.

Weight. An adjustable gain associated with a connection between nodes in a neural network.

References

[1] Firebaugh, M. W., *Aritifical Intelligence*, Boyd and Fraser, Boston, 1988.

[2] Bose, B. K., "Power electronics—recent advances and future perspective," *IEEE-IECON Conf. Rec.*, pp. 14–16, 1993.

[3] Bose, B. K., "Variable frequency drives—technology and applications," *Proc. IEEE Int. Symp.*, Budapest, pp. 1–18, 1993.

[4] Vedder, R. G., "PC based expert system shells: Some desirable and less desirable characteristics," *Expert Syst.*, Vol. 6, pp. 28–42, February 1989.

[5] Texas Instruments, Inc., "Personal consultant plus getting started," Dallas, August 1987.

[6] Texas Instruments, Inc., "Personal consultant plus user's reference manual," Dallas, November 1988.

[7] Tam, K. S., "Application of AI techniques to the design of static power converters," *IEEE-IAS Annual Meeting Conf. Rec.*, pp. 960–966, 1987.

[8] Daoshen, C., and B. K. Bose, "Expert system based automated selection of industrial AC drives," *IEEE-IAS Annual Meeting Conf. Rec.*, pp. 387–392, 1992.

[9] Chhaya, S. M., and B. K. Bose, "Expert system based automated design technique of a voltage-fed inverter for induction motor drive," *IEEE-IAS Annual Meeting Conf. Rec.*, pp. 770–778, 1992.

[10] Chhaya, S. M., and B. K. Bose, "Expert system based automated simulation and design optimization of a voltage-fed inverter for induction motor drive," *IEEE-IECON Conf. Rec.*, pp. 1065–1070, 1993.

[11] Chhaya, S. M., and B. K. Bose, "Expert system aided automated design, simulation and controller tuning of AC drive system," *IEEE-IECON Conf. Rec.*, pp. 712–718, 1995.

[12] Debebe, K., V. Rajagopalan, and T. S. Sankar, "Expert systems for fault diagnosis of VSI-fed AC drives," *IEEE-IAS Annual Meeting Conf. Rec.*, pp. 368–373, 1991.

[13] Debebe, K., V. Rajagopalan, and T. S. Sankar, "Diagnostics and monitoring for AC drives," *IEEE-IAS Annual Meeting Conf. Rec.*, pp. 370–377, 1992.

[14] Sugeno, M., ed., *Industrial Applications of Fuzzy Control*, North-Holland, New York, 1985.

[15] Pedrycz, W., *Fuzzy Control and Fuzzy Systems,* John Wiley, New Yrok, 1989.

[16] Zadeh, L. A., "Fuzzy sets," *Informat. Contr.*, Vol. 8, pp. 338–353, 1965.

[17] Zadeh, L. A., "Outline of a new approach to the analysis of systems and decision processes," *IEEE Trans. Syst. Man and Cybern.*, Vol. SMC-3, pp. 28–44, 1973.

[18] Mamdani, E. H., "Application of fuzzy algorithms for control of simple dynamic plant," *Proc. IEEE*, Vol. 121, pp. 1585–1588, 1974.

[19] Mamdani, E. H., and S. Assilian, "A fuzzy logic controller for a dynamic plant," *Int. J. Man-Machine Stud.*, Vol. 7, pp. 1–13, 1975.

[20] Lemke, H. R. Van Nauta, and W. J. M. Kickert, "The application of fuzzy set theory to control a warm water process," *Automatica*, Vol. 17, pp. 8–18, 1976.

[21] Kickert, W. J. M., and H. R. Van Nauta Lemke, "Application of a fuzzy controller in a warm water plant," *Automatica*, Vol. 12, pp. 301–308, 1976.

[22] King, P. J., and E. H. Mamdani, "The application of fuzzy control systems to industrial processes," *Automatica*, Vol. 13, pp. 235–242, 1977.

[23] Ratherford, D., and G. Z. Carter, "A heuristic adaptive controller for a sinter plant," *Proc. 2nd IFAC Symp.*, Johannesburg, 1976.

[24] Ostergaard, J. J., "Fuzzy logic control of a heat exchanger process," *Internal Report*, University of Denmark, 1976.

[25] Tong, R. M., "Some problems with the design and implementation of fuzzy controllers," *Int. Report*, CUED/F-CAMS/TR127, Cambridge University, Cambridge, UK, 1976.

[26] Tong, R. M., "A control enginering review of fuzzy systems," *Automatica*, Vol. 13, pp. 559–569, 1977.

[27] Takagi, T., and M. Sugeno, "Fuzzy identification of systems and its applications to modeling and control," *IEEE Trans. Syst. Man and Cybern.*, Vol. SMC-15, pp. 116–132, January/February 1985.

[28] EDN-Special Report, "Designing a fuzzy-logic control system," *EDN*, pp. 79–86, 1993.

[29] Li, Y. F., and C. C. Lau, "Developmnt of fuzzy algorithms for servo systems," *IEEE Contr. Syst. Mag.*, April 1989.

[30] da Silva, B., G. E. April, and G. Oliver, "Real time fuzzy adaptive controller for an asymmetrical four quadrant power converter," *IEEE-IAS Annual Meeting Conf. Rec.*, pp. 872–878, 1987.

[31] Sousa, G. C. D., and B. K. Bose, "A fuzzy set theory based control of a phase-controlled converter DC machine drive," *IEEE Trans. Ind. Appl.*, Vol. 30, pp. 34–44, January/February 1994.

[32] Won, C. Y., S. C. Kim, and B. K. Bose, "Robust position control of induction motor using fuzzy logic control, *IEEE-IAS Annual Meeting Conf. Rec.*, pp. 472–481, 1992.

[33] Miki, I., N. Nagai, S. Nishigama, and T. Yamada, "Vector control of induction motor with fuzzy PI controller," *IEEE-IAS Annual Meeting Conf. Rec.*, pp. 342–346, 1991.

[34] Kirschen, D. S., D. W. Novotny, and T. A. Lipo, "On-line efficiency optimization control of an induction motor drive," *IEEE-IAS Annual Meeting Conf. Rec.*, pp. 488–492, 1984.

[35] Sousa, G. C. D., B. K. Bose, et al., "Fuzzy logic based on-line efficiency optimization control of an indirect vector controlled induction motor drive," *IEEE-IECON Conf. Rec.*, pp. 1168–1174, 1993.

[36] Cleland, J., B. K. Bose, et al., "Fuzzy logic control of AC induction motors," *IEEE Int. Conf. Rec. Fuzzy Systems (FUZZ-IEEE)*, pp. 843–850, March 1992.

[37] Rowan, T. M., R. J. Kerkman, and D. Leggate, "A simple on-line adaptation for indirect field orientation of an induction machine," *IEEE Trans. Ind. Elec.*, Vol. 42, pp. 129–198, April 1995.

[38] Sousa, G. C. D., B. K. Bose, and K. S. Kim, "Fuzzy logic based on-line tuning of slip gain for an indirect vector controlled induction motor drive," *IEEE-IECON Conf. Rec.*, pp. 1003–1008, 1993.

[39] Sousa, G. C. D., *Application of Fuzzy Logic for Performance Enhancement of Drives*, Ph.D. Thesis, University of Tennessee, Knoxville, December 1993.

[40] Simoes, M. G., and B. K. Bose, "Application of fuzzy logic in the estimation of power electronic waveforms," *IEEE-IAS Annual Meeting Conf. Rec.*, pp. 853–861, 1993.

[41] Lawrence, J., *Introduction to Neural Networks*, California Scientific Software Press, Nevada City, CA, 1993.

[42] Miller, W. T., R. S. Sutton, and P. J. Werbos, *Neural Networks for Control*, MIT Press, Cambridge, MA, 1992.

[43] Uhrig, R., "Fundamentals of neural network," University of Tennessee Class Notes, Knoxville, 1992.

[44] Simoes, M. G., and B. K. Bose, "Fuzzy neural network based estimation of power eletrornic waveforms," *III Brazilian Power Electronics Conf.* (COBEP '95), São Paulo (accepted).

[45] Horikawa, S., et al., "Composition methods of fuzzy neural networks," *IEEE-IECON Conf. Rec.*, pp. 1253–1258, 1990.

[46] "Neural network resource guide," *AI Expert*, pp. 55–63, December 1993.

[47] Hammerstrom, D., "Neural networks at work," *IEEE Spectrum*, pp. 26–53, June 1993.

[48] Connor, D., "Data transformation explains the basics of neural networks," *EDN*, pp. 138–144, May 12, 1988.

[49] Narendra, K. S., and K. Parthasarathy, "Identification and control of dynamical systems using neural networks," *IEEE Trans. Neural Networks*, Vol. 1, pp. 4–27, March 1990.

[50] Hunt, K. J., et al., "Neural networks for control systems—survey," *Automatica*, Vol. 28, pp. 1083-1112, 1992.

[51] Dote, Y., "Neuro fuzzy robust controllers for drive systems," *IEEE Proc. Int. Symp. Ind. Elec.*, pp. 229–242, 1993.

[52] Anaskolis, P. J., "Neural networks in control systems," *IEEE Control Syst. Mag.*, Vol. 10, pp. 3–5, April 1990.

[53] Kim, M. H., M. G. Simoes, and B. K. Bose, "Neural network based estimation of power electronic waveforms," *IEEE Trans. Power Electronics*, March, 1996.

[54] Harashima, F., et al., "Application of neural networks to power converter control," *IEEE-IAS Annual Meeting Conf. Rec.*, pp. 1086–1091, 1989.

[55] Buhl, M. R., and R. D. Lorenz, "Design and implementation of neural networks for digital current regulation of inverter drives," *IEEE-IAS Annual Meeting Conf. Rec.*, pp. 415–423, 1991.

[56] Simoes, M. G., and B. K. Bose, "Neural network based estimation of feedback signals for a vector controlled induction motor drive," *IEEE Trans. Ind. Appl.*, Vol. 31, pp. 620–629, May/June 1995.

[57] Weersooriya, S., and M. A. El-Sharkawi, "Identification and control of a DC motor using back-propagation neural networks," *IEEE Trans. Energy Conversion*, Vol. 6, pp. 663–669, December 1991.

[58] El-Sharkaw, M. A., et al., "High performance drive of brushless motors using neural network," *IEEE-PES Summer Conf. Proc.*, July 1993.

[59] Theocharis, J., and V. Petridis, "Neural network observer for induction motor control," *IEEE Control Systems*, pp. 26–37, April 1994.

[60] Trzynadlowski, A. M., and S. Legowski, "Application of neural networks to the optimal control of three-phase voltage controlled inverters," *IEEE Trans. Pow. Elec.*, Vol. 9, pp. 397–404, July 1994.

[61] Kamran, F., and T. G. Habetler, "An improved deadbeat rectifier regulator using a neural net predictor," *IEEE Trans. Power Electronics*, Vol. 10, pp. 504–510, July 1995.

[62] Wishart, M. T., and R. G. Harley, "Identification and control of induction machines using artificial neural networks," *IEEE Trans. Ind. Appl.*, Vol. 31, pp. 612–619, May/June 1995.

[63] Bose, B. K. "Expert system, fuzzy logic, and neural network applications in power electronics and motion control," *Proc. IEEE Special Issue on Power Electronics and Motion Control*, Vol. 82, pp. 1303–1323, August 1994.

[64] Simoes, M. G., B. K. Bose, and R. J. Spiegel, "Fuzzy logic based intelligent control of a variable speed cage machine wind generation system," *IEEE-PESC Conf. Rec.*, pp. 389–395, 1995.

[65] Fodor, D., G. Griva, and F. Profumo, "Compensation of parameters variations in induction motor drives using neural network," *IEEE-PESC Conf. Rec.*, pp. 1307–1311, 1995.

[66] Bose, B. K., "Fuzzy logic and neural network applications in power electronics," *Proc. of Int. Pow. Elec. Conf.*, Yokohama, pp. 41–45, 1995.

Index

Biography of Dr. Bimal K. Bose

Bimal K. Bose currently holds the Condra Chair of Excellence in Power Electronics at the University of Tennessee, Knoxville, where he has been responsible for organizing the power electronics teaching and research program for the last nine years. He is also the Distinguished Scientist of EPRI–Power Electronics Applications Center, Knoxville; Honorary Professor of Shanghai University, China; and Senior Adviser of the Beijing Power Electronics Research and Development Center, China.

He received a B.E. degree from Calcutta University, India, an M.S. degree from the University of Wisconsin, Madison, and a Ph.D. degree from Calcutta University in 1956, 1960, and 1966, respectively. Early in his career for eleven years, he served as a faculty member at Calcutta University (Bengal Engineering College). For his research contributions, he was awarded the Premchand Roychand Scholarship and Mouat Gold Medal by Calcutta University. In 1971, he joined Rensselaer Polytechnic Institute, Troy, New York, as Associate Professor of Electrical Engineering and organized the power electronics program at that institution. In 1976, he joined General Electric Corporate Research and Development, Schenectady, New York as research engineer and served there for eleven years. During this period, he also served as Adjunct Professor at Rensselaer Polytechnic Institute. He has worked as a consultant in various industries and for such companies as General Electric R&D Center, Lutron Electronics, Bendix Corporation, PCI Ozone Corporation, EPRI, Research Triangle Institute, Honeywell, Reliance Electric Co., Delco Remy, and Motion Control Engineering.

Dr. Bose's research interests extend across the whole spectrum of power electronics and specifically include power converters; AC drives; microcomputer control; EV drives; and expert system, fuzzy logic, and neural network applications in power electronics. He has published more than 125 papers and holds eighteen U.S. patents (two more are pending). He is author of bestseller *Modern Power Electronics* (1992), *Microcomputer Control of Power Electronics and Drives* (1987), *Power Electronics and AC Drives* (Prentice Hall, 1986, translated into Japanese, Chinese, and Korean and several low-cost editions) and *Adjustable Speed AC Drive Systems* (1981). He has also contributed chapters and articles in a number of books. He was the Guest Editor of the *Proceedings* of the IEEE Special Issue on Power Electronics and Motion Control (August 1994). He received the GE Publication Award, the Silver Patent Medal, and a number of IEEE prize paper awards. He is listed in Marquis' *Who's Who in America* and *Who's Who in Electro Magnetics Academy*.

Dr. Bose has served the IEEE in various capacities that include Chairman of IAS Industrial Power Converter Committee, IAS member in Neural Network Council, Chairman of IE Society Power Electronics Council, and Associate Editor of IE Transactions and as a member of various professional committees. He is on the Editorial Board of *Proceedings* of the IEEE and *Asia-Pacific Engineering Journal*. He has also served on the National Power Electronics Committee and the International Council for Power Electronics Cooperation and has participated in a large number of national and international professional organizations. In 1995, he initiated Power Electronics for Universal Brotherhood (PEUB), an international organization to promote humanitarian activities of the power electronics community.

Dr. Bose received the IEEE Industry Applications Society's Outstanding Achievement Award for "outstanding contributions in the application of electricity to industry" in 1993; the IEEE Industrial Electronics Society Eugene Mittelmann Award in "recognition of outstanding contributions to research and development in the field of power electronics and a lifetime achievement in the area of motor drives" in 1994; the IEEE Region 3 Outstanding Engineer Award for "outstanding achievements in power electronics and drives technology" in 1994; and the IEEE Lamme Gold Medal (with $10,000) "for contributions to the advancement of power electronics and electrical machine drives" in 1996. He is a Life Fellow of the IEEE.